MW00813986

RUNNING DOWN

WATER IN A CHANGING LAND

RUNNING DOWN

WATER IN A CHANGING LAND

MARY E. WHITE

Cartography: Barbara Eckersley
Professional landscape photography: Reg Morrison
Foreword by Graham Harris, Chief, CSIRO Land & Water

Kangaroo Press

RUNNING DOWN: WATER IN A CHANGING LAND
First published in Australia in 2000 by Kangaroo Press
An imprint of Simon & Schuster (Australia) Pty Limited
20 Barcoo Street, East Roseville NSW 2069

A Viacom Company
Sydney New York London Toronto Tokyo Singapore

National Library of Australia
Cataloguing-in-Publication data

White, M.E. (Mary E.).
Running Down: water in a changing land.

Bibliography.
Includes index.
ISBN 0 7318 0904 1.

1. Rivers - Regulation - Environmental aspects - Australia.
2. Streamflow - Australia. 3. Nature - Effect of human beings on - Australia.
4. Conservation of natural resources - Australia. 5. Environmental protection - Australia. I. Title.

551.4830994

Cover photo: *Black Flag Lake, about 30 kilometres from Kalgoorlie, Western Australia. An ancient landscape, rusty red; salt-rimmed water cool in a hot, dry world—epitomising Australia, its ephemeral waters, its intriguing mysteries…*
PHOTO BILL VAN AKEN, SCIENTIFIC & TECHNICAL PHOTOGRAPHER, CSIRO LAND & WATER, PERTH.
(the Author expresses her deep gratitude to Bill van Aken and the CSIRO Land & Water for making this special photograph available)

Cover and internal design: Linda Robertshaw
Cartographer: Barbara Eckersley

Set in 9.5/11.5 pt Minion
Printed by Imago Productions (FE) Pte Ltd

10 9 8 7 6 5 4 3 2 1

Sponsorship acknowledgment

Kangaroo Press and Mary E. White acknowledge, with deep gratitude, the sponsorship that has enabled the recommended retail price of this book to be not more than $50, including GST.

To have CSIRO Land & Water, through its Chief, Dr Graham Harris, sponsoring this book has been an honour and a privilege beyond the wildest hopes of the author. That Graham Harris has written, as a Foreword, such a generous testimonial to *Running Down* in addition to the sponsorship, leaves her feeling humbled and at a loss as to how to express her gratitude adequately. To have the support of the CSIRO—an Australian icon, and one that has the power, the wisdom and the will to make the changes necessary to implement sustainable management of our land and water resources—is 'the stuff that dreams are made of.'

Mr R.D. (Bob) Walshe, Chairman of the Sutherland Environment Centre, has made a generous contribution to offset the print costs of this book, enabling the sale price to be kept within the range of all the readers for whom it is intended. He has been personally involved for many years with issues aimed at maintaining the health of the three main rivers in Sutherland Shire and is interested in, and concerned for, the sustainable use of Australia's resources. Because of his belief in the importance of education for valuing the environment, he has supported Mary White's contribution through this book, and she is deeply grateful for his support.

Above: Tanami Desert flooding: Mulga, termites and instantaneous brief flooding typify this land of extremes. REG MORRISON

Page iii: Flooded mulga scrub in the Tanami Desert N.T., April 1989. Termite mounds become islands and water brings life to the desert. REG MORRISON

Acknowledgments

I acknowledge, with deep gratitude, the contribution of colleagues to this book.

While the concept, the research and the writing has been a one-woman effort, the final product could not have emerged without the help and expertise of an extraordinary number of others. In particular, the contribution of two has been of inestimable value—my cartographer daughter, Barbara Eckersley, who created more than 200 maps and diagrams; and Reg Morrison, who made available slides from his professional landscape photography collection, enabling me to include a very large number of highest quality illustrations in this book. Presentation of the otherwise daunting amount of scientific information in a form which makes it easily and, I hope, enjoyably available to a wide spectrum of readers, relies heavily on the visual presentation.

The scientists on whose published papers the book has been based have generously given their time to check my manuscript and to advise, as well as supplying the essential photographs needed to illustrate their sections. There would have been no book without their original work and their generous assistance. The bibliography and the photo credits give some idea of the number of people to whom I am indebted in this regard—too numerous to thank individually, but all essential contributors to the final product.

Thank you all, most sincerely, for your help and encouragement.

River red gums, symbols of survival, beautify our unreliable rivers and ephemeral watercourses—each tree a work of art. M.E.W.

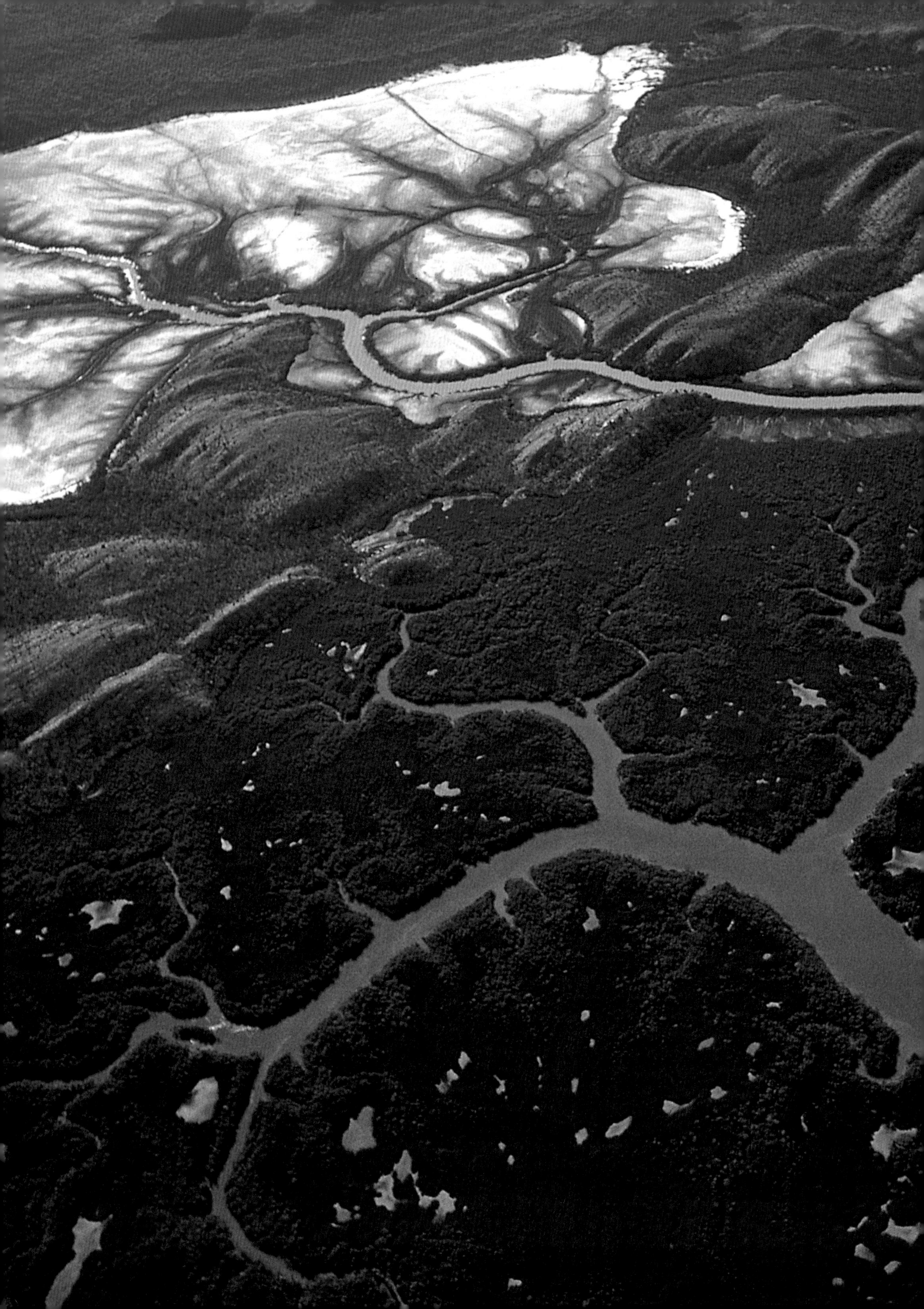

Contents

REG MORRISON

M.E.W.

Opposite: Tidal drainage patterns in mangroves, Kimberley region, Western Australia. REG MORRISON

SECTION THREE: THE TROPICAL NORTH 79

REG MORRISON

SECTION FOUR: THE GREAT ARTESIAN BASIN

M.E.W.

SECTION FIVE: EASTERN AUSTRALIA 107

M.E.W.

SOUTH–EASTERN AUSTRALIA 112

M.E.W.

REG MORRISON

R. BURNS

REG MORRISON

Foreword

There can be no doubt that water is a vital resource in Australia. If we continue to use it in our present fashion, national economic growth may well become limited by its availability. Our rivers show severe signs of degradation through extraction, regulation by dams and other forms of habitat destruction. We continue to extract unsustainable amounts of water from our surface and ground waters – and new dams and bore fields are still planned.

It is pretty clear that Australia faces a major series of decisions in the next decades. Business as usual is not an option. How do we make sustainable use of resources and really attain the fundamental goal of ecologically sustainable development: to leave the continent to our children the way we found it?

One fundamental requirement is to obtain a true sense of place. What Mary White has done in this book, once again, is to give us that sense of that place and to remind us of the great antiquity of this continent. These are not the rivers of Europe or North America, they are very old and they function differently. We are given a sense of "deep time" – the working out of geological and ecological processes over the millennia. What we see now has a very long and unique history, a history that we are destroying in the blink of an eye.

Australia is a world leader in the use of resource economics for natural resource management. Water reforms have gone further than in almost any other country, prices charged are more realistic and water use efficiency has improved, with some benefit to the environment. We have begun to tackle the issue of water for the environment and the trade-offs between water for production and water for environmental flows. These are difficult issues and require us to balance competing interests.

We need information and wisdom, such as is provided here, to manage the complex interactions contained within the concept of the 'triple bottom line' – economics, sociology and ecology – the only true meaning of inter-generational equity. We need to appreciate the basic biophysical constraints of living on this driest of continents and to achieve a true vision of a sustainable future.

Through her books Mary White has contributed much to that vision and has been a major contributor to our knowledge. It is books like this that will help us to find our own vision, and our own solutions to our uniquely Australian problems. She has taught us to appreciate this land in its true form and to look at it through new eyes. I congratulate her for producing another magnificent piece of scholarship. This book will make a major contribution to a most important national debate.

Dr Graham Harris
Chief
CSIRO Land and Water

Roaring Meg Creek, Wilsons Promontory, Victoria. REG MORRISON

Preface

WATER—ALWAYS RUNNING DOWN—ACROSS LANDSCAPES, THROUGH THE EONS AND THE AGES, CARVING CREASES AND WRINKLES ON THE FACE OF ANCIENT CONTINENTS, LIFE-GIVING, THE ESSENTIAL RESOURCE FOR SURVIVAL OF EARTH AS A LIVING PLANET—YET RUNNING DOWN, DEGRADING, UNDER OUR SELFISH HUMAN STEWARDSHIP...

I have spent the last fifteen years assembling the big picture of Australia's evolution through geological time. The stories of the co-evolution of the land and its biota, which I told in *The Greening of Gondwana* and *The Nature of Hidden Worlds,** were the product of my profession as a palaeobotanist and my fascination with the ancient world in which my fossil plants had lived. Those books led naturally into *After the Greening: The Browning of Australia* because I wanted to find out how the Australia-in-Gondwana, which had been my first interest, had become Australia the Island Continent and how it had changed from a green, well-watered Gondwanan fragment to become the driest vegetated continent.

I did not plan another book, but it became clear to me that the story I had documented in *After the Greening* accounted for so many of the environmental problems which face Australia today, and in my travels so much land degradation was clearly visible that I knew that there was another important subject to tackle. The full realisation that the geological past of the continent had pre-determined the problems that would arise as a result of the imposition of European-style agriculture and land-use practices on an ancient, fragile land was a revelation to me, and *Listen ... Our Land is Crying* was the result.

Writing, and the research which precedes it, are addictive, and I was thinking all the time while I was absorbed in finding out about changes to the land just how little I knew about water, our most vital, limiting and rapidly degrading resource. Just enough information on ancient palaeodrainage systems, prior streams detected below the modern landscape, artesian water and the like, had come my way to whet my appetite. So *Running Down: Water in a Changing Land* had been gestating for several years. As often is the case, a catalyst was needed to precipitate matters and, by chance, one appeared. *What if there was no River System?*[1] was a paper presented at a Murray–Darling Basin Conference by Robert Vincin, who had made a study of early Surveyor General's records, explorers' and early travellers' and settlers' accounts of

The Murray River, 20 kilometres west of Tocumwal, NSW. M.E.W.

Opposite: The Murray, near Gunbower, Vic. Severe bank erosion has exposed the roots of river red gums. REG MORRISON

The Darling, near Bourke, NSW, as it looks when it runs dry. REG MORRISON

rivers. These had shown how different today's rivers are from those early accounts—how many of the rivers which now form well-defined, incised systems used to 'die in the plains', or end in reedbeds and swamps; how the connections from one to another were across floodplains during big rainfall events.

This book in a way documents my search for the answer to the question about there being no river *system*—and only near the end of its collation did the full realisation come that the wrong question had been asked. It should have been **'What constitutes a river system in Australia?'**—and answering that explained the early records and showed that the explorers, naturally enough, had no idea and were looking with eyes and minds informed by their European background. The attitudes of subsequent generations, including earth scientists until recently, have continued to be coloured by the same unrealistic premises. (My researching has revealed that many scientists have been investigating early historical records in their recent work on rivers as a basis for their understanding of what constitutes an Australian river system and what changes have resulted from 200 years of our management.)

The very real, and often serious, changes that European-style land use has had on the uniquely Australian river systems, which over most of the continent are so unlike stereotypic examples in Europe, are a main theme in this book. Failure to understand that **floodplains are essential components of Australian river systems** has been the most significant omission. Before our changes to catchments occurred and many rivers became deeply incised channels, reaches of rivers on the flat floodplains were often characterised by anabranching, sometimes into many ill-defined channels. The massive floodplains, their types of sediment, the extraordinary flatness of landscapes, are all products of the geological past.

Once again, it was clear to me that an understanding of how our rivers and landscapes and their ecosystems had evolved through *geological* time was a pre-requisite. Added to the information available on their status and function at the time of European settlement, before they were changed by our activities, this provides the key to a fuller comprehension of problems. This deep time background can indicate what was required for natural, balanced hydrological systems generally, and it provides guidelines which may enable us to achieve sustainable use of our water resources.

I offer this book as a background to this understanding, believing that responsible stewardship of land and water resources—our life-support systems—is our individual and our national duty.

---

*The Nature of Hidden Worlds *has been re-issued by Kangaroo Press as* Reading the Rocks—Animals and Plants in Prehistoric Australia and New Zealand.

Portrait of an old, old land

A CONTINENT ERODED DOWN THROUGH UNIMAGINABLE LENGTHS OF GEOLOGICAL TIME, MAINLY BY WATER ... FLAT, WEATHERED ... WRINKLE CREASES WHERE WATER HAS CUT INTRICATE PATTERNS ... HOW VENERABLE AUSTRALIA LOOKS DEPICTED BY SATELLITE IMAGERY IN THE COLOURS OF OLD AGE! AND HOW FAITHFUL THIS DEPICTION IS TO THE ESSENTIAL TRUTH. OLD, WORN OUT, DRIED UP, YET YOUNG IN THE UNIVERSAL SCHEME OF THINGS. A UNIQUE PART OF THE BIOSPHERE, ITSELF THE PRODUCT OF THE CO-EVOLUTION OF LIFE AND THE ENVIRONMENT THROUGH FOUR BILLION YEARS OF TIME. TO BE TREATED WITH REVERENCE; 'MANAGED' WITH UNDERSTANDING OF THE REQUIREMENTS OF ITS LIFE-SUPPORTING SYSTEMS WHICH SUSTAIN THE RESIDENT BIOTA, INCLUDING US. WHEN WE IMPOVERISH AND DEGRADE THIS AGE-OLD LAND WE IMPOVERISH OURSELVES AND ARE DEGRADED.

REMEMBER, WE DO NOT INHERIT OUR BIT OF THE EARTH FROM OUR FOREBEARS—
WE HOLD IT IN TRUST FOR OUR CHILDREN AND GRANDCHILDREN.

Landsat imagery provided by Australian Centre for Remote Sensing (ACRES), Australian Surveying & Land Information Group, DIST, Canberra, and digitally enhanced and produced by Satellite Remote Sensing Services, Department of Land Administration, Perth, Western Australia.

A TWENTY-FIRST CENTURY VISION FOR AUSTRALIA
R. D. WALSHE, OAM

The Bungle Bungles, East Kimberley, Western Australia—a water-worn landscape. M.E.W.

Erosion by water has created the fantastic Bungle Bungles, where rivers run between the sandstone towers in the wet season. M.E.W.

Truly, we Australians are living these last days of the twentieth century in a turmoil of concern as to our identity in the Pacific and our sense of national purpose.

Uncertainty has been evident throughout the 1990s, a decade of argument and policy shift by governments and parties. Painful enough in our external relations, these have been no less painful internally as we have disputed the condition of our continent, the degree of its wounding by European settlement and whether or not the wounds are healing.

Problems crowd in, adding new concerns to old. We sense that we need a great purpose to inspire and unite us, but so far it has eluded us. This year, however, a vision has been hinted to us by a foreign observer whose words carry weight in the counsels of the United Nations. Australia, says Maurice Strong, is 'an environmental super-power'; it is 'the only island constituting an entire continent... home to a huge assemblage of plants and animals found nowhere else on Earth... rich in mineral and biological resources... [with] an important responsibility as the dominant actor in the environmentally sensitive South Pacific.'*

Strong, a Canadian, is currently Chairman of the United Nations Earth Council; he has held the post of Secretary General to the two most significant international environmental events in modern times, the 1972 UN Conference on the Human Environment and the 1992 UN Conference on Environment and Development. 'I am persuaded that the 21st Century will be decisive for the human species,' he says.

Australia's political parties have thus far been unable to provide the vision that is needed. Yet the moment we grasp the possibility that a government of our island continent has the potential to pursue with conviction a role as steward of this environmental super-power, the solutions to many problems begin to take shape.

We must first face the reality that most of the continent's land mass, for all its beauty and diversity, is extraordinarily fragile. Just look how a short period of European settlement has stripped so many forests, eroded so many soils and diminished so many rivers. It means we have a duty of care and should direct our efforts, local and national, towards a healing. What could be better? In a role of responsible stewardship we would be celebrating our national distinctiveness.

Continental stewardship, leaving behind the exploitative approaches of the past, could refresh many of our present economic, technological and scientific endeavours. For example ...

- In Science, we would stress science-for-an-arid-continent, and broaden from that to all kinds of landcare practices, together with applications in renewable energy and implementation by local governments and local groups of the range of activities called for by the United Nations' Agenda 21. And we would invite scientists worldwide to visit us and share our experience.

- In Tourism, we would not only promote all the present attractions but would add an invitation to visit the unique island continent—this environmental super-power—to observe its national program of enthusiastic care.

Reaching out thus to all countries, welcoming their visitors, offering our science, and explaining that we see ourselves as stewards on behalf of the whole world, we would be acting in the best possible way for the defence of this one and only nation continent; for we would be continuously engaged in briefing the world on how the reality of the continent's fragility dictates our environmental priorities and our population restraint. We would be making an unarguable case for a unique and admirable identity.

---

* *From the Inaugural Jack Beale Lecture on the Global Environment, 'Towards a Sustainable Civilisation', University of New South Wales, 11 February 1999.*

Bob Walshe is currently Chairman of Sutherland Shire Environment Centre. In 1972 he founded the NSW Total Environment Centre and was its first Secretary. In 1998 he was awarded the Order of Australia Medal (OAM) for his services to education and the environment.

DEEP ECOLOGY AND THE PLACE OF HUMANS IN THE SCHEME OF THINGS

The Norwegian philosopher Arne Naess formulated the concept of Deep Ecology in the 1970s.

It embodies a major division in ecological thought. On the one hand, 'shallow' ecology is anthropocentric with humans regarded as outside Nature and with Nature having only instrumental or 'use' value. On the other, 'deep' ecology does not separate humans, or anything else, from the natural environment. It sees the world not as a collection of isolated objects, but as a network of phenomena that are fundamentally interconnected and interdependent. It recognises the intrinsic value of all living things and views humans as just one particular strand in the Web of Life.

In Deep Ecology, as in the Gaia concept of the Earth acting like an enormous living organism, lies hope for the world. Ultimately, deep ecological awareness is spiritual awareness in the deepest sense. The feeling of belonging, of connectedness to the cosmos as a whole, is sustenance for the human spirit and provides the inspiration for our stewardship of the Earth. An emerging new vision of reality based on deep ecological awareness is consistent with the so-called perennial philosophy of spiritual traditions, whether it concerns the spirituality of Christian mystics, that of Buddhists, the philosophy of cosmology underlying Native American traditions, or Aboriginal Australians' identification with the natural landscape. A Deep Ecology philosophy underpins the Declaration of Interdependence which is the credo of the David Suzuki Foundation of Canada. This Declaration begins:

> *We are the earth, through the plants and animals that nourish us. We are the rains and the oceans that flow through our veins. We are the breath of the forests of the land, and the plants of the sea. We are human animals, related to all other life as descendants of the firstborn cell. We share a common present, filled with uncertainty. And we share a common future, as yet untold.*

Deep Ecology acknowledges the co-evolution of Life and the environment through the unimaginable lengths of geological time from the first living organisms 3.5 billion years ago, which started the oxygenation of the atmosphere. It sees the living modern biota as part of the first DNA which has grown and expanded, passed on by generation after generation down the ages in the Continuum of Life—we are all here because each of our ancestors back through time was a successful breeder! That original DNA went forth and multiplied! Our vertebrate evolutionary history explains our salty plasma and the gill-slits which scar the necks of our developing foetuses—our fishy, aquatic beginnings. We are not different from other life and it is only our relatively recent evolution within the last 2 million years or so which has made us think we are. That evolution has been the development of the big brain with its left and right hemispheres able to accommodate the completely rational and the irrational, intuitive and 'spiritual' at the same time, making us human (and fallible).

As a result of the enlargement of the human brain, our technology evolved to compensate for our being ill-equipped to survive on the savannas of Africa when the equatorial forests were decimated during glacial stages of the Pleistocene ice age. And it is this technological evolution which has brought us to the present dilemma which faces us and the world—for the first time in the history of the Earth we, an animal species, can change the environment on a scale which threatens the survival of the Living Planet. We have seen ourselves as having a god-given right to dominate the world, believing that everything is here for our benefit and for us to exploit.

As the Suzuki Foundation manifesto concludes: *'At this turning point in our relationship with Earth, we work for an evolution from dominance to partnership, from fragmentation to connection, from insecurity to interdependence.'*

From words to actions—from having a vision of what would enable sustainable use of the Earth's life-supporting resouces, in particular its precious soil and fresh water which are in short supply and much degraded as a result of human over-population globally—to the goal of a balanced biosphere.

INTRODUCTION

EUROCENTRIC ATTITUDES AND EXPECTATIONS: THEIR INCOMPATIBILITY WITH OUR LAND OF EXTREMES

Running Down: Water in a Changing Land is a companion volume to *Listen ... Our Land is Crying*, intended to complement it and complete the big picture of land and water resources in Australia, their origins, history, usage and present-day situation. Water-related subjects treated in *Listen ... Our Land is Crying* are not repeated in detail in this book, and text references direct the reader to where to find the information.

A main theme in *Listen ...* is that we have been mis-reading the land and failing to take into account the prehistoric history which predetermined how it would respond to European-style agriculture and land-use practices. Sustainable land management is only possible when practices are based on an understanding of the parameters set by the ancient, time-worn land itself, and by the climatic variability which adds an extra dimension of difficulty to achieving sustainability. It has become painfully obvious that Eurocentric attitudes and expectations have been at the root of the problems of land degradation and productivity loss which affect or threaten much of the small proportion of our large continent which is suitable for intensive use.

The same mis-reading, inappropriate use and unrealistic expectations applies to our surface and underground water supplies, in this driest of all the vegetated continents. The seriousness of the consequences of failing to understand the nature of our waters and what is required for their sustainable management cannot be over-estimated. Life itself depends on water—every element of the food chains which sustain the Planet, from microbe to mammal, from lowly green cell to forest tree. Nowhere is water a more vital commodity than in Australia, where 75 per cent of the land is acutely arid (40 per cent is desert); a further 10 per cent is under arid regimes for much of each year and acutely arid in droughts; and only 15 per cent is reasonably well-watered. Added to this is the El Niño–Southern Oscillation (ENSO) climate-warper which makes Australia a wide brown land, a land of drought and flooding rains (as so evocatively described by poet Dorothea Mackellar). With global climate change on the horizon, it is all the more essential that we have the knowledge and wisdom to deal with the future problems which will arise.

This climatic unpredictability is something which no one can change and with which we must learn to cooperate. To have any hope of succeeding we have to have a wide vision, see the 'big picture' and understand the co-evolution of this continent, its water resources and its biota through geological time. Then we will know, perhaps, a bit more about how to achieve sustainable, living ecosystems in our river systems, lakes, billabongs and wetlands, and

Monsoon rain comes to the Tanami Desert, NT. REG MORRISON

on the vast floodplains of this, the flattest and most poorly drained continent.

A re-think, a change in mindset, is the first prerequisite. The first settlers could only see the native vegetation as 'useless bush' and the native animals as 'vermin'. They immediately set about to rectify the situation by land-clearing and introducing foreign livestock and rabbits, and they established agricultural landscapes to which they could relate, using recognised European agricultural and land-use practices. Following generations trod in their footprints with the same expectations. Many became increasingly frustrated as their best intentions and the recognised-as-best farming practices led to loss of productivity; to the need for larger acreage to provide financial return; and as the increasing costs of inputs threatened to defeat their hard work and dedication. To a large extent, in spite of some attitudinal change and modification of methods, much the same attitudes persist today in relation to the land.

In the case of water—the disparity between the real Australian situation and how it appeared and was to be used from that European perspective was possibly even more marked. And even now, in the year 2000, just about no consideration is being given to what makes for real ecosystem sustainability in our surface waters. When it comes to groundwater, and palaeo-waters in particular, our ignorance is even more profound, and attitudes to its usage are even less acceptable. The situation which existed when European settlement occurred was the product of a dynamic co-evolution of life and the environment through great lengths of geological time. This had created a unique continent which is also the driest, flattest, most poorly drained, and in fact largely inward draining, land on Earth.[2]

Australia has greater variability in climate and river flows than any other land. Adaptation of all the life forms to the variability of climate and conditions had occurred over long periods. Today, we see, for instance, how kangaroos adapt their breeding cycles according to rainfall patterns; how birds have a-seasonal breeding patterns; how the breeding of fish and other aquatic animals is triggered by patterns of flood and drought in their habitats; how estuarine life burgeons when floods bring sediment and nutrient into that domain.

The attitudes and expectations which we still bring to management of land and water resources are deeply embedded in our culture. In the Northern Hemisphere lands from which Australia's settlers have mostly come, the deep soils and the landscapes are the product of Pleistocene ice age conditions. During the last 2 million years, great ice sheets ground their way across much of Europe (including Great Britain), North America and Asia during glacial stages, while interglacials brought deglaciation and more benign conditions. During this present interglacial in which we are all living, stable seasonality has been established in Europe's climates—spring, summer, autumn and winter have set the rhythm of life and life-cycles with built-in predictability, year upon year. It is said that in Europe any rain water that enters a major river reaches the sea within a week, at the longest. The very idea of what a river is and how it should behave is conditioned for most of us by our heritage: rivers should run swiftly to the sea, be permanent, well behaved, and stay within their banks, and should provide for all our requirements, predictably, throughout the year.

How, then, to reconcile rivers which never reach the sea, or whose waters take inordinately long to do so; which are ephemeral; or which don't go anywhere in particular and may end in reedbeds or billabongs; and others where floodplains are temporary connectors to larger river systems when flooding occurs in headwaters, frequently at enormous distance from the floodouts? Australia's Pleistocene ice age history, on top of its unimaginably long evolution as a land mass through geological time, explains why things are so different in this continent, why it is so flat and so much drains internally, and why our European attitudes have landed us in the present situation where our water and land-use practices are unsustainable.

It is no exaggeration to state that European land management has completely changed the hydrological situation in those parts of the continent subjected to agricultural and pastoral use. For the rivers, the basic patterns of change have involved deep incision of the stream bed; bank and floodplain erosion; siltation; loss of reedbeds

Debris-strewn bed of the Barwon River (the upper Darling River) at Walgett, NSW. REG MORRISON

Eucalypt woodland drowned by a holding reservoir for cotton irrigation near Bourke, NSW. REG MORRISON

The Chaffey Dam on the Namoi River NSW, with an algal bloom greening the water. BRAD SHERMAN

and swamps; drying out at depth of those floodplains which are in a quasi-natural state, because deeper river channels mean less over-bank flooding; and the creation of connected drainage systems where once many rivers had meandered on plains and ended in reedbeds or swamps, to run on beyond them only in times of flood. The changes can be described as 'turning rivers into drains', and case studies in this book will demonstrate the process. Overlaid on this basic scenario are the effects of rising regional water tables with waterlogging and salinity problems where the hydrological balance has been upset by removing deep-rooted, perennial vegetation and replacing it with vegetation which allows excessive recharge. This situation is exacerbated by irrigation. These matters are discussed at length in *Listen ... Our Land is Crying.*

The standard concept of the hydrologic cycle which applies to many other lands applies in full to only a small fraction of Australia. The 'typical' hydrologic cycle involves the following processes:

> The sun evaporates water, mainly from the sea; wind transports the vapour-laden air masses inland where they condense into clouds; rain falls and the water is dispersed in various ways—it may evaporate, filtrate into the ground or flow overland to a stream channel. The infiltrated portion may be stored as soil moisture, ready to be taken up by plants and transpired, or it may evaporate directly; or else it can move again—laterally as interflow to join a stream, or vertically to unite with the groundwater store. The groundwater often supplies a baseflow to streams and the streams discharge into the ocean to complete the cycle.[2]

Only the fringes of the Australian continent (and Tasmania) have substantial enough rainfall and run-off for the hydrologic cycle to be 'typical'. Aridity with extreme rates of evaporation afflict more than two-thirds of the continent all the time and another 20 per cent part of the time; most of our rivers are ephemeral, some very rarely have any water in them; most rivers do not run to the sea; our rivers are subject to floods which are among the world's most extreme and, because of the low gradients and vast inland plains, the area inundated by flooding can be very large; their variable flow has greater fluctuations than seen in rivers in any other part of the world.

It is not hard to see why early settlers mis-read the nature and capacity of the land. They started off in regions around Sydney without persistent water shortages or climatic extremes, and it was only when expansion occurred across the Divide that they were exposed to the droughts and flooding rains, and complete climatic unpredictability. The further west they went the more foreign the conditions they found. Pastoralists had taken up almost the whole of the arable part of eastern Australia by the middle of the nineteenth century, damming streams and creeks to ensure water supply, digging wells and, from the 1880s, tapping the artesian waters of the Great Artesian Basin.

Enormous environmental damage was done in the early days of settlement by the introduction of vast numbers of sheep and cattle, over-grazing—compounded by rabbit plagues—and by the systematic clearing of land to increase pasture area and later to produce crops. The early pastoralists greatly over-estimated the carrying capacity of the native pastures and by the 1890s the increase in sheep population was imposing considerable strain on the environment. The increase in watering places led to the build up of kangaroo numbers, which added pressure to the over-grazing by introduced animals. At the turn of the century, droughts between 1897 and 1902 halved the sheep populations, and the inland areas have never again been able to carry so many sheep. A brief boom in wheat growing followed slightly behind the early pastoral boom, when good rains in the 1870s saw expansion from the well-watered eastern coastal regions into low rainfall areas of the inland, but the bubble soon burst and wheat-growing belts were not consolidated until the introduction of dryland farming techniques.

The changes that this early period and the continuing European-style agriculture and land-use practices have brought to rivers, streams and drainage patterns are described in this book. It is amazing to find just how much change has occurred as a result of land use. It is normal for people of each generation to accept things as they are in their time, from their personal experience, as though history does not count. It is only when one takes a long view, longer than a generation, that one can see something of the true picture.

Australia's difficult climate became better understood by the early 1900s and a network of recording stations was established. Knowledge did not bring more realistic expectations, however, and exploitation of land and water resources—'mining' them—continued, and continues, little modified, today. This was excusable in the early times because of ignorance, but can no longer be tolerated.

There are always those who do not let facts interfere with the formulation of a good theory or a dream, and we continue to see proposals for grandiose schemes to impose massive change on the land, using engineering skills and technology. From early times, 'visionaries' wanted to dig a canal to join Lake Eyre to the ocean and create an inland sea, such as the early explorers were searching for before they found the central deserts. Others wanted to turn coastal rivers backwards to run into the arid centre, assuming that the increased moisture content of the air would guarantee rain. (This would not occur without new technology to lift the water vapour and create other specific meteorological conditions. More than just water vapour is needed to make rain fall.) Only 50 years ago it was proposed to feed coastal rivers into the Diamantina and the Cooper so that they would flood Lake Eyre permanently, supposedly opening up vast areas for agricultural production. The proponents did not acknowledge that rates of evaporation are such that salinity would inevitably follow watering of deserts, infertile sands need exorbitant inputs to produce crops, and fertilisers on a grand scale soon bring their own problems. Irrigation in arid and marginal lands has, in fact, a very limited life span.

The Bradfield schemes of much the same time proposed turning back the Clarence and other Queensland rivers to water areas on the other side of the Great Divide. Fortunately, because surely we must aim to leave Australia habitable for future generations, none of these schemes were, or are, financially viable. Their implementation would have resulted in hydrological disasters of an unimaginable scale.

To our national shame we still hear the same sorts of grandiose ideas being proposed at the start of the new millennium, sometimes espoused by people in government. Even in the less extreme realm of our dam building and river regulation we refuse to look at what has been learnt in other parts of the world (or what has often been found to be the case in Australia as well) and we continue to do things for present gain, disregarding the long-term consequences.

If we look honestly at the mighty Snowy Scheme and admit what it has done, apart from the economic angle which will ultimately be negated by the enormous expense of trying to remedy the environmental disasters which are now becoming clearly visible, we should admit that it was a monumental act of environmental vandalism. At the time, and in the context of that time, it did much for 'nation building' and defining Australia's identity—and it was, for all time—a remarkable engineering feat. Yet it will ultimately have contributed substantially to desertifying vast areas of our already limited agricultural land by its contribution to the inexorably rising saline water tables in the Murray Basin caused by the volume of water it made available for irrigation. It has virtually killed the Snowy River and contributes to the salinity problems in the Murray River system. (Sadly, it looks as though the regulation and exploitation of arid-land tributaries of the Darling, and of the Darling itself, for irrigating cotton will leave another set of irreversible problems in a few years' time.)

The freshwater crisis situation in which we find ourselves now has many aspects which will be addressed in this book. Some of the problems are reasonably amenable to management changes—sources of pollution can be controlled, some harmful practices minimised. But major and possibly intractable conflict exists in the agricultural domain, in particular between the requirement for stable supplies of water for socio-economic enterprises in spite of the natural variability, and the requirements of the environment to maintain ecosystem sustainability, which depends on this very variability—compounding the dilemma.

The challenge is to maintain a flexible balance between sustainable aquatic ecosystems and sustainable socio-economic enterprises by adaptive management practices which are based on the fundamental principle that variability is essential for sustainability—an extremely difficult task. (This acceptance of the necessity to maintain variability for the environment has been the missing concept in current management. River regulation, dams and all engineering works are designed to achieve stability and reliability and they have largely disregarded the requirements of aquatic, riparian and floodplain life.) If it is accepted that a radical change in management is required to address the problems caused by current 'regulation', and if realistic action follows, there is unlimited opportunity for social and economic benefit. The engineering fraternity has the challenge to come up with world-leading technologies and there is unlimited opportunity for innovative agriculture and land use, developing new crops and products, repairing the damaged wastelands we have created—by, for instance, planting some of the newly developed saltbushes that look like being the remedy for advancing deserts.

A real *vision* of the possibilities of changed water management and the wonderful benefits it can bring to the environment should replace the current 'technological fix' mentality. It would also solve the economic problems which are with us because of the land and water degradation which has come from not having that vision and understanding.

Equally importantly (and essential for the framing of that vision), we need to go back to early records and determine what was the status of rivers, groundwater, floodplains (which are our large agricultural and pastoral areas), the whole hydrological balance, at the time of settlement—before we started to make the changes which have altered everything in this context in a most dramatic way in only 200 years, and far less in many cases. In fact, we will find that serious changes to river function and all its attendant hydrological ramifications often occurred within a couple of decades of settlement, land-clearing and introduction of foreign animals and plants. Unfortunately, we will also find that the modern systems are still out of equilibrium and the impact of our land use has continued to result in rapid degradation.

Recent findings[3] that rates of siltation in our rivers and on their floodplains are datable by reference to an atomic-fallout zone in sediments, which corresponds to the years when Woomera was an atomic bomb test site, have shown clearly that there is no room for complacency. Metres of mud and sand deposited on river floodplains, which were assumed to be the results of hundreds or even thousands of years of erosion, are now known to represent only 30 or 40 years of deposition. In the case of the Murrah River on the New South Wales south coast between Bermagui and Bega, which received sudden and disproportionate media attention only because of the link with nuclear warfare, a third of the deep deposits was in fact dumped in one massive flood event, in 1971.

MAP OF GONDWANA, THE GREAT SOUTHERN SUPERCONTINENT

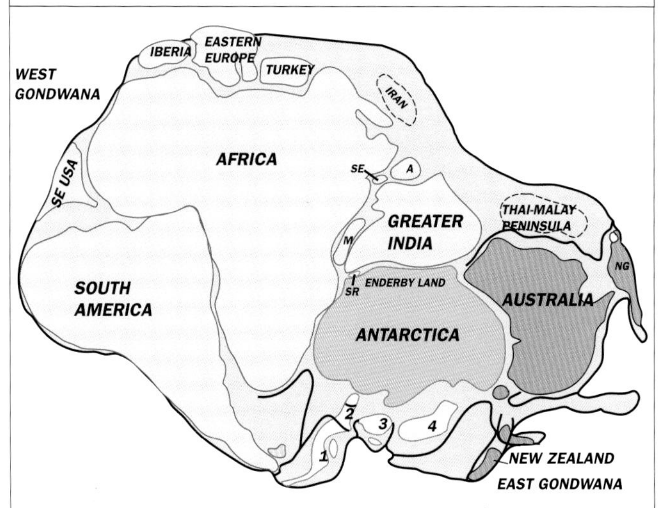

1, 2,3,4 Small plates that formed the Antarctic Peninsula and West Antarctica
M. Madagascar, SR. Sri Lanka, A.Afghanistan, SE Seychelles, NG New Guinea

I invite you to come with me on an exploration of the role of water in Australia through the ages, to assemble the background and big picture, so that we may understand the present situation (and because the story is fascinating in its own right). Water has shaped global environments since the beginning of Time, when the first rain fell on the cooling crust of the new Earth and started to erode its surface: creating drainage lines, wearing down the high parts, filling the low with sediments, and then reworking it all again and again, co-operating with tectonic and geological events to create the world as we know it.

We will see a slice of the evolution of the modern Australian

THE UPS AND DOWNS OF SEA LEVEL

Sea level changes through geological time, and even during the geologically recent climatic swings of the Pleistocene ice age, have had a profound effect on drainage patterns and landscapes. In a continent where some incredibly ancient landforms have been preserved, major changes in global sea level in the Cretaceous and any changes throughout the Tertiary are relevant to our story, as are marine incursions into sinking continental basins which did not necessarily involve a global sea level rise. Where blocks of continental crust are uplifted or where areas of crust sink to form basins, new drainage patterns follow. The processes of erosion through time are constantly altering the nature of drainages—lowering the headwaters, decreasing the gradient of streams and rivers, stripping to bedrock in places (which is a factor in the preservation of ancient patterns such as are seen on the ancient craton blocks of Australia) and filling up low areas with sediment.

In Australia, much of the continent is inward draining, and has been so for great lengths of time. Only the narrow fringes of the continent now drain to the ocean and it is therefore the coastal rivers which have been affected by rises and falls in sea level in comparatively recent geological times. Global times of high sea level, resulting in the formation of epicontinental seas in the past, have been significant in the production of modern landscapes. The limestone Nullarbor Plain was created under epicontinental seas which invaded in the Eocene and again in the Miocene—and was subsequently uplifted; marine invasions of the sinking Murray Basin left a series of marine sedimentary rocks which date the times of inundation, and the most recent left strandlines which stand out in the modern landscape as the sea retreated.

The sea level fluctuations of the last 2 million years, the Pleistocene ice age, have resulted in the sculpturing of our modern coastlines, creating the deeply incised valleys which are now drowned (like Sydney Harbour) and the canyons on the continental shelf through which the rivers ran to the sea when the coastline was many kilometres further out during low sea level stands.

PALAEOGEOGRAPHIC MAP OF AUSTRALIA IN THE EARLY CRETACEOUS

Extent of the epicontinental sea

GREATER INDIA
ARGO ABYSSAL PLAIN
DARLING SCARP
ANTARCTICA
EROMANGA SEA
SOUTH POLE

Rift / Rifted margin ***Stretched continental crust***
Volcano ***basaltic seafloor*** ***Spreading ridge***

continent by tracing the development of the ancient drainage patterns, which are like character lines or wrinkles on the face of our land, and by tracing the ancient history of our present-day rivers. Australia has some of the most ancient landscapes preserved anywhere on Earth. In Western Australia, some landscapes in the Kimberleys may date back an astounding 600 million years; the drainage patterns on the Yilgarn Craton can be traced for nearly 300 million years—and it is possible to reconstruct the changes to landscapes and climate up to today. Because the ancient history is so well recorded in the western half of the continent, that aspect of our story will be covered in detail by reconstructing the changing Western Australia landscapes through time.

The chosen examples of our rivers and waters will, where possible, show their prehistoric record, and then their status before European activities impacted upon them. Records of the early explorers, travellers and surveyors, and their maps and pictures, provide this base line—and often present amazing information on just how much change there has been.

In a subject as huge as this there will be many omissions, often of well-known subjects and aspects that readers might have expected to find included. The overall picture of *Water in Australia* presented is constructed like a mosaic, from bits of all shapes and sizes, seen from all sorts of different angles and perspectives, which, when assembled, make a whole.

GROUNDWATER

Australia's groundwater resources are very large but quality and quantity vary greatly in different parts of the land. Over about 60 per cent of the continent, the population is entirely dependent on groundwater, mainly in the arid and semi-arid regions. In other areas it is important in supplementing unreliable surface supplies. About 14 per cent of water used in Australia is groundwater.

Four classes of groundwater quality are recognised: *fresh*, which is generally suitable for drinking, domestic use, irrigation and stock; *marginal*, suitable for some irrigation uses and stock; *brackish*, which is suitable for stock and some industrial uses; and *saline*, which is generally unsuitable for use.

Aquifers on the east coast (North-east Coast and South-east Coast divisions), where population is largely concentrated, are already substantially exploited, as they are in the main foodbowl area of the Murray-Darling Division. There are large amounts of relatively untapped groundwater in northern Australia (Timor Sea and Gulf of Carpentaria divisions) and Tasmania (Tasmanian Division). In arid and semi-arid regions (Lake Eyre and Western Plateau divisions) most groundwater is of marginal quality.

DRAINAGE DIVISIONS OF AUSTRALIA
AUSTRALIAN WATER RESOURCES COUNCIL

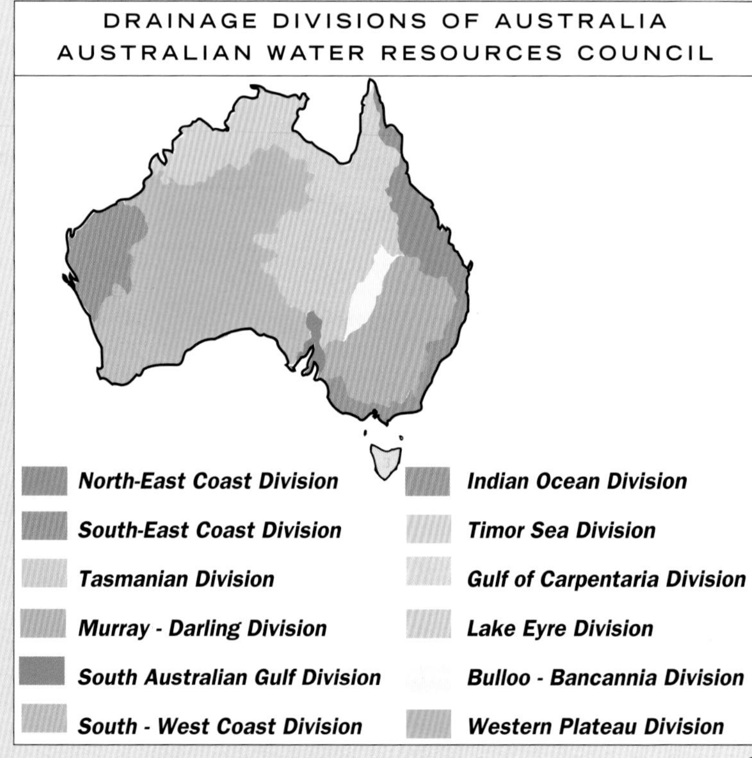

After McMahon *et al*[2]

Aquifers are of three main types:

- ***Surficial aquifers*** occur in alluvial sediments in river valleys, deltas and basins, and in dune sands accumulated by wind or water. These are major sources of fresh groundwater, and are easily exploited. About 60 per cent of groundwater used comes from surficial aquifers. (Perth and other places on the Swan coastal plain obtain much water from unconsolidated sand aquifers; and Fraser Island, the largest sand island in the world, is a huge surficial aquifer feeding perennially flowing rivers.)

- ***Sedimentary aquifers*** occur in consolidated sediments (sedimentary rocks) like porous sandstones, conglomerates and limestone. These are continuous over large areas and contain the greatest reserves of groundwater, though much of it is of marginal quality. The Great Artesian Basin, which underlies most of the eastern half of Australia, has many water-bearing sedimentary strata, only some of which have good quality water. Thirty per cent of used groundwater is from sedimentary aquifers.

- ***Fractured rock aquifers*** occur in igneous and metamorphosed rocks which have been subjected to disturbance. They may extend over large areas, but contain less available water. Groundwater extracted from fractured rock aquifers amounts to 10 per cent of that used.

For groundwater use to be sustainable, the rate of extraction must be less than the rate of recharge. This is hard to achieve in Australia where much of the groundwater is old, especially that in sedimentary aquifers. In these the rate of lateral movement is slow, recharge often takes place in far distant regions, and it may take very long times for water to travel. In the Great Artesian Basin, water moves at only a metre a year, and the distance from major recharge areas on the Great Divide is such that water emerging at mound springs near Lake Eyre is up to 2 million years old. (See page 97-104 for information on the GAB.) In the arid and semi-arid zone, very little recharge of localised smaller aquifers takes place at all, and the water is fossil water which resulted from recharge when rainfall in the area was much higher than today. Using such groundwater is obviously an unsustainable practice.

DAMS AND WATER DIVERSIONS

In Australia, this driest of all the vegetated continents, substantial rainfall and run-off occurs only within a narrow belt along the northern, eastern, south-eastern and south-western coasts of the continent, plus Tasmania. Large variations in quantity exist even within these sectors, and availability does not coincide with human settlement and activity in a proportionate manner. About two-thirds of the run-off occurs north of the Tropic of Capricorn, where only 5 per cent of the population lives and which contains only a small share of the productive land. This imbalance between locations of need and availability has predictably resulted in a high level of human interference with the natural hydrologic cycle, even in a country so recently and sparsely settled.

The situation has been compounded by the variability of flow in our rivers, which exhibit extreme highs and lows, problems again addressed by regulating them to reduce the variability. Engineering and technology have been used to make the hydrology fit our lifestyles and expectations. We have not modified our practices to suit the natural hydrological systems

From the start of European settlement, dam building has been an ongoing activity to deal with the problems of water scarcity and variability. Small farm dams exist in untold numbers, probably in excess of 500 000, with a great increase in recent years aided by the ready availability of earth-moving machinery. The water extracted from rivers to fill these is not fully taken into account when calculating how much water is being allowed for irrigation and how much for 'environmental flows' in rivers. Little information is available on these small river water storages and their capacity and management, but large dams with enormous water storage capacity have been well documented.

The greatest numbers of major dams are in the north-eastern and south-eastern coastal divisions, Tasmania and the Murray-Darling division. Although the greatest number of storages are for water supply, the greatest share of the water is used for irrigation. There has been a dramatic expansion in storage capacity over the last three decades.

Between 1971 and 1980 seven enormous reservoirs were commissioned—the Gordon, Ord River, Dartmouth, Serpentine, Fairbairn, Copeton and Wyangala dams—with a combined capacity of about 40 per cent of Australia's total storage.

Dams alter the flow regime, tending to distribute flow below the dam more evenly throughout the year, increasing low flows and reducing high flows. Small floods are eliminated, but large floods are not greatly affected. The hydrology of the dammed area changes; where the original vegetation was maintaining balance by evapo-transpiration, there is now evaporation from the dam surface.

The Hume Dam. DANIEL CONNEL, MDBC

The Hume Reservoir. DANIEL CONNEL, MDBC

THE GROWTH OF STORAGE CAPACITY

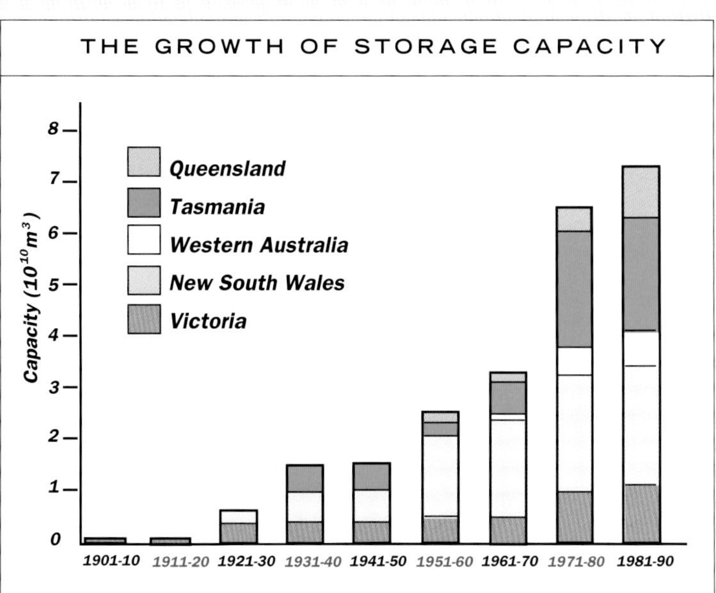

After McMahon *et al*[2]

INTER-BASIN WATER TRANSFERS

The damming and diversion of rivers induce significant changes to the hydrologic cycle with potential for environmental degradation and ecological change.[2]

- The Shoalhaven and Thomson Schemes transfer water to Sydney and Melbourne respectively.
- Water from the Lower Murray is diverted to Adelaide and towns at the head of the Spencer Gulf.
- In Western Australia, water from the Swan-Avon basin is diverted eastwards to supply Kalgoorlie and other towns.
- In Queensland, the headwaters of the Barron River are diverted to the Mitchell River Basin for irrigation.
- In western Victoria, the headwaters of the Glenelg River are diverted into the Wimmera-Mallee basin for irrigation.
- The Snowy Scheme diverts between 1 and 2 million cubic metres of water annually. Practically all Snowy River water above Jindabyne is transferred to the upper Murray and Murrumbidgee rivers. On the lower Snowy River, the mean annual flow has been reduced by 45 per cent, although the catchment area

above Jarrahmond has been reduced by only 14 per cent. Low flows at Jarrahmond are now about one-third of what they were before diversions, while a 1-in-10 year 7-day low flow, used as a base for some water quality standards, has been reduced by 75 per cent. Low flow intervals on the lower Snowy have become more frequent, showing a 14-fold increase; they now extend for longer and build up greater water deficits. There has been a reduction of 40–80 per cent in typical flushing flows in each month of the year. However, the Scheme has had only a small effect in reducing flood flows.

The hydrologic consequences of inter-basin transfers may be expected to be more severe than those arising from within-basin re-distribution. The imposition of artificial hydrological regimes may lead to an ecosystem structure that is atypical of the river. Even the setting of 'minimum environmental flows' does not adequately protect biological values, because in-stream biota are adapted to extremes of flow where these occur naturally, and they exist in a state of dynamic equilibrium with environmental disturbances. Floods may act as cues for breeding, and by providing a productive floodplain environment for juvenile fish, increase the number which reach maturity. Floods may also improve water quality by flushing away fine deposits.

PLEISTOCENE CLIMATIC SWINGS

The climate changes of the Pleistocene ice age, in which we are living in an interglacial, have created the Australia of today—the driest vegetated continent with more than 70 per cent under arid regimes. (A full account of how this continent changed from a green, well-watered and largely forested land 45 million years ago, when it started its northwards drift as an island continent, to become what it is now, is told in *After the Greening* and will not be repeated in this book.) The fluctuating climatic regimes of the last 2.6 million years have had a profound effect on the surface and below-surface waters of our continent. Some surprising facts emerge when a study is made of river behaviour through this small slice of geological time, and a picture emerges of a very different, and at times much less arid, Australia, even as recently as when the Aboriginals first came here.

Alternating dry and wet climatic regimes across central and eastern Australia during the past 300 000 years have greatly affected Australia's rivers, lakes and dune fields.[3A] Fluvial conditions dominated part of the last two interglacials, resulting in large sand loads in rivers in the Simpson Desert and south-eastern Australia.

The palaeochannels of central Australia were highly competent sand-load rivers during the last interglacial. There, fluvial activity peaked about 110 000 years ago, 5000 to 10 000 years behind world temperature and sea level maxima. Then aridity spread from central Australia towards the margins, with a peak at the last glacial maximum. A less widespread wet phase between 55 000 and 35 000 years ago is associated with high lake levels and palaeochannel activity in south-eastern Australia.

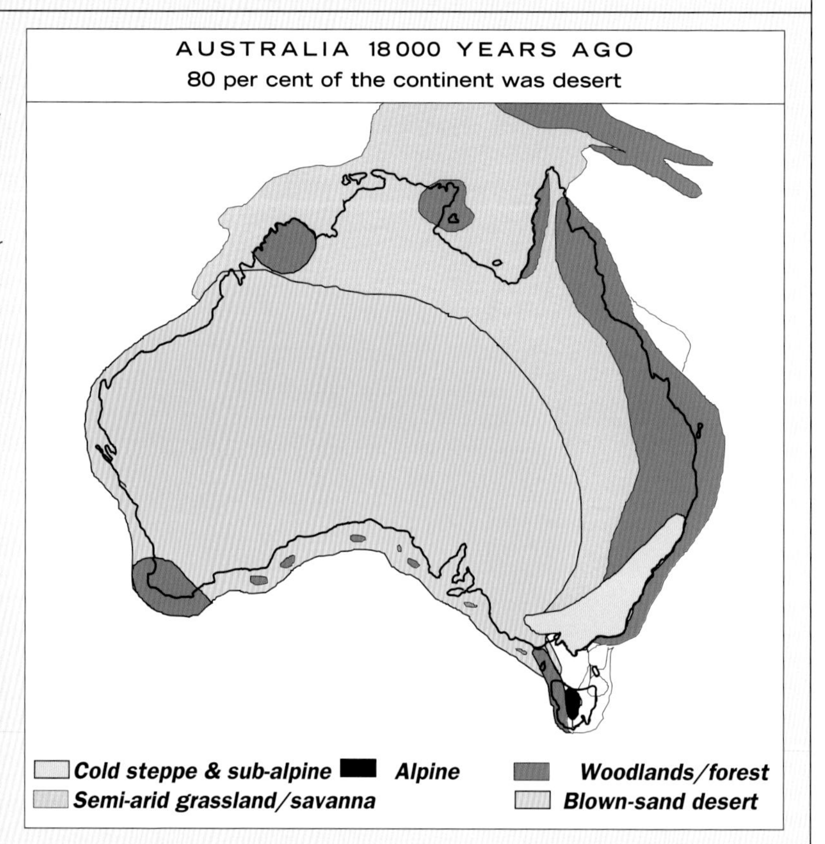

The coastward spread of aridity in Australia is recorded in the sedimentary record of rivers and lakes, and in the chronology of dune formation, as we shall see in accounts of rivers and landscapes in this book. For example, on the Riverine Plain the peak in river activity in prior streams occurred at about 90 000 years ago, and operated until 85 000 years ago. Cranebrook Terrace on the Nepean River, closer to the coast, gave an age of 80 000 years for its peak flow; and coastal rivers transported sand and gravel until about 70 000 years ago.

Dunes started to form in the Eyre Basin in the last interglacial from about 95 000 years ago. On the Riverine Plain, midway between the Centre and the coast, dunes bordering streams and lakes started to form in the period between 70 000 and 50 000 years ago. In the Lake George and Shoalhaven Basins, closer to and on the coast, dunes only started at about 20 000 years ago, during the last glacial maximum. At the peak of the last glacial maximum Australia was twice as dry (had half the rainfall) and twice as windy as it is now, and up to 80 per cent of the land was covered by wind-blown sand. The major dunefields assumed their present form then, and much of the Murray Basin was a salt-sand desert.

In the Holocene, the last 10 000 years, temperature and sea level have both varied. A mini-Greenhouse at 9000 years ago saw temperatures higher than today's over the continent, and increased rainfall. Maximum Holocene sea level occurred between 7500 and 6000 years ago, and sea level has been more or less stable since.

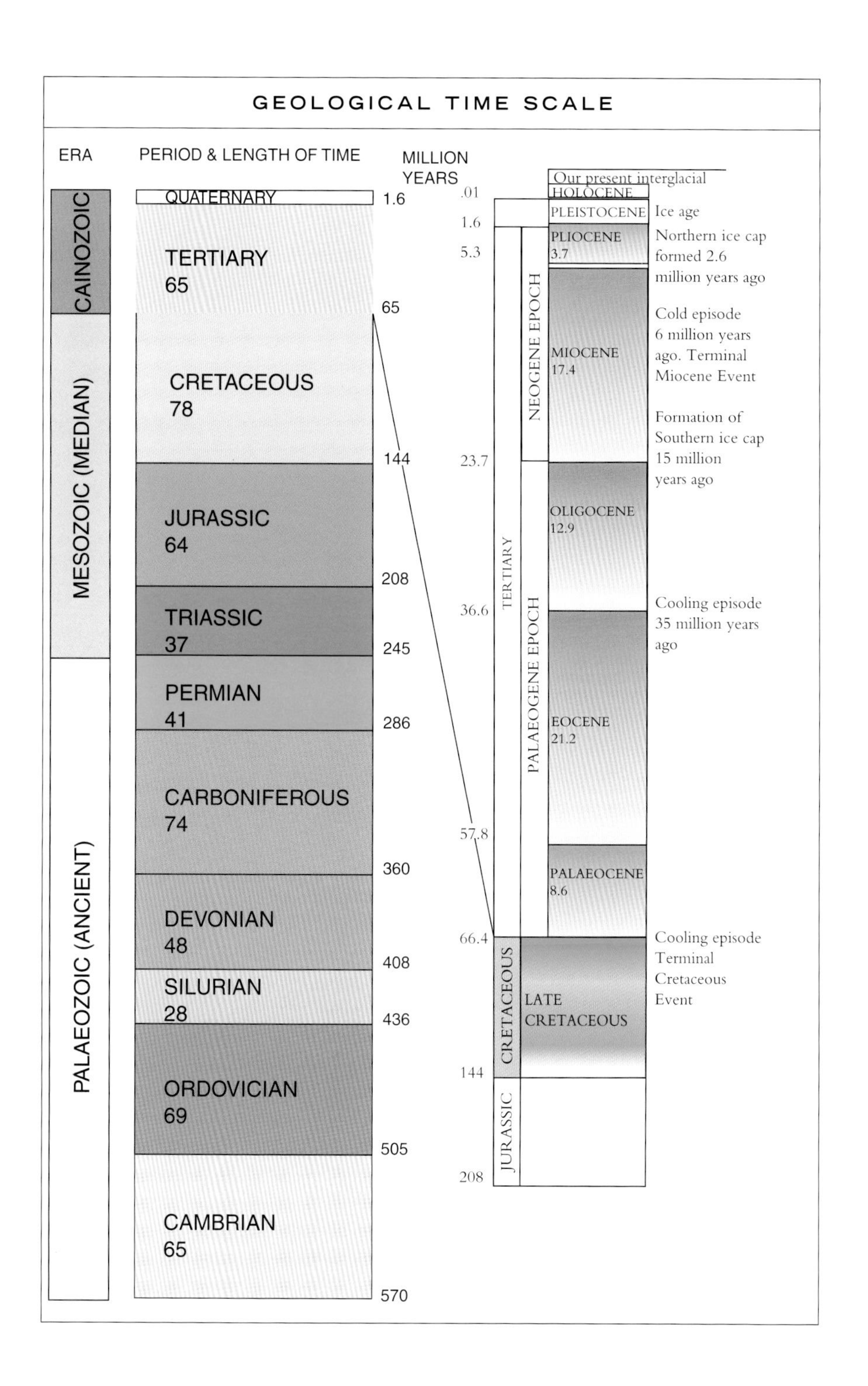

GEOLOGICAL TIME SCALE
ERA
PERIOD & LENGTH OF TIME
MILLION YEARS
CAINOZOIC
MESOZOIC (MEDIAN)
PALAEOZOIC (ANCIENT)
QUATERNARY
1.6
TERTIARY 65
65
CRETACEOUS 78
144
JURASSIC 64
208
TRIASSIC 37
245
PERMIAN 41
286
CARBONIFEROUS 74
360
DEVONIAN 48
408
SILURIAN 28
436
ORDOVICIAN 69
505
CAMBRIAN 65
570
.01
1.6
5.3
23.7
36.6
57.8
66.4
144
208
TERTIARY
NEOGENE EPOCH
PALAEOGENE EPOCH
CRETACEOUS
JURASSIC
Our present interglacial
HOLOCENE
PLEISTOCENE
PLIOCENE 3.7
MIOCENE 17.4
OLIGOCENE 12.9
EOCENE 21.2
PALAEOCENE 8.6
LATE CRETACEOUS
Ice age
Northern ice cap formed 2.6 million years ago
Cold episode 6 million years ago. Terminal Miocene Event
Formation of Southern ice cap 15 million years ago
Cooling episode 35 million years ago
Cooling episode Terminal Cretaceous Event

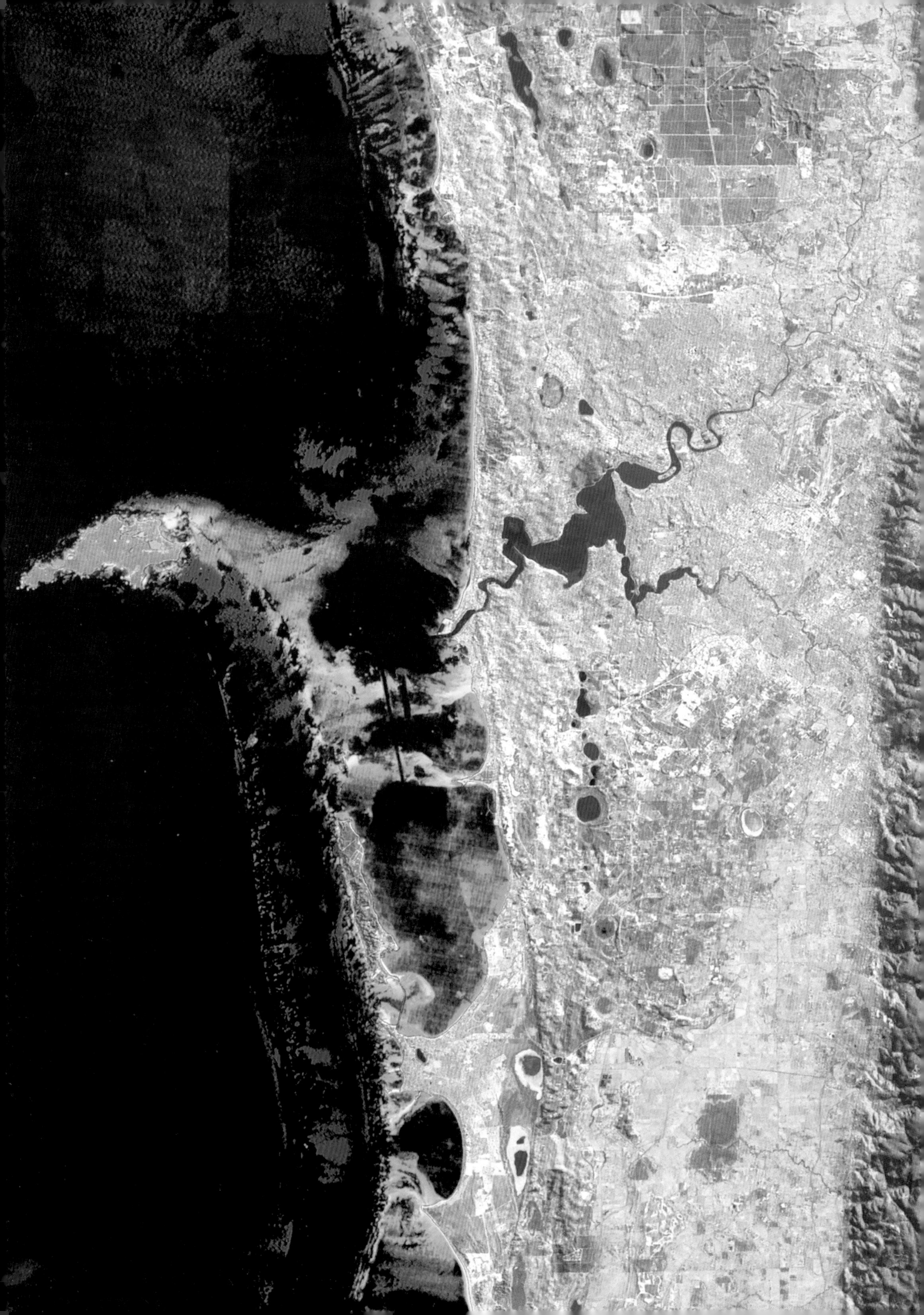

SECTION ONE

RIVERS RUNNING THROUGH TIME

WESTERN AUSTRALIA—WHERE ANCIENT LAND SURFACES ENABLE A STUDY OF DRAINAGES THROUGH GEOLOGICAL TIME

Satellite imagery is a wonderful new tool which reveals much more than large-scale snapshots of the Earth. The geological history of the landscape in this portion of Western Australia, with the Swan Estuary in the centre, is laid out before us and can be read like a book. One of the most ancient landscapes preserved anywhere on Earth—the wrinkled, ancient Yilgarn Block—forms the eastern half, and a clear-cut line, the Darling Scarp, separates it from the geologically young Swan Coastal Plain along its western margin. The Scarp tells us about the opening of the Indian Ocean and the creation of Western Australia's coastline during the split-up of Gondwana. It marks the edge of the rift which developed between India and Australia in the first stage of our continent's separation. The low, flat plain with its dune ridges and drowned off-shore features, tells of the advance and retreat of the sea during sea level fluctuations caused by polar ice melting and building up again in interglacial and glacial stages of the Pleistocene ice age. Only 18 000 years ago, sea level was 130 metres lower than it is today, and a wide stretch of the continental shelf was exposed. The offshore islands—Rottnest, opposite the Swan Estuary, and Garden Island to the south—were raised areas on the exposed continental shelf and the coastline was many kilometres further west. Rivers cut deep channels across the continental shelf to reach the ocean. Sea level stabilised only about 6000 years ago and then the estuaries and other features of the modern coastline developed.

The themes captured in this satellite snapshot are repeated and are described in the wider context of the whole of Western Australia in this section. The State of Western Australia comprises about one-third of the continent and encompasses much that relates to other areas, so it provides an introduction. Because ancient land surfaces

Satellite image of a slice of Western Australia from the coast at Perth to the ancient Yilgarn Plateau.
Landsat imagery provided by Australian Centre for Remote Sensing (ACRES), Australian Surveying & Land Information Group, DIST, Canberra, and digitally enhanced and produced by Satellite Remote Sensing Services, Department of Land Administration, Perth, Western Australia.

are preserved over large areas, the opportunity exists to trace the ancient drainage patterns—the *palaeodrainages*—and to see that modern landscapes retain a mixture of ancient and modern features. Many modern rivers lie in valleys which were carved out by *palaeorivers* incredibly long ago; chains of salt lakes in the arid areas follow drainage lines which have not seen perennial streams for many millions of years. The history of climates through the ages can be read in sediments preserved in the ancient drainages.

The Yilgarn Plateau

The Yilgarn Plateau is an ancient building block (craton) of Australia, with some of the oldest rocks found anywhere on Earth and some of the oldest preserved landscapes. The satellite image on page 11 shows the low-relief patterns of creases and rumples—little changed on this scale from nearly 300 million years ago. The earliest event relevant to the modern landscapes is the Late Carboniferous to Early Permian ice age of about 290–270 million years ago when continental ice sheets ground across the already ancient surface, levelling, gouging hollows and leaving low bosses between. At first, after the ice had gone, there was no organised drainage. As time went on, rain fell, water sculpted the surface and drainage patterns etched themselves across the land, deepening with time but barely keeping up with the slow erosion that was wearing the landscape down further.

Today, a central watershed runs north to south, roughly down the middle, with elevations of about 600 metres above sea level in the north decreasing to 300 metres in the south. On its western slopes, rivers are active—including the Swan, Murchison, Gascoyne, Ashburton and Fortescue. On the eastern slopes, rivers which previously drained into the Canning, Officer and Eucla Basins have dried up and today are represented by chains of salt lakes. There are no rivers of major size flowing to the south coast, where drainage is much younger, developing after the southern continental margin tilted along a hingeline, the Jarrahwood Axis (or Ravensthorpe Ramp), which created a drainage divide between the coastal region and the Plateau about 30 million years ago.

The Canning and Officer Basins were uplifted at the end of the Cretaceous and developed river systems of their own—the Canning, draining north-west, and the Officer to the south. The Eucla Basin was uplifted more than 30 million years later, after the Miocene marine incursion, to become the Nullarbor Plain (also known as the Bunda Plateau).

By the end of the Cretaceous or early in the Tertiary (65–60 million years ago), much of the Western Australian landscape was similar to today, its palaeodrainage patterns established.[5, 5A] Significant flow in the central-desert palaeodrainages stopped before the Mid Miocene—rivers have not been active there in the last 15 million years.

Seven major palaeodrainage provinces can be recognised:

- Valleys that drained via the Great Sandy Desert into the Indian Ocean. Remote sensing has enabled the detection of palaeoriver systems beneath the desert sands, where the salt lake chains of Percival Lake and Lake Gregory (in Canning Stockroute country) are the surface evidence for their existence.
- Valleys in the south-eastern Great Sandy Desert and north-eastern Gibson Desert which appear to have formed the only true internal drainage system in Western Australia.
- Valleys that drained via the Northern Territory into the Bonaparte Gulf.
- Valleys that drained eastward through the Northern Territory into the Finke River and hence into Lake Eyre (the playa chain of Lakes Hopkins, Neale and Amadeus—now the Central Groundwater Discharge Zone).
- South to north trending valleys in the south-western part of the State that drained into active rivers flowing into the Indian Ocean (the Avon—Swan, Canning) and the east to west flowing Beaufort and Darkan palaeochannels whose rivers probably discharged across the Darling Fault into the Perth Basin.
- Valleys that drained into the ancestral Great Australian Bight before its emergence as the Nullarbor. The presently isolated

DRAINAGE AND PALAEODRAINAGE OF WESTERN AUSTRALIA

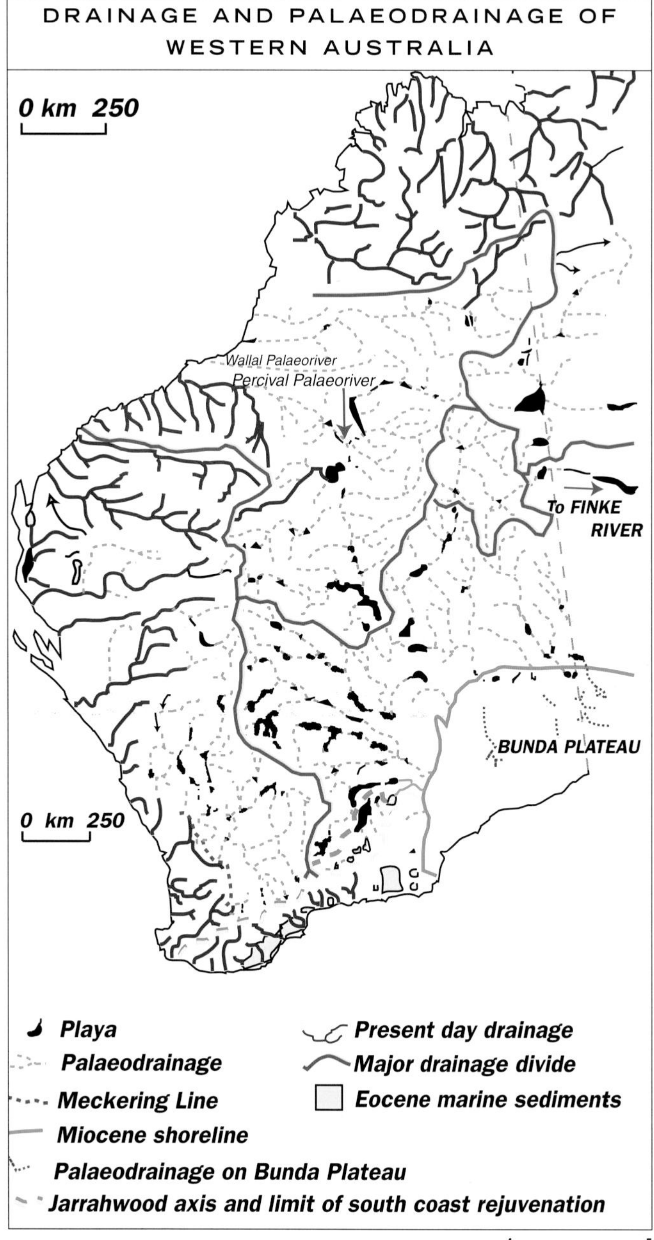

After Beard,[4] and van de Graaff *et al*[5]

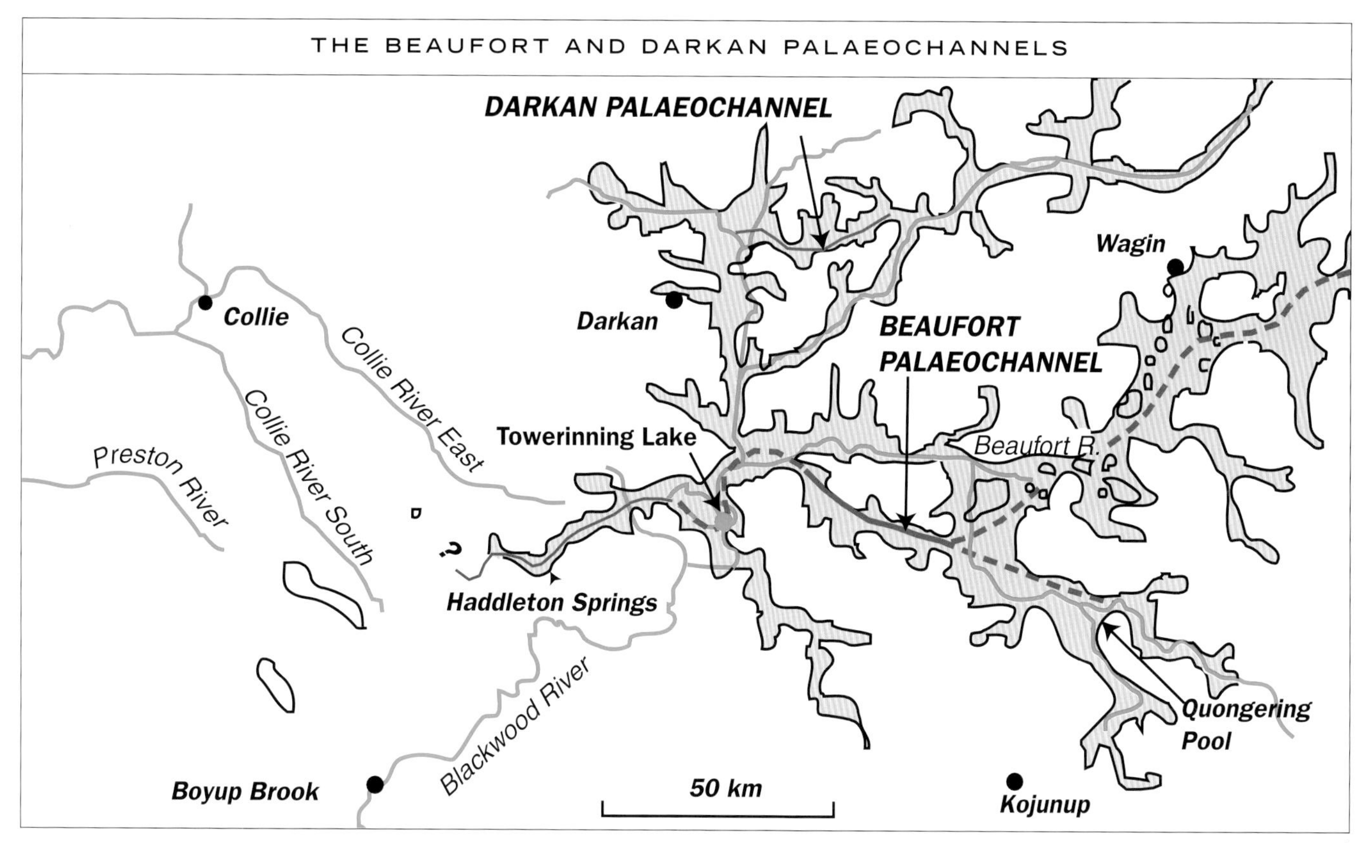

After Waterhouse *et al*[6]

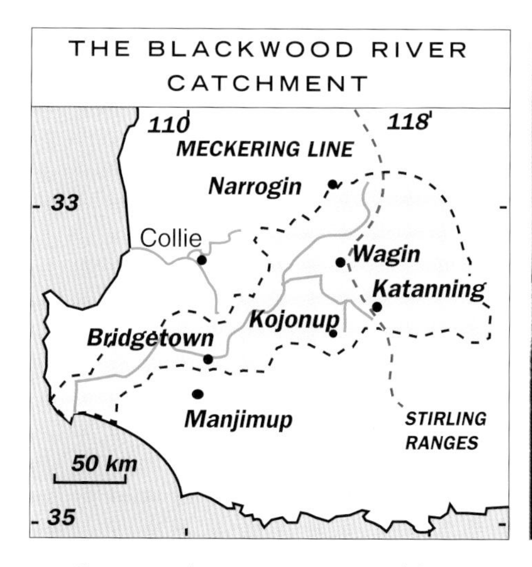

About 49 per cent of the upper catchment of the Blackwood River drains into Lake Dumbleyung, which only overflows about once every 20 years. The lake intercepts silt and salt from the saline headwater streams, which would otherwise freely enter the river downstream. M.E.W.

Cowan Drainage Basin is part of this group. These drainages can be traced for hundreds of kilometres, and stop abruptly at the edge of the Nullarbor, mostly ending in playas (e.g. Plumridge Lakes, Lake Boonderoo).

- A few isolated and discontinuous palaeodrainages on the Nullarbor. Evidence suggests that the climate on the Nullarbor has been arid since its emergence, limiting surface drainage.

The age of the palaeodrainages can be determined by dating their valley-fill deposits. In the Kalgoorlie region, Late Eocene (45–34 million years old) lignitic sediments form valley fills, like the Rollo's Bore beds which are up to 120 metres thick. Under Lakes Lefroy and Cowan such beds are overlain by 30 metres of Late Eocene marine Norseman Limestone, so the palaeodrainages were established before the Late Eocene.

The Beaufort and Darkan palaeochannels on the western margin are the first fairly complete east to west flowing palaeochannels to be recognised in the south-western sector of the Yilgarn Craton.[6] Part of the Beaufort palaeochannel, extending more than 60 kilometres, has been traced from Quongering Pool, 25 kilometres north-north-east of Kojonup, to Haddleton Springs. It probably discharged across the Darling Fault into the Perth Basin as the ancestor of the Preston or Collie River (or both).

The sediments in the palaeochannel comprise up to 65 metres of sands, silts and clays which record the Eocene history of the drainage system, from 45 to 35 million years ago. It was originally a meandering river system, and as the Yilgarn block tilted to the south during this period, the gradient of surface flow was reduced along the palaeoriver, creating lakes in the palaeovalley (as evidenced by *lacustrine* sediments). Following this tilting, the diversion of drainage which resulted saw the modern Blackwood River capturing the headwaters of the ancestral Beaufort River and beheading it by cutting through Archaean bedrock south of Lake Towerrinning. The palaeochannel west of Towerrinning has also been cut by Darlingup and Haddleton Creeks, both of which are southward flowing tributaries of the Blackwood River. (The Blackwood catchment shows no evidence in its sedimentary record of penetration by the Eocene Sea. The modern status of the Blackwood is the subject of a case study in *Listen ... Our Land is Crying.*)

The palaeodrainages are of major economic importance in arid Western Australia. The valley calcretes which occur in many trunk valleys form the major freshwater aquifers, and in landscapes where all surface waters are saline they are a vital resource. At Yeelirrie, in the Wiluna area, carnotite-bearing valley calcrete forms a major uranium deposit. Smaller deposits have been reported from Lakes Way and Maitland near Wiluna, and from near Mt Venn. In the Kalgoorlie region, alluvial gold was mined from deep leads. Most of the deep leads are infilled tributaries of the trunk valleys of the palaeodrainage systems.

THE KALGOORLIE REGION

Kalgoorlie lies within the Eastern Goldfields Province of the Yilgarn Craton. Mining of gold, which comes in tellurides and has to be smelted out of the rock, has had a widespread impact on the region. Furnaces needed wood, and wood trains radiated out into the countryside removing anything burnable. Water was always a problem in this arid region and a major pipeline was constructed from Mundaring near the coast. In recent years much drilling and exploration has taken place, mapping the palaeodrainages which contain saline water to allow expansion of the mining and towns in the region.

PALAEODRAINAGES OF THE EASTERN YILGARN

Kalgoorlie lies at the centre of the Roe Palaeodrainage,[7] a Cretaceous to Early Tertiary drainage system which discharged eastwards into the Eucla Basin. The Roe palaeoriver flowed through the existing Lake Roe area. This palaeodrainage is separated by palaeo-divides from the Rebecca Palaeodrainage to the north, and the Lefroy to the south.

The Kalgoorlie area is part of the Yilgarn Block, with some of the most ancient landscapes preserved anywhere in the world. Gently undulating, with subdued relief, low breakaways, granite tors and greenstone ridges, and with strings of playa lakes in low areas, it reflects the amazing age of its landforms and the processes of deep weathering and erosion to which it has been subjected through geological time. It is considered possible by experts[8] that the area may have been land and already well planated in Precambrian times (as much as 600 million years ago) but, ignoring such speculation, and only considering evidence for which there is some proof, a time frame of nearly 300 million years is no less awe-inspiring. Flat-lying Permian fluvio-glacial rocks on the Yilgarn are evidence of glaciation during the Late Carboniferous to Early Permian ice age. Whether the massive ice sheets which covered the land then were responsible for creating the low relief landscapes, or whether they merely contributed to the flatness by further grinding down the surface over which they moved, is not known. The flat landscapes of hollows and low ridges may have characterised the region before the ice age. In either case, palaeodrainages have since have etched their creases across the ancient planated surfaces.

GONDWANA WITH ICE SHEET
300 million years ago

The north-west to south and south-east trending greenstone belts and intervening granites determine the landforms. Two predominantly greenstone belts in the north-east and south-east of the Kalgoorlie region give rise to ranges of low hills, strike ridges and broken slopes with plains between the ridges. Granite terrain in the north-west and centre of the area consists of sinuous ridges and extensive debris fans. In contrast, granite country in the south-west consists of undulating sandy plateaux with rims of exposed ferricrete, straddling a major palaeodrainage divide between the Swan–Avon Basin (draining to the west) and the Kalgoorlie region which drains to the east. Valleys have been incised into the plateaux and plains occupy the surrounding areas. Granite hills are numerous in the south.

The chains of playa lakes occupy the palaeodrainage channels, forming a dendritic pattern. Their wide valleys drained east and south-east towards the Eucla Basin, which was a sea during the Eocene and again in the Miocene before it was uplifted to become the Nullarbor, and south towards the Bremer Basin, to the west of the Eucla, which was also a sea during the Eocene.

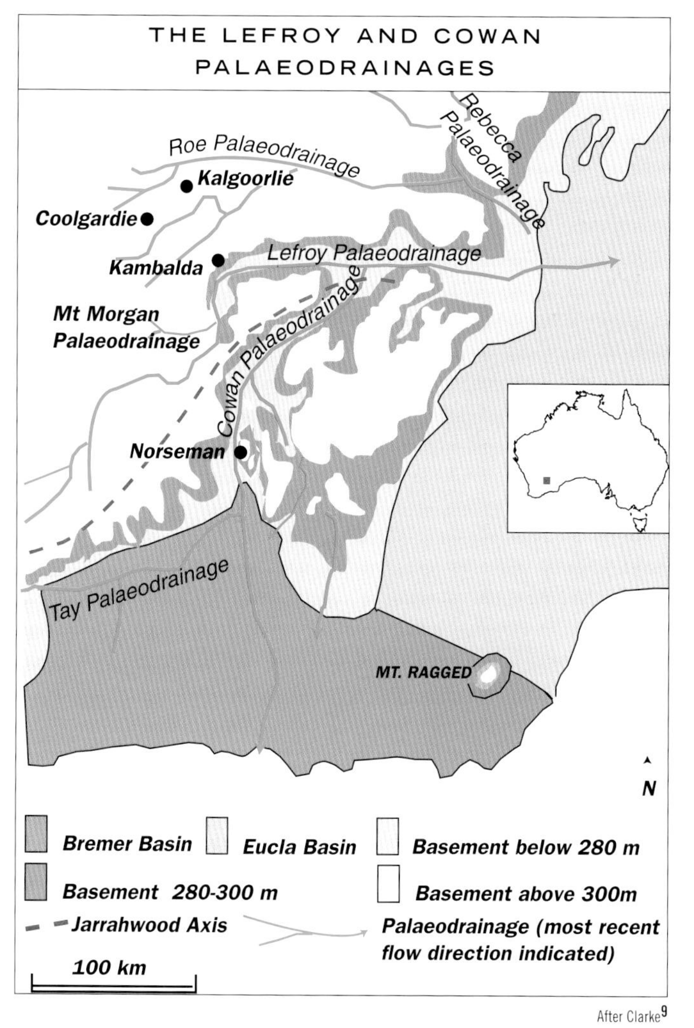

After Clarke[9]

Eocene sediments, some with plant remains and spongolite, occur as valley fill in parts of the palaeodrainage systems, penetrating as far inland from those palaeo-seas as Norseman. (Spongolite is a sedimentary rock which largely comprises siliceous sponge spicules. Sponges, which have little needles of silica in their spongy tissues, flourished in vast monoculture 'forests' in the warm waters of Eocene seas.) The valleys are older than the tectonic movements along the Jarrahwood Axis which occurred in Mid Tertiary about 30 million years ago, creating a drainage divide parallel to the south coast.[8]

A strong case has been made for the palaeodrainage systems predating the separation of Australia and Antarctica, in which case the markedly south trending valleys (and their connections south of the Jarrahwood Axis in the Esperance region) would represent palaeoriver valleys which ran from Antarctica into Australia.

A timetable of the evolution of the Kalgoorlie region of the Yilgarn shows the following stages:

- **Permian glaciation** created a palaeoplain when ice ground across the landscapes 300 million years ago (or further denuded an even more ancient planated landscape).
- In the **Mesozoic**, slow erosion of the landsurface continued. Major valleys of at least Early Tertiary age are incised into what was already a remarkably level land surface.
- **Establishment of a drainage pattern** with major valleys several kilometres wide. The modern lines of salt lakes follow these old valleys. There was a major divide between the Swan–Avon catchment and the drainage in the Kalgoorlie region.
- **The break up of Gondwana** started in the Jurassic. The broad valleys of the Swan–Avon catchment are wide at their start. They used to carry major rivers from Antarctica before rifting. A

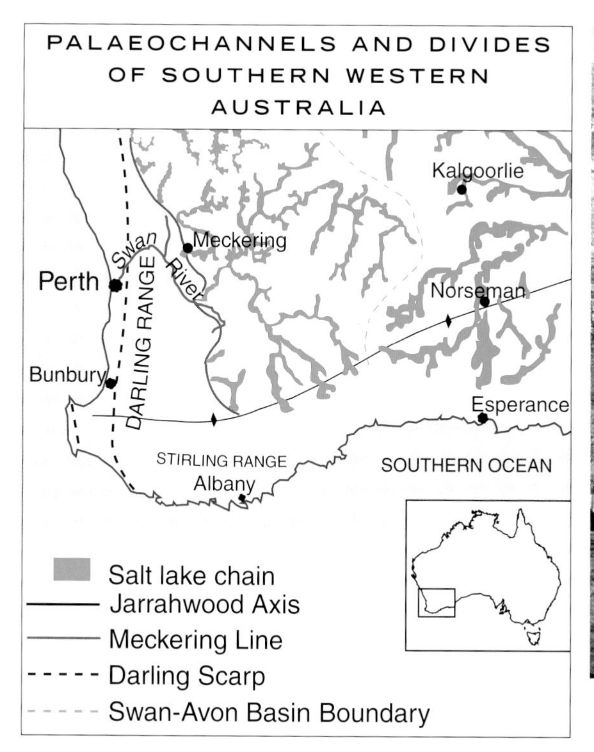

Section at the Princess Royal Mine, Norseman, showing Eocene valley fill beneath marine limestone. AGSO

EVOLUTION OF THE LEFROY AND COWAN PALAEODRAINAGE CHANNELS

SATELLITE IMAGE: DUNDAS

Landsat imagery provided by the Australian Centre for Remote Sensing (ACRES) Australian Surveying and Land Information Group, DIST Canberra, and digitally enhanced and produced by Satellite Remote Sensing Services, Department of Land Administration, Perth, Western Australia.

Lakes Lefroy, Cowan and Dundas occur within a palaeodrainage system that may have originated in pre-Jurassic times (prior to 200 million years ago).[15] The acute angles formed by convergence of the Cowan and Lefroy systems indicate that the combined system originally flowed north and east. The southern end of the Cowan palaeodrainage has been truncated by the modern coastline at Esperance. Its width there is poorly defined but probably exceeds 10 kilometres. This wide palaeovalley suggests that the original palaeoriver flowed from headwaters in Antarctica, prior to the start of rifting in the Jurassic which created the Bremer Basin.

PALAEOGEOGRAPHIC EVOLUTION OF THE SOUTH-EAST YILGARN

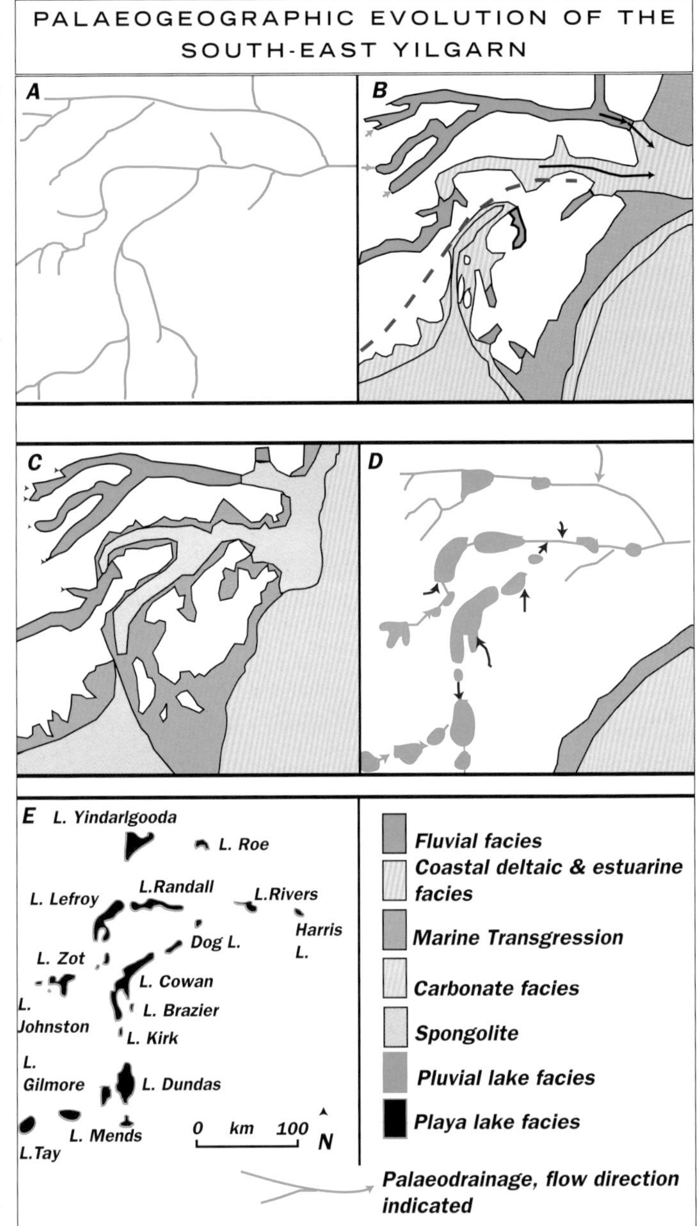

A: Palaeodrainage channel incision, post-Jurassic.
B: In the early to middle Eocene, lignitic silt of the Werillup Formation was laid down directly on Precambrian basement. The extent of the following Middle Eocene Tortachilla transgression (into the Cowan palaeodrainage from the Bremer Basin) is shown by the distribution of Norseman Limestone.
C: The more extensive Aldinga transgression saw spongolite laid down in the Lake Cowan and Dundas palaeovalleys. The seas regressed and marginal marine and freshwater sedimentary deposits followed.
D: Throughout the Miocene semi-permanent lake sedimentation took place and the connecting rivers started to dry. Lake sediments were becoming saline. Gypsum was being deposited in Lake Lefroy by the end of the Miocene.
E: The Pliocene saw aridity and salt and gypsum deposits—the Roysalt and Polar Bear Formations—in the chains of salt lakes in the wide palaeodrainage valleys.

After Clarke[15]

warm, wet climate, even in the high latitudes while Australia was connected to Antarctica, and high mountain ranges in Antarctica, provided water for these major palaeorivers which carved the wide valleys. Major south to north trending palaeovalley systems in the south-eastern Yilgarn, particularly the Cowan palaeodrainage, are also remains of the Antarctica connection.

- **Incision of valleys.** By the Eocene, rivers had cut a system of valleys into bedrock. The Late Cretaceous and Early Tertiary had been warm and wet, and rivers were active. During the Eocene, valley-fill sediments accumulated. An Early to Late Eocene pollen record in Lake Lefroy contains grains from 26 species in 12 genera, including 8 species of Proteaceae, 4 species of *Nothofagus* (southern beech), and tree ferns. Other pollen records of similar age in parts of the palaeodrainage systems indicate an abundance of Araucarian and Podocarp conifers in the vegetation, and *Nypa* palms and mangroves were present in marginal marine areas. The palaeovalleys obviously supported riverine rainforest.
- **Eocene marine incursions.** Two marine incursions occurred during the Middle to Late Eocene**.** The first, Tortachilla, led to the deposition of marine limestones in the Cowan palaeochannel; the second, more major Aldinga transgression of the Late Eocene penetrated far up valleys, and marine sediments are found at Norseman and near Lake Cowan, as much as 270 kilometres from the present coast. The Princess Royal spongolites were laid down during the Aldinga incursion.
- **Tectonic activity uplifted the area,** in some places by as much as 300 metres, while the southern edge tilted, forming the Jarrahwood Axis parallel to the coast. Some of the Eocene palaeorivers crossed this axis, as did the Cowan near Norseman, and the slope of rivers south of this Axis is reversed.
- **Landscape evolution after the Eocene.** After the filling of the valleys with Eocene sediments, weathering and erosion continued. Duricrusts were formed on surficial deposits at different times. Australia was drying out as it drifted northwards as an island continent, rivers on the Yilgarn were drying up, and those in the palaeovalleys east of the main drainage divide have not run perennially in the last 15 million years.

Lake Boonderoo, November 1996. ANDREW CHAPMAN

LAKE BOONDEROO AND THE EASTWARD-DRAINING PALAEORIVERS

Lake Boonderoo lies on the western edge of the Nullarbor Plain, about 20 kilometres south-east of Kitchener.[10, 11] Ponton (Goddards) Creek, which follows the lower half of the channel in a north-west to south-east trending palaeodrainage which extends for approximately 500 kilometres, terminates in the lake. The palaeodrainage contains the string of Raeside playas, which extends from 75 kilometres west of Leonora. A large inward-draining region west of the first lake in the chain was headwaters for the palaeoriver. A second palaeodrainage, south of the Raeside, contains the Lake Rebecca playas. It joins Ponton Creek which, like all the other drainages around the edge of the Nullarbor's limestone plain, was cut short when the plain was uplifted. The palaeochannels drain the north-eastern Goldfields.

Lake Boonderoo is normally dry. In recent times it has filled only twice, with 20 years between the events:

- **1975:** A year of above average rainfall including Cyclone Trixie which deluged the Murchison and Goldfields region in February. Water remained in the lake for eight years following this event. It was suggested that over-grazing by sheep of the eastern Goldfields as far away as Leonora contributed to this lake-filling event and the added run-off enabled the water to penetrate further downstream and reach the lake. The railway bridge was washed away for the first time since World War I, and no records exist of previous flooding there.
- **February 1995:** Cyclone Bobby brought torrential rain to the north-eastern Goldfields. Rainfall figures for February in millimetres (mean in brackets) illustrate the magnitude of the rainfall event: Cashmere Downs 367 (23), Kalgoorlie 241 (12),

Leonora 285 (25), Sandstone 271 (26), and Yundamindra 347 (22). There had been some water as far south in Ponton Creek as Goddards Bridge (crossing the railway) before the cyclone as a result of some local rainfall, and it took the water from the cyclone six weeks to reach the bridge, and another week to run out across the edge of the Nullarbor and fill Lake Boonderoo. Waterbirds appeared on the lake shortly afterwards, as they do on other desert lakes whenever they have water.

WATERBIRDS AND WESTERN AUSTRALIA'S DESERT LAKES

The playas of the Yilgarn Block in Western Australia are usually dry, salt-encrusted pans, unbearably hot in summer, freezingly cold in winter nights, and apparently devoid of life. Some of them are enormous—up to 1800 square kilometres in area. Occasionally the tropical cyclones which develop off the north-west coast degenerate into rain-bearing depressions and travel across the continent in a south-easterly direction, dumping 300–400 millimetres in three or four days. The land is flooded, and the dry saltpans are transformed into small seas.[12,13]

A miraculous transformation occurs and abundant life appears from apparently nowhere in the playas—now lakes. It is hard to imagine how small eggs, like those of the fairy shrimp, *Parartemia*, could survive the heat, the cold, the salt and the dryness over long periods—up to ten years between good drinks in some cases—and be ready to swell and hatch and populate the saline, ephemeral waters for a brief period. But this they do in their billions.

Still more miraculous is what happens next. What signal is received, what inbuilt genetic memory and response is activated, is unknown—but within days of the filling of the lakes, banded stilts in their thousands arrive, having flown up to 1000 kilometres from permanent waters near the coast. They have come to breed, and they have been waiting for the signal to do so for five or even up to ten years! Within two weeks of the end of the cyclonic rains they have formed island colonies of 20 000 or more nests, all 30 centimetres from each other, each containing two to three eggs. Three weeks of incubation follow, and within a day of the chicks' hatching they are feeding on shrimp at the lake's edge. Small parties leave the islands, paddling downwind, feeding as they go, until tens of thousands of birds are gathered in the shallows at the end of the lake, often as much as 50 kilometres away from their nests.

Lake Barlee—full in June 1992. ANDREW CHAPMAN

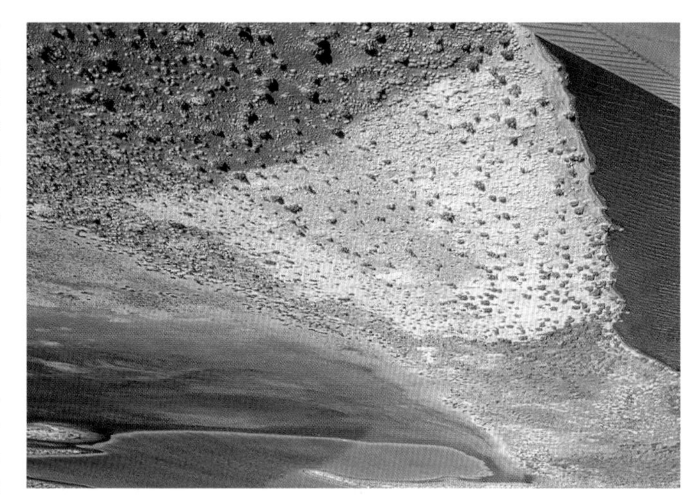

A banded stilt colony, Lake Marmion. ANDREW CHAPMAN

Stilts flying over Lake Marmion. ANDREW CHAPMAN

BANDED STILTS

Banded stilts are Australian endemics.[12,13] They are unique among the 214 species of wading birds in the world in that they nest colonially, lay white eggs (with a few black streaks and blotches), have white, downy chicks, put their young into creches, and have a special habitat requirement—recently flooded playa lakes—because they rear their young on brine shrimps. Only 20 nesting events have ever been recorded, three in inland South Australia, and the remainder in the southern interior of Western Australia.

REG MORRISON

It takes seven weeks for the chicks to be fully feathered and ready for the long flight home. The whole breeding event encompasses less than four months. Usually, the lakes dry within six months, becoming steadily more saline as they evaporate, and eventually killing all the shrimps. But the shrimps' magical drought-resistant eggs are safely in the silt and will wait for the next flood … and for the next breeding event of those wonderful birds for which they provide the energy and sustenance.

While the banded stilts and the brine shrimps have this almost-beyond-belief 'coincidental' relationship, which presumably has a very long evolutionary history, other waterfowl also make use of the lakes and wetlands of the arid interior. Unlike the stilts, which only breed when they 'know' that the food source is there, other species are opportunistic and nomadic and do not restrict their breeding exclusively to the rare major events for which they may have to wait for a decade. Whenever water arrives in the arid interior—and its persistent presence for long enough to sustain waterfowl breeding is almost randomly scattered—some birds find it and use it. In exceptional rainfall years, like 1992–93 in the Goldfields region, ducks of many sorts, grebes, herons, coots, crakes, black swans, plovers, avocets, dotterels, terns and stilts were all breeding in large numbers in ephemeral freshwater lakes and wetlands as well as in the brackish to salt playa lakes.

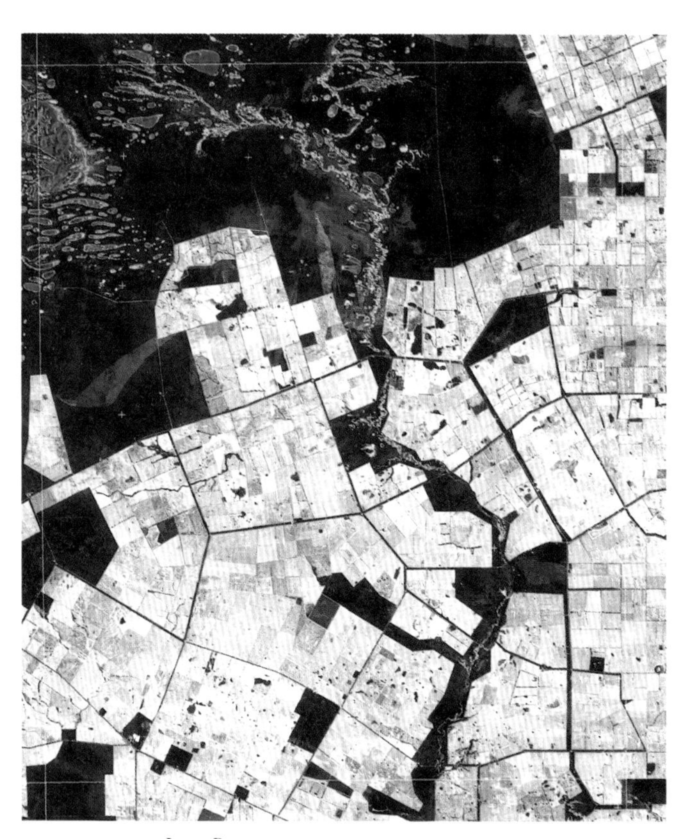

SATELLITE IMAGE: LORT RIVER
Landsat imagery provided by Australian Centre for Remote Sensing (ACRES) Australian Surveying & Land Information Group, DIST Canberra, and digitally enhanced and produced by Satellite Remote Sensing Services, Department of Land Administration, Perth, Western Australia.

THE ESPERANCE REGION: *landscape evolution of the south coast of Western Australia*

At the start of the Eocene (55 million years ago) Australia lay between 60° and 30° South, but even the southern edge of the continent had a warm, wet climate. (Climate in the Middle Eocene is believed to have been subtropical; a 5°C decrease occurred with cooling and drying in the Late Eocene. Rainfall was about 1500–2000 millimetres throughout most of the period.) A proto-Leeuwin Current down the west coast of Western Australia and around the corner into the Great Australian Bight Basin provided warm water for a proliferation of sponges. Global sea levels were high and the sea entered the on-shore Bremer Basin (which roughly corresponds to the Fitzgerald National Park, with an extension along the coast eastwards into the Esperance region), and the Eucla Basin (which was to become the Nullarbor). In the Bremer Basin, a wide embayment of the Southern Ocean resulted and marine limestones, including spongolite layers which are mainly derived from siliceous sponge spicules, were laid down. Sea penetrated far up river valleys from the main embayment and also penetrated inland in the Esperance region along major palaeodrainages northwards as far as Norseman. (Palaeodrainages flowing into the Eucla Basin epicontinental sea also have spongolite deposits in west to east flowing palaeorivers of the Lefroy and Cowan palaeochannels.)

The Barren Ranges, between Bremer Bay and Hopetoun, have a wave-cut basal platform 90 metres above sea level, formed by the Eocene Sea. Peaks of the Barren Ranges and other promontories in the Albany–Fraser Province were islands in the Eocene Sea, just like the islands of the Recherche Archipelago today. The spongolite mesas in the Fitzgerald National Park are the result of the dissection of the plateau that originated from the uplift of the marine sediments after the sea had retreated. They are now 300 metres above sea level.

The Eocene Sea was at its greatest extent in the Late Eocene. After its retreat, changes occurred to the regional drainage. A new divide (the Jarrahwood Axis) developed parallel to the south coast on a line from Augusta near Cape Leeuwin, to Kojonup and Ravensthorpe, and on to the edge of the Nullarbor. This axis resulted from the uplifting of the Darling Plateau, which began in the Oligocene, about 30 million years ago, causing tilting of the south coastal region towards the south. The elevated divide which developed along this axis, the Ravensthorpe Ramp, rejuvenated rivers to its south and reversed the flow in the southern portions of ancient valleys which had carried rivers northwards into Australia from Antarctica.

A spongolite breakaway on Twertup Creek in the Fitzgerald National Park. Flat-topped mesas in the background are remnants of the plateau created when the marine sediments were elevated and later eroded. M.E.W.

Fossils in sediments in the Bremer Basin are a key to interpreting landscapes and climate of the Eocene. The widespread forests were Gondwanan 'mixed' forests (described in detail in *After the Greening*) containing numerous species of *Nothofagus* (Antarctic or southern beech;) podocarps and araucarians; *Ficus* (fig trees); *Brachychiton* (Illawarra flame trees and kurrajongs); banksias and their relatives; *Eucalyptus, Gymnostoma* (a type of casuarina now confined to tropical regions); and *Typha* reeds. At Balladonia on the Nullarbor, *Acacia* and grasses occur in strata of the same age. The Gondwanan forests were mixed in that they comprised taxa which have since undergone sorting and sifting according to their ability to cope with increasing dryness, as global climate changed gradually from the uniformly warm and wet Eocene towards the establishment of modern climatic zones. The Gondwanan biota was undergoing this adaptation while the Australian continent moved northwards to become the driest vegetated continent. Some broadleaf elements of the mixed forest are now confined to the wet tropical and subtropical areas; some to cool temperate provinces; most to sclerophyll communities which now populate most of the continent.

Water use—water balance in the Esperance region

The main limiting factor for agriculture in the Esperance region is water. [14,14A] Problems arise because European agriculture and land-use practices upset the delicate balance in hydrological systems in this most marginal region. It is almost ironic that in this environment dominated by aridity, many of the problems are associated with too much water becoming available at the wrong times.

Native vegetation in Esperance and Ravensthorpe Shires uses 99 per cent of the rain that falls. The agricultural crops and pastures which replace the natural vegetation in this heavily cleared region use at most 75 per cent of the rainfall; the 20 per cent which is not being used causes rising water tables, seasonal waterlogging and even overland flooding. Crop yields are reduced on up to 60 per cent of the Esperance sandplain as a result. Recharge of the water table results, because the introduced systems cannot use enough of the water when it falls. Even a very small recharge change (less than 5 per cent) can have big effects. The balance is so sensitive that any small changes in surface hydrology result in waterlogging and secondary salinity. This is because the soils are mainly sands over relatively impermeable clays. Such soils have low water-holding capacity and store large quantities of salt.

A major problem is one of the incompatibility of European cropping and pastures with rainfall patterns, not to mention with the soils, or combined with the even more fundamental problem of the Eurocentric thinking that imposes such inappropriate and unsustainable use on fragile salinity-prone landscapes. After a period of good water use in the spring, the crop or pasture dries off and the water-use system fails until the next growing season, next spring.

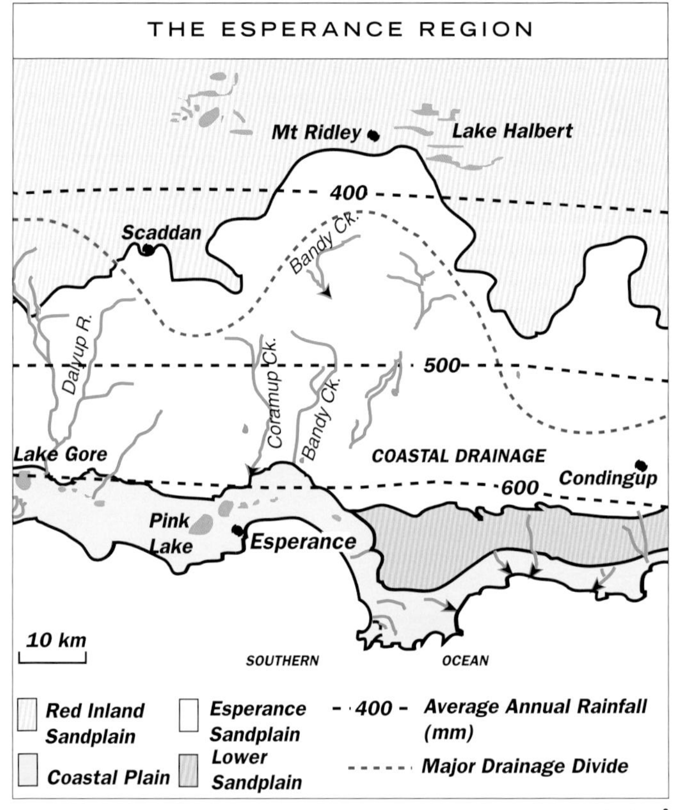

After Clarke[9]

The Pink Lake at Esperance turns this wonderful colour when an algal bloom of Dunaliella salina *transforms it.* WATER & RIVERS COMMISSION, PERTH

*A granite mound—Mt Burdett—which rises abruptly out of the sandplain. The resurrection plant (*Borya sphaerocephala*) forms attractive gardens along cracks in the massif.* M.E.W.

Wheatfield Lake, one of a string of sandplain lakes at Esperance. M.E.W.

Approximately one-third of the annual rainfall comes outside the growing season. In order to maintain water use throughout the year, perennial plants, which continue evapo-transpiration in all seasons, are needed to maintain balance. Because of the climatic uncertainties and drought, and the very high incidence of wind erosion, even planting perennials remains a high-risk proposition. One has to admire the courage and the considerable expertise of the farmers of today who are using the latest technology, keeping up with the latest scientific thinking and managing to survive financially, if precariously, in this fundamentally hostile environment.

The planting of exotic, high water-use trees is increasingly being used as a solution to some of the waterlogging and salinity problems.

A round 'Esperance dam', clay-lined to prevent seepage into the sand-plain, Oldfield catchment. M.E.W.

Native Western Australian species have often proved not sufficiently adaptable to the changed conditions, and exotics like *Pinus pinaster* and *P. radiata*, and eucalypts from other parts of Australia, such as blue gums (*Eucalyptus globulus*) and oil mallees, are being planted in enormous numbers. Attempts are being made to spread clay to overcome water repellance. (The shiny surface of the sand grains which make up the sandy soils promote run-off instead of absorption. The addition of fine clay particles promotes absorption.) Rotations are modified in order to minimise the bare ground which increases wind erosion, and minimum-till practices are used for the same reason.

Difficulty in establishing and maintaining water storages on farms is another problem. The Esperance District has low intensity rainfall. Surface flows from farmland catchments are not annual events; soils are sand; evaporation is high. Specially designed 'Esperance Dams', which are clay-lined in order to stop seepage of the water into the sandy soil, are a feature of the region. Wind erosion is an ever-present threat or problem; acidification and decreased productivity of soils results from fertiliser use. Dryland salinity has the potential to destroy 35 per cent of sandplain and 50 per cent of mallee sub-regions when groundwater levels eventually reach equilibrium.[16] Water-repellant soils, waterlogging and weeds and rabbits are additional problems. Ephemeral drainages, optimistically called 'creeks' by the locals, are usually salt-encrusted tracks across the flat landscapes. Some have drainage canals cut to channel water when rain comes and local flooding is a nuisance. Rivers like the Oldfield have billabong reaches which retain water during the dry season.

One can only wonder if any sort of cultivation of such marginal land is really justified. Some farmers have multiple properties because the overheads and inputs are so large that only by working very large areas can they make a precarious living. And yet a mood of optimism prevails among this younger-than-average farming community, and while as a conservationist one may disapprove of the risks to the environment which have been, and are being, taken, one can only be inspired by the spirit which drives people to find solutions to seemingly unsolvable problems.

(It has recently been suggested by Professor Bob Gilkes of the University of Western Australia's Centre for Land Rehabilitation that many of the waste products produced by mining and minerals processing could be used to improve the quality of soil and water resources. Vast areas of Western Australia are affected to some extent by the problems seen in the Esperance. About 90 per cent of the State's

'CREEKS' WHICH ARE NO MORE THAN DRAINAGE LINES

A salt lake at the top of the Dalyup catchment.

Disorganised, wide, saline drainage channel, top of the Dalyup catchment.

Speddingup Creek, a tributary of the Dalyup—a wide, almost flat area of salt drainage.

Speddingup Creek, drained to prevent flooding.

A drainage lake and dead paperbarks on the flat, saline floodplains near Esperance. M.E.W.

soils suffer from a lack of nutrient ions; salinity affects 30 per cent; waterlogging affects 40 per cent and water repellance in sandy soils affects another 30 per cent. The clay needed to combat water repellance is produced in quantity in processing of mineral sands and lateritic gold, both widespread activities generating a lot of the waste product in many parts of the State. Another waste product is quarry washings—the fine material produced when aggregate and other quarry products are processed. Usually ending up as land fill, this waste contains large amounts of plant nutrient elements such as calcium, magnesium, phosphorus, iron and manganese, so widely deficient in the soils. The 'red mud' by-product of bauxite mining is already used in agriculture on sandy soils where it helps to bind phosphorus and make it available for up-take by plants—which also decreases its concentration in run-off and helps to decrease nutrification of streams and waters. Professor Gilkes believes that we have the opportunity to turn two environmental problems—mining waste and land degradation—into a solution. It is only innovative thinking like this followed by action that is going to solve the massive problems confronting Western Australia and the rest of the continent.)

Bandy Creek, near Browns' Farm. M.E.W.

Tree planting near the headwaters of Bandy Creek, Wittenoom Hills. M.E.W.

THE OLDFIELD RIVER

The Oldfield River rises on the edge of the Yilgarn plateau about 95 kilometres inland, at about 300 metres above sea level. Its headwaters are in gently undulating and often saline country, cutting shallow valleys through sandstone to expose granite, before the river carves deep valleys through siltstone as it descends to the coastal plain. The Munglinup River is a large tributary of the Oldfield whose flow originates from within the sandplain.

The Oldfield catchment is approximately 248 000 hectares and the river discharges about 6900 megalitres (ML) a year on average. The upper reaches remain naturally vegetated but, with the characteristic low rainfall and low run-off, would account for less than half of the annual discharge. The rivers flow briefly each winter, or negligibly in dry winters, and are naturally brackish to saline.[16] Salinity levels have been elevated by clearing in the lower catchment, and siltation has also occurred as a result of this, as well as filling pools and clogging channels. High nutrient levels, resulting from the heavy application of artificial fertilisers, affect the river.

River sections have been evaluated for the Oldfield (and for other rivers around Esperance) as seen in the map. Landcare groups are involved in wetland management in the coastal sandplain areas

A billabong section of the Oldfield River. Marine Eocene sediments, containing spongolite, form the far bank. M.E.W.

Young River upstream and downstream from the highway. The river runs into Stokes Inlet on the coast. M.E.W.

CHANNEL EVALUATION OF RIVERS OF THE ESPERANCE SANDPLAIN

Lort River
Young R.
Oldfield R.
Munglinup R.
Lort R.
Dalyup R.
MALLEE-SANDPLAIN
Coramup Ck.
Bandy Ck.
STOKES-BARKER
ESPERANCE
Esperance
OLDFIELD
0 km 50

A : Near Pristine: river section & upstream catchment in natural bush
B1: Relatively natural - river section in natural bush but part of catchment cleared or other land uses
B2 : Corridor River - river section in substantial corridor of vegetation
B3: Habitat river - river section retains significant riparian vegetation
C : Agricultural drain
Catchment boundary
Wetland group

After Pen(in 16)

(Oldfield, Stokes–Barker and Esperance) and in the Mallee–Sandplain. The Munglinup River and the lower reaches of the Oldfield lie generally within a vegetated corridor, with some sections and tributaries passing through or from large bush blocks. South of the highway there are some large, deep pools which are thought to be spring-fed. These represent a very important habitat and there is concern that they are being degraded through the effects of high nutrient levels from catchment land use. The Oldfield River is estuarine for 8 kilometres, from the coast almost to Springdale Road. It discharges into the Oldfield Estuary, which forms a 3-kilometre-long meandering basin with a maximum width of 500 metres. Although it is up to 5 metres deep, it has extensive shallow areas which often dry out. The mouth is 200 metres wide and until recently may have been very much wider. The long, high bar breaks only every three or four years after heavy rain falls in the catchment, and stays open only briefly.

Gabions in Bandy Creek to stop flood erosion from the road culvert. M.E.W.

Sulphur and iron sulphate (white) in Sulphur Springs, in the Pilbara. REG MORRISON

THE NULLARBOR AND ITS WATER RESOURCES

Drainage lines from terrain surrounding the Nullarbor extend out only a short distance onto the flat limestone plains, and surface water is limited, except after rain. The extensive cave systems within the limestone contain considerable reserves of fresh water, however. The region has been arid ever since the Miocene when it was uplifted to become a plateau. (It was under the sea in the Eocene and again during the Miocene, when its marine limestones were laid down.)

Weebubbie Cave, beneath the Nullarbor. REG MORRISON

Abrakurrie Cave. REG MORRISON

The Oldfield and Munglinup Rivers provide important vegetated corridors between the coastal strip and large tracts of natural bushland in the upper catchment. Road reserves, 20 to 200 metres wide, also provide important corridors between bush remnants.

Wetlands, in the form of permanent and intermittent lakes and swamps, occur on the sandplain. Lake Shaster, just west of the Oldfield catchment and adjacent to the coast, is one of the largest wetlands in the district. It is monitored by CALM and Birds Australia for waterbird usage. It is hypersaline and open, supporting few birds and little breeding. A number of small wetlands are present on farmland. The majority have little or no remnant vegetation, other than swamp yate and paperbark.

THE PILBARA

PALAEODRAINAGES IN THE HAMERSLEY RANGES

The courses of palaeorivers can frequently be traced by mapping features capped by duricrusts, which were formed when iron, silica or calcium-rich waters hardened valley sediments along river courses long ago, and which now form the high features in landscapes. These features have been produced by 'inversion of relief' processes in which the softer regolith materials around them have eroded away through time while their duricrusted capping has protected the sequence below them.

In the iron-rich Hamersley Ranges in Western Australia, spectacular inversions occur due to the transport of ferruginous material in rivers, such as the palaeo-Robe River system, from the upland towards the coast. Such detritus weathers and breaks down and the free iron cements the local sediments in the river channel, forming ferricrete (which is pisolitic in form, when the iron nodules are like small marbles or peas in the iron-cement matrix). The alluvial Robe River pisolite was reconstituted from weathered detritus in Eocene times (about 45 million years ago). It protected the valley bottoms from erosion, and today they stand proud as the sinuous lines of ferricrete-capped mesas, dramatic features in the west of the Hamersley Basin.

The ancestral courses of some tributaries of the Hardy, Beasley, Cane and Duck Creek palaeorivers are also preserved and left in local positive relief as a result of inversion. Some have no modern descendants, but many have migrated laterally and remain active a short distance from their precursors. Some of the inverted palaeochannels form landscape features which have persisted for 60 million years.

THE ROBE RIVER PISOLITE DEPOSITS

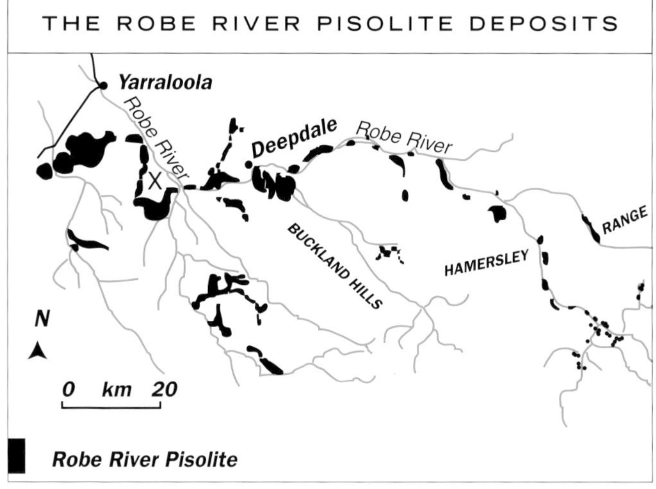

After Twidale[17]

THE FORTESCUE RIVER: *the effects of a dam and of mining operations on the river's floodplains, and on the viability of local pastoral leases*

The Fortescue River rises in the Opthalmia Range at the eastern extremity of the Hamersley Ranges.[20] The river's upper reaches derive their waters from two sets of headwater streams. First are the uppermost tributaries which include the Spearhole, Warrawanda, Whaleback and Kalgan systems. A second set of tributaries, the Jimblebar, Carramulla and Jiggalong Creeks, joins the main stream lower down in the vicinity of Ethel Creek Station. The Fortescue enters the Indian Ocean at James Point, midway between Dampier, 70 kilometres to the north, and the Robe River mouth a similar distance to the south.

The Fortescue and its headwater tributaries flow erratically depending on episodes of heavy rain during the December to May summer rainy season, and from the passage of rain-bearing depressions which are the aftermath of Indian Ocean cyclones. During the winter, when there is less than 20 millimetres of monthly rainfall, parts of the river system may still contain water in pools or swamps.

The Opthalmia Dam has been constructed at the head of the gorge where the river leaves the ranges and curves east before swinging west and flowing along the broad valley between the Chichester and Hamersley Ranges. It captures the waters of the upper Fortescue, Whaleback, Spearhole and Warrawanda Creeks systems, but not Kalgan Creek. These upper tributaries used to provide most of the floodwater for the floodplains before the dam was constructed. The Upper Fortescue floodplain may be inundated for as much as 160 kilometres, when it comprises a complex pattern of rapidly flowing river channels, anabranches, creeks and gullies, spreading floodwaters right across the valley until it meets higher ground. There, mulga and spinifex replace the characteristic coolibah trees of the floodplain country. In high flood, the Upper Fortescue drainage system is so complex that it can only be appreciated from the air.

The Opthalmia Dam was constructed in 1981 because of a growing shortage of water at Newman, where alluvial aquifers were being tapped. It was considered at the time that the Fortescue and its floodplains below the dam would be little affected because there was believed to be considerable run-off from the Opthalmia Range. However, the dam has not only drastically altered the frequency of flooding but also, by reducing the volume of water moving downstream, has altered the spread of floodwater across a floodplain whose ecology is adapted to and sustained by irregular, but in the long run reliable, floods.

The position is further complicated by the way interruptions to the flow below the dam changed the balance between the volumes flowing from the Fortescue and its undammed upper tributaries. When this tributary floodwater flows into an almost dry Fortescue channel it causes erosion, and the outward spread of floodwater is also reduced because of the absence of the backing-up effect which the main channel flood would have supplied.

Since the construction of the dam there has been widespread tree-stress and death. In addition, perennial native pasture grasses have been replaced by annuals and the quality of grazing fodder has declined. Reports commissioned by the Department of Agriculture concluded in 1991 and 1992 that the lack of regular flooding caused by the presence of the Opthalmia Dam was to blame. The dam has only overflowed, very briefly, three times between 1982 and 1990. The authors of the reports concluded that without returning to unimpeded flooding, the

THE HAMMERSLEY RANGES

The Hamersley Range. REG MORRISON

Hancock Gorge. The banded ironstone forming the walls of Hancock Gorge is about 2.5 billion years old. The ironstone was formed by oxygenation of the iron in ocean waters by early photosynthetic life forms. REG MORRISON

The brilliant red of a banded ironstone cliff emphasises the white bark of Eucalyptus papuana. REG MORRISON

vegetation already under stress will continue to die until either a lower population of the current species is reached, or the present species become dominated or replaced by species with a lower water requirement. There has also been a tendency to dismiss changes to

PALAEODRAINAGE OF THE GREAT SANDY DESERT SEEN BY REMOTE SENSING

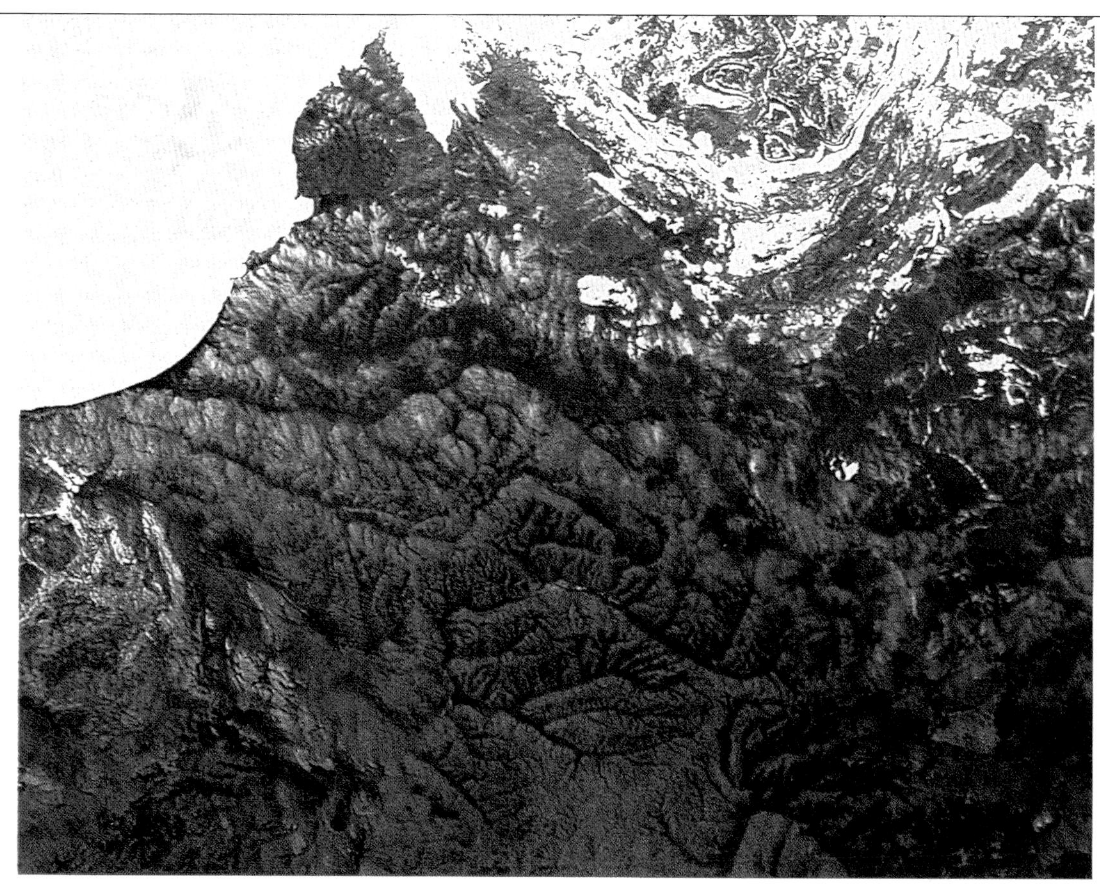

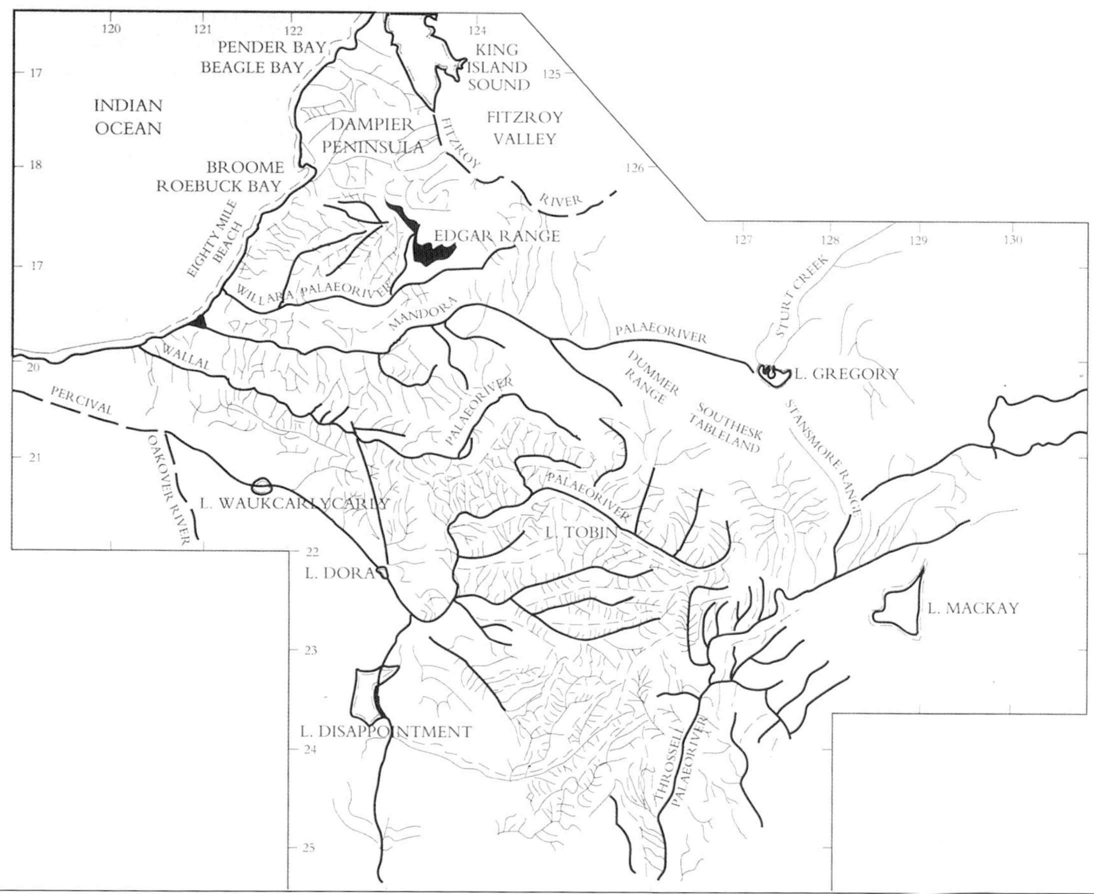

After Tapley[18]

The North Pole stromatolitic reef in the Pilbara is one of the oldest records of life found anywhere on Earth. It is dated at 3.5 billion years. The cyanobacteria which formed the reef were photosynthesisers; organisms of this sort produced the oxygen which oxidised the iron in the oceans—which in turn created the banded ironstone formations. REG MORRISON

RIVERS OF WESTERN AUSTRALIA

Exmouth Gulf
Fortescue River
HAMERSLEY RANGE
Lyndon R
Minilya R
Lake Macleod
River
Shark Bay
Gascoyne
Murchison R
Meekatharra
Greenough R
Geraldton

After Tapley[18]

INVERSION OF RELIEF: tracing ancient drainages by the silcrete and ferricrete cappings on lines of mesas

Mesas capped by Robe River pisolite. PROFESSOR C. W. TWIDALE

Inversion of relief refers to a process in landscape evolution when a former valley bottom becomes a ridge, bounded by newly formed valleys on either side.[19] It is well known in the context of lava flows and also occurs in landscapes with duricrusts.

In the context of rivers and palaeodrainage lines, inversion of relief occurs when materials on valley floors are, or become, more resistant to erosion than the adjacent valley slopes. This happens when silica, iron or lime in solution have indurated valley floor sediments. As erosion proceeds, the softer surrounding terrrain is stripped, leaving the valley floor and the section of regolith below it as higher features. When one looks at aerial photographs of mesas in parts of our arid continent, a dendritic pattern sometimes emerges when they are linked, showing the branching of the palaeodrainage to which they owe their hard, protective cappings.

pasture as having been caused by over-grazing, even where it has been shown that grazing pressures had been much reduced and the causes of degradation could not be attributed to this or other normal causes.

The Fortescue is a river whose ecology is based almost entirely upon episodic river flooding. The plight of the floodplains below the dam emphasises that arid and semi-arid ecologies are marginal ecologies in which comparatively slight but prolonged changes in water supply can bring about much larger biological responses. Little is known about subterranean flows in the Upper Fortescue but there is some anxiety that the saline water tables at depth below the floodplains could start to rise when water abstraction continues and reduced flooding does not compensate. Recently, sink-holes have appeared near Roy Hill Homestead and also in the Tom Price borefield, where a massive cavern collapse is suggested in the underlying dolomitic limestone (which forms the substrate below the Quaternary and modern alluvium in the Fortescue valley).

The direct effect of mining operations has to be considered as a contributor to tree-stress and death on the floodplains. The mines in the Hamersley Ranges might be the source of pollutants, either airborne or transported by water, or both. In fact, it is inconceivable that mining on the scale which is current in the Hamersleys would not pollute, given the well-documented evidence for pollution associated with mining activities elsewhere in Australia, and globally.

An interesting indicator that something was wrong with the water flowing across the floodplain during the 1995 flood, which was the first since 1981, was the way in which exposed nodules of calcrete were bleached white by the passing water. (The 1995 flooding was the result of Cyclone Bobby—see page 231 in *Listen … Our Land is Crying*.) In addition, coolibah trees, which would have been expected to benefit from the return of floodwater, have been dying in their thousands since the floodwaters passed. A November 1995 air reconnaissance by Murray Kennedy found large patches of dead and dying mulga situated away from the main floodplain. Though local drought might be the cause, a connection to mining operations is not ruled out.

Another long-term effect of the Mt Newman and associated mining operations involves the extremely large-scale blasting and subsequent removal of millions of tonnes of overburden to extract, crush and transport, in open trucks, enormous quantities of iron ore to Port Hedland. The Mt Whaleback deposits, from which the iron ore is mined, contain pyritic Mt McRae Shales which are exposed as the mine deepens. Fifty million tonnes have been mined and put into tailings dumps. The shales emit sulphur dioxide gas and acidify mine water, of which large quantities have to be pumped to allow mining below the regional water table. When the Shale is exposed and subject to weathering, rapid oxidation of the pyrite takes place, which can be so fierce that spontaneous combustion can occur. Emission of additional sulphur dioxide results from the combustion.

Blasting releases large quantities of pyritic shale and iron ore dusts into the air on a daily basis. In addition, the ammonium nitrate and fuel oil mixture which is used in blasting is converted into nitrous and nitric oxides as well as carbon dioxide. The connection of all these factors with 'acid rain' is obvious. While scientific testing has still to be done, experts consider that acid rain is an unlikely cause for the tree deaths because the climatic and other factors make its production on a large enough scale unlikely. Dust, however, is another matter. The very fine dust which is generated by blasting can travel up to 200 kilometres and if deposited on the leaves of plants can interfere with function as well as being directly toxic.

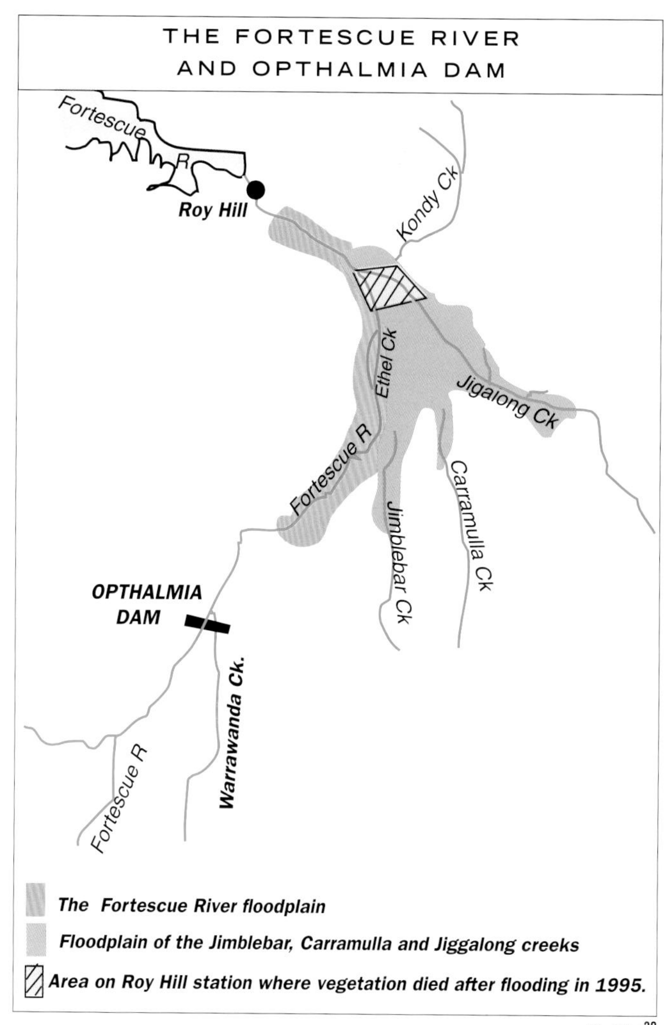

After Webb[20]

Battle of the station owners affected by the Opthalmia Dam

The pastoral lease Roy Hill Station has suffered severely as a result of the construction of the Opthalmia Dam. Ramon and Murray Kennedy, whose company holds the pastoral lease, estimate that the carrying capacity of the station has been halved. The Fortescue River floodplain has become seriously degraded since it was deprived of the floodouts which are essential for its survival. They have been able to show conclusively that it has not been grazing mismanagement which has caused the amount of degradation seen on the floodplain. Since the first signs of trouble started to appear shortly after the dam was built, they have been keeping a photographic record and carefully documenting the changes that have been occurring. They have used every occasion to canvass politicians, officials concerned with water and grazing management, scientists and the media, to try to get something done about the ever-worsening situation. The Department of Land Administration has reduced their land rent by 20 per cent in recognition of the reduced carrying capacity.

The Kennedys have lived and worked in the Pilbara for thirty years and their on-the-ground knowledge is enormous and valuable. They should be heeded and action should be taken by government and BHP and everyone else involved to do something about the situation. They point out that the recent changes, which include the escalating rates in death of trees and pasture, represent 'a huge environmental disaster affecting up to 12 million hectares of grasslands, and millions of native trees and shrubs are either dead, dying or under extreme stress'. The more cynical among us wonder what hope there is of doing what is necessary to restore the floodplain when big money and iron ore exports are on one side of the equation while only biodiversity and environmental matters are on the other. (Newman has alternative options for water and the Opthalmia Dam is not essential for its supply.)

Is there a lesson to be learnt from the Opthalmia Dam–Fortescue floodplain situation?

The effects of a dam which severely restricts flooding of the floodplain below it in a semi-arid to arid environment (even if there is no added effect from mining operations) is so clearly seen in the case of the Fortescue River and the Opthalmia Dam that it should serve as a warning for future dam building. A case in point is the proposed dam on the Fitzroy River in the Kimberleys where vast floodplains downstream of the dam would have their widespread flooding greatly reduced in volume and extent. This would undoubtedly mean stress and death to floodplain ecosystems, and a great reduction in carrying capacity. The grazing industry in the area involved is not limited to one or two stations, as is the case in the Fortescue, and the impact of a Fitzroy dam would have wide and serious economic consequences.

The floodplains have evolved from river floodouts, bringing silt and nutrients. They require long periods of inundation to saturate the soils and activate the gilgai clays. Prolonged inundation saturates the soil to greater depth, providing a much longer growing season for grasses, shrubs and trees. Denied this saturation, the river plains will degenerate, losing perennials which will be replaced by annuals. What has happened on the Fortescue floodplains clearly warns of what will happen to the Fitzroy if the dam goes ahead.

Millstream, with Afghan cameldrivers' date palms. REG MORRISON

14 March 1990: waist-high perennial grasses in a paddock on Roy Hill Station three months after the Jiggalong floodout of 20 January 1990. MURRAY KENNEDY

26 April 1995: ankle-high annuals and dead and dying trees after the Fortescue ran for the first time in 13 years and its floodwaters flowed across the Jiggalong floodplain, carrying who-knows-what contaminants. MURRAY KENNEDY

The Kimberley Plateau

The Kimberley Plateau is a plateau cut across a range of Precambrian rocks of the Kimberley Group, older than 2 billion years.[21,22] The Kimberley Block is an 'exotic terrane' which docked against Australia, and it has retained its independent drainage system. It is rimmed by younger Halls Creek Belt rocks between 2 and 1.75 billion years old. The term 'High Kimberley Surface' has been given to the oldest planated surface; the 'Lower Kimberley Surface' comprises the younger surrounding surface, which is usually duricrusted, like the Mitchell Plateau.

ANCIENT LAND SURFACES

The land surface of the Earth and its landscapes are the product of cycles of erosion, mainly by water, through time. Most modern land surfaces are the product of geologically recent and continuing processes which cover those that preceded them, but in some parts of Australia, areas of the ancient cratons are exposed where all the cover which had accumulated on them has disappeared.[21]

Ancient land surfaces are of three types:

- **Epigene** or **subaerial** land surfaces have been shaped by processes of weathering and erosion at the earth's surface. Most are of riverine origin. Deep weathering has taken place beneath some stable epigene surfaces, and thick regoliths (soils and sub-soil layers) have been developed, separated from the fresh bedrock by the *weathering front* or lower limit of effective weathering. Such weathering created a marked physical contrast between bedrock and the covering layers, and where the cover has later been stripped over wide areas to expose the weathering front, the result is *etch* surfaces. Both epigene and etch surfaces have been buried and later exposed as *exhumed* surfaces. Glacial, marine or fluvial burial is usually accompanied by erosion of the pre-existing regolith, so that most exhumed surfaces are also of etch type. In some cases, old regolith is not stripped before a land surface is buried. For example, burial by wind-blown calcareous dune sand has allowed a deep regolith developed on sandstone and granite to be preserved on the west coast of the Eyre Peninsula in South Australia.
- **Exhumed** land surfaces can be dated by reference to the youngest rocks in which they are cut, and the greatest age of those that buried them, the two dates bracketing the maximum age range. Etch forms have two ages: first the period during which weathering took place and when the shape of the weathering front was determined by the interaction of groundwaters and bedrock; and second, the period during which the regolith was stripped to expose the weathering front.
- **Epigene** surfaces are more difficult to date. Planation can sometimes be correlated with deposition, and ages of faults and volcanic activity can be useful. Duricrusts can be useful markers.

Determining the age of strata and events can be done by working out the *stratigraphy* or sequence of deposition of sedimentary strata; or by detection of palaeomagnetic reversals in volcanic rocks; or by dating techniques using radioactive carbon or breakdown rates of other radioactive materials such as argon; or by new techniques using thermoluminescence.

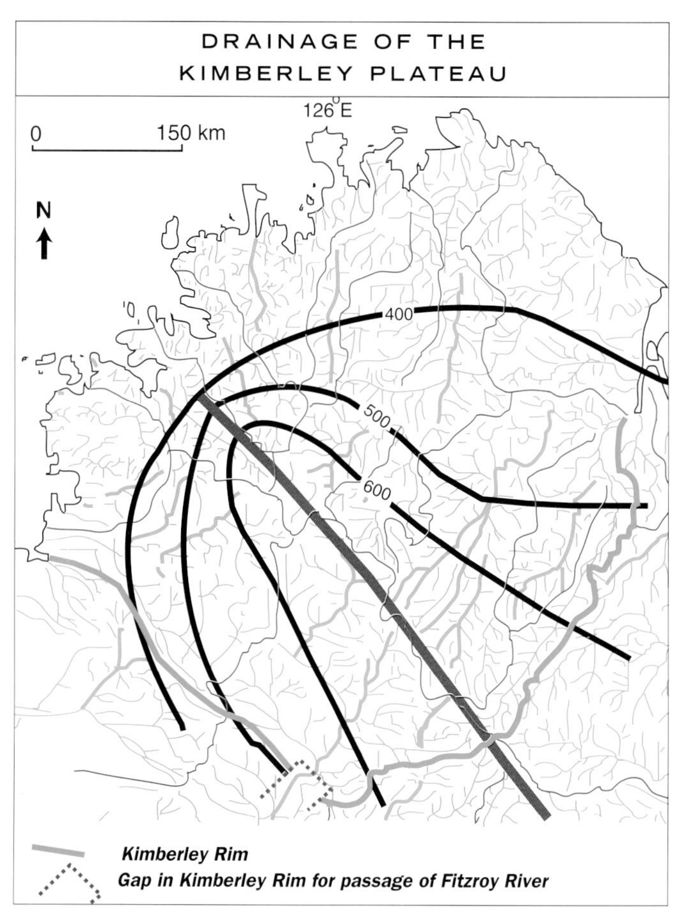

After Ollier *et al*[22]

The plateau is surrounded on all sides except the east by sedimentary rocks which were laid down in younger (less than 700 million year old) Phanerozoic basins.

A large area of glacial sediments is preserved in the south-western part of the plateau, in palaeovalleys between ridges on which the Kimberley Surface is developed. The glacial sediments are dated at 700 million years, remnants from the major Precambrian ice age which nearly froze the whole world between 900 and 600 million years ago, and they date the Surface as having been in existence for at least 700 million years! (The Canning Basin, on the south-western side of the Kimberley Block, has sediments which have been dated as Ordovician (460 million years) and its younger Silurian and Middle Devonian sediments are believed to have come from the erosion of the Kimberley Block.) No bedded terrigenous (accumulated on land) sediments have remained on the Block in the last 750 million years—the land surface on the Block is the oldest found anywhere on Earth.

Drainage of the Block is due to warping of the ancient surface, forming an elongated dome with an axis running north-west to south-east. A marked feature, the Kimberley Rim, virtually isolates the drainage of the ancient terrane from the rest. Only one river, the Fitzroy, crosses the rim, and it does so at a structurally weak point.

THE FITZROY RIVER

The Fitzroy River has been in the news lately because of a proposal to build a major dam and bring waters in a giant canal 500 kilometres long to irrigate cotton in an area 150 kilometres south of

SATELLITE IMAGE: THE FITZROY RIVER FLOODING

Landsat imagery provided by Australian Centre for Remote Sensing (ACRES), Australian Surveying & Land Information Group, DIST, Canberra, and digitally enhanced and produced by Satellite Remote Sensing Services, Department of Land Administration, Perth, Western Australia.

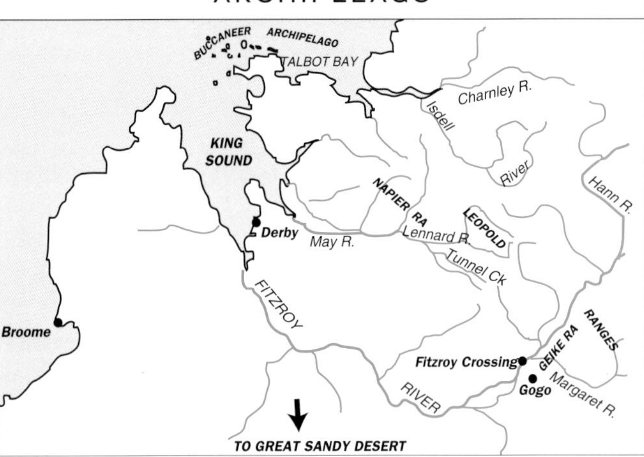

The Buccaneer Archipelago, Kimberley Coast, WA. REG MORRISON

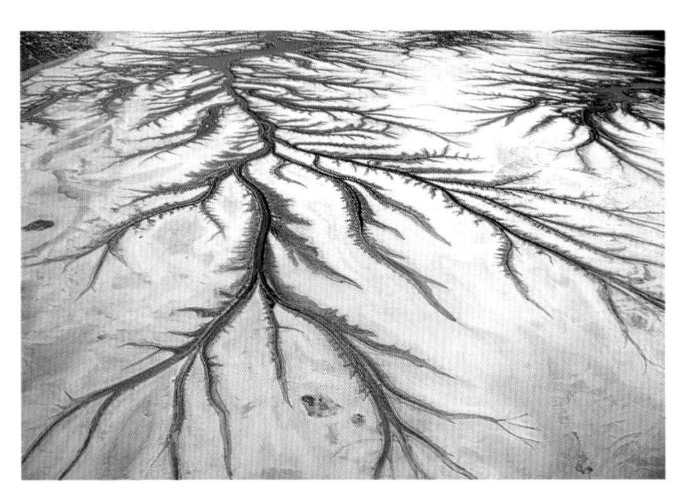

Tidal drainage patterns near Derby. REG MORRISON

Broome. The plan envisages 175 000 to 250 000 hectares being irrigated when the dam and canal are finished; and a first stage of 25 000 hectares would be irrigated with groundwater before dam water became available. Such a scheme would be nothing short of environmental vandalism.

No thought has been given to the enormous loss to evaporation along such a canal; to the degradation of the vast floodplains which depend on annual widespread flooding which would be greatly reduced; to the damage to pastoral and tourism interests which would easily counterbalance the supposed economic advantages of the cotton in a region where all sorts of potential problems and unpredictable influences, such as cyclone damage, might make such an enterprise non-viable—not to mention the enormous initial expense of the engineering projects involved.

The **Fitzroy estuary** experiences a long dry season which restricts the presence of mangroves.[23] Between 7400 and 6000 years ago, tall mangroves colonised the area and the mangrove swamps were far larger than they are today. A longer and heavier wet season characterised the area then. The onset of drier climatic conditions in the Pleistocene led to the formation of floodplains along the Fitzroy and Lennard Rivers and their tributaries and to extensive alluviation marginal to uplands in the Fitzroy Basin. During the glacial maxima there may have been an absence of cyclone activity, due to reduced sea temperatures or a change in pathways. (The exposed continental

Tide race in the McLarty Range. REG MORRISON

Mudflats and mangroves, Talbot Bay, Kimberleys REG MORRISON

shelves would have affected climate.)

The Fitzroy estuary enters King Sound just south of Derby; the Meda, into which the Lennard runs, flows into Stokes Bay north-east of the town. Vast tidal floodplains characterise the southern end of King Sound and amazing dendritic drainage patterns are visible from the air.

Across the northern end of King Sound lies the Buccaneer Archipelago, an amazing collection of small islands. During low sea level stands in glacial stages of the Pleistocene ice age, rivers cut deep channels across the wide, exposed continental shelf to the sea. When sea level rose and finally stabilised about 6000 years ago, the deeply incised valleys were invaded by the sea and the surrounding low hills remained above water as islands. Some sinking of the continental shelf and coastal land in that region has exaggerated the topography.

In Talbot Bay is found the world's biggest tide race, where the sea rushes in through a gorge in the McLarty Range. A tide range of 10 to 13 metres makes this a wild and awesome place. A proposal to generate electricity by using the tidal power in this region is currently under consideration.

RIDGE-FORM ANABRANCHING RIVERS—UNIQUELY AUSTRALIAN

As if it were not enough to be able to walk on land surfaces which are as unimaginably ancient as those on the Kimberley Plateau, researchers at Wollongong University have discovered and described a unique type of river there, which is so different as to be almost as unbelievable. On reaches of the Durack River which are alluvial, and alternate with narrow, bedrock-confined stretches, roughly parallel, steep-sided and heavily vegetated sandy ridges divide the flow system into remarkably straight, canal-like, anabranching channels.[24] The ridges, supporting a dense growth of trees, are of approximately floodplain height. There is an abundance of riparian vegetation, in places growing chaotically over wider sections of stream bed.

The vegetation in the study area, which is near Karunjie Homestead, is grassy savanna woodland. Tall trees with a dense understorey occur on the sandy alluvial soils adjacent to the streams. Rainfall is summer monsoonal, about 700 millimetres annually. Potential evaporation is high, about 3000 millimetres annually. Tropical cyclones periodically dump heavy rain and cause severe flooding. Above and below the study reach, the channel is cut into bedrock and deposits of very large boulders are evidence of occasional high-magnitude floods.

Streams in the region frequently occupy bedrock and boulder-bed channels, especially along their steepened and incised downstream reaches. In the middle reaches, stream gradients are much lower and sandbed alluvial channels dominate, although they alternate with bedrock reaches. Alluvial stretches are wider, frequently straight, and are laterally and vertically stable, with bedrock outcropping in places along their beds. Floodplains are discontinuous, generally restricted to the widest sections of valleys and often associated with tributary junctions. Within-channel growth of shrubs and large trees is widespread, especially *Melaleuca leucadendron*, establishing on the river beds and banks during the dry season, tapping the groundwater that lies beneath the channel.

Along larger streams, channels formed in alluvium typically reveal a multi-channel planform, often of the ridge-form anabranching type, with up to 10 parallel anabranches. At the downstream end of many alluvial channel reaches, large permanent waterholes are found. They represent transitional areas separating alluvial channels upstream from bedrock channels downstream. The well-defined anabranches between the ridges are on average 10 to 15 metres wide, can be continuous in length for several hundred metres, have flat channel beds, and are generally free of tree growth.

How and why this planform develops and how it functions is still open to some speculation. The formation of ridges is believed to cause a reduction of flow resistance with an increase in depth. The

CROSS SECTION OF THE DURACK RIVER: Floods of different magnitudes I—IV

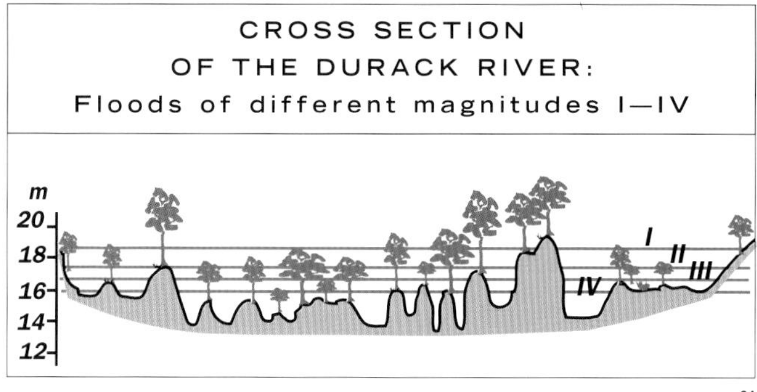

After Wende & Nanson[24]

RIDGE FORM ANABRANCHING RIVERS

The formation of alluvial ridges to form anabranches within Fine Pool on the Durack River near its confluence with the Chapman River, the Kimberley region, Western Australia. GERALD NANSON

Aerial view of alluvial ridges dividing the channel to form separate anabranches on the Marshall River, Northern Territory. GERALD NANSON

A RIDGE-FORM ANABRANCHING REACH OF THE DURACK RIVER

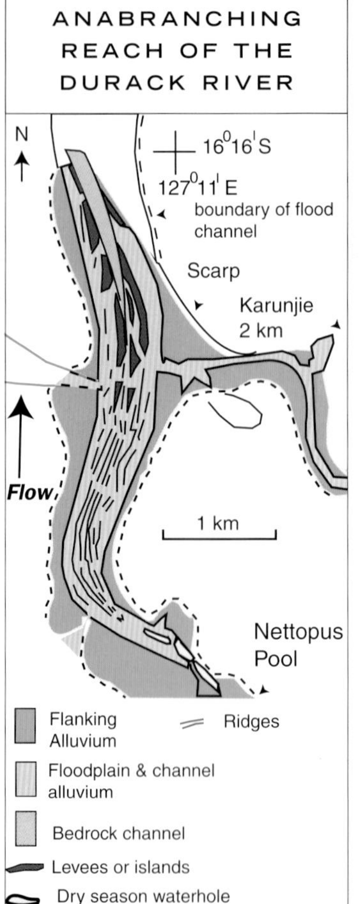

After Wende & Nanson[24]

An alluvial ridge dividing the channel to form separate anabranches on the Marshall River, Northern Territory. GERALD NANSON

An early stage in the development of an alluvial ridge that will divide the channel to form separate anabranches on the Sandover River, Northern Territory. GERALD NANSON

An alluvial ridge dividing the channel to form separate anabranches on the Sandover River, NT. GERALD NANSON

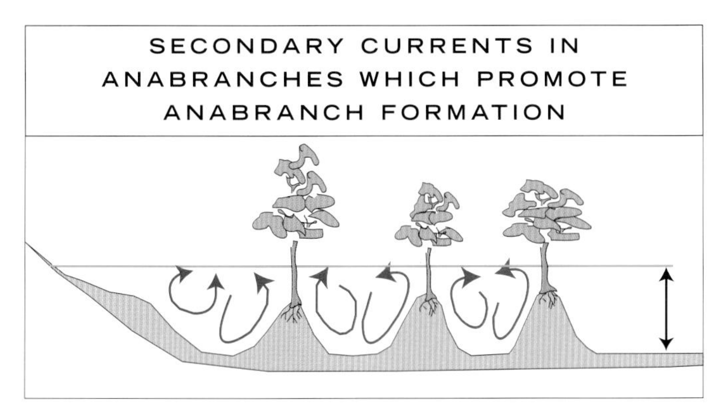

After Wende & Nanson[24]

ridges act to concentrate or squeeze the flow between them, producing higher velocities and greater sediment transport than would otherwise occur. This compensates for the less efficient flow conditions associated with these well-vegetated alluvial reaches; trees generally do not survive on the beds of higher-energy channels that form between the ridges. The formation of the channels and intervening ridges may be aided by the development of double flow helices in each anabranch. Some of the anabranches were observed to have formed by channel avulsion, when new channels were scoured across adjacent floodplains, particularly in the widened sections of valleys near tributary junctions (where the flow from adjacent tributaries may be the cause).

It appears that ridge-form channels organise riparian vegetation and sedimentation into a linear pattern and limit flow resistance, maintaining sediment transport—and the unique conditions in tropical northern Australia have seen the evolution of this very different type of system.

Rock slabs stacked like giant dominoes on steps in some stretches of bedrock channels in the Durack River are another unusual feature.[25] They have been investigated in detail at Jack's Hole and also on an unnamed creek between the Durack and the Ord Rivers by Raine Wende, a post-graduate student of Wollongong University's Geoscience department.

This sort of feature occurs where the bedrock is jointed and layered slabs dip gently downstream relative to the channel slope. Steps develop in the bedrock and domino-like slabs accumulate singly or in clusters leaning against the steps, or extending upstream and downstream beyond the bedrock step. Some of the slabs in the channels at Jack's Hole and the other creek are enormous, with an intermediate axis of more than 8 metres. To erode and transport

Jack's Hole, Durack River. Huge blocks swept from the exposed channel pavement and stacked together during extreme floods. GERALD NANSON

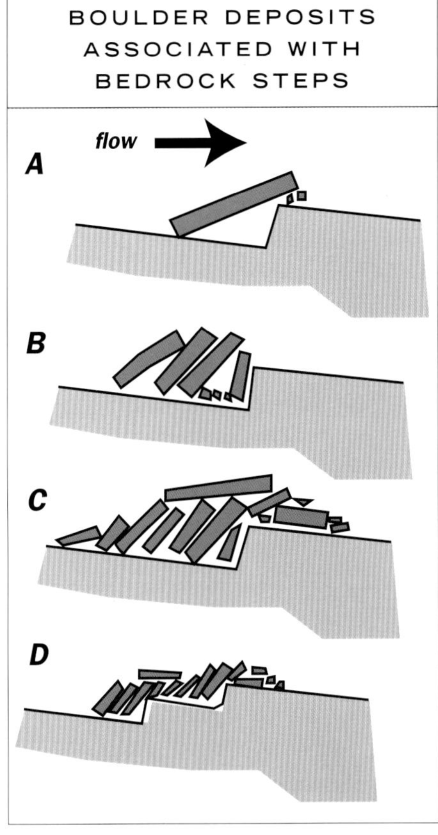

After Wende[25]

Enormous blocks stacked together on steps in the Durack River during extreme floods. The white scale is 5 metres in length. GERALD NANSON

such large rock slabs high magnitude floods and high flow velocities are required. Because they only move rarely, under exceptional circumstances, the slabs act to slow bedrock erosion of the channel beneath them.

Numerous successive upstream-facing steps which extend more or less transverse to the channel are typical of channel reaches where the strata dip is downstream relative to channel slope. Individual steps are produced by hydraulic plucking of blocks from upstream of these steps, while the succession of steps is the result of the dip of the strata. The height of the individual steps and the spacing of the steps along the channel in a downstream direction is strongly influenced by the spacing of the truncated horizontal joints. The planform of bedrock steps is controlled by the vertical joint pattern of the rock mass forming the channel bed. If only a single, roughly parallel set of vertical joints traverses the channel, straight or linear steps are the result, while intersecting sets of vertical joints result in non-linear steps, frequently following a zigzag pattern.

THE ORD RIVER

The **Ord River Scheme** is the subject of detailed study in *Listen ... Our Land is Crying* (pages 223–7). The Ord River is the only north coast river significantly impacted upon by agriculture. It is regulated by two dams—Lake Kununurra, constructed in 1963, and the larger Lake Argyle, about 50 kilometres upstream, which was completed in 1972.

The Ord runs into the sea in Cambridge Gulf. Its palaeochannel used to flow into the Bonaparte Gulf via what is now the Keep River, and the detection of this fact has been important because of the potential for alluvial diamonds in the gravels. Exploration is under way in the estuary of the Keep. (The Fitzroy River estuary and the wide continental shelf westward of it are potential alluvial diamond areas as well, because of the occurrence of a number of volcanic events in the west Kimberley when diamond pipes like the Argyle were formed.)

Recent success with sugar cane in the Ord River Scheme has led to expansion of the project, which had been a disappointment and has only involved about one-third of the originally planned area until recently. It is now proposed to re-introduce cotton. (Agriculture has been constrained by infertile, leached soils, a harsh climate and an abundance of vertebrate and invertebrate pests.) Cotton plus invertebrate pests adds up to pesticides in abundance. Floodplain wetlands of the region are vulnerable. If expansion of agriculture, especially cotton, occurs and other rivers are dammed

THE DURACK AND ORD RIVERS

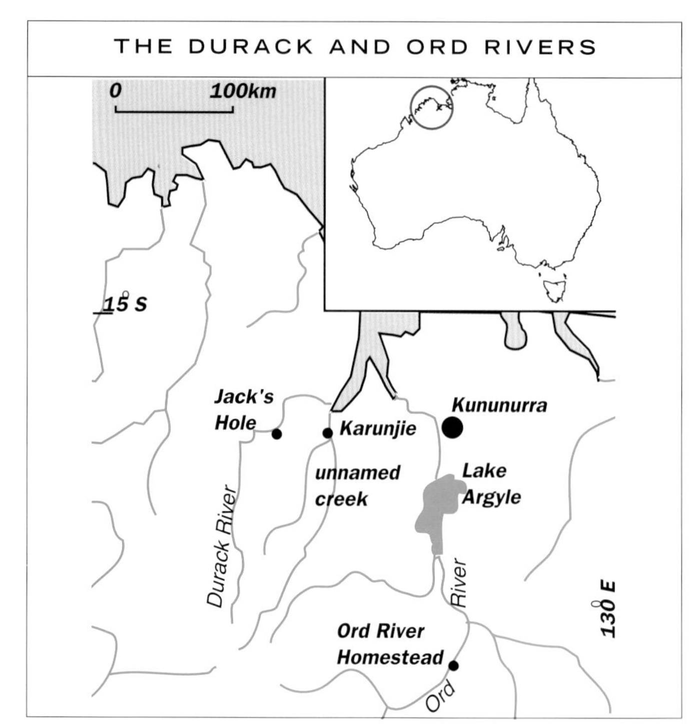

The Argyle Dam, Ord River. M.E.W.

Irrigation channel, the Ord River Scheme. M.E.W.

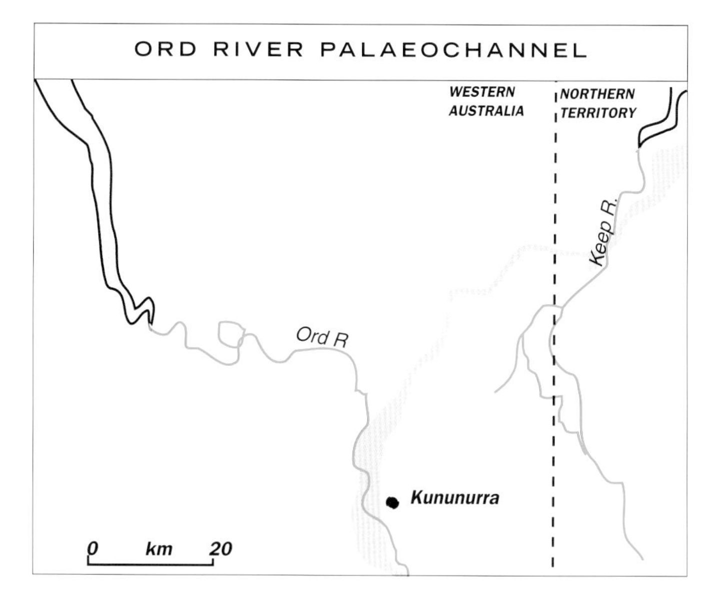

and other projects are started there are important biological and biodiversity problems waiting to manifest themselves. The wide spread of introduced pasture grasses into the floodplains which would follow development of the area for more intensive grazing use would threaten waterfowl and other vertebrate habitats. The ponded pasture 'weed' species which have taken over on the Mary River floodplains and elsewhere bring monoculture conditions, loss of biodiversity and irreversible change.

The Carnarvon Basin

THE GASCOYNE RIVER

The Gascoyne River catchment suffers more acute rainfall variation than most of the rest of the Australian arid zone as it lies in a 200-millimetre rainfall zone, which here reaches the coast, and at the same time is tracked across from the north and north-west by remnants of tropical cyclones.[26] Carnarvon, at the mouth of the Gascoyne, has winter-dominant rainfall patterns with infrequent summer extremes from cyclones. Run-off is significant but irregular in the Gascoyne Basin; an average annual volume exceeding 550 million cubic metres discharges into Shark Bay and the Indian Ocean. The over-grazed catchment provides massive amounts of sediment which is transported by the river and out to sea during floods. The almost complete absence of adequate storage sites means that water resources development is constrained to exploitation of local shallow unconfined aquifers associated with the lower course of the Gascoyne River.

The Gascoyne River system has one major tributary, the Lyons River, which enters from the north about 160 kilometres from the mouth. All streams are low gradient with wide, ill-defined sandy beds

Satellite image of the Gascoyne River in flood

Landsat imagery provided by the Australian Centre for Remote Sensing (ACRES), Australian Surveying & Land Information Group, DIST, Canberra, and digitally enhanced and produced by Satellite Remote Sensing Services, Department of Land Administration, Perth, Western Australia.

THE GASCOYNE RIVER CATCHMENT

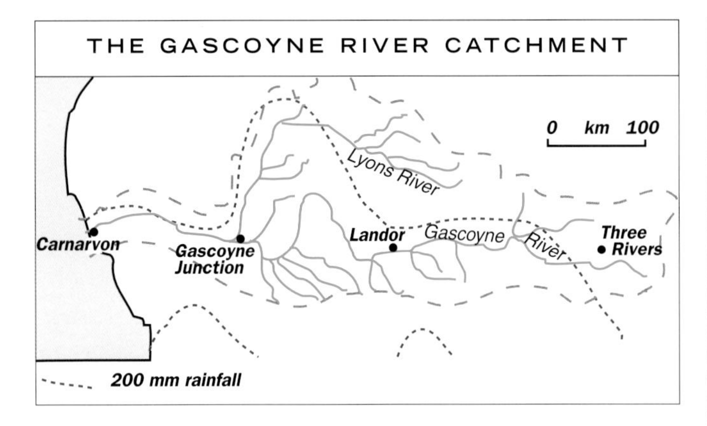

in their lower courses. Water quality is good, and in the lower course there is an extensive sandy bed which acts as an unconfined aquifer. The streambed aquifer is connected to a regional aquifer of lower quality, partly confined and spreading laterally away from the river.

The streambed aquifer along the lower 50 kilometres of the river is utilised for irrigation of sub-tropical horticultural crops and for domestic purposes for Carnarvon. There is no direct storage or use of surface run-off. All supplies are drawn from sand aquifers and the storage available in the bed aquifers is limited. Pumping restrictions apply and are lifted during occasions of river flow. Major flow events are strongly correlated with ENSO, with positive Southern Oscillation Index values. From January 1990 to November 1994, when negative indices also brought recurring drought to eastern Australia, there was a total of 1440 days of no flow, including one of 537 days, and four flow periods totalling 265 days, including one of 125 days commencing in April 1992.

The town water supply for Carnarvon before 1971 was dependent on a small well-field on Water Supply Island, which continued as a partial source until its collapse in 1986. Horticultural plantations, mostly bananas, obtained their own supply from bores, mostly close to the river. The 1958–60 drought, when the river did not flow for 546 days, caused a serious deterioration in water quality and quantity, and government assistance was sought.

In 1962, a pilot scheme was introduced to tap the river bed aquifer upstream of the plantation area, and in 1972 this was extended some 40 kilometres upstream, tapping 10 designated aquifer basins and 63 production wells. Most plantation owners have other licensed bores on their own properties drawing on the deeper prior stream aquifers. All bores are metered and the local Advisory Committee sets the limits for safe annual extraction.

The Gascoyne River, a river of sand in the dry season. REG MORRISON

THE STROMATOLITES OF SHARK BAY

Living stromatolites in the Hamelin Pool Marine Nature Reserve in Shark Bay are created by descendants of the stromatolite-constructing early life forms which made the 3.5 billion year old North Pole reefs of the Pilbara. The cyanobacteria which form stromatolites are photosynthesisers and produce oxygen as a by-product.

REG MORRISON

THE MURCHISON RIVER

The Murchison River runs through the southern Carnarvon Basin. Most of its flow originates from headwaters in the hard-rock Gascoyne Province far inland. It enters the sea in Gantheaume Bay at Kalbarri (not to be confused with the other Gantheaume Bay near Broome).[17, 27] Inland it flows through the Victoria Plateau which is developed on a well-jointed Silurian sandstone. Joint control has resulted in development of rectangular patterns and in the so-called Z-Bend where the river is deeply incised to form the Murchison Gorge. The gorge was in existence prior to the Early Cretaceous, when shallow epicontinental seas flooded the valley, leaving behind marine strata. The ancestral Murchison which cut the gorge thus predates the Early Cretaceous sea.

In the Murchison Gorge a Silurian rock platform is exposed. It has 'tyre tracks' across it, made by metre-long eurypterids (large crab-like creatures) which were among the first animals to leave their footprints on the land, about 425 million years ago.

The Perth Basin

Z-bend, Kalbarri National Park. REG MORRISON

A Z-bend in Tumblagooda Sandstone, Murchison Gorge. REG MORRISON

The Perth Basin extends for 800 kilometres along the west coast of Australia; onshore, it ranges from 15 to 90 kilometres wide. The Darling Fault and Scarp form the eastern boundary of the Basin.

Groundwater supplies in the Perth Basin are largely from unconfined aquifers (up to 70 metres, but usually 30 metres deep). These aquifers comprise Quaternary Complex sand, silt and clay sediments which accumulated during the time when sea level was rising and falling during the Pleistocene ice age and sand was being blown during glacial stages. Shallow bores are used to tap the aquifers. Perth's large unconfined aquifer is a major source of water supply for the city and also has an important role in the conservation of wetlands. It is mainly recharged by rainfall, and partly recharged from upward leakage from the confined aquifers below in Upper Jurassic and Lower Cretaceous sandstone units. The water in these confined aquifers can only be directly accessed by deep artesian boreholes. Dams on the Darling Plateau augment Perth's water supply.

Mean annual rainfall in Perth is 869 millimetres and the annual potential evaporation is 1972 millimetres. On average, about 11.5 per cent of rainfall recharges the Perth aquifer. Groundwater fluctuates in response to recharge, withdrawal rate and land-use management. The water table is high in spring and low in autumn. The quality of groundwater is, on the whole, good but pollution of unconfined aquifers would be all too easy, and has in fact occurred in some regions. Privately owned bores and wells abound.

The south-western corner of the continent, the winter rainfall region, has experienced declining rainfall during the second half of the twentieth century (while the opposite has been the case along the eastern continental margin). With Greenhouse global warming from the greenhouse effect and climate change a further 20 per cent decrease is predicted by 2040, accompanied by a rise in average temperatures and higher evaporation. These changes have serious implications for water supply and for salinity problems. (A full account of Western Australia's problems with salinity and related issues is given in *Listen … Our Land is Crying.*) Rising sea level will cause more intrusion of seawater into estuaries. Salt water already penetrates many kilometres up the Swan River.

Many wetlands, swamps and shallow lakes are present along the

Eurypterid tracks on a rock platform in the Murchison Gorge. These footprints were made about 400 million years ago, when living creatures were first emerging from the water and starting to colonise the land. REG MORRISON

Bubble experiment on the Swan River, mixing the water layers by forcing air under pressure through a perforated pipe lying on the river bottom. M.E.W.

coastal plain, occurring where the unconfined groundwater table intersects the surface. Many have been much reduced, modified or destroyed by urban sprawl and agricultural development.

The degraded state of the Avon–Swan River is discussed in *Listen … Our Land is Crying*. Experiments have been carried out to try to develop methods of oxygenating the water in the lower reaches to prevent algal blooms. Forcing air through a wide pipe with small perforations which lies across the river on the bottom makes a curtain of bubbles, mixing the water layers.

THE SWAN COASTAL PLAIN

The Swan Coastal Plain was largely a wetland in 1829 at the start of settlement. Winter rains caused flooded streams to flow out onto the plains, filling an interconnected chain of swamps many kilometres wide. In the Peel–Harvey catchment only the Murray and the North and South Dandalup Rivers had permanent stream beds, and even they spread out over the plains during floods. The other main rivers, the Serpentine and the Harvey, were well-defined watercourses in their upper and lower sections but their middle reaches were a maze of swamps, with paperbarks, flooded gums and sedges. Here, the rivers spread out in winter and joined with the flow from all the other streams, so that the plain was flooded from the scarp to the long ridge of dunes close to the coast. Some water would eventually seep into the lower reaches of the rivers and via estuaries into the sea.

Many of the major wetlands of the coastal plain lie in troughs between the parallel dune systems. Lake Cooloongup and Lake Walyungup at Baldivis lie between the Quindalup and Spearwood systems, and the long saline Lakes Clifton and Preston are constricted between limestone ridges of the Spearwood system.

Away from the floodplains and major wetlands, the coastal plain is still a maze of open wetlands and large swampy areas, even after a hundred years of human effort in digging thousands of kilometres of drains. In the early days of settlement, as the Swan River colony spread out and the plain was progressively farmed, the annual flooding was a daunting problem.

As settlement proceeded and land-clearing intensified, the flooding was found to worsen. Rising water tables due to clearing increased the winter flooding. By the late 1890s, large areas of jarrah forest on the Darling Range had been logged, increasing streamflow, and the situation was becoming critical for farmers. Many abandoned their land. Major drainage works were undertaken, the Harvey Main Drain being constructed at the turn of the century to channel the water in the middle reaches to the Peel–Harvey estuary. Other major draining projects followed.

THE PEEL–HARVEY ESTUARY

Three major rivers flow into the estuarine system of the Peel–Harvey—the Serpentine, the Murray and the Harvey.[28] The Murray is the largest with its upper tributaries rising 150 kilometres from the estuary, on the Yilgarn Plateau, where they occupy broad valleys. These tributaries meet to become two small rivers, the Hotham and the Williams, which in turn join to become the Murray, flowing through narrow, forested valleys between the hills of the Darling Ranges. There, the jarrah forests, endemic to Western Australia, grow on the thin, lateritised soils.

THE PEEL–HARVEY CATCHMENT

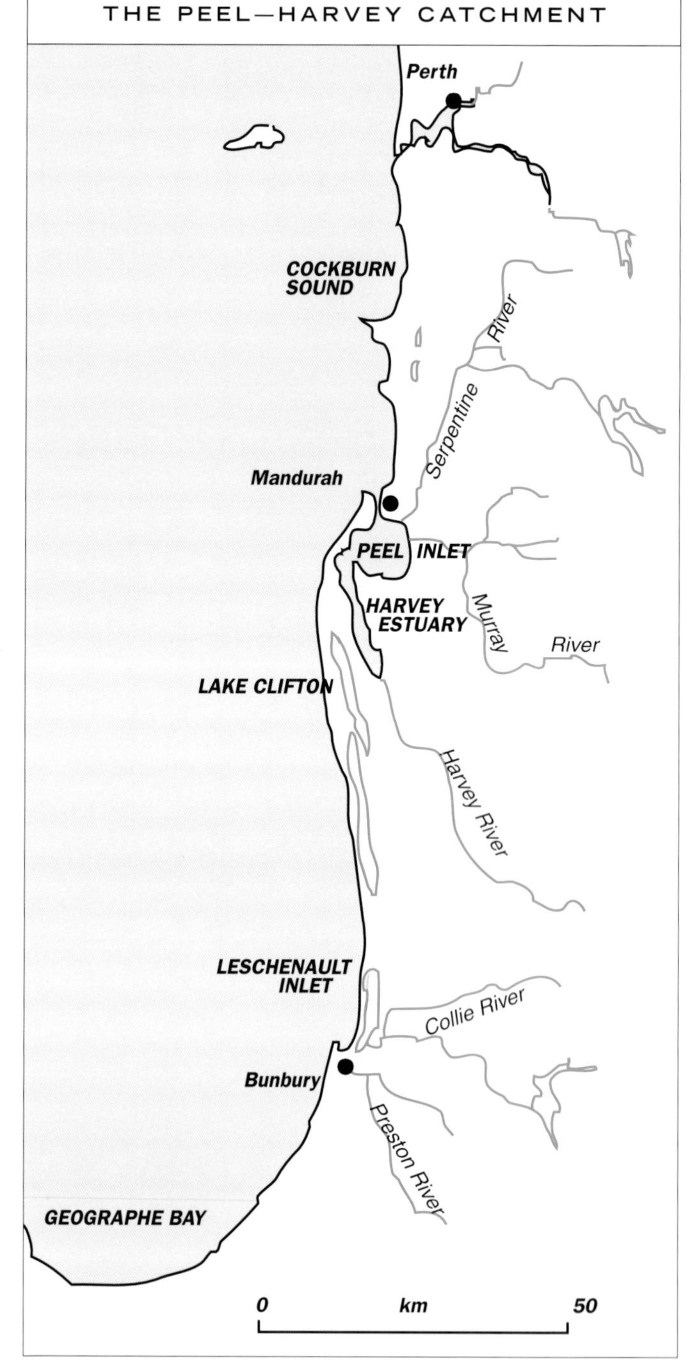

The Serpentine and the Harvey rise in the Darling Range, as do the North and South Dandalup and a number of other small streams. All this water flows through the jarrah forest to the Darling Scarp, crossing it in a succession of rapids and small waterfalls and emerging onto the coastal plain within 5 kilometres of the scarp. The rivers have no direct outlet to the sea because the Spearwood dune ridge runs along the western edge of the plain—so they used to form a system of interconnected wetlands. While the Harvey made its way northwards across the flat plain, the Serpentine was deflected southwards. Its wetlands lie between the Bassendean and Spearwood dune systems. Today, both the Harvey and the Serpentine flow in constructed channels along much of their length.

The Peel–Harvey Estuary consists of two shallow lagoons, the tidal reaches of three rivers and a narrow channel to the sea. The daily tide range on the south-west coast is only just under a metre (in contrast to the tide range on the Kimberley coast, where range in spring tides averages 8 metres). The entrance to the estuary is narrow, slowing down the entrance of tides, and before the Dawesville Channel was opened in 1994, the tidal range was only about 10 centimetres—10 per cent of the ocean tide. Water levels in the sea and estuary also rise and fall with the regular passage of high pressure and low pressure barometric systems, and with the annual changes that occur in the overall sea level. The estuary is subject to extremely variable conditions as a result of the restricted tidal exchange and the marked seasonality of rainfall which results in 70 per cent of river flow occurring in the four months June to September. Water is almost fresh in winter but is saltier than the sea in summer.

Origins of the estuary

Like all Australia's estuaries, the Peel–Harvey is recent, having attained its present form after sea level stabilised 6000 years ago, following its rise during deglaciation after the last glacial maximum. At 18 000 years ago, sea level was 130 metres lower than it is today. Peel Inlet was a shallow valley across which the Murray River flowed to a channel carved through the sand dunes near Halls Head, then across the wide coastal plain (the exposed continental shelf) to the sea. Harvey Estuary was a long valley between the dunes through which river water flowed to meet the Murray. Over the next 11 000 years sea level rose, flooding the valleys, and spreading out over low-lying areas inland of the present coast when it attained a height 1 to 2 metres above its present level at about 7000 years ago. The estuary then would have been 7 to 8 metres deep. Since stabilising of sea level

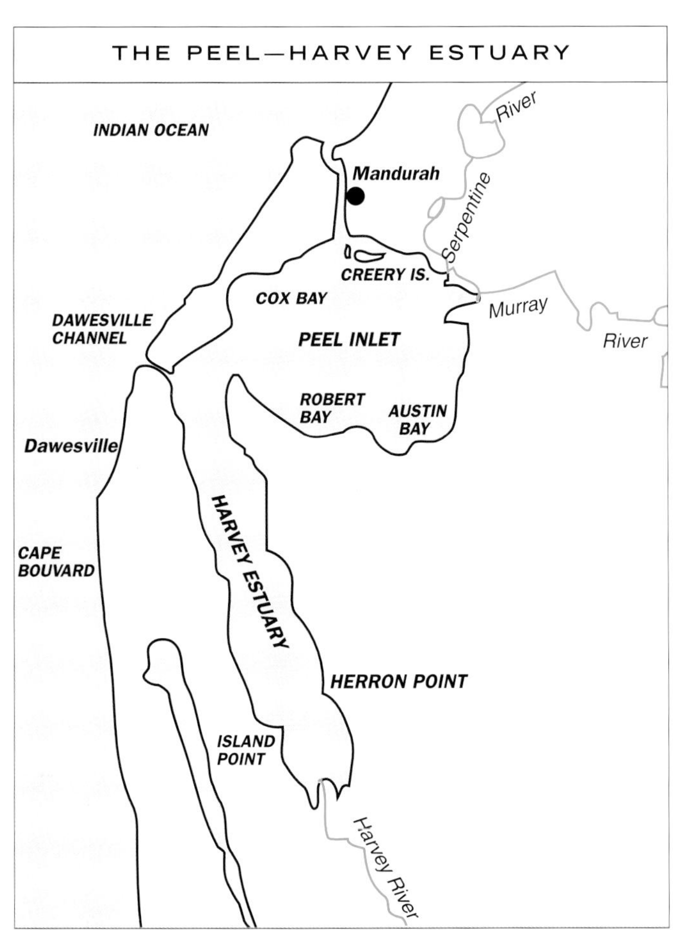

After Bradby[28]

at the present lower level and siltation of the estuary, the depth has been no more than 2 metres. The sea level drop decreased tidal exchange between the estuary and the sea.

When the estuary first formed, the entrance channel was wide and at least 12 metres deep, and the environment was essentially marine. Sand washing into the channel with the tides made the channel shallower and flood tides dropped sand where the channel met the open water of Peel Inlet. There it built a large delta almost 4 kilometres across. The slight drop in sea level at 4000 years ago stranded much of the delta at water level, forming the Creery Wetlands on the east shore. Sedimentation of the Estuary continues up to the present.

Bottom sediments in the estuary have recorded some of its past history. Marine shells of cockles and other molluscs in deeper sediments have been radiocarbon-dated and shown to be between 4200 and 7000 years old, showing that the Peel Inlet was once a bay where marine animals lived, just as they do in Oyster Harbour on the south coast today. In the upper sediment, materials deposited in the last 4000 years contain only estuarine species of molluscs that still live in the estuary.

Changes to the estuary since settlement

The healthy and resilient estuarine system that existed when the first settlers arrived lasted until the end of the nineteenth century.[28] The estuary was highly productive, supporting large numbers of fish and waterbirds. Kingfish regularly made a spectacular sight as they made their way across the bar and into the rivers. By the late 1890s a decline in fish and bird numbers was already noticed and the estuary had silted visibly. Many attempts were made to reopen the mouth, where a large sandbar had formed, and a wall was built to deflect sand, but was not successful.

Between 1900 and 1939 massive changes occurred to all the wetland systems on the Swan Coastal Plain. Most large wetlands had been drained to some extent; most trees along rivers and streams had

Exit from the Peel–Harvey estuary to the sea—the mouth of the Dawesville Channel. M.E.W.

Bridge over the Dawesville Channel. M.E.W.

been cleared; channels had been de-snagged to allow free flow of water from the new farmlands; deep pools in rivers had gone, filled with silt; drains had replaced rivers; and floodplains had been deprived of their life-giving supply of water.

The consequence of all the draining and channelling was that more fresh water entered the estuarine systems, carrying with it soil, sediment and extra nutriment. In an extra wet year in 1910, fish kills and disease found among the fish caught for marketing were blamed on the heavy influx of contaminated fresh water and sediment.

Fish populations were managed in the estuary, using the Serpentine Lakes as nurseries by impounding schools of fish there, particularly mullet, when water levels were high in winter and releasing them into the lower reaches of the lakes and rivers to replenish stocks later.

The bar at the ocean entrance to the estuary closed completely many times over the years, affecting water quality and salinity. Crisis conditions existed in 1935 when the estuary remained blocked even after heavy rains and slimy weed started to grow in the murky waters. There is no doubt that the wetlands, the rivers and the estuary had suffered permanent and serious damage by the start of World War II.

The Green Revolution agricultural boom of the post-war years saw vast expansion of farming on the plains and little attention was paid to the declining health of the estuary, although prolific weed growth in the wet winters of 1946 and 1947 began to alarm the fishing industry, whose nets became encrusted with pink slime that stuck firmly. (The 'weed' which affected nets is believed to have been the fine red alga *Corynospora australis*, an epiphytic alga which can grow on the leaves of seagrasses and restrict their ability to photosynthesise.) The weed problem improved in the 1950s but was back again in the 1960s. The Fisheries and Public Works Departments decided that better through-flow would help, so the Serpentine was cleared of 'obstructions' and many new drainage canals were dug. Within a couple of years of drains being cut, lakes filled with weed. Fish catches improved over the next few years, between 1960 and the mid 1970s, in spite of the weed, partly because more effort was put into fishing because of better prices, but partly because the weed provided extra food for food webs. However, the weed problem was really only starting to emerge.

Several species of green algae were becoming established in the estuary. One of these, *Cladophora*, known locally as goat weed, grew as unattached balls that lined the estuary bed. These balls would float to the surface and drift into the shallows. By 1976, the condition of the Serpentine Lakes had deteriorated further and the schools of mullet which used to feed and grow there for months stayed only a few weeks; seagrass beds were disappearing and the swans which fed on them were also disappearing.

The construction of training walls in the mouth of the estuary in 1967, allowing free flow of tidal water and emigration of fish and shellfish, should have improved matters in the estuary, but it had no visible effect on the impact of 'weed'. Floating pink slime was prevalent; goat weed was accumulating on the shores and causing not only unpleasant odours but also so much rotten egg gas that people were nauseated and tourists ceased to visit. In the southern end of the Harvey Estuary 'a type of weed that grows from the bottom and almost reaches the surface in five feet of water' was reported by fishermen to be causing loss of fishing grounds; and other green algae were present. In November 1970, a thick green sludge covered more than 6 kilometres of the Serpentine River at Barragup—a toxic cyanobacterial infection which drifted like an oil slick into the estuary.

Various methods of collecting and removing weed and improving water through-flow were tried in the 1970s. In 1976, the EPA commissioned a major study of the Peel–Harvey Estuary, which ran from 1976 to 1980.

By 1978, researchers had established that diatoms play an important part in ecosystems in the estuary. They flourish from July to September, scavenging the nutrients which wash in from the rivers and drains. They can be so numerous as to give the water a cloudy appearance. While the diatoms feed on the nutrients, microscopic predators (copepods and nematodes) feed on the diatoms. The waste products from this activity fall to the bottom of the estuary where they decay with the help of large masses of bacteria. The bacteria use up so much oxygen in the bottom water that phosphorus in the sediment is freed and released into the water in spring when the water is warm. (The source of the phosphorus in the sediment was believed to be fertilisers in farm run-off into drains and rivers.) The availability of phosphorus creates conditions ideal for the growth of algae. The rate of growth of *Cladophora* was found to correspond to the amount of phosphorus available.

In 1978, another major cyanobacterial bloom occurred, and in 1980 the whole estuary was blanketed with green scum. The conclusion of the study was that the amount of nutrients, and phosphorus in particular, in the estuary must be reduced.

By the 1980s, dense growths of *Chaetomorpha* and *Enteromorpha* replaced *Cladophora* as the dominant weed in the Peel Inlet. Rotting weed was creating health problems and things were so bad that workers clearing weed wore gas masks. More studies were being carried out and the situation was becoming more and more political. The Concerned Citizens of Mandurah were becoming militant.

Management of the problem involved several different approaches:

- Measures to decrease the nutrients making their way into the estuary. Advice was given to farmers on the use of fertilisers after research had shown that on many soils little of the fertiliser was being used by the pasture or crops to which it was applied. Often, more than 30 per cent was lost to drainage, and more than 50 per cent was unaccounted for in the surface soil, or the pastures, or the run-off waters. Using less fertiliser and applying it at times when it was less likely to be washed away made great savings in the amount of fertiliser used. The growing Landcare movement can take much credit for the improved attitude of the farming community and its cooperation with the Departments and extension officers who were orchestrating the changes.
- The discovery that **bauxite residue**, known locally as 'red mud' and produced by Alcoa as a by-product of its aluminium production, stops the leaching of phosphorus when added to soils, has been useful. The iron in the product binds the phosphorus and makes it available for uptake by plant roots, increasing productivity while helping with the problem of too much phosphorus entering waterways. This method works particularly well for the coastal sandy soils which lack iron.
- **The use of wetlands as filters** to remove nutrients from run-off has been shown to be effective. Many of the wetlands, however, had been destroyed and those that were left were insufficient to deal with the problems faced by the estuary, although they were capable of greatly reducing the phosphorus load in water passing through them. The most effective wetlands were those where paperbarks and sedges had been left around the water.
- **Engineering options** of many sorts were considered and a final decision was reached to construct a channel between Harvey Estuary and the sea. The channel was to be 2.5 kilometres long, 220 metres wide and between 4.5 and 6.6 metres deep. At the estuary end, the channel was to extend through the tidal flats to deeper water to maximise the flushing of the system. The first earthworks for the Dawesville Channel began in 1992 and it was completed ahead of time and within budget in 1994.

SOUTH COAST RIVERS: *estuary and stream rehabilitation involving the Green Skills Program*

Western Australia's South Coast estuaries and rivers face mounting problems of salinity and increased nutrient levels.[30]

The South Coast Region comprises about 5.4 million hectares and extends from the Frankland River in the west to Cape Arid in the east. Development of the region has mostly occurred over the last 50 years, with agriculture the main industry. Wool, beef and cereal crop production have been the main land uses, and lately horticulture, forestry and viticulture have increased. Tourism is a growing industry.

High conservation values exist in the region, which contains one-third of the State's plant species, many of which are endemic. Landscape values include mountains, dramatic coastlines and remnant biodiversity. This is a fragile region, subject to wind and water erosion, rising saline water tables and waterlogging. It has been estimated that 1 million hectares of agricultural land could be subject to dryland salinity in 30 years—a figure which is even worse than at first appears because a fair proportion of the region does not comprise usable agricultural land. The need to protect remnant vegetation and to manage the area so that declining water quality and other issues are addressed is becoming increasingly urgent.

Green Skills is a non-profit incorporated community organisation whose aim is the creation of employment in the environmental field. It has offices in the towns of Denmark, Albany and Fremantle. Its main areas of work include providing environmental training and educational programs to increase skills and raise awareness; helping people find meaningful work and fostering the development of ecologically sustainable enterprises; and carrying out practical conservation work in partnership with landholders, managers and community groups. Its involvement in projects in Oyster Harbour, Wilson Inlet and the Kent, Frankland and Pallinup catchments have included fencing, revegetation, weed management, foreshore surveys and broader promotion projects.

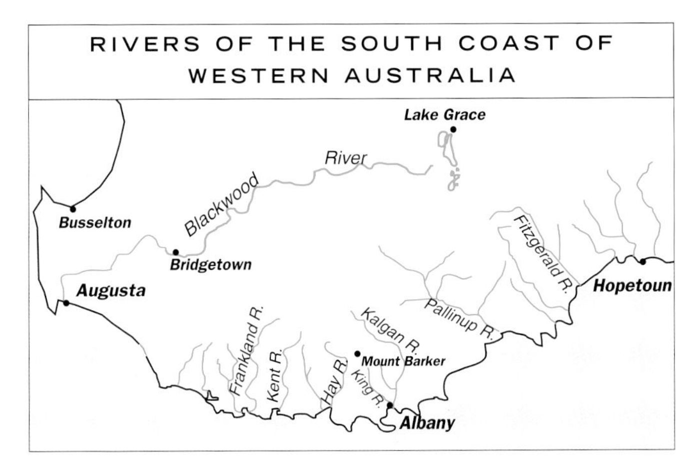

Green Skills has used a number of programs to carry out its riparian projects including LEAP (Local Environment Action Programs); NOW (New Work Opportunities); Gordon Reid Foundation Grants Projects; and Green Corps. Close cooperation has been maintained with the Albany Waterways Management Authority and the Wilson Inlet Management Authority which support projects like foreshore surveys, grant assistance for fencing

Mt Shadforth Lookout to Wilson Inlet, South Coast, Western Australia. M.E.W.

and revegetation, development of demonstration sites and community watercare projects.

Successful joint projects have included surveys and riparian rehabilitation of the Kalgan River, where demonstration foreshore revegetation projects have been established; surveys of the Hay and Denmark Rivers; the Little River catchment; foreshores of the Oyster Harbour catchment; the Frankland and Gordon Rivers; the Scotsdale catchment; and the upper Kent catchment. The surveys have been important in guiding on-ground work. The Oyster Harbour and Kalgan surveys have allowed agencies and Landcare groups to target fencing and revegetation works. The Hay and Denmark Rivers survey enabled the Wilson Inlet Management Authority to allocate its fencing subsidies as well as to recognise that some of the smaller coastal catchments are key 'hot spots' and need more attention than the main rivers. The entire Hay River frontage of the Mount Barker Research Station was fenced and revegetated in a joint project between the Wilson Inlet Management Authority, Green Skills, the Research Station and two landholders as a result of the initial Green Skills survey.

The Little River report allowed the Denmark Environment Centre to access Natural Heritage Trust funds to appoint a catchment officer to coordinate riparian repair projects. The Kent and Frankland surveys have led to greater allocation of funds for riparian fencing and rehabilitation. The Oyster Harbour survey has led to major weed management work focussed on the lower King River.

All this successful activity has been a wonderful example of what can be achieved when individuals, community organisations and government agencies work together to achieve worthwhile goals.

The Donnelly River, a pristine stream in the forested south-western corner of Western Australia. M.E.W.

WHEATBELT RIVERS AND THEIR PROBLEMS WITH SALINITY

The tributaries of the Avon River down to York, the Blackwood River down to Bridgetown and the upper reaches of the Kent and Pallinup Rivers all originate within the Western Australian Wheatbelt, in an annual rainfall zone of less than 600 millimetres. Salinities in these four streams, as in others in this region, have reached high levels and are rising continuously. The trend in salinisation is forecast to at least triple, possibly to quadruple, the area of land currently affected, with associated rises in stream salinity. Large areas of stream, riverine and wetland ecosystems have already been devastated.[29]

The current estimate of the area of these catchments which will ultimately be affected by salinity at equilibrium is 40 per cent; at present it is approximately 10 per cent. The impacts on the biota are most obvious in lower parts of the landscape, but the relief of the Wheatbelt is so low that a much larger area will be affected. At risk, or already affected, is 80 per cent of susceptible remnant vegetation on farms and 50 per cent on public lands. To date, degradation is focussed on wetlands and drainage lines. The beds and banks of 80 per cent of the region's rivers and streams are seriously degraded; the degradation of wetlands is well advanced but largely unrecorded.

Salt is a natural, integral part of the Wheatbelt landscape—it is estimated that a staggering 10 000 tonnes lie below each hectare. Changes in hydrology due to land-clearing have led to rising water tables, as in the Murray Basin and elsewhere in Australia. To remedy the situation, Landcare and other community groups and organisations, and landholders, have been active, planting deep-rooted perennials, rehabilitating riverine corridors, fencing remnant vegetation and stream banks and seeing, in some cases, some reward for their labours. (There are several case studies on salinised land and rivers in *Listen ... Our Land is Crying.*)

A recent assessment of the hydrological effects of tree planting in the Wheatbelt is interesting, if depressing.[29] It was found that:

- Trees planted in discharge areas have only a small impact on water levels, and only when the groundwater is reasonably fresh.
- Water levels in local aquifer systems only fall significantly in areas with a significant fraction (50–80 per cent) of the land planted to trees.
- Trees have little or no effect (yet) on water tables more than 10–30 metres from the planted area.

There is no doubt that tree planting has local effects on groundwater, but major impacts on catchment scale behaviour require significant areas of high density plantings, and time measured, perhaps in decades, to show results in terms of reduced salt loads and stream salinity in the Wheatbelt.

What of engineering options for combating salinity? Once again it has been shown that draining and pumping, on a large scale or small, reduces groundwater levels and salinity, but the overall effect is small, local, and can take a long time. What to do with the saline water drained or pumped out of the landscape is a serious problem.

The study concluded that:

- is unlikely that salinisation can be significantly reversed with revegetation, at least within human time scales. The hydro-geological systems of the Wheatbelt are too saline, too flat and too untransmissive to respond quickly to tree planting.
- Engineering can have positive local effects, but to substantially change the salinity of drainage lines would require pumping over very long time scales, with associated problems of disposal of saline water. (It has recently been seriously suggested that the only viable option is to sacrifice a couple of rivers and turn them into salt-removing drains. The two most suitable rivers are the Avon–Swan and the Blackwood! Horrifying indeed—but this is an indication of the gravity of the salinity situation in the Wheatbelt.)
- We may have to accept that some changes in the hydraulic and hydro-chemical characteristics of the system are irreversible. Geological time, perhaps geomorphological change, and massive climate change may be required to change the situation fundamentally or even superficially.

We have an ethical responsibility, however, to do what we can and manage the situation to the best of our ability. This may require reforestation over large fractions of the landscape and expensive and ongoing engineering solutions simply to buy time.

THE CENTRAL AUSTRALIAN GROUNDWATER DISCHARGE ZONE

An ephemeral watercourse in the Amadeus Basin. REG MORRISON

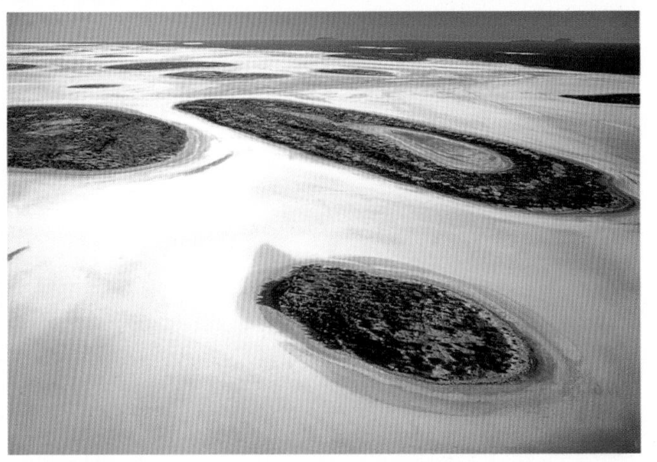

A great valley runs from Lake Hopkins, on the Western Australian side of the Northern Territory border, extending 500 kilometres eastwards between the Cleland Hills—George Gill Range to the north and the Petermann and Musgrave Ranges to the south. Today this valley is a major groundwater discharge zone, comprising a chain of playas running from Lake Hopkins through Lake Neale, Lake Amadeus and several smaller playas to the dune fields along the Finke River at the edge of the Simpson Desert. It is a remnant of an ancient palaeodrainage system whose rivers flowed into the Finke and ran on into Lake Eyre, prior to the most recent onset of aridity and dismemberment by drifting sand.[5, 17] Today the Finke disappears into the Simpson Desert and its flood-outs end about 200 kilometres from the Lake.

The great valley is devoid of surface drainage, and the playas and their landforms and chemical sediments are the result of groundwater transmission, and discharge from a large regional flow system.[37, 37A, B, C, D, E] The valley contains extensive groundwater calcrete deposits, which are commonly silicified. The playas contain gypsum and glauberite deposits resulting from concentration of discharging groundwater by evaporation. Gypcrete crusts have developed on the margins of playas and on islands as a result of induration of gypsum dunes.

The catchment for the groundwater discharge zone extends over 90 000 square kilometres. Several intermittent streams drain the ranges along the northern and southern divides and flood out into dune fields, recharging groundwater, but no surface water reaches the playa chain.

The groundwater system is two-layered.[37] Cainozoic sedimentary aquifers overlie fractured bedrock aquifers. In the centre of the basin, Cainozoic sediments are more than 100 metres thick and consist of clay and sand of alluvial and aeolian origin, with interbedded calcrete, gypsite and gypcrete chemically precipitated from the groundwater. The main aquifers are alluvial sand units and calcrete. Age of the waters has been dated as from modern up to 1300 years old. The base of the Cainozoic sediments has been dated as Middle Palaeocene (60 million years) at Ayers Rock, and the original surface of the basin was probably Late Cretaceous in age. (The Cretaceous Gosses Bluff meteorite impacted with the ancient land surface in the northern Amadeus Basin, dating that area, and similar ancient Mesozoic land surfaces have been recognised in north-west Queensland, the Flinders Ranges, the Gawler and Macdonnell Ranges.)

Bedrock hills, including Uluru (Ayers Rock) and Kata Tjuta (the Olgas), protrude through the Cainozoic sediments. Run-off from these protruding

Islands in the glittering salt crust in the Amadeus Groundwater Discharge Zone. REG MORRISON

areas, particularly the bare rock of Uluru and Kata Tjuta, provide local recharge of groundwaters. (The flow-on zone around Uluru has a vegetation zone distinct from the surrounding plains.) A very ancient drainage system exists between Kata Tjuta and Uluru.[224]

The chain of small playas near Curtin Springs receives groundwater flow from the flanks of the valley (north and south) as well as some eastwards down the valley. On the bare playa beds, brown areas of heaved gypsum ground form 'ploughed field' topography where groundwater discharges and evaporates; in occasionally flooded areas where groundwater outcrops, the flatter, white areas are encrusted with salt. The playa beds have small dune-sand islands, gypcrete ringed. Quartz sand dunes, some up to 30 metres high, stable and vegetated, surround the playas.

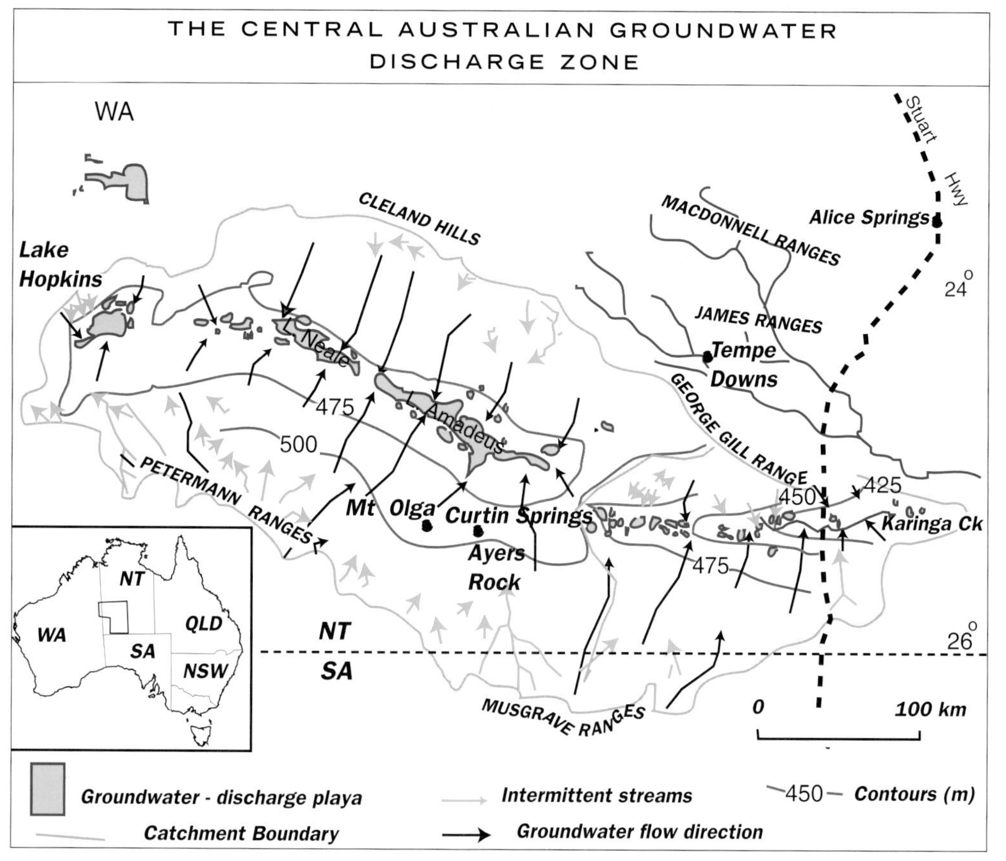

After Jacobson *et al*[37]

The age of the calcite and gypsum layers in the playa sediments gives a record of the climatic changes in the ancient palaeochannel.

- Between 75 000 and 35 000 years ago there was a long period of low velocity groundwater regime, similar to that of the present day but with greater hydraulic head, and a great deal of calcrete was precipitated.
- Between 27 000 and 22 000 years ago, the nature of the calcrete deposits indicates episodes of high intensity rainfall.
- The gypsum deposits which indicate much increased aridity are dated within a period from 16 000 to 8000 years ago.

The 27 000 to 22 000 year wetter period is believed to be the time when humans colonised the arid zone—estimated as just prior to 22 000 years ago.[223] The maximum lake-full stage for Lake Woods in the Northern Territory, where evidence of early human occupation is found, is 23 000 years ago. (Two older dates on calcrete stringers are 183 000 and 200 000 years, providing evidence of other episodes of calcretisation in the palaeodrainage system.)

After Jacobson et al[37]

SECTION TWO

CENTRAL AUSTRALIA

FLOODPLAINS AND ARID CENTRAL AUSTRALIA—TWO CONCEPTS WHICH CONJURE UP SUCH DIFFERENT IMAGES IN THE MIND WOULD SEEM TO BE INCOMPATIBLE! YET MANY STUDIES OF EPHEMERAL RIVERS IN THE ARID ZONE SHOW THAT DURING VERY LARGE FLOODS, WATER CAN COVER MOST OF THE LANDSCAPE AND SEDIMENTS ARE WIDELY SPREAD OVER GREAT AREAS, FAR AWAY FROM THE MAIN CHANNEL SYSTEMS.

Most of Australia's fluvial landscapes occur in the dry inland of the continent [17] and their formation, and the nature and behaviour of the river systems which formed them, are very different from those of the eastern 'green crescent' of our land with which we are reasonably familiar. While New South Wales coastal rivers, for instance, have short courses, relatively steep gradients and alluvial plains dominated by fairly recent features, the situation in the Centre is completely different. In this flat, inward-draining mega-region, landscapes and their drainages are often incredibly ancient. The Finke River, for instance, may have followed some sections of its course since Devonian and Carboniferous times, when the Alice Springs orogeny (mountain building event) occurred, starting about 380 million years ago and continuing for 100 million years,[31] resulting in major folding, faulting and uplift across central Australia.

Topography of the Alice Springs region is characterised by the east–west trending Macdonnell, Chewings and other ranges developed by folding of Precambrian and Palaeozoic strata during major orogenies.[225] Most central Australian river systems originate in these central Australian ranges where the intricate ridge and valley topography, determined by the resistant folds in the ancient rocks, controls river patterns. The ranges are surrounded by an extensive area of plains with surface characteristics of varying origin and history. They include pediments at the base of the ranges (fans of sediment on the steep slopes); sandplains and dunefields of wind-blown origin; floodplains; stony desert; calcrete rises; claypans; and ephemeral lakes—all of which occur, as we shall see, in other arid parts of the continent as well.

Some rivers, like the Finke and Hugh, are not constrained by the east–west ridge pattern of the ranges.[17] Instead, they flow across several of the individual ridges, transversing the structural grain and in some cases breaching the snouts of folds. Relics of Cretaceous surface (100 million years old) exist in the high tops of the Macdonnells, but most of the crests are duricrusted Miocene

A flooded spinifex plain in the Tanami Desert, NT. Even the deserts flood in this land of contrasts. REG MORRISON

Redbank Gorge, a water-worn slot canyon in the Macdonnell Ranges.
REG MORRISON

Glen Helen Gorge in the Macdonnell Ranges, N.T.
REG MORRISON

ANOMALOUS COURSES OF THE FINKE AND HUGH RIVERS

CHEWINGS RANGE
Glen Helen
Alice Springs
Finke R.
Ellery Creek
Owen Springs
WATERHOUSE RANGE
Palm Valley
JAMES RANGE
Hugh R.
Finke River
0 km 30
Strike line
River

After Twidale[17]

MODERN CENTRAL AUSTRALIAN RAINFALL PATTERNS

Average rainfall decreases from 250 millimetres a year at Alice Springs, in the foothills of the Macdonnell Ranges, to less than 125 millimetres a year to the south in the Simpson Desert and the central Lake Eyre Basin, where south-flowing drainages end. However, rainfall is highly variable.[38] At Alice Springs, annual rainfall varies from 82 millimetres to 783 millimetres. Monsoon storms in the summer months can cause spectacular floods in the canyons of the ranges. Where the floods escape from the ranges they spread out across the plains, creating remarkably wide shallow flows that ultimately dissipate in the central deserts.

A summer storm front moves across the central plains south of Alice Springs. REG MORRISON

surfaces. Thus the modern Finke and other transverse drainages of Central Australia can be dated as at least 20 million years old.

Central Australian floodplains contain landforms which suggest three scales of activity in their formation during the last 10 000 years:

- **A set of large sandsheets, sand threads and megaripple-covered channels**, related to a few enormous floods, dominate the landscape, setting the basic configuration of the floodplain. They are much larger or more widespreading than those associated with currently active river systems and they have not changed significantly since air photos became available in the 1950s. Evidence for the major floods can be found in the bedrock gorges as well as in the floodplain sediments.
- **The contemporary floodplain** which consists of channels and levees, flood-outs, unchannelled floodplains and flood basins—systems which have been growing upwards and outwards from the mountain ranges since the last mega-flood. Considerable

The Finke River and the Goyder Creek (below). The township of Finke is marked. South of the river, gibber plains (eroded Jurassic pebbly sandstone) with red linear dunes are overlain by moderately well-spaced linear dunes orientated northwards. Source-bordering dunes occupy a zone 2 to 5 kilometres wide north of the floodplain. North of this zone are regional linear dunes characteristic of this part of the Simpson Desert. Erosion cells are clearly seen north of the township, north-east along the Finke, and in the median sector between the rivers west of Finke.

PHOTOGRAPH KINDLY SUPPLIED BY DR JOHN MAGEE, ANU, CANBERRA

changes are seen in the aerial photo landscape record between the 1950s and 1980s.

- **A mosaic of erosion cells** which result from local redistribution of sediment. This has intensified since the arrival of Europeans 100 years ago, which caused disturbance of systems by grazing and other activities, and also as a result of increased flooding due to variations in rainfall.

HISTORY OF THE FINKE RIVER

The Finke River rises on the ancient Arunta Block, about 75 kilometres west of Alice Springs and more than 350 kilometres north-west of the township of Finke. It transects a series of east–west ridges and ranges as it maintains an ephemeral southward course into duricrusted plains and sand-ridge deserts, heading for Lake Eyre which lies 15 metres below sea level and about 800 kilometres away, as the crow flies, from the river's source. (It vanishes into the sand dunes of the Simpson Desert about 150 kilometres south-east of Finke, joining floodplains of the Alberga and Macumber and other streams draining along the western edge of the desert.[36])

The Finke carries a large sand load in a well-defined channel when it is flowing. Spectacular water gaps and gorges are developed through the resistant sandstone ridges of the Macdonnell, Krichauff and James Ranges where the river is deeply entrenched into bedrock. There are, in fact, two gorges in the section of river north of the James Range. One is currently occupied by the river; the other is not as deep and looks like another meander train intertwined with that of the main gorge. Deposits in the unused gorge are ancient, and the palaeochannel probably dates to the Miocene. (The climate at that time is believed to have been subtropical, with semi-arid conditions developing during

LANDFORM ZONES OF THE FINKE RIVER SHOWN IN THE AERIAL PHOTOGRAPH

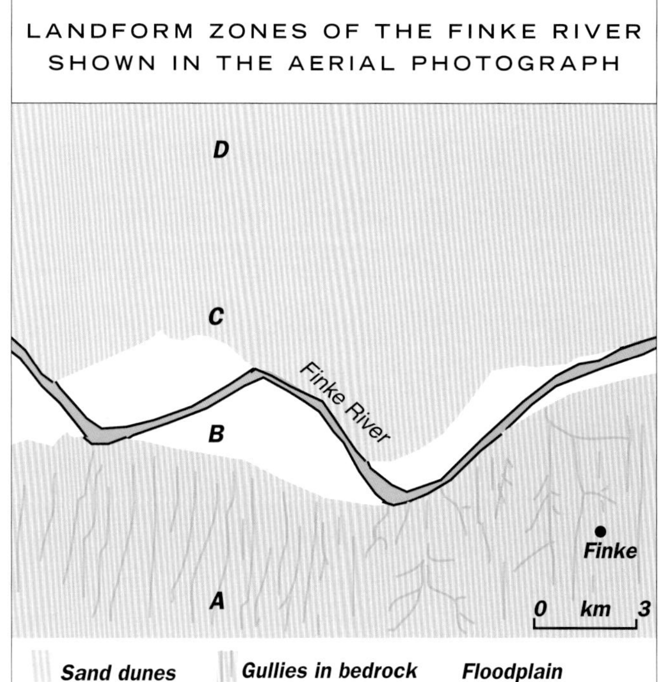

A: Gibber plains with linear dunes, oriented northwards
B: Finke River channel and floodplain
C: Source-bordering dune zone, 2 to 5 kilometres wide, north of floodplain
D: Moderately vegetated dunes north of source-bordering dunefield are part of the Simpson Desert regional dunefield

After Nanson *et al*[32]

PALAEO-MEANDER SYSTEMS OF THE FINKE GORGE

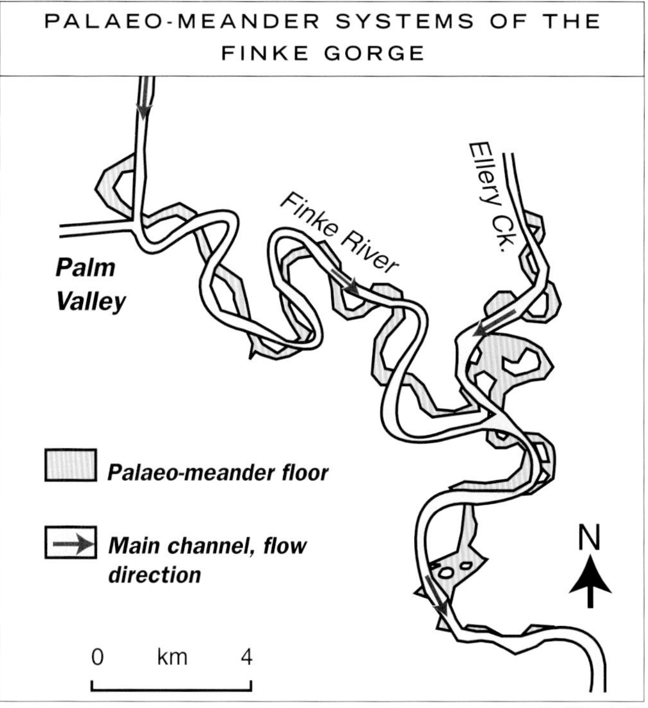

After Pickup et al33

The Finke River, James Ranges in the background. REG MORRISON

The Finke River, near Finke, Northern Territory. GERALD NANSON

Pancakes of green algae in Palm Creek, Finke Gorge National Park, N.T. REG MORRISON

the Pliocene, and aridity following in the Pleistocene.)

The entire Finke River system is ephemeral, flowing only after intense rain periods, when a high velocity, turbulent flood rushes through the gorges, carrying a sand load in suspension. In eddies and backwaters some of the sediment may be deposited, leaving *slack-water deposits.* These can often be dated and used as evidence of previous flood heights.[33, 34, 35]

HEADWATERS OF THE FINKE RIVER

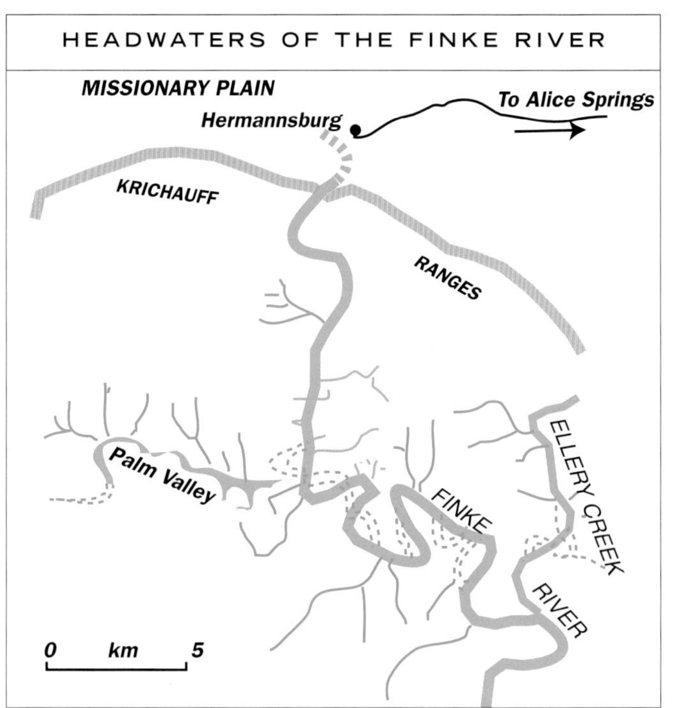

After Baker *et al*[34]

The Finke Gorge slack-water deposits are the best sequence so far discovered in Australia. These deposits have enabled the dating of seven flood events, all of which occurred in the last 850 years—supporting the case for a marked increase in the frequency of extraordinary floods during the Late Holocene, and the extreme variability which is in-built into the hydrology.

For a flood to occur which passes along the full length of the river, heavy rainfall over a long period and over a very large area is required. Lesser events in the headwaters occur but do not travel far downstream because of high infiltration losses into the river's sandy bed during its long track across the Missionary Plain, and because of evaporation. So when there is a flood, it is likely to be a large one, and because such events are rare, their effect on channel morphology is long-lasting.

The 1967 floods in the Finke River resulted from Tropical Storm Gwen which deposited 150 millimetres in 11 hours on some stations. Where the Finke emerged from the gorge, it spread out 11 kilometres wide and averaging 0.8 metres deep, then split into several long fingers where it flowed down north–south-trending flats between the longitudinal dunes of the Simpson Desert. Some interdune flats were flooded to a depth of 6 metres and water stayed in these flood-outs for up to 19 months. The 1967 floods were considered to be the greatest historical floods, or at least the greatest since 1895 when the Hermannsburg Mission recorded 240 millimetres in an episode in mid-January. But subsequent larger floods occurred in 1971, 1974 and 1988—the most concentrated recorded cluster of flood events ever in a seven-year period.

Sediments near the township of Finke have provided information on changing climate in the western Simpson Desert in the Late Quaternary. The alluvium in the Finke valley here has been dated as having been deposited prior to about 90 000 years ago, following which there is a gap in the record of depositional activity until the Holocene (the last 10 000 years). The activity which took place in the Early to Mid-Holocene was probably the result of the reactivation of the northern monsoon, which had been interrupted during the last

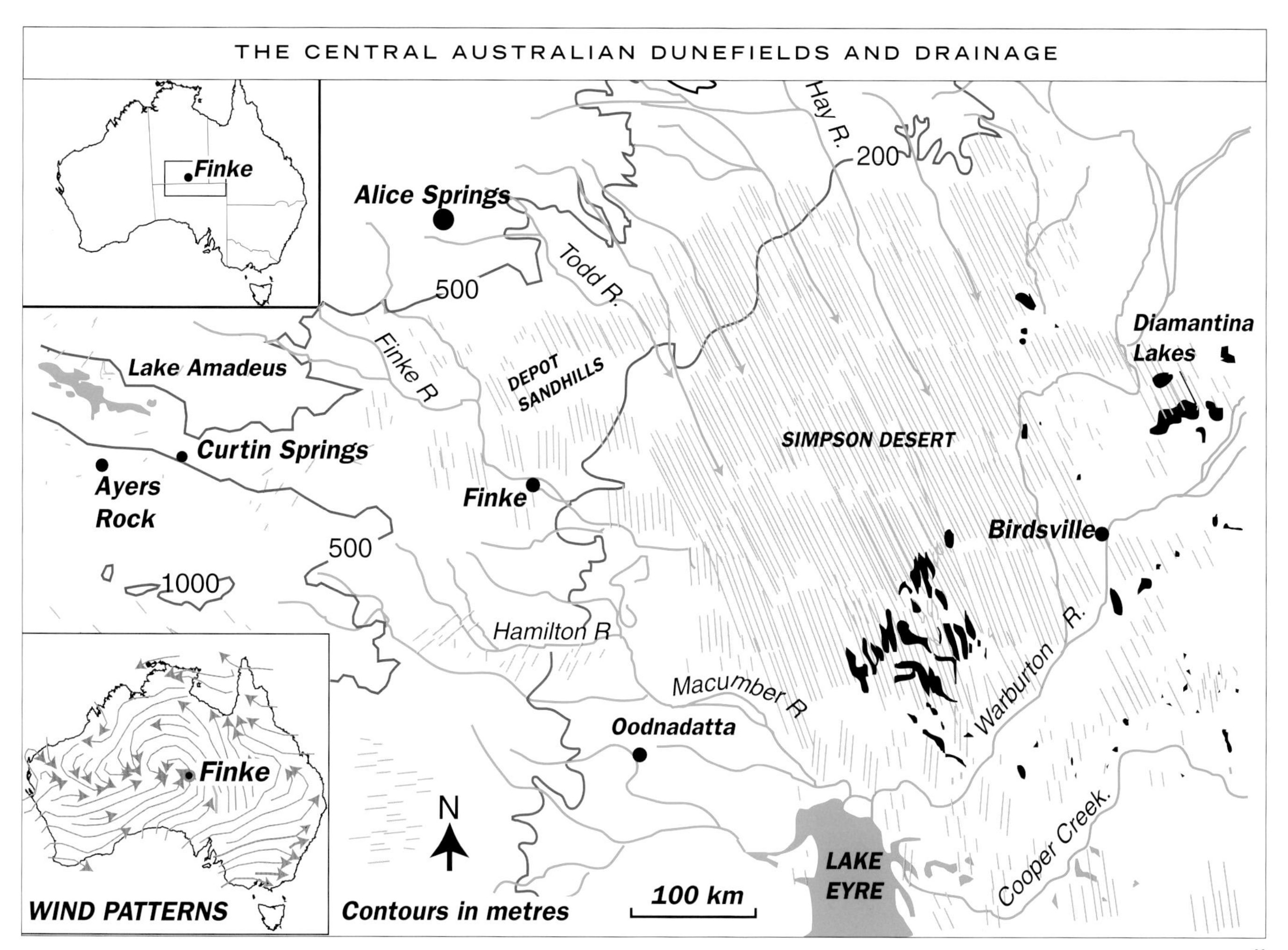

After Nanson *et al*[32]

Salt-encrusted watercourses south-west of Lake Eyre in the Simpson Desert. REG MORRISON

Pale coloured source-bordering sand dunes blown from the Finke River in the western Simpson Desert of the Northern Territory about 18 000 and 5000 years ago. GERALD NANSON

glacial stage of the Pleistocene ice age, and its penetration into central Australia. Rivers in the eastern part of the Lake Eyre Basin do not show this active phase, possibly because they were much lower-energy and less responsive.

The regional dunefield near Finke consists of linear dunes largely reworked and aligned during the last glacial stage (30 000 to 12 000 years ago) as part of the great anti-clockwise whorl of dunes in the Centre. The oldest source-bordering dunes from the Finke River are bright red in colour, and they were deposited at about 100 000 years ago and are now buried below paler source-bordering dunes. The latter consist of two units: a lower one of unknown orientation which dates between 17 000 and 9000 years, and younger paler dunes which appear to have had a different origin and to have extended only recently into the dunefield.

The 30 000 to 18 000 years old regional linear dunes near Finke are aligned almost due north; their cross-sectional asymmetry with steeper eastern slopes suggests a response to south-westerly or westerly sand-transporting winds between 18 000 and 10 000 years ago; these winds appear to have shifted to their present south-east orientation during the last 5000 years. A northward shift in the wind pattern at Finke by about 100 to 150 kilometres (1 to 1.5° of latitude) during the last glacial period can be postulated, but the dune pattern has remained remarkably stable.

THE ROSS RIVER

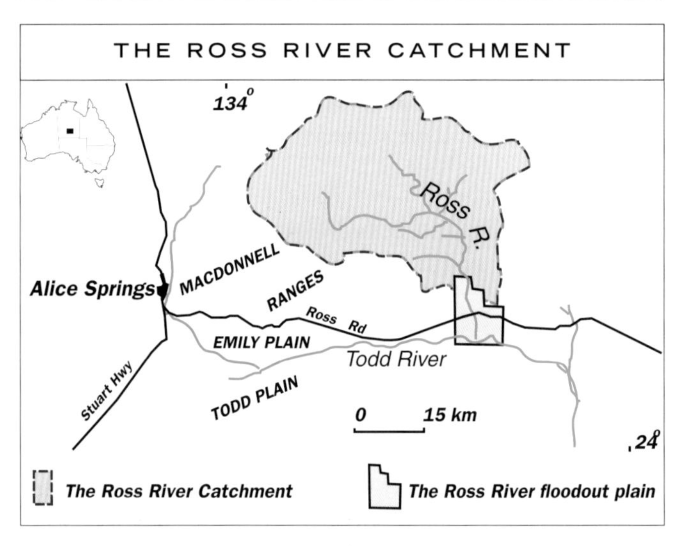

The Ross River drains an area of 1570 square kilometres in the eastern Macdonnell Ranges, making it the single largest tributary of the Todd.[38] The headwaters are in the ancient Proterozoic Arunta Block, which comprises the high ranges, and the Ross flows through the surrounding low ranges and hills before it emerges onto the broad low-relief plains and travels south to join the Todd. The Ross River floodplain begins where the river leaves the southern margin of the hills. It has a well-defined channel, approximately 300 metres wide, where it crosses the plain to its junction with the Todd River. The active channel is incised up to 5 metres into the floodplain deposits. The northern section of the incised river consists of a single meandering channel, but the southern reaches comprise braided channels separated by large diamond-shaped bars.

The floodplain is underlain by alluvial sediments and well-preserved palaeochannels are clearly visible on aerial photographs.

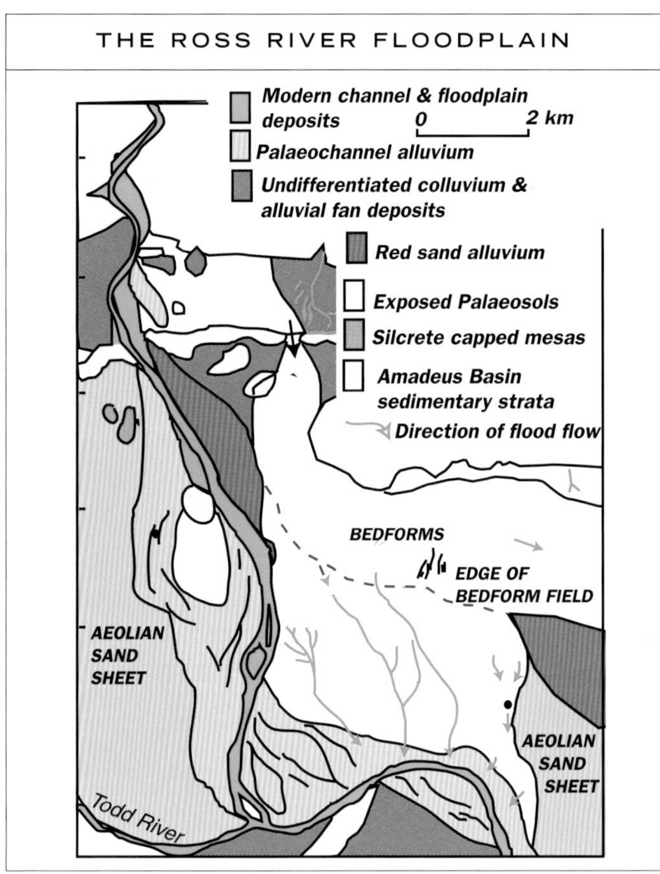

After Patton *et al*[38]

The western edge of the floodplain is defined by the contact between a major palaeochannel up to 1000 metres wide and the adjacent aeolian dune field. This wide palaeochannel swings to the south-east and becomes part of the Todd River floodplain. Palaeochannels of the Todd River upstream of its confluence with the Ross are obscured by sand sheets which have not been disturbed by floods during the Holocene.

A record of palaeofloods is preserved in the alluvial deposits and palaeochannels of the Ross River. The stratigraphy of the floodplain records the areal extent and frequency of Holocene floods. The plain is underlain by alluvial deposits, characterised by red earth soils dated at more than 59 000 years, sediments so deeply weathered that they retain no evidence of their alluvial history. This alluvium is covered by a sheet-like deposit of very silty sand of probable aeolian origin dated at 9200 ± 900 years, from the Late Pleistocene arid period. (This arid period is believed to correspond to the Younger Dryas event, when increased aeolian activity occurred, as recorded in the Gulf of Carpentaria[39] between 11 375 and 10 430 years ago. The 2500-year cycle of sunspot activity is involved.)

The oldest Holocene alluvium occurs as broad low-relief bars and levee deposits flanking the modern channel, and as low-relief, long-wavelength bedforms that fan out across the plain. This deposit resulted from a flood flow, up to 10 kilometres wide, that covered the entire plain. It has not been possible to determine the age of this deposit closely—it falls within the range of less than 10 000 years ago, and older than the newer deposits. Evidence for several large floods between 1500 and 700 years ago is also preserved in a palaeochannel 500 to 1500 metres wide. The most recent phase in the river's evolution has been its entrenchment into the older alluvium on the plain, resulting in the present narrow, deep channel.

The pattern of events has been one of long periods of stability punctuated by a few large flood events, and it appears that these events are concentrated in the Late Holocene (as in the Finke River).

THE TI-TREE BASIN IN THE NORTHERN TERRITORY

The Ti-Tree Basin is a Cainozoic structure covering 5000 square kilometres which lies between the Plenty Highway and Barrow Creek, due north of Alice Springs. It is presently the subject of groundwater withdrawals for one of the largest arid zone irrigation projects in Australia. About 1 million cubic metres of water are withdrawn for horticulture and agriculture annually.[40]

At Ti-Tree, annual rainfall averages 300 millimetres but it is very variable and potential evaporation is about 3200 millimetres. Nearly all the incident rainfall is used in transpiration.

The Basin is rimmed by deeply weathered Precambrian rocks and contains up to 300 metres of Cainozoic sediments. Eocene lacustrine mudstone is overlain by Late Miocene to Early Pliocene limestone and sandstone, referred to as the Waite Formation. Above this are Pliocene and Pleistocene sands that comprise the main aquifers of the Basin. Quaternary calcrete, aeolian sand and alluvium form a covering in parts of the Basin.

Detailed mapping has been carried out, using data from 100 water bores. The general passage of groundwater is from south to north, down the hydraulic gradient, The depth to standing water level is variable over the Basin, being 40 to 50 metres in the south-east and near the Woodford River in the east, 10 to 20 metres in the centre, and less than 10 metres in the north. Water in the centre of the Basin in a linear zone below the Allungra Creek is fresh. There, water flows in the floodout after exceptionally heavy rainfall events, such as that of January 1974 when 405 millimetres was recorded, but such floodout events probably occur only every 20 years. Elsewhere in the Basin the water has varying amounts of dissolved salts, magnesium and calcium, and high nitrate levels are common.

Isotope investigations have shown that all shallow groundwater in 1991 was more than 50 years old.[40] The age of groundwater increased with the depth of the aquifers, with radiocarbon dating indicating that the age range was between 1730 and 8900 years; thus, recharge from infiltration of rainfall had taken place episodically in the Holocene. The isotope 'fingerprint' indicated that recharge had occurred during exceptionally heavy rainfall events, with evaporation prior to, or during, infiltration. A disturbing finding was that very little recharge is taking place in the modern climate regime. It occurs only in areas of concentrated run-off.

Therefore, the irrigation at Ti-Tree is using a resource which is only partly renewable (and a small part at that) and is mining sub-fossil waters.

THE TI-TREE BASIN

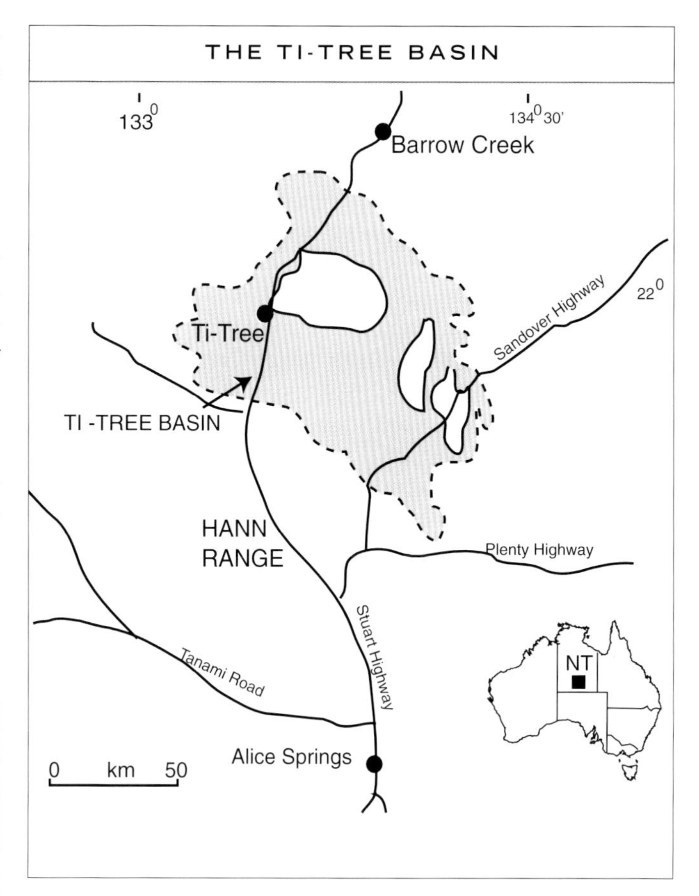

THE TODD RIVER FLOODPLAIN

The active floodplain of the Todd and its tributaries, south-east of Alice Springs, occupies a large area and includes a major flood-out. It is confined to higher ground, most of which shows evidence of very great floods in the form of megaripples, overflow channels, discontinuous sand deposits and the Undoolya bar field. This feature is 5 kilometres long, 2.5 kilometres wide, and consists of a system of linear bars up to 2 metres high and sometimes several kilometres long, separated by swales 100 to 200 metres wide. The bars are built of coarse sand and gravel similar to bed material in the Todd and appear fresh, but no flood of the last 100 years has come even close to overtopping them. They are persistent remnants of a megaflood.[36]

THE TODD RIVER FLOODPLAIN

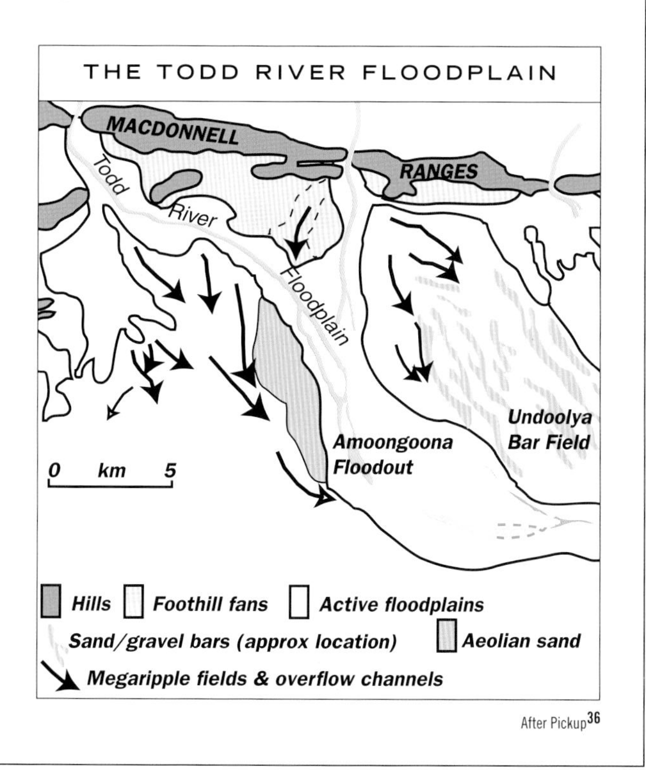

After Pickup[36]

HYDROLOGY OF THE ULURU–KATA TJUTA NATIONAL PARK

(Ayers Rock and the Olgas)

A recent comprehensive investigation of the Uluru–Kata Tjuta region of the Northern Territory, using all the latest technology, has revealed some remarkable information about the region.[31] Who would have imagined that just below the surface, in the Dune Plains between the two inselbergs, another Kata Tjuta lies buried, and a major river system once flowed northwards to Lake Amadeus?

Uluru National Park is a World Heritage Area and a Biosphere Reserve. It lies within the major geologic province of the Amadeus Basin and is bounded to the south by the Musgrave Complex. The landscape comprises a flat sandy plain and undulating dunefields, punctuated by the steeply protruding inselbergs of Uluru and Kata Tjuta, and by lesser outcrops of bedrock. The plain slopes gently south to north, descending about 3 metres every kilometre to Lake Amadeus.

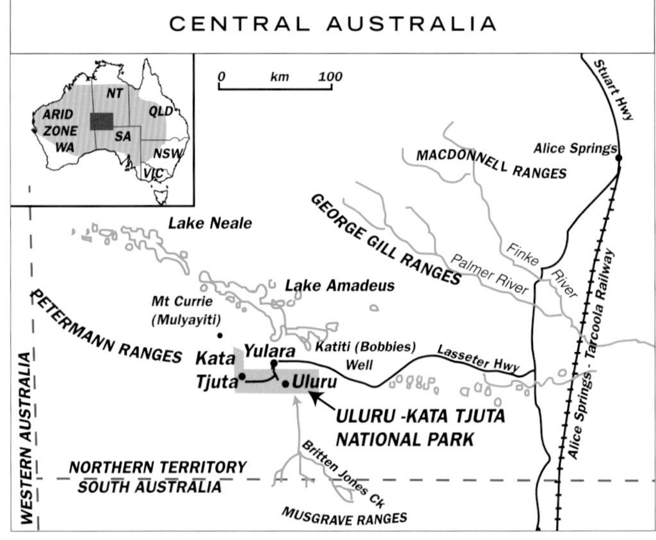

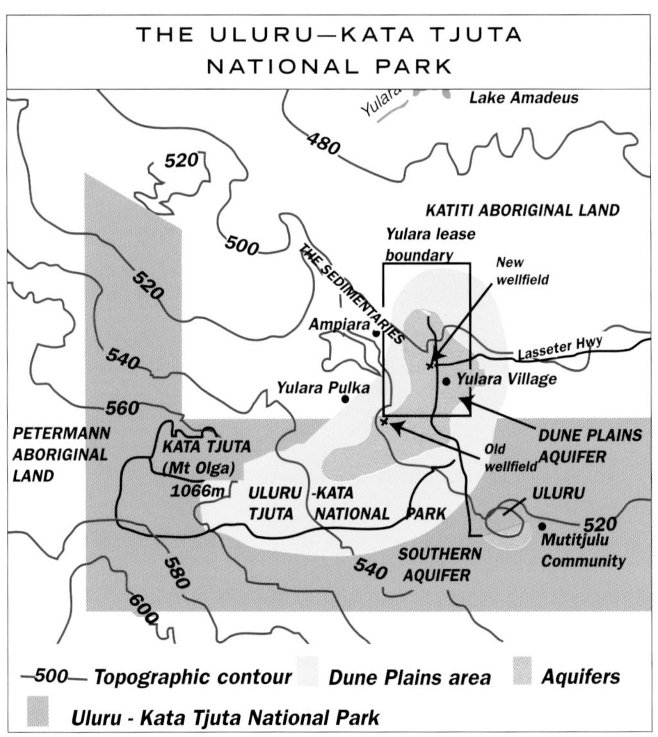

After English[31]

The mulga zone at the foot of Uluru receives the run-off from the inselberg. This zone is biodiverse and supports more animals than the surrounding plains. M.E.W.

Kata Tjuta has a mulga zone around the base of the towers which supports more species than occur in the surrounding spinifex plain. M.E.W.

Two important aquifers provide community water supplies: the Southern Aquifer system supplies the Ayers Rock borefield at Uluru; the Dune Plains aquifer system supplies the Yulara (Dune Plains) borefield. Both comprise bedrock and Cainozoic aquifers. Much of the annual 300 000 cubic metres of groundwater extracted from the Dune Plains aquifer system for Yulara is saline. Raw, untreated water is used for some irrigation and domestic services and for fire fighting; the remainder is desalinised to make it fit for human consumption. (Waste water from Yulara is treated and the effluent used for irrigation in the village.)

The Dune Plains system is an arcuate, essentially north-trending zone about 20 kilometres long that passes east of the Sedimentaries and beneath the Yulara area. It is bounded by east-north-east striking basement faults. It occupies a section of the palaeovalley of a former watercourse, the Dune Plains palaeoriver, between Uluru and Kata Tjuta.

The palaeovalley is filled by up to 100 metres of sediments, mainly Cainozoic, but with some dated as Late Cretaceous. The age of the palaeovalley and the ancient landsurface in which it was incised is thus Cretaceous, and may even be Jurassic, like the age of surfaces determined in the Yilgarn of Western Australia. It was originally thought to flow east to join the proto-Finke River. It is now known to flow north to Lake Amadeus. The palaeovalley was originally a closed valley with discrete depocentres in which lake and alluvial fan sediments accumulated. Later, a river evolved and flowed north, creating a broad deltaic braidplain as it ran to the Lake.

EARLY TERTIARY PALAEOGEOGRAPHY OF THE DUNE PLAINS PALAEOVALLEY

Yulara Industrial Area
Alluvial fans & sheetwash
THE SEDIMENTARIES
Yulara Village
0 km 2
Domes or protruding ridges
Swamp
Alluvial fans
Lake
Alluvial fans & sheetwash
Dome
Lake
Swampy valley floor
Lake
Domes
Dome
W
E
Lacustrine & alluvial fan sediments
Bedrock

After English[31]

Lakes and swamps developed in the centre of the palaeovalley, around the protruding domes; alluvial fans and sheetwash from the surrounding hills accumulated on the valley sides.

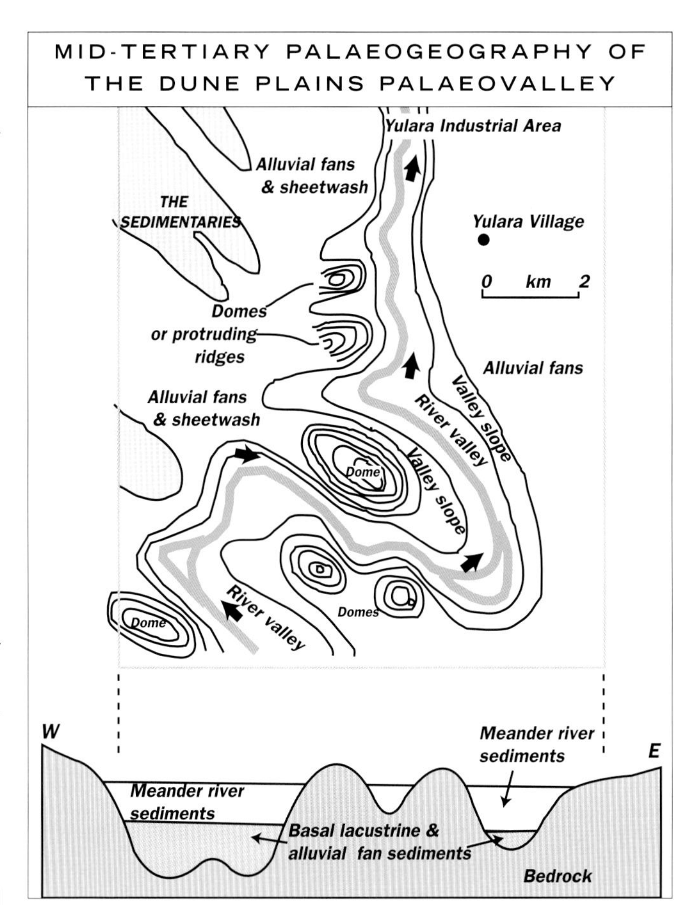

After English[31]

The broad meanders of the Dune Plains River were confined by the relict basement topography of domes and depressions. A bedrock constriction in the north (beneath the Yulara area) was breached and the river flowed northwards to Lake Amadeus.

Uluru and Kata Tjuta were shaped by millions of years of weathering and erosion which occurred while they were deeply buried in the Amadeus Basin. This subterranean, chemical weathering, accelerated where seeping rainwater was channelled along joints and microfissures and along primary bedding planes, created the curved and ridged forms which characterise the now-exposed inselbergs. Sub-aerial erosion has continued since the rock massifs, of which the two features are a small remnant, were exposed by the Late Cretaceous. (See also pages 100–2 in *After the Greening.*)

During the Late Cretaceous to Late Eocene, sediments were laid down, forming the lowest 20 metres in the valley depressions. They contain organic material derived from swamp vegetation. The valley was rapidly filling with sediment eroded from the surrounding mountains under the wet climatic conditions which prevailed. The lakes were filled up progressively and a 'brim-full' threshold was reached at about 60 metres below the present ground surface. Some time during the Mid-Tertiary, either a river or an overflow lake established a narrow outlet northwards for free-flowing drainage.

The initial Dune Plains palaeoriver probably traced a meandering course as it followed subtle depressions that were once deep sub-valleys, and curved around the summits of still-protruding domes in the valley. Eventually river sediments aggraded to 40 metres below the present surface, domes were buried and the landscape flattened out. The river was no longer constrained by the domes and wandered about on its floodplains. The 40-metre deep Mid-Tertiary valley

LATE TERTIARY PALAEOGEOGRAPHY OF THE DUNE PLAINS PALAEOVALLEY

After English[31]

The late Tertiary river flowed northwards between Uluru and the Sedimentaries as a broadly braided river, contracting in the region of Yulara where there is a bedrock constriction, and fanning out again in a broad deltaic floodplain to Lake Amadeus.

which developed on a buried prior landscape can be regarded as the Dune Plains palaeovalley. (This palaeovalley is now the setting for a compound bedrock-Cainozoic sediment aquifer system which is a major source of water supply for the region.) The river course then developed over a distance of 55 kilometres from south of Kata Tjuta to Yulara and then north, as a broad, deltaic distributary system which flowed to Lake Amadeus. Such a system does not imply that many channels flowed simultaneously. Through time, flow would have switched haphazardly and, as drier conditions became established, become ephemeral. The increasingly arid climate during the latest Pliocene, 3 to 2 million years ago, would have favoured accumulation of aeolian sediments in the Dune Plains palaeovalley, choking surface drainage and leading to infiltration of water into the evolving groundwater system below. The braid plain has received recent episodic floodwaters that have disrupted the Quaternary dune fields.

Aridity intensified during the Pleistocene, although wetter and drier regimes oscillated. By 75 000 years ago, Lake Amadeus was saline and was contracting and rapidly becoming a groundwater discharge zone. Aridity and fluctuating water tables in the Dune Plains river system (and others) led to the formation of groundwater calcrete deposits, particularly in areas where groundwaters were constricted. Where such bodies have been subject to dissolution and have become porous, they can be useful aquifers.

Where Uluru and Kata Tjuta protrude through the surrounding plains they create a localised zone—a sheetwash landscape unit composed of red earth—forming a broad, gently sloping, encircling apron. The soil in the zone is calcretised and during rainfall, surface run-off from this hardened unit constitutes a distinctive 'sheetflow recharge' mechanism that maximises water conservation and infiltration for the underlying aquifer. The sheetwash landscape unit supports banded mulga shrubland. The transition zone to the surrounding sandplain–spinifex associations has been found to be a most significant zone for biodiversity. The hydrodynamic processes of the sheetwash unit carry concentrated nutrients to the base of the slope, where infiltration occurs, and the influx of water benefits the surface biota as well as replenishing the deeper aquifers. In the Uluru National Park the endangered mulgara, a small marsupial, is dependent on the transition zone for survival.

The Southern Aquifer at Uluru had a similar history to that of the Dune Plains palaeovalley and river system. In its case, the outlet for surface water was probably established through an east-north-east-flowing tributary of the palaeo-Britten Jones Creek, and may have flowed round the base of The Rock.

LAKE FROME

Lake Frome is a large playa south-east of Lake Eyre which has a local catchment in the Flinders Ranges as well as connection via Lake Blanche, Lake Callabonna and the Strzelecki Creek to the Cooper System. High lake levels in Lake Frome require considerable input from the Cooper system, thus establishing the likelihood of water-level records coeval with those in Lake Eyre.

A late-Quaternary salinity record from Lake Frome,[45] derived from analysis of gypsum deposits, is consistent with much of the

Strzelecki Crossing, S.A.: the dry bed of Strzelecki Creek which floods occasionally and carries water to Lake Frome. M.E.W.

Filamentous green algae in a hot artesian bore outflow, Strzelecki Desert, S.A.. REG MORRISON

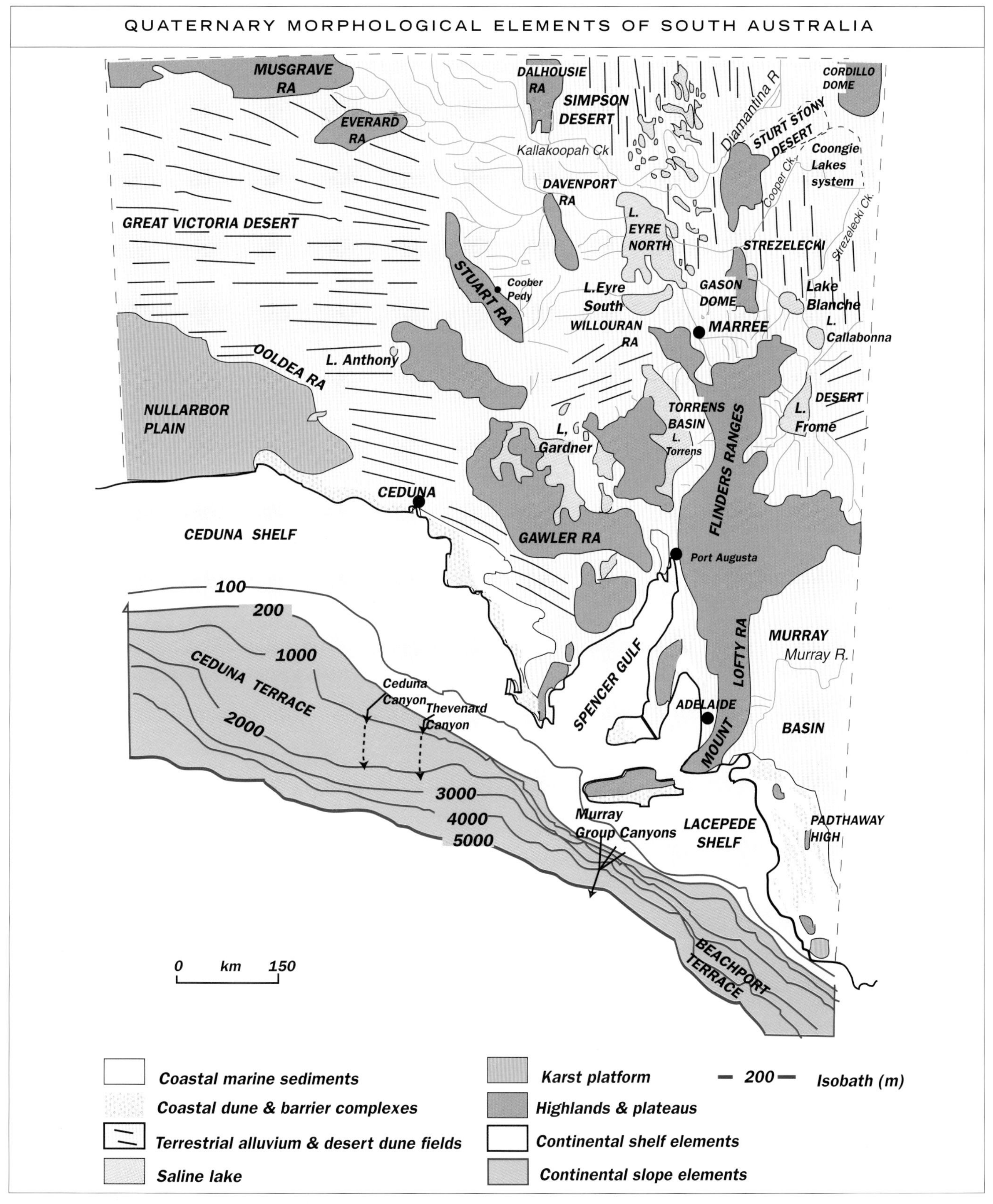

QUATERNARY MORPHOLOGICAL ELEMENTS OF SOUTH AUSTRALIA

After Drexel & Preiss[42A]

inferred climatic history of south-east Australia. Three periods of lake-filling are indicated: at about 17 000 years ago; from 15 000 to 13 500 years ago; from 13 000 to 11 000 years ago.

The lake was dry prior to 17 000; between 19 500 and 15 000 years ago; at 13 000 years ago; and between 6000 and 4000 years ago.

THE COOPER CREEK CATCHMENT

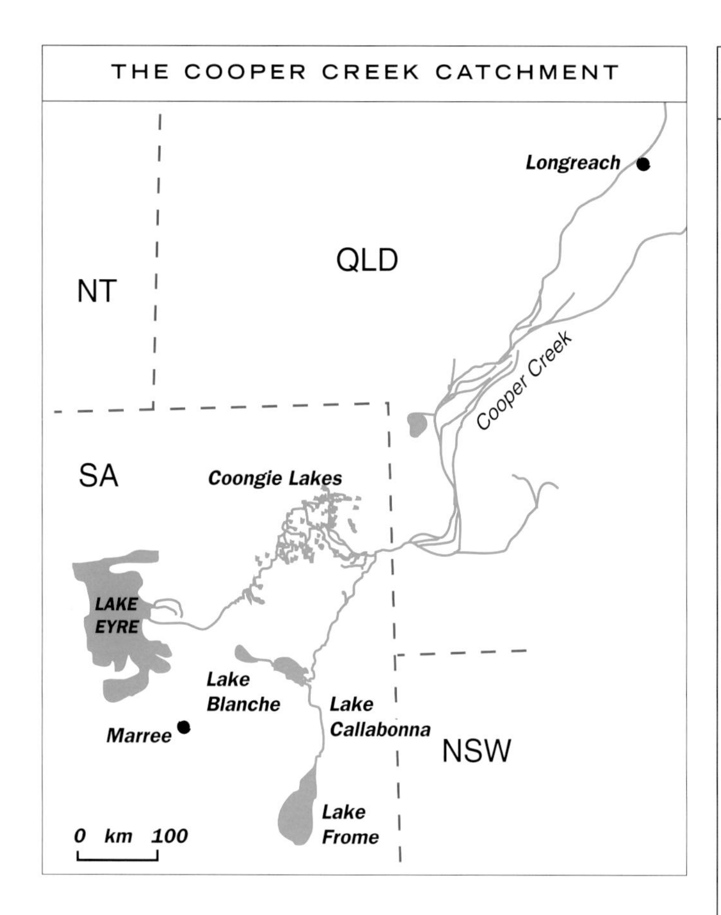

THE STRZELECKI CREEK CONNECTION

Floods in Cooper Creek bring life-giving water to the Coongie Lakes and down the Strzelecki Creek to Lakes Blanche, Callabonna and Frome. The Coongie Lakes are one of Australia's most important wetland areas, vital for maintenance of biodiversity, and the breeding place for millions of waterbirds.

Among ephemeral rivers which feed into Lake Frome from the Flinders Ranges are Arkaroola Creek and others draining the mountainous Arkaroola Station; Chambers Creek, which flows through the spectacular Chambers Gorge before crossing the plain to the Lake; and Italowe Creek in the Gammon National Park.

The Strzelecki Creek floodplain at Merty Merty, S.A. M.E.W.

THE COONGIE LAKES

An enlarged Lake Appanburra spreading among the coolibahs during the Cooper floods of 1989. REG MORRISON

South Australian University researchers navigate flooded desert during the 1989 Cooper floods. The Coongie Lakes were much enlarged. (Hawk's nest in foreground.) REG MORRISON

Rain brings new life to the bare dunes in the Coongie Lakes area of the Strzelecki Desert during the 1989 floods of the Cooper. REG MORRISON

Senecio gregorii, *yellowtops, growing in a flooded swale in the Strzelecki Desert during Cooper Creek floods in 1989.* REG MORRISON

THE MT POOLE REGION, FAR NORTH-WESTERN NEW SOUTH WALES
EXPLORER STURT'S COUNTRY

Mt Poole is the only prominent feature in the flat, gibber-strewn landscape in Corner Country, about 40 kilometres south-south-west of Tibooburra. Climbing its stony slopes and looking out across the empty landscape from the cairn built by Sturt's party, who were trapped in its vicinity by drought for six months in 1844–45, one can only wonder at the fortitude of the early explorers. That ill-fated expedition, in which James Poole died of scurvy, was looking for the inland sea which, in those days, everyone expected to be there—because, from the point of view of people of those times, rivers ran to the sea, not to the sunken centre of a continent.

By constructing an 'underground room' in Depot Creek (a tributary of Evelyn Creek) where there was a permanent waterhole, the party managed to survive the drought. 'To give the men occupation and to keep them in health', Sturt had his men march daily to 'Red Hill' (now Mt Poole) and construct the cairn. The only trees to be seen on the gibber plains round the hill are corridors of river red gums along the ephemeral stream beds. When one enters the deeply incised Depot Creek channel, it is to arrive in a surprising microcosm—a living artery where the roots of trees are tapping underground water in the channel to survive until the next rain which will make the river run and bring brief greening to the gibber plains.

Mt Poole across the Stony Desert. Some of the gibbers are white quartz, on the underside of which are green and blue-green algae, actively photosynthesising when the soil on which they lie is wet. They rely on reflected light from other gibbers to penetrate to their under surfaces. M.E.W.

The permanent waterhole in Depot Creek. M.E.W.

The wide, stony bed of Evelyn Creek. M.E.W.

View from Mt Poole—flat gibber plains, corridors of eucalypts along ephemeral water courses. M.E.W.

Inside the deeply incised channel which sheltered Sturt and his party when they were stranded for six months by drought. M.E.W.

LAKE EYRE: the last 130000 years: Sedimentation and Hydrology

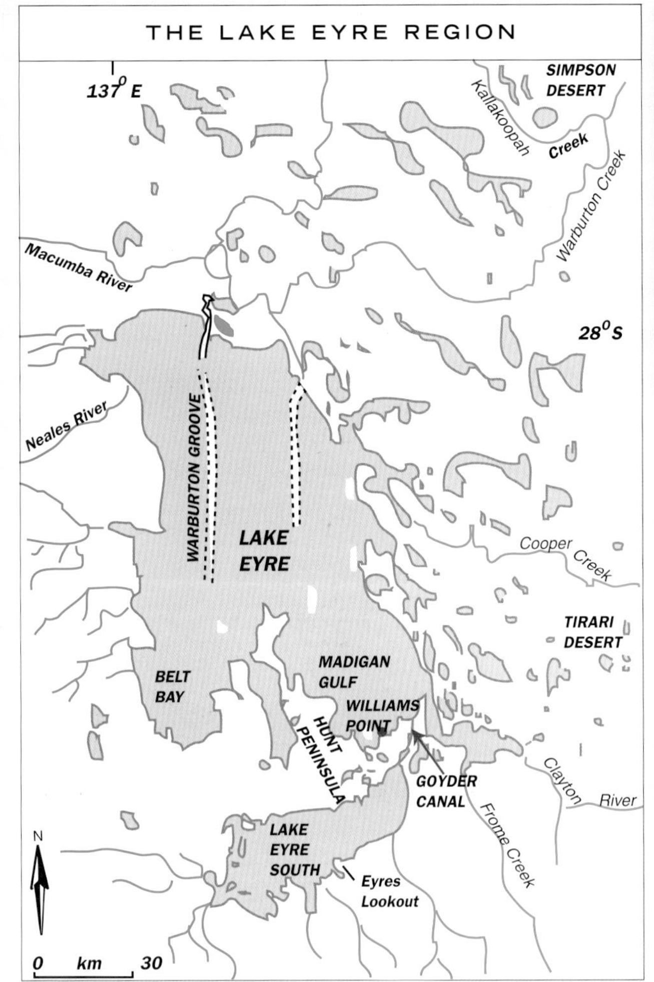

After Magee *et al*[43]

Lake Eyre. REG MORRISON

Lake Eyre North, full during the 1989 Central Australian floods. REG MORRISON

The Goyder Canal, looking north towards Lake Eyre. REG MORRISON

The Causeway across the Goyder Canal. M.E.W.

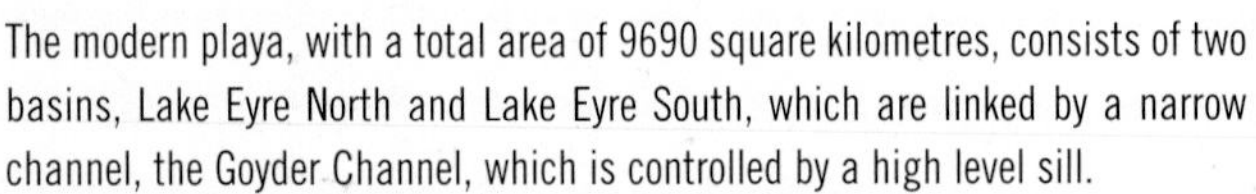

The modern playa, with a total area of 9690 square kilometres, consists of two basins, Lake Eyre North and Lake Eyre South, which are linked by a narrow channel, the Goyder Channel, which is controlled by a high level sill.

Lake Eyre North contains the lowest point in Australia, almost 16 metres below sea level, and both lakes have strong salt crusts. The playa is normally dry or has ephemeral surface water of limited extent and depth, and infrequent major floodings. In recent years Lake Eyre North has spilled into Lake Eyre South and they became one lake between 1974 and 1976; in 1984 and 1989, Lake Eyre South filled and spilled over into Lake Eyre North. The 1974 to 1976 flooding was the greatest since European settlement.

The floor of Lake Eyre North slopes towards the south and intersects the more nearly horizontal water table at the southern end, where a salt crust occurs. Floodwaters from streams entering from the north and flowing across the lakebed floor to the deeper southern bays have excavated a number of shallow grooves in the sloping northern playa floor. The Warburton Groove is the most remarkable of these, extending in a straight line some 85 kilometres from the mouth of the Warburton Creek towards Belt Bay.

During the Quaternary Period, the major climatic oscillations associated with glacial–interglacial cycles have resulted in an alternation of wetter and dryer episodes. During these fluctuations, Lake Eyre has varied from being a perennial lake, significantly larger than the present playa and more than 25 metres deep, to drying up and suffering extensive deflation of basin sediments.[43, 44] This deflation, controlled by groundwaters as the water table was lowered by evaporation, produced the morphology of the modern playa basin. The present playa–ephemeral lake phase is in a relatively stable state with sediment input balancing deflation.

Madigan Gulf, the largest of three large bays at the southern end of Lake Eyre North, has been the subject of an extensive sedimentation study, and the history of the Lake's reaction to climatic changes during the last full glacial cycle (130 000 years) has been elucidated.

- **Early stage 5 lacustral phase** (not yet conclusively dated but falling within the range of 130 000 to 90 000 years ago): The lake was permanent, but mostly saline, throughout this interval, until it dried at Williams Point. Some deflation may have occurred briefly.
- **Later stage 5 lacustral phase** (90 000 to 70 000 years ago): The lake refilled and oscillated between ephemeral and more permanent saline conditions, with occasional brackish to fresh interludes. It eventually dried again and soil formation occurred before about 70 000 years ago, when conditions approximated those of today. It refilled at about 70 000 years ago and probably fluctuated until about 65 000 years ago.
- **Dune-building and probable deflation phase** (60 000 to 50 000 years ago): Aeolian sediments derived initially from beaches, and later with gypsum and pelleted clay derived from playa deflation, characterise this phase.
- **Possible stage 3 lacustral phase** (50 000 to 25 000 years ago): A shallow, saline lake during this period is suggested by laminated sediments in some drill cores, but more investigation of this stage is required. (This stage would correspond to the lake-full stage during this interval in the Willandra Lakes, and a less certain episode in Lake Frome.)
- **Playa phase** (about 25 000 to 10 000 years ago): At some stage prior to the Holocene, groundwater-controlled deflation removed previously deposited lacustrine sediments and excavated the basin to 17.4 metres below sea level, producing the present playa morphology. When water table lowering stopped and deflation ceased, deposition of sediment occurred in an ephemerally flooded playa environment. The sediment was oxidised and soils formed during dry stages. A thick salt crust was formed, probably during the very arid glacial maximum.
- **Early Holocene shallow lacustrine phase** (10 000 to 4000 years ago): By at least 10 000 years ago the lake had filled, probably semi-permanently, and the halite crust partly dissolved before it was sealed by a laminated gypseous clay layer which prevented further dissolution.
- **Modern ephemeral playa lake** (3000 years ago to present day): The fraction of the lower halite salt unit, which was dissolved during the onset of the early Holocene wet phase, forms the relatively thin modern salt crust. During modern major ephemeral floods, the entire surface of the salt crust is dissolved and re-precipitated as the lake dries, always migrating upwards through any sediment deposited during the flooding. The present interglacial is seen to be drier than the previous one, which saw a semi-permanent lake between 130 000 and 70 000 years ago. The monsoonal influence in the northern catchments was obviously much stronger then.

Today Lake Eyre generally fills from the Diamantina and the Cooper, but in 1984 and 1989 it filled from local rainfall to the south, delivered by Frome Creek.

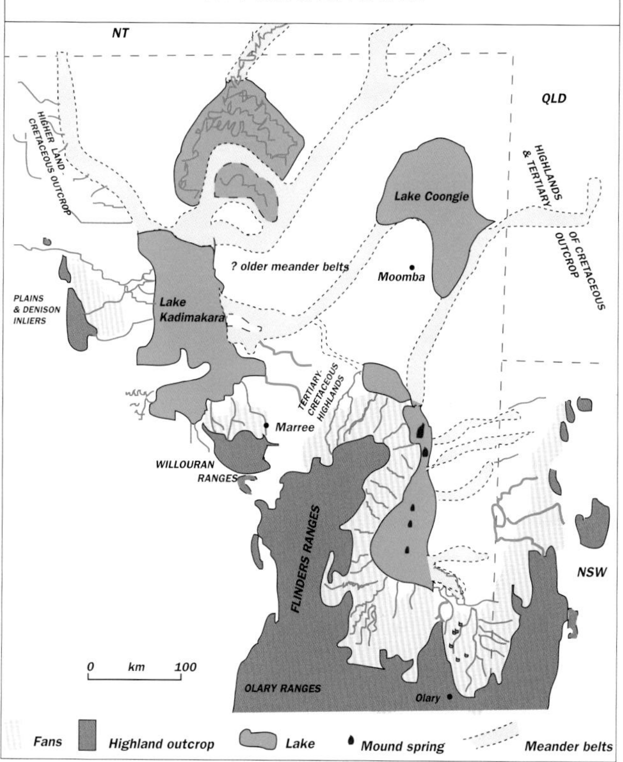

After Belperio *et al* in Drexel & Preiss[42A]

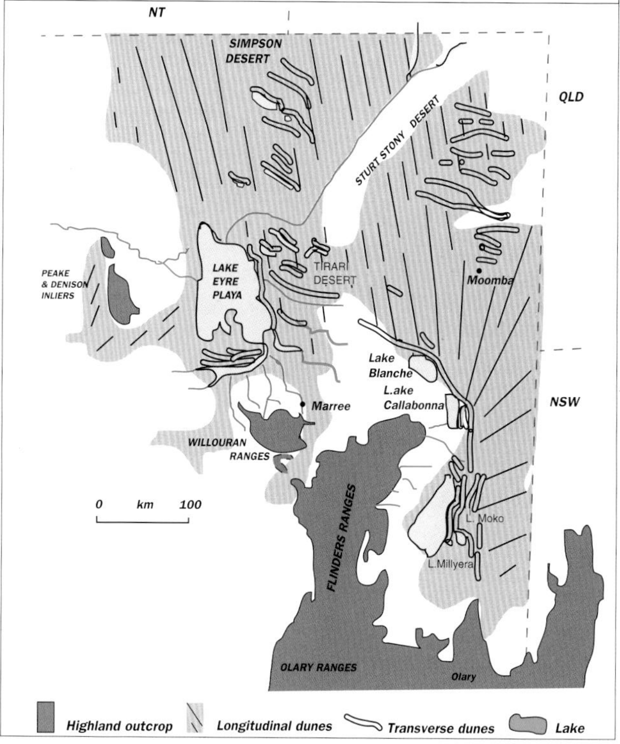

After Belperio *et al* in Drexel & Preiss[42A]

THE LAKE EYRE BASIN

Salt heaving in the Goyder Canal, S.A., near the Causeway. M.E.W.

Tourists on the salt crust of Lake Eyre South in 1988. REG MORRISON

Salt patterns in the Goyder Canal. M.E.W.

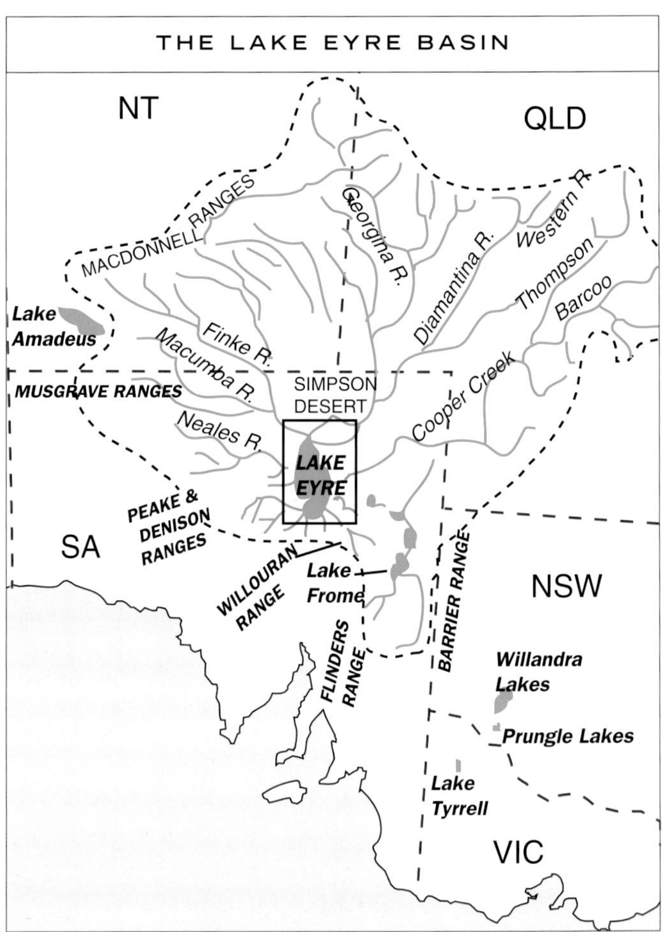

The Lake Eyre Basin is the largest internal drainage basin in Australia, covering 1.3 million square kilometres. The main river systems on the east side of the Basin are the Cooper (downstream of the Barcoo–Thomson confluence), the Diamantina–Warburton and the Georgina, which arise in the humid subtropics and discharge towards Lake Eyre, about 1000 kilometres away in the sunken centre of the continent. Shorter streams drain the Stuart, Everard and Musgrave Ranges to the west, entering the lake via the Neales and Macumber Rivers; others drain the Flinders Ranges to the south, entering the lake via Frome Creek. The Margaret River runs to Lake Eyre South from hilly terrain to the south-west, and Screechowl Creek from due south. To the north, ephemeral streams draining the Macdonnell Ranges, like the Finke and others described above, are prevented from entering the lake by the Simpson Desert dunefield. The Finke retains a connection through the edge of the dunefield to the Macumber system, but contributes little if any flow to the lake. All the ephemeral streams have extremely variable discharge, both seasonally and from year to year.

The climate in the Basin is hot and dry with short cool to cold winters. Annual rainfall varies from close to 500 millimetres in the far north-east to less than 150 millimetres in a large area in the vicinity of the lake. Mean annual pan evaporation varies from 2400 millimetres in the north-east to 3600 millimetres at Lake Eyre, exceeding rainfall in all months over the whole Basin.

The Lake Eyre Basin started to form in the Early Tertiary when the south-eastern sector of South Australia started to sink to form a large, shallow structure which became a depocentre for river and lake sediments. Sedimentation, and sinking, continue up to the present. The Basin is bordered on the west by the Peake and Denison and Dalhousie Dome provinces; in the south by the Willouran, Flinders and Olary Ranges; and in the east by the Barrier Ranges. In the north, the Basin extends into the Northern Territory, Queensland and New South Wales and is bordered by domes of Mesozoic rocks. The southern margin is deeply defined and faulted as a result of Late Cainozoic uplift.

The Eyre Formation, the oldest sedimentary unit in the Basin, contains plant fossils of Palaeocene to Eocene age. Some of these

Screech Owl Creek,S.A., cutting down through a duricrusted desert surface. M.E.W.

Margaret River, S.A., runs into Lake Eyre South. M.E.W.

Frome Creek, S.A., a wide, sandy ephemeral river bed typical of rivers of the Centre. M.E.W.

fossils are described in *After the Greening*, which gives an account of the drying out of the Australian continent and how it changed from a green, well-watered and largely forested land to become the driest vegetated continent. Nowhere have the changes been more profound than in the Lake Eyre region of the Basin, which today constitutes the 'dead heart' of the continent. Between the Late Oligocene and Mid-Miocene, large, relatively shallow lakes covered much of the Basin, fed by meandering rivers with extensive floodplains. Contraction of lakes and drying of the area proceeded through time, but desertification only occurred in the Pleistocene ice age of the last 2.6 million years when dry glacials and more benign and wetter interglacial stages alternated.[234]

RIVERS OF THE CHANNEL COUNTRY

The rivers in the north-eastern sector of the Lake Eyre Basin contribute the largest discharge and sediment load and deposit most of their sediment in south-west Queensland and northern South Australia, in the Channel Country.[46, 46A, 47] Virtually all the alluvial area lies below 150 metres elevation, and little of the catchment is above 300 metres. The rivers mainly transport mud. Sand is present in minor anastomosing channels, but the dominant channel pattern is braided. The braid bars and channel floors are composed of mud which comprises 10–20 per cent sand, 25–35 per cent silt and 55–65 per cent clay. The clay includes extremely fine fractions.

River gradients are very low and river planforms are predominantly anastomosing, although the Cooper and the Diamantina–Warburton flow into Lake Eyre as single-thread, incised channels that flow intermittently. The average channel gradient of Channel Country rivers in western Queensland is 1 in 6000, or 17 centimetres per kilometre. (The inward-draining Murray–Darling system has similar low gradients. For example, Walgett, on the Darling River in northern New South Wales, approximately 2000 kilometres from the river mouth, is only 140 metres above sea level, which represents an average gradient of only 1 in 14 000, or 7 centimetres per kilometre.)

Floodplains in upstream parts of the rivers and in tributary

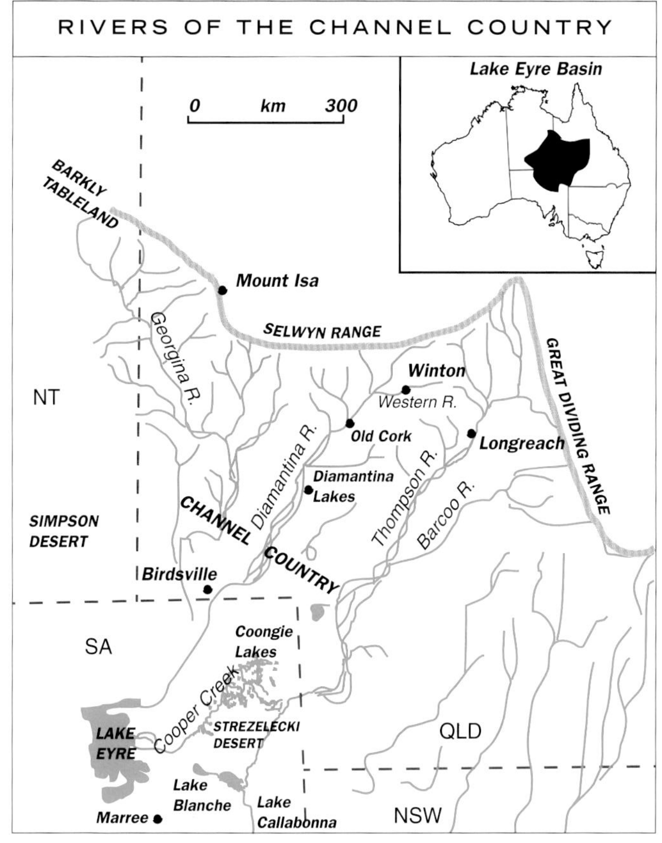

After Gibling *et al*[46]

ANABRANCHING RIVERS

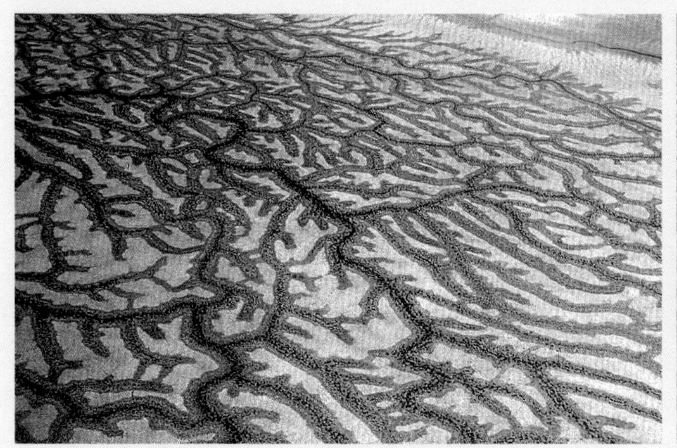

Anabranching—the Diamantina shows us how it is done. GERALD NANSON

Anabranching of Cooper Creek, South Galway. GERALD NANSON

Most rivers flow in single channels with adjustable slopes in order to maintain efficient transport of water and sediment. Less common worldwide, but widely represented in Australia, are anabranching rivers, consisting of multiple channels separated by vegetated, semi-permanent islands and ridges.[41, 42] The multiple channels are, in combination, usually substantially narrower and deeper than if the same discharge were to occupy a single channel.

Over the last decade, studies of Australian anabranching rivers by geoscientists at Wollongong University have unravelled the mysteries surrounding anabranching. Analysis of basic hydraulic relationships for alluvial channels that include flow continuity, roughness and several sediment transport functions, showed:

- Where there is little or no opportunity to increase channel slope, conversion from a wide, single channel to a semi-permanent system of multiple channels will reduce total width and increase average flow depth, hydraulic radius and velocity.
- Water and sediment throughput is maintained or enhanced by multiple channels, even over-riding moderate increases in channel roughness.
- Anabranching rivers appear to be closer to exhibiting the most efficient sections for the conveyance of both water and sediment than are equivalent wide, single channels at the same slope.
- Not all anabranching systems are finely tuned to hydraulic efficiency. Once formed, some channels may continue to operate in spite of increasing inefficiency and the channel's decreasing function with time. Others may form a distributory system for dispersing and storing water and sediment across extensive low-gradient floodplains.
- Anabranching rivers are usually characterised by flood-dominated flow regimes, banks that are resistant to erosion relative to stream energy, and mechanisms that lead to channel damming and avulsion.
- In some cases, anabranching is a chance event, a single channel dividing round an obstruction; in others it is clearly a general condition.

The very nature of the Australian continent and its climate predisposes it to the formation of anabranching rivers. Very low relief, intensely weathered and cohesive fine-grained sediment in an arid environment, with declining flow discharges and increasing sedimentation downstream, all encourage sediment storage and the development of anabranches. Tree-lined rivers with muddy, cohesive banks promote the formation of narrow channels which maintain water and sediment conveyance for the maximum distance downstream. In parts of arid and semi-arid Australia, anabranching rivers are the most dominant type, and they are of many different kinds because anabranching is a condition locally superimposed on stable or laterally active straight, sinuous, meandering, or braided systems.

Floods of 1990 inundated the Cooper floodplains; only the large coolibahs on the main anabranches protruded through the waters. GERALD NANSON

Many rivers in the two major drainage basins, the Murray–Darling and Lake Eyre, exhibit anabranching, as do many rivers in other smaller basins. This abundance, not seen on other continents, is clearly the product of Australia's geological history. Our continent is an old, eroded land mass, the flattest, with an average elevation of about 340 metres above sea level—less than half the world average. Away from their headwaters, rivers have little opportunity to increase their gradients if required to do so. Becoming the driest vegetated continent has involved declining river discharges and increased sediment accumulation and storage downstream. Deep weathering and low gradients have resulted in much alluvium being fine-grained and cohesive, particularly in parts of the Murray–Darling Basin and the Channel Country of western Queensland—the region of the infamous bull-dust.

In addition to these geological reasons, a botanical one contributes to the frequency of anabranching. Even in the most arid regions, a variety of riparian-adapted trees grow on stream banks. The rivers frequently have low flows or are ephemeral, and the trees can grow right down the banks, protecting them even at the base, stabilising the channels. (This is in contrast to large rivers in humid areas where the lower banks are often relatively unprotected by vegetation because of their constant inundation.)

Aerial views of the larger anabranching channels of Cooper Creek near South Galway, QLD showing the floodplain channels flowing over the top of the main anabranches. GERALD NANSON

Numerous waterholes, 8 kilometres north of Durham Downs, QLD. GERALD NANSON

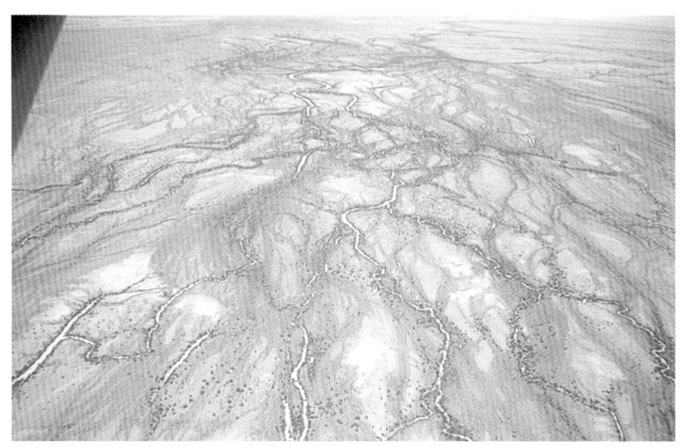

The Cooper Creek floodplain. GERALD NANSON

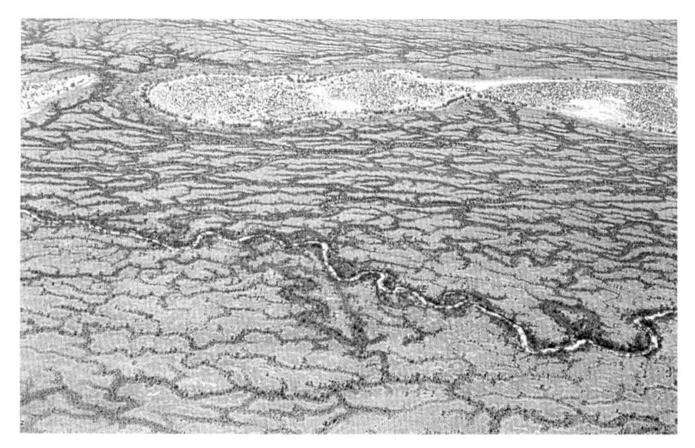
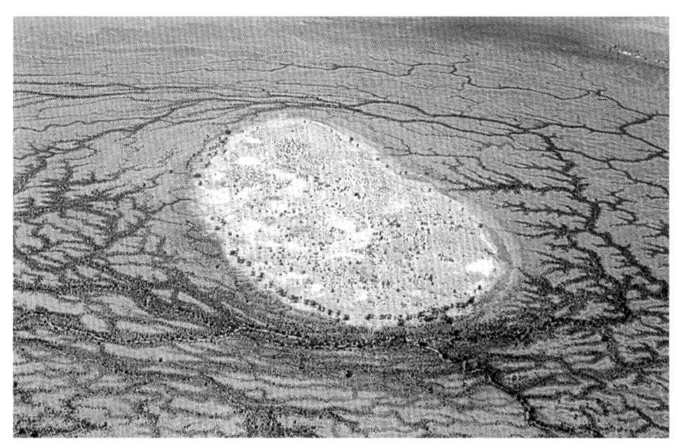

Aeolian dunes surrounded by floodplain mud and reticulate channels, indication of past periods of exceptionally dry climate (glacial stages of the Pleistocene ice age). In places, the present floods of the Cooper are squeezed between dunes, forming scour channels. GERALD NANSON

valleys are typically a few kilometres wide, but the Cooper floodplain south of Windorah widens to 70 kilometres.

Annual rainfall in the arid to semi-arid Lake Eyre Basin ranges from 400 to 500 millimetres in the headwaters to about 120 millimetres a year in the Simpson Desert. The rivers show extreme flow variability. The coefficient of variation of annual flows for the Diamantina system is among the highest recorded, and flood discharge of a single week can greatly exceed a river's mean annual discharge. During floods, the rivers expand to extraordinary widths, resulting in sheets of water 70 kilometres wide on the Cooper below Windorah and up to 500 kilometres wide on the Diamantina and adjacent channels above Birdsville. 1990 floods in the Lake Eyre and Murray–Darling basins submerged 220 000 square kilometres. The flood waves move slowly down the long low-gradient rivers, taking several months to reach Lake Eyre, which has been filled only a few times in the twentieth century. If the headwaters of several tributaries receive heavy rain, downstream locations can receive multiple and reinforced flood waves. Flood volumes decrease downstream due to evaporation and losses by seepage into underlying sediments. Such losses have been estimated, over a 400-

The Cooper at Innaminka, S.A. Dense lignum forms an impenetrable barrier on the opposite bank. M.E.W.

Eucalypts standing on their stilt-like roots are a common sight on floodplains of the Cooper. REG MORRISON

The Georgina River, a major Channel Country river. M.E.W.

A billabong on the floodplain at Bedourie, QLD—milky water, opaque with fine clay in suspension. M.E.W.

The Cooper flowing at Innaminka, May 1999. M.E.W.

kilometre stretch of the Cooper, as averaging more than 75 per cent, but vary greatly depending on the magnitude of the flood.

Crayfish burrows are widespread in all studied reaches of the rivers during and immediately after flows, and occur also in dune troughs, in channel walls up to flow-level marks along the channel-ward side of benches and locally on bank tops. Burrows are believed to extend down to the water table several metres below the stream beds; presumably the crayfish (*Euastacus serratus*) survive deep down when river channels are dry. Other larger fauna which live in the rivers are crabs and tortoises, while bivalves and gastropods are also common.

Flow in the Channel Country rivers ceases during droughts and no discharge is reported in some years, but stagnant water is retained in the expanded, waterhole segments of channels. More than 300 waterholes are present in the Cooper system between Windorah and Nappa Merrie, and they are generally two to three times wider than associated channels. In planform they are relatively straight but may have large meanders.

The geology of the Channel Country has a great deal to do with the nature and planform of its river systems. The Rolling Downs Group of mudstones, siltstones and sandstones, of Cretaceous age, are ubiquitous in the Queensland section, overlain by the quartzose Eyre Formation (Tertiary). A silcrete duricrust, developed on the Tertiary sediments, covers kaolinised (weathered down to clay) bedrock which is highly erodable and contributes fine clay-rich sediment to the lower rivers. The anabranching of rivers is related to the cohesive, clay-rich sediments which form the stable banks of the multiple channels. During dry periods, the soils of the Cooper and Diamantina floodplains are characterised by complex networks of cracks up to 1 metre deep. When floodplains are inundated, the cracking-clay soils are the source of the mud which lines the channels, the clay which is carried in suspension, and the silty sand-sized aggregates which form the surface layers on the floodplains. The distributory floodplains of the Channel Country have a plentiful reserve supply of fine sediments.

Aeolian dunes flank the Diamantina and Cooper for much of their lower reaches, invading the floodplains in many cases. The dunes are easily mobilised by the wind. Sediments do not leave internal drainage basins except when transported by the wind. Where rivers are ephemeral, large quantities of sediment are left in river channels and floodplains. They dry out quickly and are vulnerable to wind erosion.

AQUATIC FOOD WEBS IN TURBID ARID-ZONE RIVERS

The rivers of the Channel Country, with their extensive floodplains and network of small channels that connect only during episodic floods, exist for most of the year as strings of very turbid waterholes.[48] It would be logical to assume that their aquatic food webs would be driven by fluxes of energy and nutrients derived from the riparian zone and from the floodplains after floods, as is the case in 'normal' rivers elsewhere. It would also be logical to predict that aquatic plant production would be extremely limited in turbid waterholes because of low light penetration, and that in consequence the waterholes would support very little life.

Research by scientists from Griffith University, Queensland, and the University of Western Australia has shown that none of these assumptions apply, however, and that a most interesting and different system has evolved—with algae, the driving force supporting the web of abundant life that the waterholes, surprisingly, contain. Large and thriving populations of snails, crustaceans and fish, defy logical expectations.

The research focussed on permanent waterholes in the main channel of the Cooper near Windorah, and small ephemeral pools in claypans on the floodplain of the Cooper were also sampled. The aim was to measure the amount of organic carbon produced and consumed within the system, and to trace the fate of terrestrial and in-stream sources of organic matter in the aquatic food web. Stable isotope analysis of carbon was the scientific tool used. It enabled the sources of the organic carbon assimilated by the consumers, such as invertebrates, fish and waterbirds, to be pinpointed. The carbon isotope signature of a consumer is determined by diet alone—and the scientists commented that this proved 'you are what you eat'. The very abundant thiarid snails (*Notopala*), large shrimps (*Macrobrachium*) and crayfish (*Cherax*), and several species of native fish were sampled. Very few aquatic insects were collected (in contrast to forested streams where they are abundant). Even though most of the riverbed organic matter in the waterholes was of terrestrial origin (more than 50 per cent in diagram 1), much of the biomass carbon of the inhabitants was of algal origin.

The shallow margins of the permanent waterholes were characterised by a conspicuous band of algae, largely filamentous blue-green Schizothrix, growing on the mud surface and forming a 'bath-tub ring'. This algal zone showed extremely high rates of primary production and respiration. The high rate of respiration results from high algal respiration at night, rather than from decomposition of organic matter. The deeper water had very low primary production, not surprisingly, because of the turbidity and low light. The littoral region was a major producer of organic carbon, and the mid-channel habitat was a net consumer. Although the shallow littoral

A 4-metre deep excavation into the floodplain immediately adjacent to a waterhole (figure gives scale). The exceptionally cohesive and watertight nature of the clays was illustrated by no leakage of water into the pit.
GERALD NANSON

FOOD WEBS IN COOPER CREEK

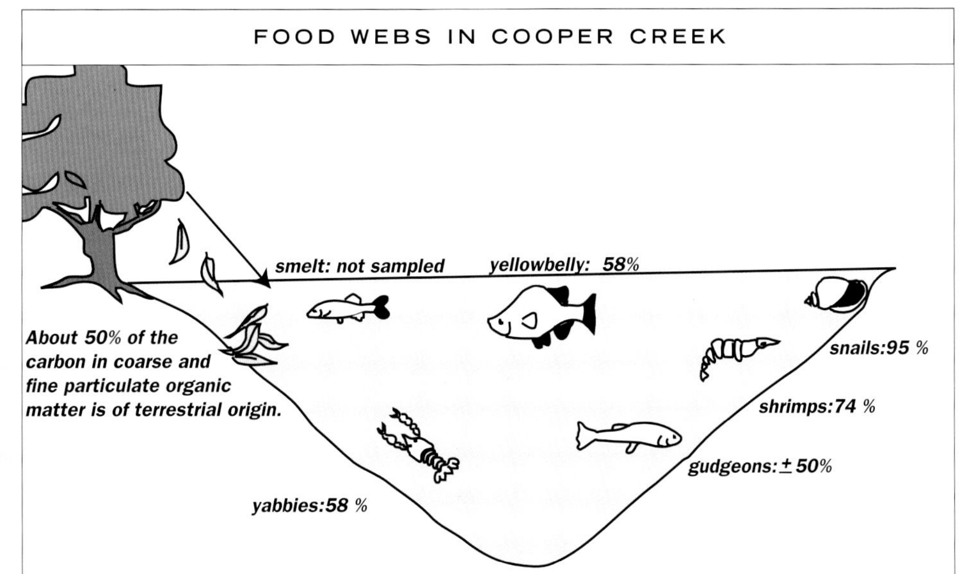

The aquatic food web for Cooper Creek, showing mean carbon-13 isotope signatures of primary sources and consumers. Most animals have signatures similar to algae showing that they obtain their carbon from that source. Ninety-five per cent of the biomass of snails, 74 per cent of the biomass of shrimps and 58 per cent of the biomass of yabbies, is derived from algal carbon. At least half of the biomass carbon of small gudgeons was derived from algal carbon, probably because they feed on small invertebrates; the yellowbelly has 58 per cent algal-derived carbon, also because of the food chain. Smelt, believed to feed on plankton, were not sampled, and rainbow fish, a surface feeder whose diet is based on insects and not leaf litter, had values closer to terrestrial sources. (Since the original research was carried out in 1996, new work has revealed that zooplankton are an abundant component of the waterhole food web and are well represented in the diets of several fish and in filter-feeding mussels. The role of terrestrial carbon in the food web, as shown in the diagram, is therefore believed to be far less important, and it is possible that the food web is almost entirely driven by benthic algae and plankton—-an even more bizarre situation than that originally described.)

After Bunn & davies[48]

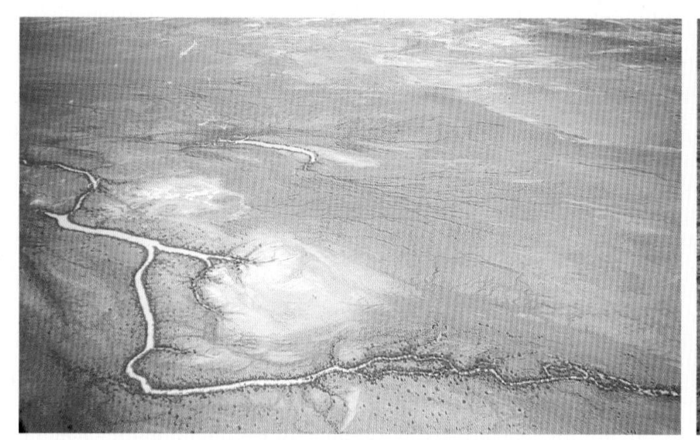

Pritchella Waterhole, east of Durham Downs. GERALD NANSON

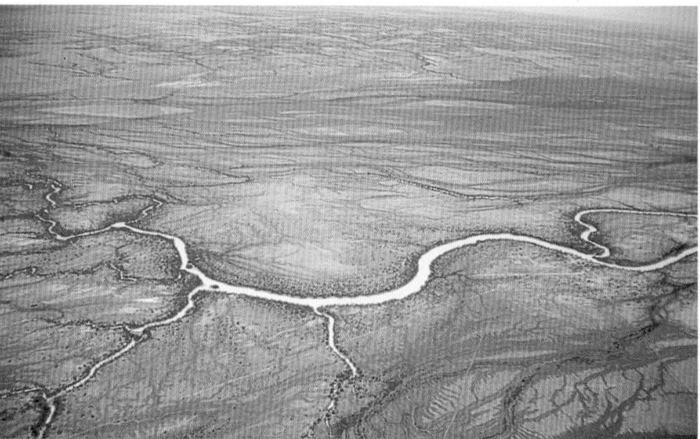

Meringhina Waterhole on the Cooper Creek floodplain near Durham Downs. GERALD NANSON

zone represented less than 8 per cent of the total riverbed habitat, primary production was so high that the entire waterhole was a net producer of organic carbon. The highly productive algal zone is a potentially important source of available food for aquatic invertebrates and other consumers.

The shallow pools in the claypans on the floodplain, which had filled after recent rain, contained numerous shield shrimps (Triops), some smaller crustaceans and snails. The ooze on the bottom of the claypans proved to be of algal origin, and it was shown that the biomass carbon in all these creatures came from this algal source, and not from terrestrial organic matter.

Despite the extensive connectivity with the riparian zone in anastomosing channels, and the well-known massive floods of the Channel Country, terrestrial carbon has been shown to be unimportant to aquatic consumers. While floods redistribute large amounts of terrestrial material in the channels, it is not important in the food web. The value of floods remains for mobilising nutrients; for provision of habitats by filling floodplain wetlands, lakes and billabongs; and in recruitment and dynamics of populations of fish and invertebrates.

The same sort of situation in regard to food webs is likely to be found to apply to other arid zone rivers, including the Paroo. Ecologically sustainable usage of the rivers will have to take this new knowledge into account.

If irrigation or other uses results in rapid draw-down of water in the deep waterholes, the all-important algal ring that fuels the in-stream ecosystems will be damaged or destroyed. The fragility of systems that depend on algae and also, as now known, on zooplankton, for healthy survival is clearly evident. What we see as green scum and stagnant water is actually a complex web of life which is obviously the product of co-evolution of life and environment through great lengths of geological time.

THE AQUATIC FOOD WEB IN FLOODPLAIN CLAYPANS

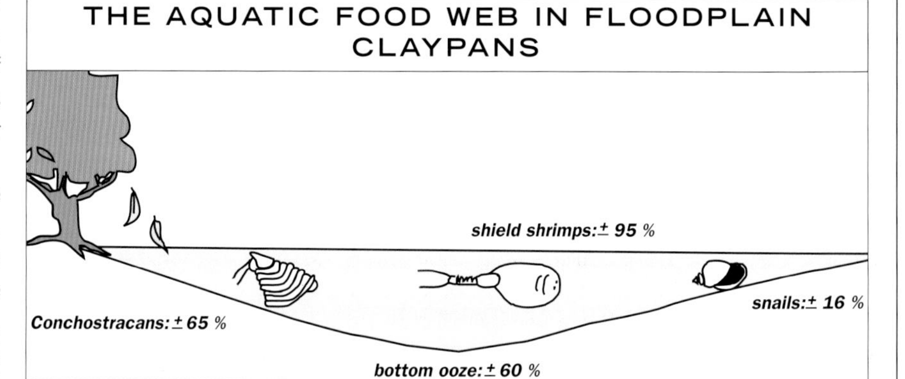

A simple food web for the floodplain claypans shows the mean carbon signature of primary sources and consumers.

After Bunn & davies[48]

Krefft's turtle (Emydura krefftii) *in the Coongie Lakes during the 1989 Cooper Creek floods of the Strzelecki Desert.* REG MORRISON

The yabbie or freshwater crayfish (Cherax destructor), *in Cooper Creek, Strzelecki Desert.* REG MORRISON

A flooded claypan, during the Cooper Creek floods of 1989. REG MORRISON

The self-mulching grey clays of the Cooper floodplain are locally known as 'bull dust'. Dust storms, according to meteorological records, occur about three times a year in the Channel Country. But Birdsville, during 1960–61, experienced 55 dust storms, compared with an average frequency (1960–84) of 6.6. Dust storms are more frequent when dry periods in the Channel Country follow flooding.

CHANNEL COUNTRY PALAEOCHANNELS

Along the middle and lower reaches of the Cooper, aerial photographs and shallow augering have revealed palaeochannels and bank deposits scaled to river meanders far larger than any present in the system today; subsequent research using thermoluminescent dating techniques has elucidated the history of sedimentation in the Channel Country. Extensive alluvial sand deposits are present beneath 2–3 metres of mud along the channels of the Cooper (and also along the Diamantina).[46, 46B]

It has been shown that there are two distinct ages of sandy alluvium.[3] The oldest of these alluvial strata were laid down by large meandering sand-load channels that correspond broadly in age to the penultimate interglacial (oxygen isotope 7) period between 260 000 and 220 000 years ago. The climate at that time must have been much wetter than today. A period of limited river activity followed—a drier phase—and then a second major pluvial period occurred in which sediment dated at about 110 000 years (stage 5) was deposited. It was following this period of active rivers and sediment deposition that the present river regime was initiated. As the pluvial period declined, the large meandering sandbed channels which characterised wetter times were replaced by numerous low-energy anastomosing and braided streams which transported mud over extensive floodplains. Dates for the mud range from 85 000 years ago to modern, and correspond to the period of increasing aridity and dune building which peaked during the last glacial stage (30 000 to 10 000 years ago).

In Australia, the period between the last two interglacials was probably largely arid. A dune in the upper Diamantina has been dated at 274 000 ± 22 000 years. Few older dunes have been dated because dune sands have mainly been reworked with each successive dry period. The transition to arid conditions in the Lake Eyre Basin is also dated by gypcretes, which form only under conditions of severe aridity. (On the inland plains of south-eastern Australia it has been clearly shown that climatic changes in the last 40 000 years or so have led to major shifts in the hydraulic regime of the Murray–Darling system.[91] But the chronology emerging for the Channel Country indicates that last glacial and Holocene climatic changes were of little consequence: it seems that major changes in hydrologic regime long predate them. The present interglacial in the Lake Eyre Basin is not as humid as the preceding interglacial.)

The channels on the Cooper floodplain are mostly fringed with coolibah trees (*Eucalyptus microtheca*). Away from the active channel belts, near-level floodplains up to tens of kilometres wide have a few distinct sinuous channels and networks of small channels which are commonly anastomosing. Gradients are extremely low, and gilgai formation in the soils (see page 203) has resulted in mounds which may have influenced drainage networks. The Channel Country floodplains have suffered some modification by humans. Aboriginal burning may have altered the vegetation to some extent, and cattle grazing has certainly had an effect and caused widespread erosion.

INTENSIVE, IRRIGATED AGRICULTURE AND THE COOPER

It's on again! The thin edge of the wedge?

On page 178 of *Listen … Our Land is Crying* is an account of a proposal to develop Currareva near Windorah as an irrigated cotton business. There was huge local opposition and the proposal was dropped (or, as it now appears, shelved). The main arguments against the idea were the effect on the river and its floodplains, and on the important wetlands downstream all the way to Lake Eyre, and the impact on the local 'organic' beef industry. In a region without large-scale agriculture, no pesticides, herbicides or chemicals from fertilisers are present in the environment to contaminate water and soil and impact on the completely 'green' image of beef fattened on the rich floodplains after the annual Cooper floodings. The more-than-$100 million industry depends on this guarantee, with overseas markets increasingly turning to such clean products as the rest of the world becomes more and more contaminated by human activities. (An account of the problems with chemicals from irrigation areas entering the wider environment of the Liverpool Plains appears on pages 207-8, giving some idea of this situation.)

Now the problems of irrigation and Currareva have resurfaced, and in a softer, more insidious form which to the cynical is designed to create an open-slather situation for water extraction and irrigation—which would be the beginning of ecological vandalism on a scale not to be tolerated by any responsible Australian. To let a few people get rich at the expense of fragile ecosystems in the arid Centre for which the Cooper's floods are the lifeblood is unthinkable. Even from a purely economic standpoint, the possible damage to the cattle industry; the irreversible damage to the potential for tourism that the Coongie Lakes and other wetlands present; the risk to biodiversity in the Cooper system itself; and many other issues, make this a proposal to be quashed without argument. This is a case where so little is really known about so many of the aspects which are involved that caution has to be the watchword. There is great danger in setting a precedent which makes it difficult to control future expansion, and the picture elsewhere has been of a rush by other enterprises to be part of any new Development (with a capital D) and the inevitable over-exploitation of the resources involved.

The new Currareva proposal is to produce tea-tree oil and various fruit crops which can deliver fruit during traditional off-season periods. The company involved already has water licences, granted by a careless previous Queensland Government without full understanding of the environmental impact. Water is said to come from pumping from the Cooper during flow events and from groundwater from 'sub-artesian aquifers … believed to be recharged from floodwater'.[49] The proposal admits that 'some impact will occur as a result of harvesting water from Cooper Creek' and there is some uncertainty about water licences until the Draft Water Management Plan appears. The tea-tree plantation will be watered by furrow irrigation and water will be stored in 'several above ground storages'. The orchard will be watered using micro-irrigation. 'If local groundwater resources are available, water will be pumped directly from the bores. If only limited groundwater is available, Cooper Creek water will be pumped to the orchard.' The design of the project will include levees to keep all water on the property because of 'the use of various chemicals that may be carried in run-off'.

What is not mentioned is the evaporation which occurs in open supply channels and storage dams under the climatic conditions at

Windorah. This could be as much as 80 per cent during the dry times (low flow or no flow in the river during these times as well). It is clear from the many uncertainties mentioned in the proposal just how little real research and understanding of the hydrological situation exists at this stage.

A Draft Impact Assessment Study (DIAS) of the proposal has been prepared, and it in turn is waiting on the acceptance of the Queensland Government's Draft Water Management Plan (DWMP) for Cooper Creek. The draft is open to criticism and suggestions from any interested parties.

That these assessment studies and management plans have not fully taken into account the downstream effects, ecological and socio-economic impacts, is argued by the Australian Society of Limnology in their critical review of the DIAS.[50] This review emphasises that:

- Although the DWMP claims that the variability of the river system makes it impossible to quantify downstream environmental impacts, there is in fact a great deal of information available from Australian and overseas rivers and it is well known what deleterious effects can follow.
- The modelling which has been done is not precise enough for the Cooper—each river is individual and long times are needed for collection of the data used in the modelling process.
- It has been impossible to model flows to wetlands (here and elsewhere) and most of the dependent ecosystems on the Cooper are wetlands and these are the areas that graziers rely on.
- The DIAS admits that low flows will be impacted by the Currareva proposal, and it has been shown that the low flow phase of the flow pulse is ecologically critical.[48] Reducing the permanency of Cooper Creek waterholes would reduce fish diversity. (Already, many small dams and water off-takes are imposing stress on the river in this respect.)
- Extraction of a potential 21 000 ML from the Currareva Waterhole with a capacity of 4000 ML will drastically affect natural flooding and drying patterns in the waterhole, which is one of the permanent aquatic refuges on the Cooper.
- The international significance of the Ramsar-listed wetlands has to be taken into consideration. The Coongie Lakes cover nearly 2 million hectares and are one of the largest Ramsar sites in the world, and Lake Eyre itself is a candidate for World Heritage Listing.
- The effects on floodplain flows of building banks and levees and construction of a below-ground channel which will drain floodwaters, enabling them to be pumped into off-river storages, have not been clarified.
- Groundwater investigations are not complete and little is known about aquifers except that they are recharged by river flows and floodplain flooding—both of which will be affected by the proposal.
- A serious flaw in the Development Application, and also with the DWMP, is its failure to deal with the whole catchment; the Queensland Government does not care what happens in the South Australian sector. (Oh for Federal Government control of all important matters relating to the basic life-support systems—soil and water!)
- The socio-economic impacts have not been assessed. These include the effect of the introduction of a chemically-sustained enterprise on the 'clean, green' beef exports; the downstream effects of reduced annual floodplain renewal for landowners; the eventual cost to the community of environmental repair; the intrinsic high risk character of the enterprise itself in view of the extreme natural variability of the river and the climate.

It is to be hoped that the DWMP for the Cooper will involve the revoking of the water licences for irrigation and an end to the Currareva proposal. It will be too late to express regrets when irrigated cotton appears as the next step on this property or on others in similar circumstances. The mulga–spinifex natural vegetation of the region tells us clearly what is a sustainable type of ecosystem under the harsh and highly variable conditions which apply at Windorah.

OTHER RIVERS OF THE CHANNEL COUNTRY

Thomson River At Longreach this tributary has a braided pattern of shallow channels across the floodplain surface. The Longreach Waterhole is the only large channel and permanent water body in the area. Augering revealed the same sequence of deposition as that seen in the Cooper—a capping of about 2 metres of muds on clean sand, in this case in a palaeochannel which was sinuous and probably braided. Adjacent to the waterhole, the mud unit is very thick. It is assumed that at the termination of the pluvial sand-transport phase the wide, sandy channel contracted and partially in-filled to form the present 6-metre-deep waterhole.[51] There appears to have been a considerable reduction in the cross-section of the waterhole during the last 10 000 years.

Western River This river at Winton shows a section similar to the Thomson when its palaeochannel is investigated, except that the

*Little corellas (*Cacatua sanguinea*) on the Diamantina River at Birdsville. Very large flocks of corellas occur in the Lake Eyre Basin.* REG MORRISON

A RIVER OF FAR WESTERN NEW SOUTH WALES: UMBERUMBERKA CREEK

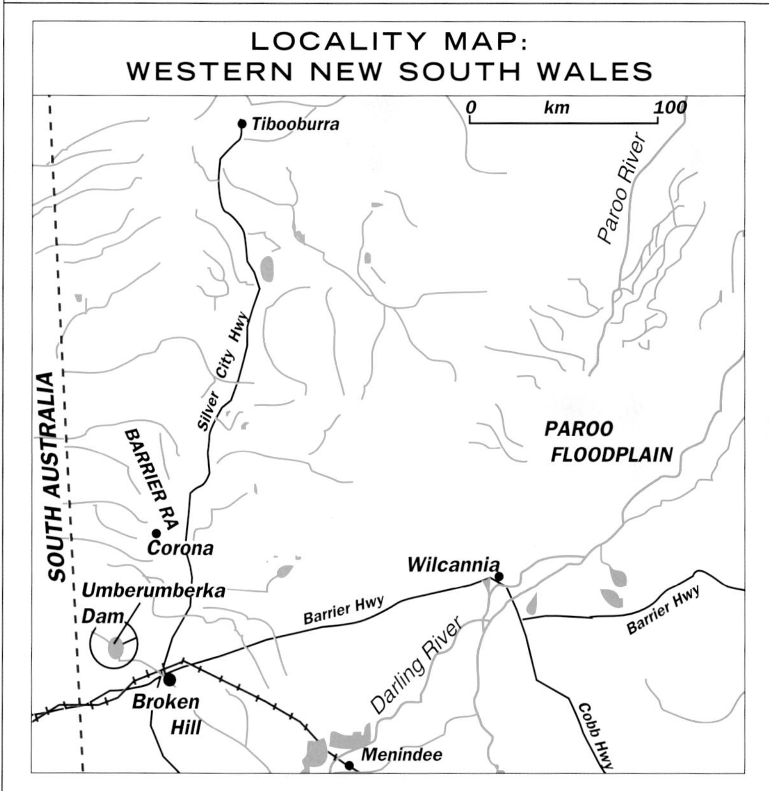

that has its apex close to the scarp. The lower end of the creek draining the fan in turn runs out onto a smaller flood-out fan that lies about 14 kilometres from the fault scarp. Sediment transported from the catchment was deposited on the alluvial fan and flood-out until the Umberumberka Dam was constructed in 1913–15. Since then it has accumulated behind the dam. As the fan does not join a stream or lake, the sediment transported to it is retained and so can provide a record of sediment yield by the catchment.

Pre-settlement yields of sediment have been estimated from the radiocarbon-dated alluvial fan, while post-settlement yields have been calculated from surveys of sedimentation in the reservoir. The results are startling. The average post-settlement yield of 1.9 cubic metres per hectare, per year is about 50 times greater than the average for the 3000 years preceding settlement. The highest recorded post-settlement yield of 3.1 cubic metres per hectare, per year, over the interval 1915–41 is 90 times greater. Sediment yield in the interval between 3000 and 6000 years ago was higher than in the interval from 3000 years to European settlement but lower than the post-settlement yield. The high initial yield into the dam is believed to represent the passage of the eroded material from the initial degradation by over-grazing and drought down the catchment since the 1800s, exacerbated by the removal of timber for the mines. It is somewhat at odds with the known increase in rainfall which has characterised the period from 1940 in the eastern half of the continent. (In *Listen ... Our Land is Crying,* pages 91–2, an account is given of the dramatic erosion in the Homestead Creek at Fowlers Gap in the Barrier Ranges about 100 kilometres from Broken Hill where a completely unstable erosional situation exists and will persist until a new equilibrium is established, emphasising the fragile balance which typifies arid environments.)

Broken Hill lies at the extreme western edge of the Murray–Darling Basin where it abuts the Barrier Ranges, which are mined for their silver, tin and lead deposits. Rivers which rise on the Barrier Range and flow westwards are outside the Basin and flow towards the Centre. The township of Umberumberka, near Silverton, was the focus of a prospecting rush in 1881 when silver was first discovered there, and the Umberumberka Creek and Reservoir have recently been a focus for some interesting and telling research into rates of sedimentation prior to and following European settlement in the catchment.

The widespread excessive erosion which followed the introduction of sheep and cattle into the arid rangelands of the Western Division of New South Wales during the nineteenth century is well documented. The Royal Commission of 1901 into the plight of farmers whose land was literally blowing away following over-stocking, over-grazing and recurrent droughts is testimony to the early damage done to fragile ecosystems—but there has seldom been an opportunity to quantify the changes which occurred. Umberumberka Creek presents such an opportunity because of the presence on this one creek of an alluvial fan and a reservoir.[52]

The creek rises in the Barrier Ranges just west of Broken Hill and flows generally north-west, towards the Lake Eyre Basin. After crossing the steep Mundi Mundi Fault Scarp it debouches onto an alluvial fan with an area of about 200 square kilometres

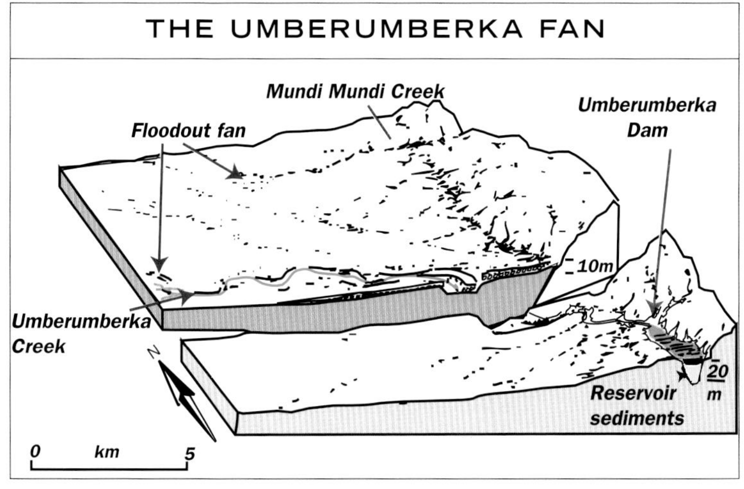

After Wasson & Galloway[52]

mud unit is separated from the underlying sand by a zone of interbedded sand and mud with a median age of about 38 000 years. This seems to indicate a fluctuating dry and wet regime during the transition from wet to dry regimes.

Bulloo River The Bulloo runs parallel to the Cooper but degenerates into uncoordinated drainage south of the Bulloo Overflow near Tibooburra. Waters reach the overflow in summer and occasionally penetrate further west towards Lake Frome.

EMU EGGS AND CLIMATE CHANGE IN THE PLEISTOCENE

Who would ever think that you might be able to read climate change by studying emu eggs? Using radiocarbon-dated emu eggshell fragments from the continental interior of Australia, three earth scientists working at the ANU in Canberra have been able to reconstruct low-altitude subtropical temperatures for the past 45 000 years (using temperature-dependent amino-acid racemisation) and some surprising findings have resulted.[53]

This period is within the last glacial cycle and, perhaps because there is so little evidence of glaciation in Australia (where only about 50 square kilometres was glaciated on Mt Kosciusko at the last glacial maximum 18 000 years ago) it was assumed that temperatures generally, though cooler, were not markedly decreased. (Tasmania had a small ice cap on the Central Plateau, and minor glaciation occurred on New Guinea's high peaks.) We knew from the records of intense aridity and windiness that perhaps there was half the rainfall—twice the dryness—and twice the windiness, and it was a time of blowing sand and salt-laden dust. The dominant environmental indicators are vegetation change, assumed to relate to water balance; episodic wet/dry cycles across the semi-arid interior; and tree-line lowering on highlands. Because of the nature of the Earth's orbit it was assumed that during Northern Hemisphere glaciations, southern lands would experience summer insolation maxima, and thus temperature change across the interior lowlands would be minimal.

This new research has shown that millennial-scale average air temperatures were, in fact, at least 9°C lower between 45 000 and 16 000 years ago. The results also show that warming was rapid after the 16 000-year point.

Australia's geographical situation has rendered it ideal for recording clear base-line information on global climate change. It has been a passive recorder of global-scale climate events as a consequence of its distance from polar ice, and its low relief without any real mountain ranges—factors which preclude feedback mechanisms which would mask responses.

A temperature change of the magnitude indicated by the emu egg research must reflect global processes, and the answer is believed to lie in the enormous reduction of water vapour in the atmosphere that characterised the times when so much of the Earth's water budget was tied up in polar ice caps. Water vapour is a powerful greenhouse gas. Its global decrease in the atmosphere was a factor in global cooling, along with changed ocean circulation, increased terrestrial ice, decreased ocean area and reduction in carbon dioxide concentration.

In addition, as a result of this research, the original concept of a relatively sharp drop in temperature at the glacial maximum, with stable conditions before and after, is no longer acceptable. Temperatures throughout the period covered by the emu egg research were consistently at least 6°C lower than present, so the glacial maximum must have been much colder than the average 9°C reduction.

SILCRETE FORMATION AND THE INVERSION OF RELIEF

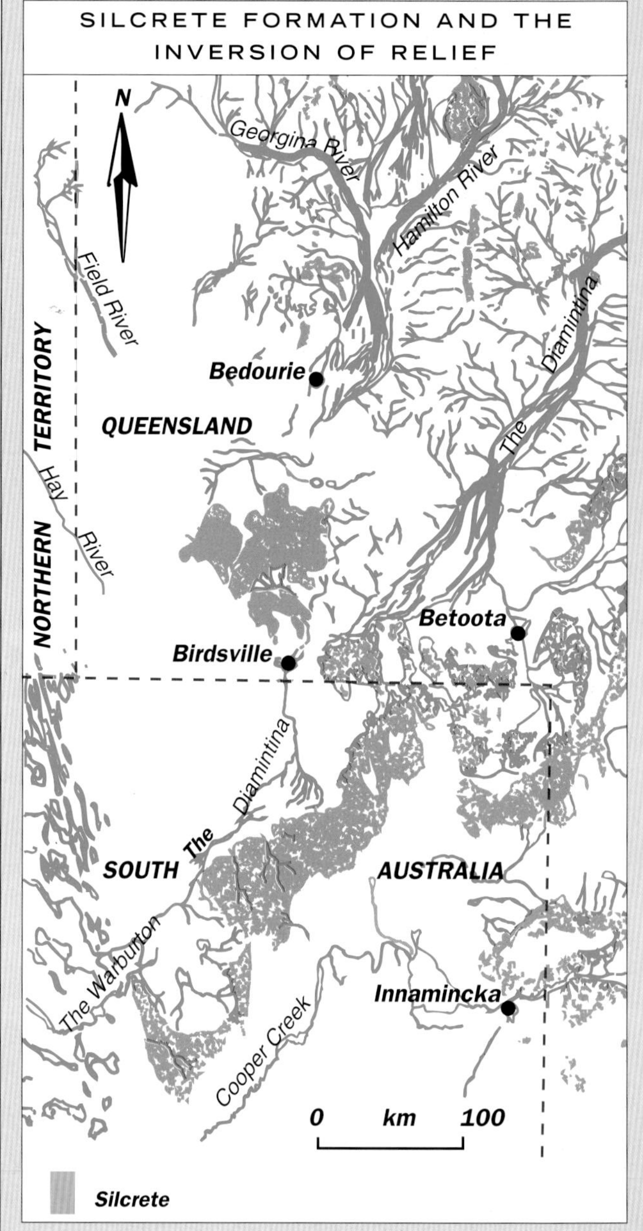

After Twidale[17]

Some palaodrainage channels floored by silcrete, which was produced when silica-rich water hardened sediments below the river channels, now stand high in the local topography.[17] The silcrete duricrust protects the channel floor, and the sediments below it, and when the surrounding landscape is levelled by erosion, the palaeochannels, which may involve wide palaeovalleys, stand proud. Simple palaeochannels with silcrete linings are mostly reduced to the strings of flat-topped mesas which are so characteristic of some desert landscapes. Where wide valley deposits of siliceous alluvia occur, such as those of the Diamantina and Warburton in south-western Queensland and north-eastern South Australia, the modern rivers flow between plateau-like upstanding old river palaeo-floodplains. In the Ooraminna Ranges, south-east of Alice Springs, silcrete has developed on Miocene lake sediments which have withstood weathering and erosion and now cap prominent mesas and plateaux.[17B]

THE FLINDERS RANGES

The Flinders Ranges, beginning 300 kilometres north of Adelaide, extend for 400 kilometres as a series of parallel north–south trending strike ridges, intersected by many gorges. Rising abruptly from the arid plains, the ranges are biodiversity-rich refugia for flora and fauna. Their present-day hydrology reflects the regional low and variable rainfall, which averages about 200 millimetres a year and falls mainly in winter. Average evaporation of more than 2000 millimetres prevents the formation of permanent water bodies

Geological time has seen great changes in the landscapes of the ranges, and even in comparatively recent times, when Australia was already inhabited, there have been significant changes.

A high energy ephemeral river of the Northern Flinders Ranges

Some rivers arising in the Northern Flinders Ranges flow east towards Lake Frome; others flow westward towards Lake Torrens in the southern sector or towards the Frome River and other systems draining towards Lake Eyre in the northern sector.

ARKAROOLA CREEK

Tree trunks stacked 2 metres high by floods in the stony creek bed indicate the force of the floods. M.E.W.

Large trees within the stream bed, and a passing emu. M.E.W.

Italowe Creek in the Gammon Ranges National Park is another high energy ephemeral stream carrying boulders and large logs in floods. M.E.W.

Echo Creek Waterhole—permanent water through the dry season, but green with algae. M.E.W.

Stubbs Waterhole—a pool barely surviving in the dry season. M.E.W.

THE CENTRAL FLINDERS RANGES

Brachina Gorge, in the central Flinders Ranges, is a well-known tourist attraction. River red gums, in all their glory of coloured trunks and wondrous shapes, adorn the creek beds; white cypress pine and gum-barked coolibah and red mallee cover many of the ridges; and grasses and low shrubs, and ephemeral herbs after rain, cover the plains. Brachina Creek rises in the centre of the ranges and flows westward towards Lake Torrens, winding through the 10-kilometre stretch of the gorge as a meandering sandy channel in the wider sections, and confined to a narrow rocky bed in the 'bottleneck' section of the gorge.

Recent mapping and dating of creek-bank sediments has provided evidence of the different hydrological conditions which prevailed in

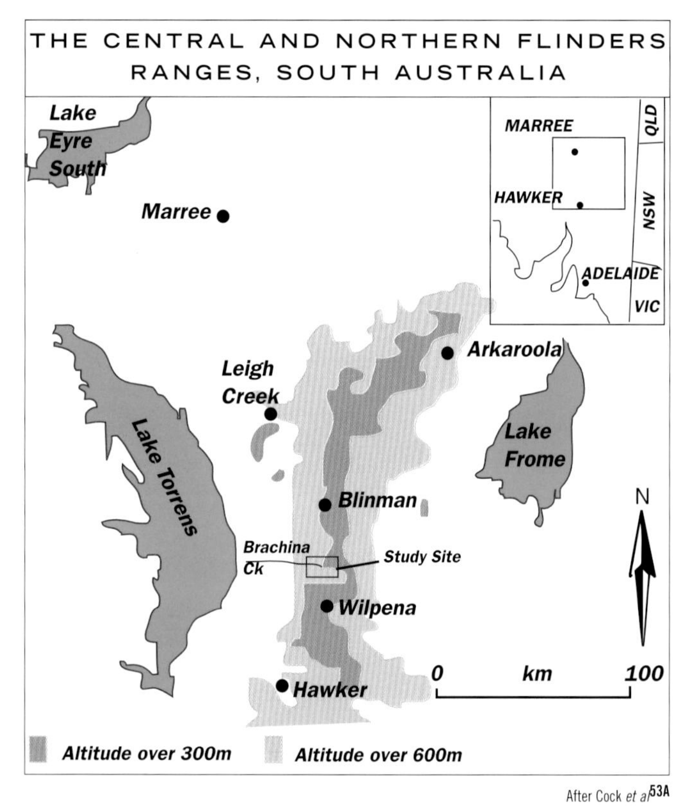

After Cock *et al*[53A]

Brachina Creek, Flinders Ranges, with young river red gums. Lacustrine sediments are found on the right-hand side of the creek at this point.
DON ADAMSON

Brachina Creek, with Archaeocyathid limestone exposed in the stream bed.
M.E.W.

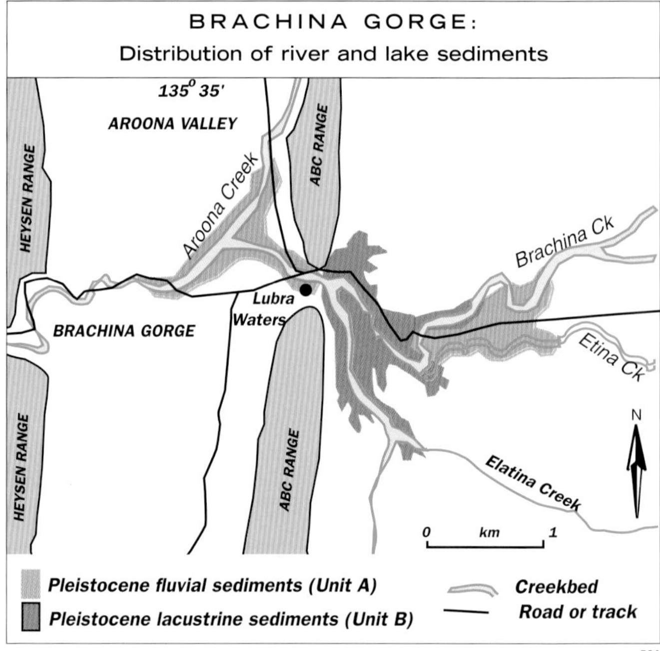

After Cock *et al*[53A]

the past, and the study has also shown that a lake existed in the area.[53A] The sedimentary column consists essentially of a basal, probably lacustral, clay-rich deposit separated by a weakly developed palaeosol horizon from a zone of sediments laid down in a lake. This study is of particular interest as it shows the influence of local conditions which modified the response to the regional climatic changes occurring in surrounding territory. A comparatively mountainous region within a flat, arid region to some extent creates its own local climate, and the history of sedimentation recorded in Brachina Creek documents a timetable which is somewhat anomalous.

The sedimentary record shows that prior to about 28 000 years ago, rapidly developing dust mantles altered local hydrology by reducing run-off and increasing infiltration. The fine-grained, clay-rich sediment in this lower layer is interpreted as being reworked aeolian dust, transported during the dust storms which resulted from the increasing aridity on the rundown to the last glacial maximum. Dust mantles blanketing the ridges and slopes were remobilised and deposited in the regional drainage channels as thick, uniform deposits of silt and clay. The timing of this sedimentation accords with evidence from other inland sites, and the termination of this phase of deposition at 28 000 years ago is synchronous with low fluvial activity recorded at other sites, representing a response to regionally decreasing rainfall.

However, the palaeosol zone which follows indicates that despite decreasing regional moisture, conditions locally remained wet enough to support modest soil-forming activities at some time between 28 000 and 21 000 years ago. (The lead-up to the glacial maximum did not bring a dramatic increase in aridity locally.) The lake sediments which overlie the palaeosol result from a blockage of the bottleneck in the ABC Range, causing the formation of a lake which persisted through the 4500 years which span the glacial maximum. (Whether the blockage resulted from tectonic activity, a landslide, or a sediment or vegetation plug is not known, but the evidence for a sudden event is seen in the sequence of depositional layers of lake sediments.)

Lake Brachina is not the only lake which broke the rules and was in existence at the extraordinarily dry time of the glacial maximum—Lake George on the Southern Tablelands of New South Wales has a similar record, admittedly in a much less arid climatic zone. The suggestion in the case of Lake Brachina is that during the glacial maximum, evaporation was lower, the winter cloud base was lower, and the winter rainfall somewhat higher than in the surrounding areas and also higher than it is today.

CHANGES IN THE CENTRAL FLINDERS RANGES 30 000 to 15 000 YEARS AGO

Prior to 28,000

WIND BLOWN DUST

ABC RANGE

Formation of dust mantles alter local hydrology by reducing runoff

Regionally more effective precipitation

Fluvial deposition widespread

N

28,000 - 20,500

WIND BLOWN DUST

ABC RANGE

Regionally less effective precipitation

Locally cessation of fluvial deposition

Period of stability allows soil development in fluvial sediments

Input of windblown dust continues

N

20,500 - 16,000

WIND BLOWN DUST

ABC RANGE

Regionally lower evaporation increases effective precipitation

Locally enhanced orographic rainfall

Sudden damming event

Deposition of lacustrine sediments

N

After Cock *et al* [53A]

THE SOUTHERN FLINDERS RANGES

The Willochra Plain is an intermontane topographical basin developed when faulting resulted in a block of terrain being lowered between adjacent ridges.[17] It is drained by Willochra Creek. The Willochra Basin is an important source of underground and artesian water. (The Walloway Basin is another intermontane basin in the Southern Flinders.) During Middle Eocene times the northern Willochra Plain was a large lake, with arms extending up valleys of the Mt Arden and Kanyaka Creeks as well as Boolcunda Creek.

THE WILLOCHRA AND WALLOWAY BASINS IN THE SOUTHERN FLINDERS RANGES

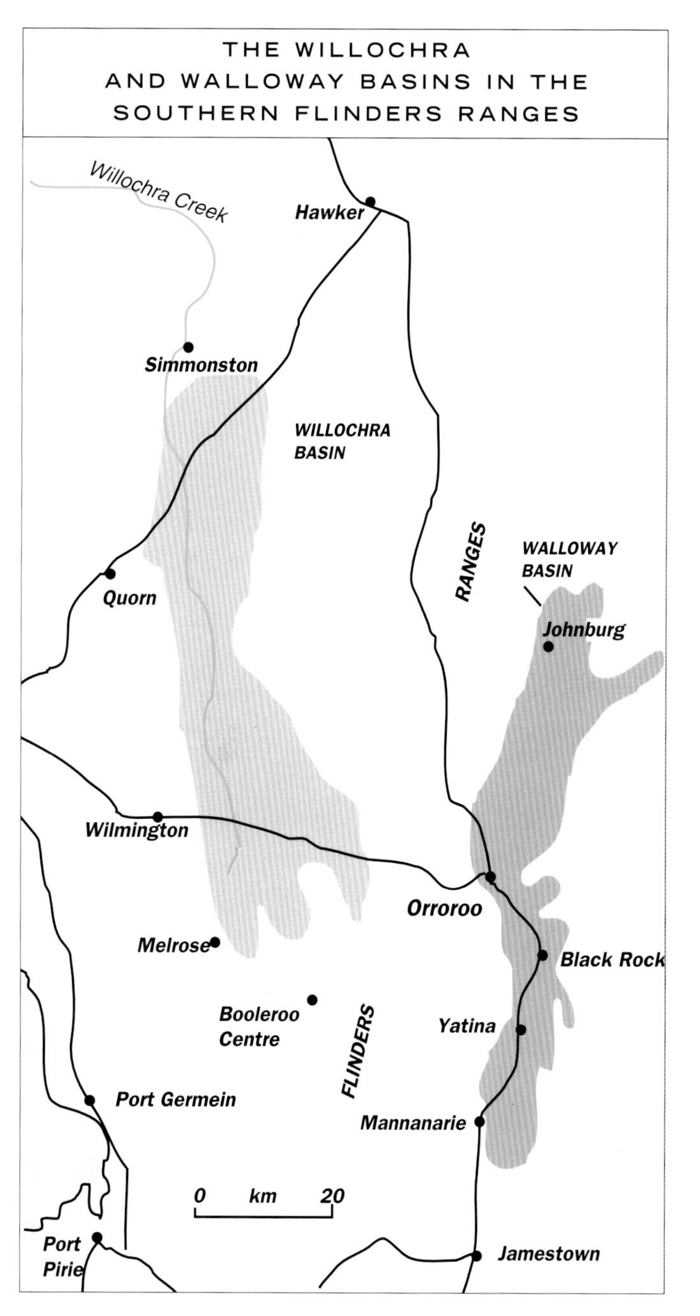

SECTION THREE

THE TROPICAL NORTH

Extensive low-lying coastal and riverine plains characterise northern Australia. These plains are geologically young and have mostly developed to their present form only in the last few thousand years.[54] Sea-level changes in the Quaternary have profoundly affected both south-eastern Asia and northern Australia, where the extensive low-gradient continental shelves on the Sunda platform and Sahul shelf occur. Continental shelves were exposed during times of low sea level (glacial stages of the Pleistocene ice age), Australia was joined to New Guinea, and the Gulf of Carpentaria was land with a central lake—Lake Carpentaria.

The sea encroached on the land at times when sea level was higher than it is today in previous interglacials and has been shown to have fluctuated rapidly during the last 240 000 years. (There is at present no evidence of the higher sea stands during the last interglacial from localities along the northern Australian coastline. Such evidence is found at 10 metres above present sea level at Ningaloo in Western

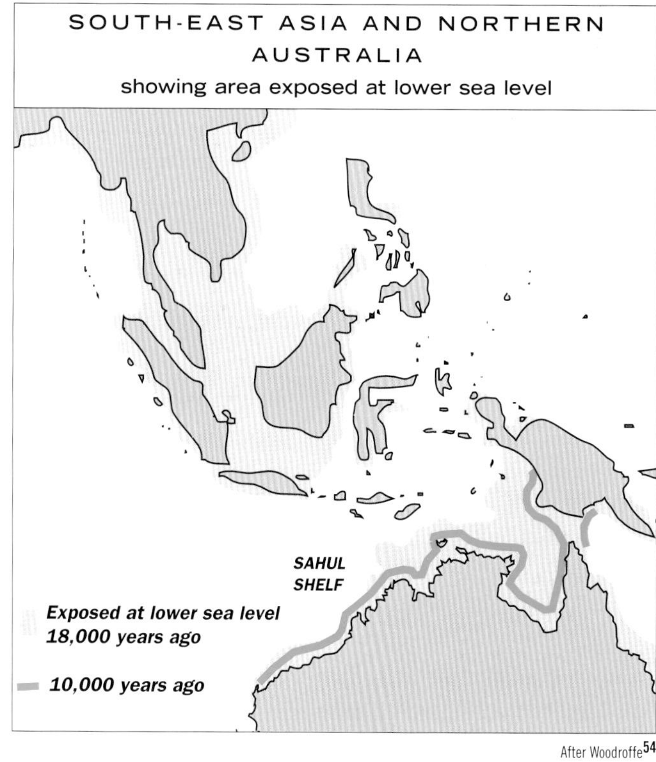

After Woodroffe[54]

Tidal drainage patterns fringed by mangroves, on the western side of Cape York Peninsula. REG MORRISON

Australia, while on the opposite side of the continent evidence of a higher stand occurs in the Great Barrier Reef region, suggesting that the northern part of the continent has gone down over that time. Some experts argue that the sea level was not in fact higher and that on some shorelines the land has gone up during isostatic adjustments and we therefore see the fossil shoreline as higher.)

SEA LEVEL CHANGES OVER THE LAST 240 000 YEARS

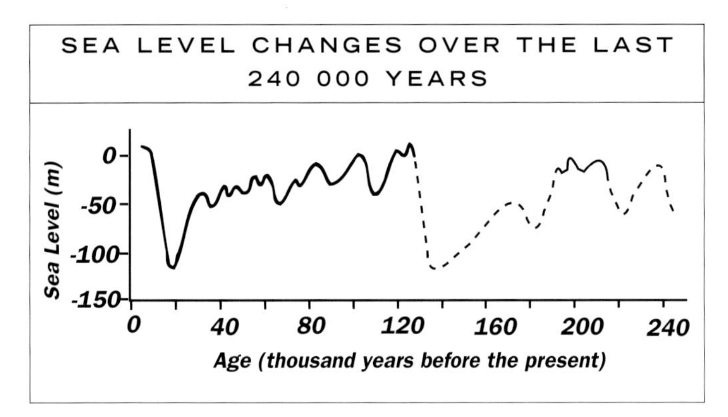

The last interglacial shoreline was assumed to have been 6 metres above present-day height at its peak; sea level reached its lowest at the last glacial maximum which is now known to have been between 21 000 and 22 000 years ago, instead of the 18 000 years which had been suggested by the less sophisticated dating techniques previously available. Off the north coast of Australia, the lowest sea level reached at the glacial maximum was between 150 and 165 metres below today's level. (The depth varied at different points round the continent.)

Since the last glacial maximum, sea level rise has been rapid, up to 30 millimetres a year, which means that the shoreline was advancing across the low-gradient continental shelves at several tens of metres a year—a very visible advance to the Aboriginal inhabitants of northern Australia. Sea level is believed to have risen steadily to above its present height at between 6500 and 6000 years ago (the 'big swamp' phase) and then to have decreased and become stabilised at close to its present levels. (The rising, peaking and fluctuating before stabilising of sea level is seen in the South Alligator River where a big swamp phase with spread of mangroves characterises the peak.)

THE GULF COUNTRY OF NORTHERN QUEENSLAND

Today, the Gulf of Carpentaria is a shallow epicontinental sea, less than 70 metres deep, which lies on continental crust between Australia and New Guinea, is connected by the Torres Strait, a seaway only 12 metres deep, and to the Arafura Sea across a sill 53 metres below average sea level.[55]

Lake Carpentaria

Late Quaternary fluctuations in sea level, controlled by polar ice build-up during the last 40 000 years, exposed the shallow Carpentaria Basin. Information from bore cores of sea-floor sediments in the Gulf indicate that after drying out when sea levels dropped, a shallow lake with brackish to fresh water up to 10 metres deep and with an extent of 29 000 square kilometres occupied the

LAKE CARPENTARIA

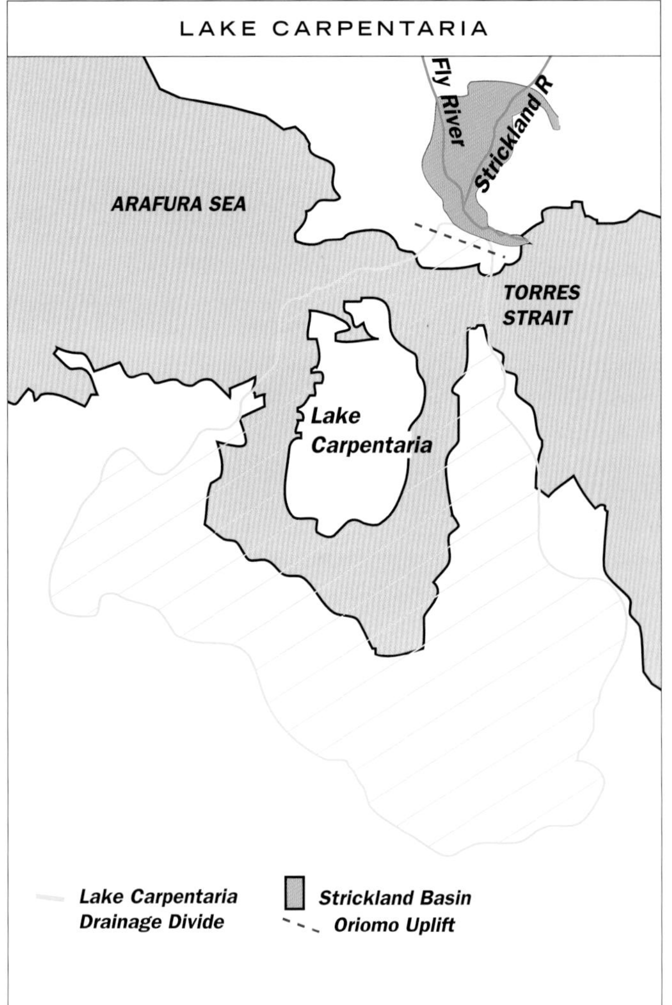

After Torgersen *et al*[55, 39]

Basin between 35 000 and 12 000 years ago.[55, 39, 93] Between 35 000 and about 26 000 years ago a wetter than present-day climatic regime operated. During the last glacial cycle, the environments of the Basin were influenced by the level of the sea; climatic variations (the monsoonal patterns in those latitudes were changed and finally interrupted during the build-up to the glacial maximum); and the tectonic diversion of the Fly River. The Fly–Strickland Basin lay to the north of the Lake Carpentaria catchment boundary, with the Fly and Strickland Rivers flowing southwards above their junction. When the Oriomo Uplift occurred prior to 35 000 years ago it resulted in the diversion of the lower Fly River to the east, so that it no longer ran straight across the Torres Strait land bridge and into Lake Carpentaria.[56]

Rivers that today run into the Gulf drained then into the Carpentaria Basin, and extensive swamps, expanding the floodplain swamps which exist today, would have surrounded the central lake.

Artesian springs found along the southern perimeter of the Gulf arise from the northern marine extension of the Great Artesian Basin.[57] They are clearly visible in the intertidal zone at low tide, their presence distinguished by clouds of fine sediments stirred up by submarine spring water emanating from the sea bed. Others can be found in deeper water. Prawn fishermen try to avoid these disturbed areas, known locally as 'wonky holes', because the suspended sediment can foul their nets. Whether these littoral springs influence local biodiversity, especially in the seagrass communities which provide important nursery habitats for juvenile prawns, is currently unknown. It is also unknown whether the declining pressure in the rest of the Great Artesian Basin as a result of water extraction has affected these springs. When the Gulf was Lake Carpentaria, however, the springs would have been located on the sloping flats round the southern edge of the lake, which occupied the lower middle of the region, and they may well have been small mound-springs.

Pollen analyses of the cores of lake floor sediments show that the region during the last 35 000 years supported much the same vegetation as the black soil plains south of the Gulf do today—now sedge-grass communities with aquatic taxa in the low-lying swamps, woodlands on the higher, better drained slopes. Eucalypts, cypress pines and casuarinas comprise the woodlands and the Basin, apart from the swamps, was grassy savanna. Data about the consistent nature of the vegetation through at least 40 000 years across this northern sector of the continent is important information. It has often been postulated that rainforest was ubiquitous across the tropical north. Riverine vine forests would have extended along the major rivers southwards from the edge of the Gulf and during the 35 000 to 26 000 years ago wetter phase would have extended further into the dry interior of the continent south of the present monsoon-dominant zone.

From about 12 000 years ago, sea-level rises started to affect the lake, with a permanent open connection to the sea. Estuarine conditions became established, and the salinity is believed to have remained low both because of limited water exchange over the Arafura sill, and because the climate was wetter and tropical cyclones kept up the freshwater supply. At 8000 years ago the Torres Strait was flooded and fully marine conditions returned to the Gulf.

The bore cores of lake sediments which span the 35 000 to 10 000 years period are interesting for more than their proof of the existence of the lake, which was permanent throughout the period and did not have dry stages like other Australian lakes. As well as the pollen record which they contain, they also provide a record of cyclic aeolian activity with layers of wind-blown sand and dust, indicating dry climatic conditions, occurring every 2250 years and persisting for 600 years.[39] One of these dry, windy events falls between 11 375 and 10 430 years ago, coinciding with the Younger Dryas Event in the Northern Hemisphere. This Event represents a cold spell of nearly 1000 years when glaciers advanced and conditions were very unpleasant in northern Europe. Glaciers also advanced in New Zealand, and in Australia there is evidence of aridity over the north-eastern sector and reactivation of dunes in central deserts. (The cyclic colder, dryer fluctuations are also documented in oxygen isotope records from Greenland and Antarctic ice cores, and changes in deep-sea foraminiferal faunas. Cores from the Great Salt Lake in Utah show the same fluctuations, as do patterns of growth in coral reefs in the Barbados—so the phenomenon was global.)

The cyclic dry and windy periods in Australia which resulted in drought and dune reactivation over 600 years, followed by about 1500 years of more stable rainfall conditions, are thought to represent intensification of the eastern trade winds over northern Australia. This would have forced the Intertropical Convergence Zone further north than it is today, implying a temporary failure of the monsoon over the Gulf of Carpentaria and an intensification of the Southern Oscillation over the Pacific at the same time. The 2500-year cycle is believed to be governed by sun-spot activity.

MODERN RIVERS OF THE GULF OF CARPENTARIA

The monsoon turns the flat coastal plains of the tropical north into a drowned landscape. NEIL LONERAGAN

Rivers in the Gulf of Carpentaria are laying down alluvium ranging in size from fine gravel to mud, very slowly building extensive fan deltas into the Gulf.[58] Under conditions of intense tropical-monsoon weathering in northern Australia, these non-cohesive sediments can become welded into duricrust, the cement being limonite and calcite. In geological terms, this process is rapid, occurring over a period of less than 30 000 years. In some reaches these rivers are like any others, re-working soft alluvial sediments to form compliant channels with

Cemented alluvium on the Nicholson River near Burketown in the Gulf of Carpentaria, QLD. GERALD NANSON

The Mitchell River, western Cape York Peninsula, one of the major rivers of the Gulf Country which forms a huge delta. REG MORRISON

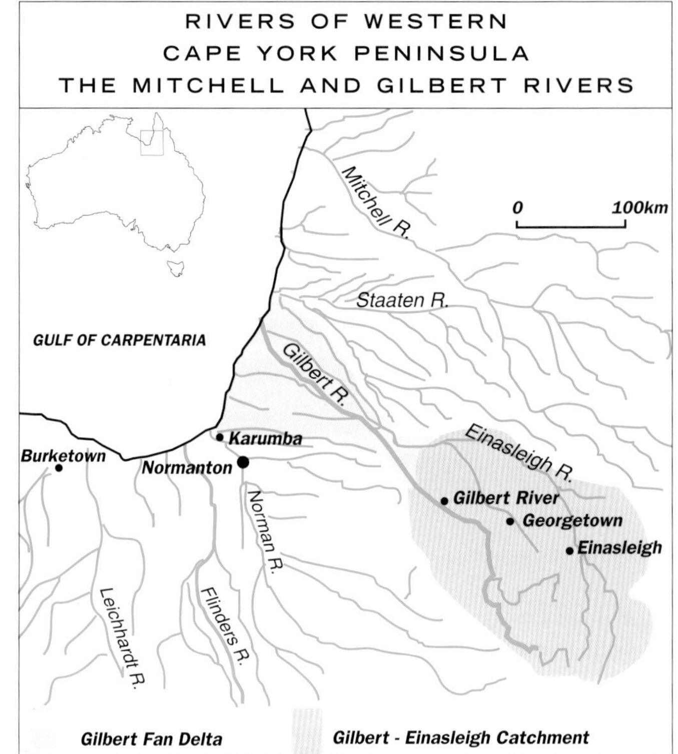

After Nanson *et al*[60]

Cemented alluvium on the Gilbert River in the Gulf of Carpentaria. It is the weathering and cementing of alluvium into a soft rock along these northern rivers that inhibits the ability of the channel to adjust to variations in climate and flow regime. As a consequence, the rivers periodically abandon their existing channels to form new ones in more erodable areas of the plains surrounding the Gulf. Clearly visible is a small gorge subsequently cut by the river into its cemented sediments. GERALD NANSON

gently sloping banks and uniform gradients, free to adjust to gradually changing hydraulic conditions. Yet where the channels incise into these alluvial duricrusts, the sides become cliff-like and waterfalls and rapids form; thus *the river becomes constrained by its own alluvium.* No longer free to adjust gradually, such channels will sometimes avulse, breaking out of their regular path and cutting a new course elsewhere across the floodplain. This erratic behaviour causes the development of multiple channels at various stages of erosion and infilling. This is yet another example of Australian rivers showing a radically different type of response to what might be expected.

The Gilbert River

The Gilbert River flows some 500 kilometres from the highlands of north-eastern Queensland to the eastern shore of the Gulf of Carpentaria.[60] Mean rainfall over the catchment varies from 800 to 1000 millimetres, and is highly concentrated in the summer.

The Gilbert River is an ephemeral, sand-dominated, low sinuosity channel that breaks up into anabranches below its confluence with the Einasleigh River. Sand is largely restricted to the anabranches, with extensive floodplains formed of thick muds and muddy sands deposited gradually by extensive over-bank flooding during the monsoon each December to March. Over the lower 200 kilometres of its course, the river has deposited a very large low-angle fan and associated deltaic deposits. The upper 8 to 12 metres consist of interbedded muds and muddy sands beneath which occur coarse sands and gravels. The mud/sand alternation repeats itself with depth. The uppermost 10 metres of the fan deposits, visible in the channel banks, have been extensively weathered and variably indurated with calcium and iron compounds. In some parts the induration is in the form of nodules, in others fusion results in calcrete and ferricrete horizons. The present channel system and some of the more recent palaeochannels are incised into this upper, fine-textured indurated unit. Extensive over-bank flooding across the entire delta fan occurs several times each century.

A major study, using both thermoluminescence (TL) and uranium-thorium (U/Th) dating, has determined the chronology of the delta formation in the Gilbert River. The findings apply to other rivers and their deltas in the monsoonal tropics of the Gulf. The

Cemented alluvium on the Gilbert River. The river has cut a gorge into the cemented sediments which now confine it. GERALD NANSON

Cemented alluvium on the Gilbert River. Here the river has cut down deep into the restricting material which results from its own presence. GERALD NANSON

determination of ages in the duricrust being formed in the rivers (within the last 30 000 years) is of interest as it has often been assumed that such processes ended in the Late Tertiary.

Beneath the top 5 to 12 metres, the Gilbert fan delta consists of an extensive sand body, older than 85 000 years and probably about 120 000 years old, which is highly indurated, often forming calcretes or ferricretes. (The same layer has been identified and dated below the Leichhardt River fan delta near Burketown.) This underlying old sand zone represents a period of major fluvial activity not repeated since that time. Above it are muds and fine sandy muds that extend uninterrupted to the present surface, except in the downstream fan, where they are bisected by a thin unit of medium sand that dates at 40 000 to 50 000 years ago. This represents a period of enhanced rainfall, seen also in Channel Country rivers. A system of sandy distributary channels over the fan surface represents an early Holocene fluvial phase probably more active than at present. Dates for these channels range from 7000 to 10 000 years. (A terrace in the upper catchment at Georgetown showed an enhanced fluvial period at about 9000 years which accords with these findings in the lower portion of the river.)

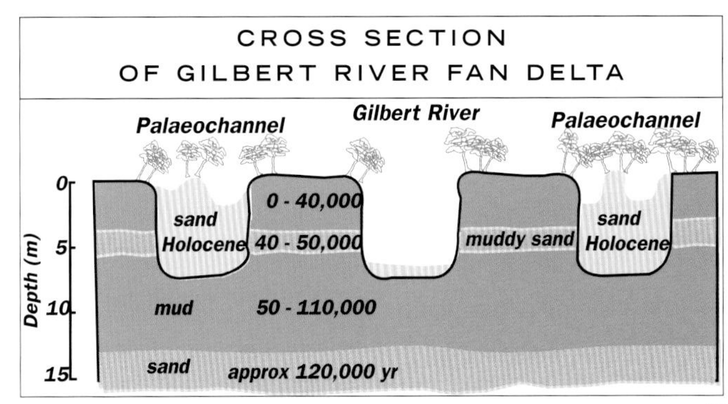

After Nanson *et al*[60]

A cross-section of the Gilbert River fan delta showing the basal sand unit, the upper floodplain muds intersected with a muddy-sand unit and the main Gilbert River channel in the centre, flanked by two abandoned Holocene distributary channels.

The McArthur River

The McArthur River drains a semi-arid monsoonal catchment with an area of 18 400 square kilometres. It discharges into a sheltered embayment behind the Sir Edward Pellew Islands.[61] The sheltered position of its exit has much to do with the silting-up of channels and formation of small sand islands, and with the size of the McArthur River delta (sub-aerial and submerged).

Annual rainfall in the catchment is around 600 millimetres, decreasing from north to south. South-easterly winds predominate in the dry season (April to November), and north-westerly winds are most frequent during the wet season (December to March).

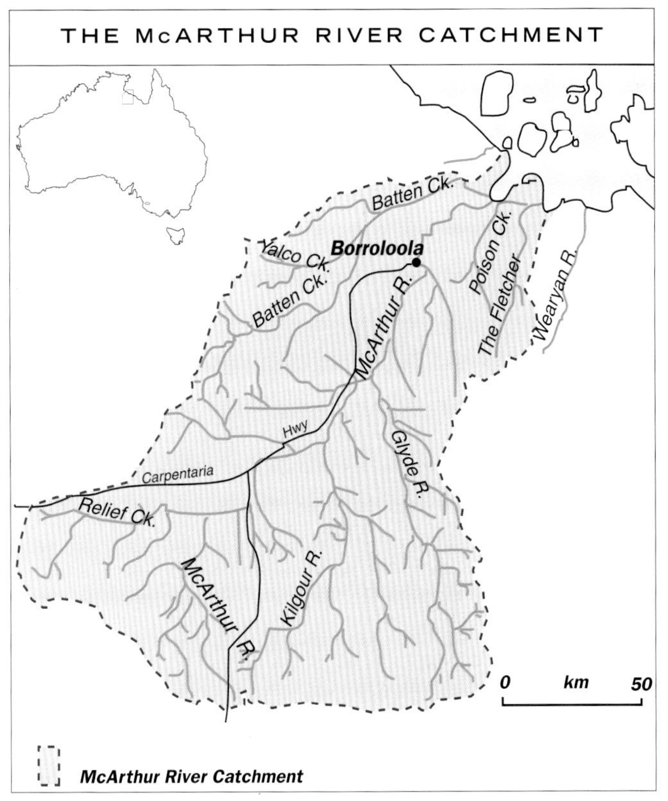

After Woodroffe & Chappell[61]

The McArthur River near Borroloola, N.T. COLIN WOODROFFE

Bessy Springs, on the McArthur River Station, N.T. COLIN WOODROFFE

Salt marsh above high tide level, McArthur River. COLIN WOODROFFE

Megaripples on a sandstone block, on the McArthur River. The ripples were created when the sediments were accumulating 1.5 billion years ago! REG MORRISON

Discharge of the McArthur River is extremely variable: at Borroloola, 68 kilometres from the coast, discharge varies from no flow to peak floods of more than 3200 cubic metres a second. Tropical cyclones affect the area and in 1984 Cyclone Kathy devastated the town and had a major destructive effect on the vegetation of the delta.

In the upper part of the catchment, where there is pronounced fluvial erosion, the river is braided and entrenched. Extensive Quaternary alluvial deposits are associated with river meanders in the lower catchment, particularly in a Pleistocene system of sandy fluvial ridges flanking the present river course downstream of Borroloola. Towards the coast, some stranded beach ridges and fossil dunes have been reported (evidence of higher sea stands in the past) and for the last 25 kilometres of its course the McArthur flows across a Holocene delta and associated coastal plain. The apex of the delta is marked by the confluence of Batten Creek and the McArthur River. Batten Creek's flow continues north through the Carrington Channel while the McArthur River dominates the eastern margin of the delta. Tidal circulation in the Gulf is complex, mainly clockwise, and at Centre Island the tidal range is about 3 metres. Tidal influences penetrate 68 kilometres up the river into the vicinity of Borroloola.

Holocene deltaic sediments form a broad, shallow wedge within the pre-Holocene river and coastal sediments. The sub-aerial deltaic plain can be divided into an upper and a lower segment. The upper plain is constrained by the local topography, bounded by the Carrington Channel to the west and the Johnson River to the south.

HOLOCENE EVOLUTION OF THE McARTHUR RIVER DELTA

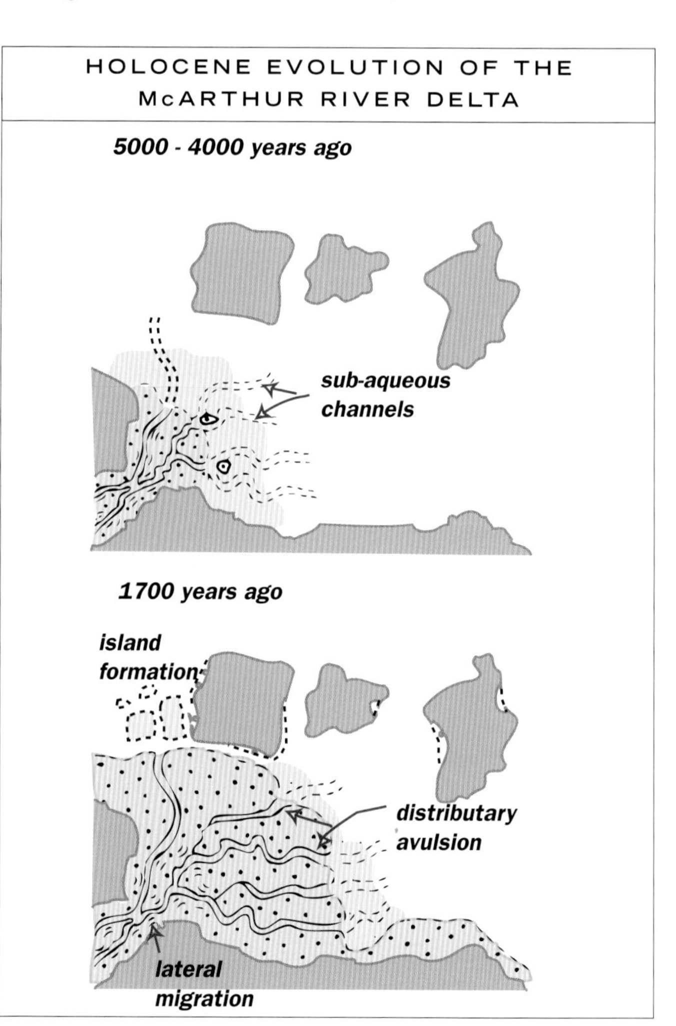

After Woodroffe & Chappell[61]

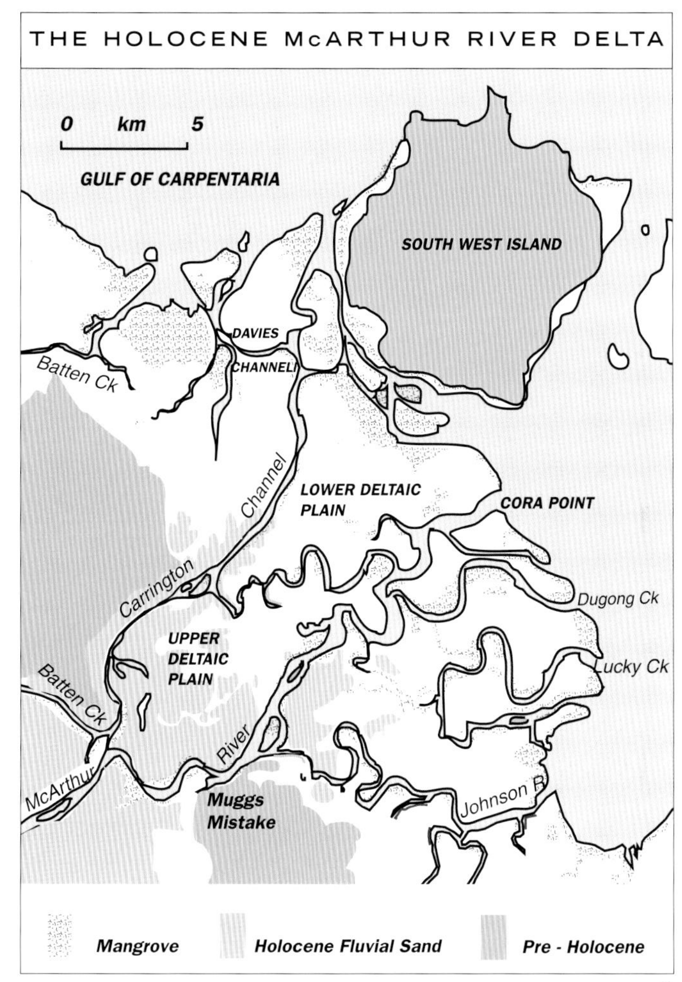

After Woodroffe & Chappell[61]

Katherine Gorge in the Nitmiluk National Park in the Northern Territory. The resistant Kombolgie Sandstone dictates the course of the river, which has cut down along joint cracks through the ages. REG MORRISON

Katherine Gorge: the reach where tourists change boats to proceed further along the river beyond the rocky bar which is exposed when the river is not in flood. M.E.W.

The lower deltaic plain is flanked by the broad coastal plain to the north-west and to the south. Extensive mudflats, bare or vegetated by samphires and grass, and a series of mangrove-lined distributaries characterise the lower deltaic plain. The shoreline, fringed with mangrove forests, divides the sub-aerial part of the delta from the sub-aqueous, which consists of a broad low-gradient delta front composed of shelly sands. The upper deltaic plain has sandy ridges left by past migrations of the river channel. The ridges support open grassy eucalypt woodland, and termitaria are abundant on the higher crests.

The McArthur River has built its delta as relative sea level dropped over the last 4000 years. Distributaries have migrated across the upper deltaic plain by lateral migration, leaving a series of fluvial ridges. In lower parts of the deltaic plain, channel migration appears to have been largely by avulsion—as each channel silted up, the river made a new channel and the old one became increasingly infilled and tidally dominated.

RIVERS OF THE TOP END, THE NORTHERN TERRITORY

The Katherine River

The modern Katherine River drains the Arnhem Land Plateau of north-central Australia. The plateau is an upland of relatively flat-lying quartzose sandstone which is extremely resistant to erosion and forms a prominent escarpment.[62, 63] Upon emerging from the plateau, the Katherine River traverses a granite plain where erosion produces a great deal of sand, available to be transported in floods. The Katherine Gorge lies immediately downstream of the granite plain. It is a narrow, deep canyon system incised into the hard and resistant Kombolgie Sandstone, 32 kilometres north-east of the town of Katherine. Most of the gorge is within a national park. The gorge exhibits a rectangular pattern of slot-like canyons which results from the pronounced vertical joint pattern in the sandstone. Beyond the gorge, the river flows south-west and joins the Daly River which runs north-west to Anson Bay.

The tropical monsoonal climate results in rare flood events of very large magnitude such as the 1998 flood which devastated the town. The confined nature of the channel results in enormous flow power during such events, and large boulders are moved, pools are excavated to great depth and the floodwaters are destructive where they exit the gorge and rush across the plains.

The Katherine Gorge from the air, during the 1998 floods. JOHN CHILDS

The force of the floodwaters is clearly seen in this photograph of the Katherine Gorge during the 1998 floods which devastated the town of Katherine downstream. JOHN CHILDS

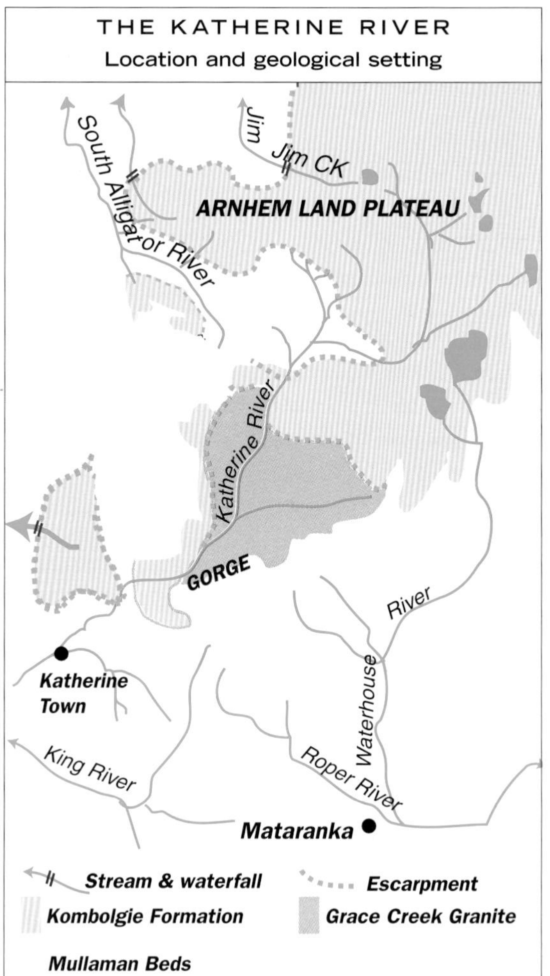

After Baker & Pickup[63]

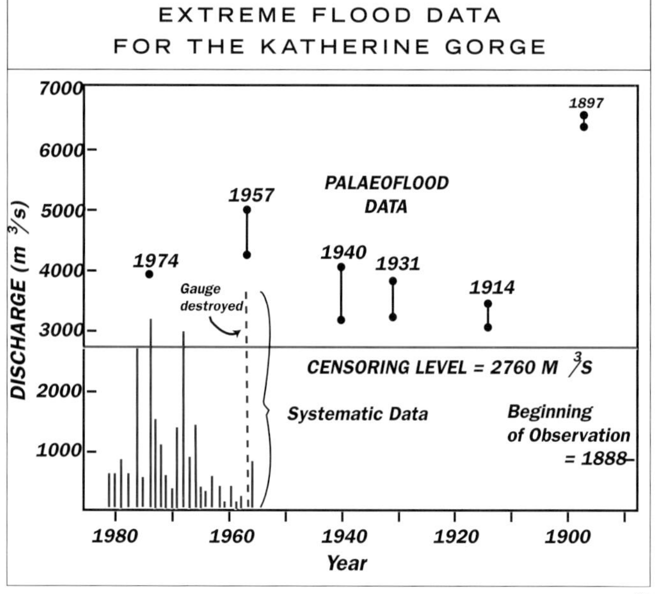

After Baker & Pickup[63]

Slack-water deposits date the extreme flood events in the Katherine Gorge since 1897.

In 1980, a depth sounder was used to map the bed profile in the gorge. It revealed a striking pool-and-riffle morphology. At low water, a series of pools and shallow rapids comprise the five gorge sections. Very deep pools occur where major joints were deeply weathered and more easily eroded. The heavy sand load carried in floods has polished some bedrock platforms, and deep potholes in bedrock often have spiral grooves on their walls—evidence of origin through abrasion by vortex flow (whirlpools) of sediment-charged water.

The average annual rainfall is about 1000 millimetres, with 95 per cent occurring in the 'Wet' (October to April). In January 1910 a record 234 millimetres fell in 24 hours at Katherine; a maximum for 24 hours of 545 millimetres was recorded at Roper Valley Station, 200 kilometres east of Katherine on 5 April 1963.

The Katherine River is ancient. The palaeodrainage system which preceded it cut down through sediments laid down when the Top End was inundated during the Early Cretaceous (about 120 million years ago), when much of the continent lay beneath epicontinental seas. All that earlier cover was removed over time as the land surface was planated, and during the Miocene the ancestral Katherine River started cutting down into the bedrock which lay below. Making a channel through the extremely hard Kombolgie Sandstone has been slow and difficult, and a river like this, running in the erodable joints in the bedrock, has not been able to change course (unlike rivers which traverse floodplains which do not confine them).

Many of the tributaries of the Katherine on the plateau occupy 'hanging valleys' and have spectacular waterfalls which only flow during the Wet. The smaller tributary rivers without the flow power of the Katherine were unable to incise their beds at a rate that kept pace with its down-cutting, hence their 'hanging' nature.

MAJOR RIVERS DRAINING INTO VAN DIEMEN GULF

The Van Diemen Gulf is approximately 120 kilometres wide from west to east, and averages 75 kilometres from north to south. It is enclosed to the north by the Cobourg Peninsula and Melville and Bathurst Islands. The tidal range averages 5 to 6 metres through most of the Gulf and macrotidal influences affect the estuaries of rivers which run into it. Several seasonal rivers with major estuaries drain the region.[64, 67]

Extensive low-lying coastal plains characterise the southern shore of Van Diemen Gulf and extend inland as estuarine plains along the lower reaches of the rivers which drain into the gulf from the undulating hinterland. Littoral mangrove flats lie seaward of saline mudflats of variable width. Sedge-grass plains, freshwater wetlands and paperbark forests lie landward of the mud flats. These coastal and estuarine plains contain some of the most productive freshwater wetlands in tropical Australia and are important wildlife habitats. They also support commercial and recreational fisheries, tourism and a pastoral industry. The climate is monsoonal and river flow is extremely seasonal. Flooding of the wetlands and plains occurs from December to April, and desiccation of the plains develops in the dry season. Saltwater intrusion into freshwater wetlands is occurring in many areas, to a greater or lesser degree. Over-grazing by buffalo, mainly feral, is associated with the penetration of salt water, which is a worrying phenomenon. The situation is particularly serious in the Mary River floodplains.

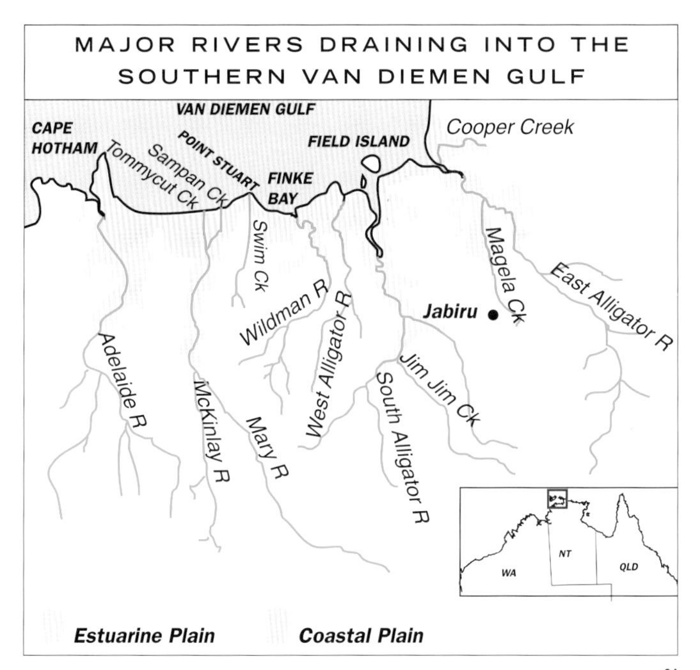

After Woodroffe *et al*[64]

The East Alligator River, Kakadu National Park. REG MORRISON

The **South and East Alligator Rivers** have their headwaters on the Arnhem Land Plateau, not far from those of the southward flowing Katherine River. They run northwards across the wide plains where enormous wetlands become widespread approaching the coast. The **Nourlangie Creek** which, with the South Alligator River, provides the major wetlands of Kakadu National Park, and **Magela Creek**, which flows past Ranger Mine and Jabiluka into wetlands round Oenpelli, also originate on the plateau. To the west of the plateau, the **Mary River** emerges and swings north to run parallel to the western margin of Kakadu National Park, into the floodouts and swamps of the coastal plain of Van Diemen Gulf. The **Adelaide River** and the Mary have their headwaters on the ancient, laterised land surface west of the Plateau.

Run-off from the landscape during the dry season is negligible, and tidal reaches of the rivers become saline. Freshwater run-off is increasingly important as the wet season progresses, and floods tend to develop towards the end of the season. The large tidal range, combined with the flatness of the plains, causes tidal effects to extend more than 100 kilometres inland in the larger rivers. The plains adjoining the tidal river (the deltaic-estuarine plains) are covered by freshwater vegetation except where brackish waters intrude via tidal creeks. Saltwater intrusion into freshwater wetlands is an ongoing and undesirable process.

THE SOUTH AND EAST ALLIGATOR RIVERS

Waterbirds and wading birds of many species, including magpie geese, collect on the floodplains of the Alligator Rivers, seeking out the persistent swamps there and on the Magela and Nourlangie floodplains and Boggy Plain (a large backswamp of the South Alligator) in the dry season.[66] Peak numbers on these floodplains have been estimated as up to 400 000 wandering whistling ducks, 70 000 plumed whistling ducks, 20 000 Burdekin ducks (radjah shelducks), 50 000 black (or Pacific black) ducks, and 50 000 grey teal. Pink-eared ducks and white-eyed ducks (hardheads) were recorded sporadically in small numbers. Few waterbirds occupy these floodplains in the wet season. Large areas of wetlands are being lost to massive invasion by *Mimosa pigra*, a noxious weed, which has taken over at least 800 square kilometres.[68] Substantial areas, including several thousand hectares of the East Alligator floodplain, have been invaded by para grass, *Brachiaria mutica*. Para grass is still being promoted for pasture improvement.

Channels of the tidal rivers in the region can be subdivided into segments on the basis of their form and dynamics. A wide entrance tapering upstream, is the *estuarine funnel*. Upstream, but within the

STAGES IN THE DEVELOPMENT OF THE SOUTH ALLIGATOR RIVER AND ESTUARINE PLAINS

TRANSGRESSIVE PHASE

BIG SWAMP PHASE

SINUOUS/CUSPATE PHASES

0 km 5

Upland
Mangrove
Palaeochannel and track of river
Floodplain (including saline mudflat and backwater swamp)

After Woodroffe *et al*[64]

Pied geese and plumed whistling ducks in the Kakadu National Park.
REG MORRISON

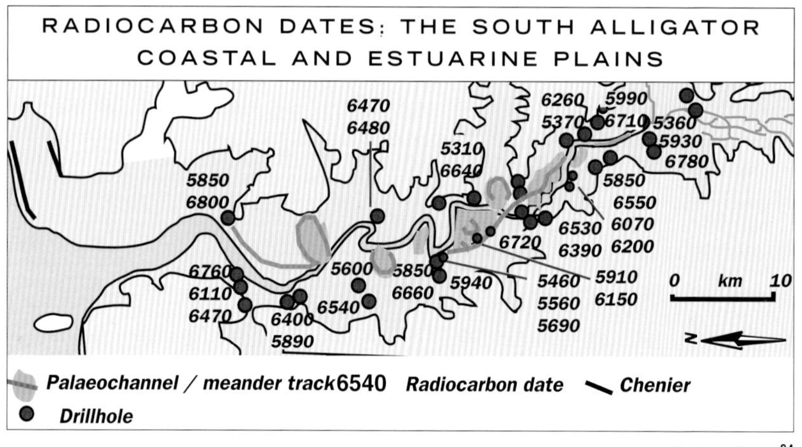

After Woodroffe *et al*[64]

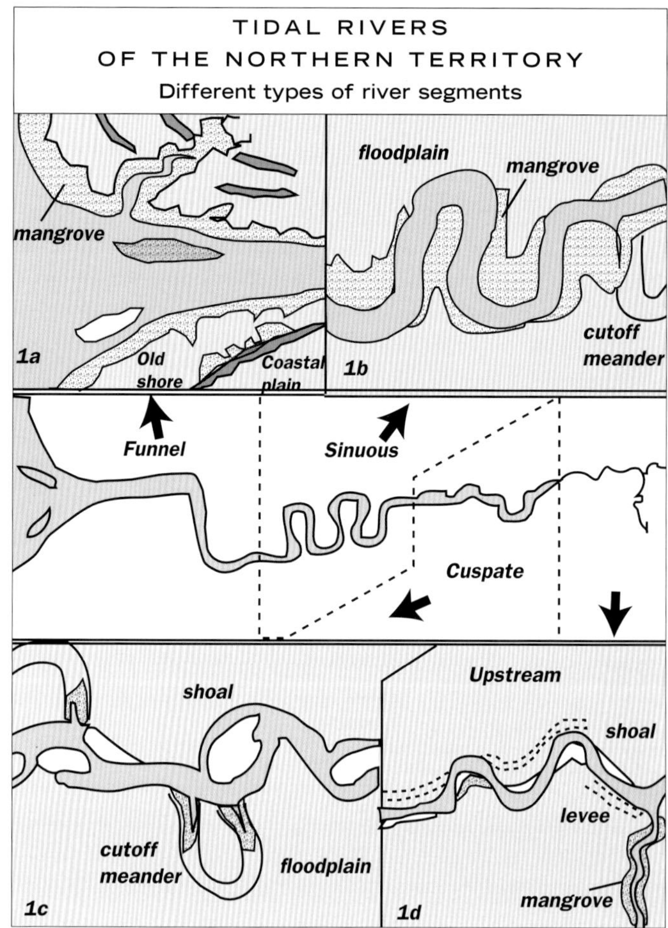

After Woodroffe *et al*[69]

tidal system, two types of meandering channel may occur: *sinuous*, resembling meanders in normal rivers, where the meanders have been inherited from prior fluvial channels which existed during Late Pleistocene times of low sea level; and *cuspate*, or estuarine, meanders generated by the interaction of two-way tidal flows and wet-season floods. Both types can occur within a single tidal river, as in the South Alligator, where the channel is sinuous between 30 and 48 kilometres of the entrance and cuspate from 48 to 70 kilometres. An *upstream* segment, characterised by long straight reaches separated by angular bends, can be recognised between the most inland meanders and the tidal limit of most of the rivers.

Detailed investigation in the **South Alligator River** has revealed that the sediments underlying these coastal and estuarine plains were deposited in Holocene times (during the last 10 000 years), through estuarine infill and the migration of the coastline. Rapid infill of the prior valleys occurred between 8000 and 6800 years ago, and was associated with development of extensive Mid-Holocene mangrove forests between 6800 and 5300 years ago as sea level stabilised. Mangroves became established throughout the estuarine plains. This *big swamp* phase is also seen in sediments in estuaries of the Fitzroy and Ord, and in King Sound in Western Australia; in the Daly, Mary and West and East Alligator Rivers; and has also been detected below the coastal plain of Princess Charlotte Bay on eastern

Cape York Peninsula, but not on the margins of the Gulf of Carpentaria.[69]

Mangrove sediment of the big swamp stage in the South Alligator is up to 8 metres thick, and pollen analysis of sediments shows that mid-tidal Rhizophoraceous mangrove communities persisted until the upper 1 to 2 metres of deposition, when they were replaced by high-tide *Avicennia*-rich communities and then the freshwater wetlands. This phase is interpreted as vertical sedimentation in the mangrove big swamp while sea level passed through the final 5 to 7 metres of its post-glacial rise. It is estimated that the mangrove area in the South Alligator alone would have been about 80 000 hectares.

The sinuous/cuspate phase of floodplain development has occurred since 5300 years ago. The river became confined into a sinuous, meandering channel, which actively migrated with development of point bars and shoals. Palaeochannels were abandoned in middle reaches by meander cut-offs; in upper tidal reaches the channel avulsed at times. Freshwater deposition of clay on the floodplains, which succeeded the mangrove big swamp, has been slow. Radiocarbon dates of Aboriginal shell middens on or slightly buried by plains sediments indicate sedimentation rates of a few centimetres per thousand years.

The coastal plain of the South Alligator River, about 4 kilometres wide east of Midnight Point, includes two discontinuous chenier ridges (former shorelines) composed of shelly sand, resting on the fine-grained mangrove sediments—evidence of prior shorelines. The present coastline has been near to its present position since about 2900 years ago, the seaward chenier ridge being built around 1600 to 2100 years ago.

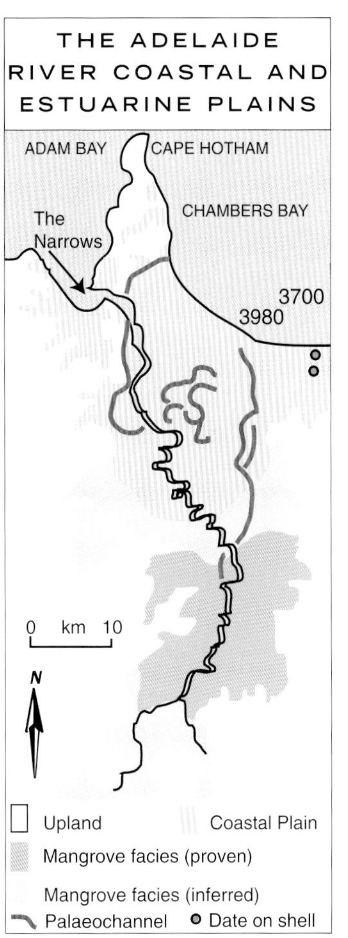

After Woodroffe *et al*[64]

THE ADELAIDE AND MARY RIVERS

The **Adelaide River** now enters to the west of Cape Hotham, but relict Holocene channels, blocked by younger coastal deposits, show that formerly it drained into Chambers Bay in Van Diemen Gulf. A narrow passage through bedrock, known as The Narrows, separates the wider lower reaches of the Adelaide from Adam Bay. Between 25 and 75 kilometres upstream, the channel follows a highly sinuous course of regular meanders, fringed on their outer banks by tall mangrove forests. Tidal effects extend to about 130 kilometres.

The same sequence of sedimentation as in the South Alligator, with a matching big swamp phase, has been identified in the Adelaide River, the only differences being that the widespread mangrove forests persisted for up to 1000 years longer in the Adelaide River than on the South Alligator plain; and the river did not meander actively after the freshwater floodplain was established. However, it flowed for a time through alternative north-eastern courses into Chambers Bay.

The **Mary River** is the most unusual of the rivers flowing towards Van Diemen Gulf, because its main channel does not reach the coast.[64] It appears from aerial photographs that before 1940 the river spread into a series of lesser channels and linear billabongs which represent palaeochannel remnants. Much of the peak wet season floodwater overtopped the river banks, flooding the plains and evaporating from the plains' surface. Over the last 40 to 50 years the situation has changed as a result of saltwater incursion into low-lying areas and rapid headward extension

The Adelaide River (spot the crocodile?). COLIN WOODROFFE

The Adelaide River snaking its way across the coastal plains. COLIN WOODROFFE

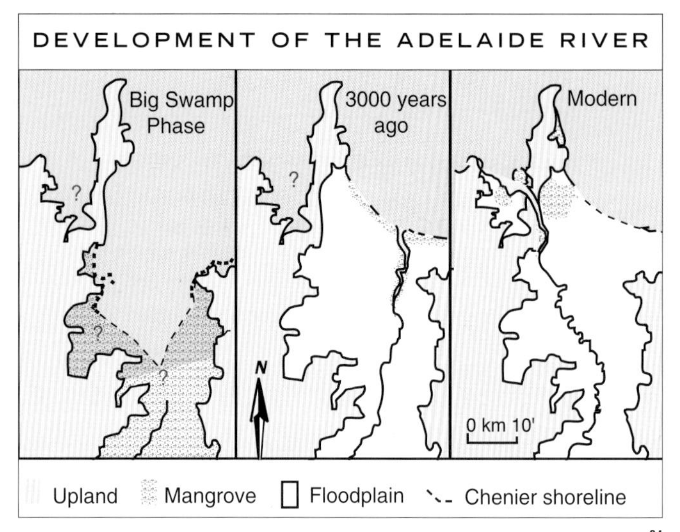

After Woodroffe *et al*[64]

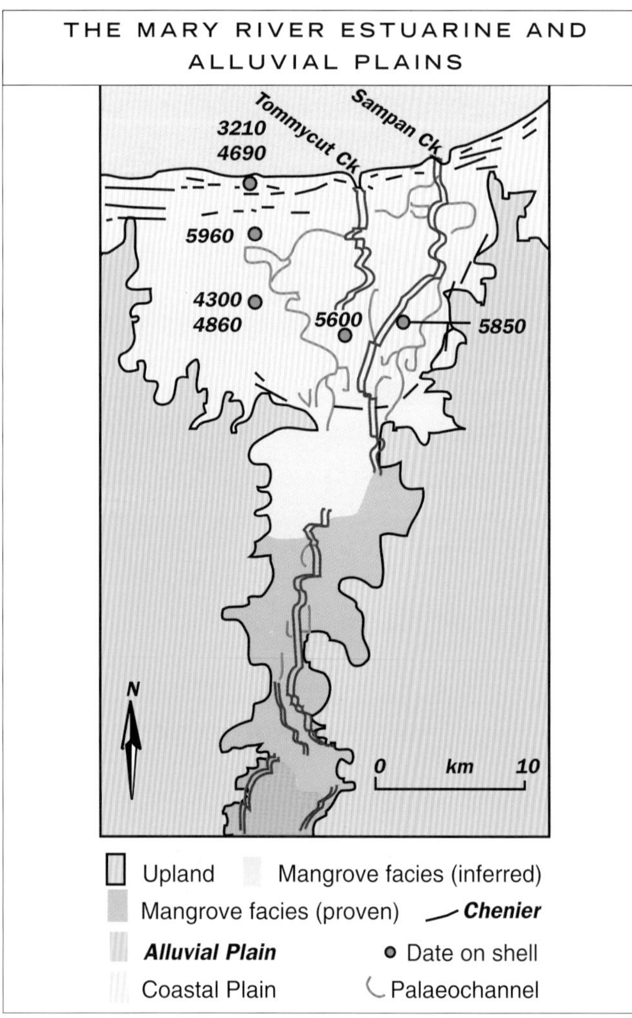

After Woodroffe *et al*[64]

of two tidal creeks, the Sampan and the Tommycut, which have expanded to accommodate larger tidal flows which currently extend 30 kilometres upstream, connecting with the freshwater Shady Camp Billabong. Over the same period the position of the coastline has also changed.

The coastal fringe of the lower Mary River is drained by many tidal creeks, most of which are limited in their inland extent by sub-parallel chenier ridges which represent former shorelines, a barrier

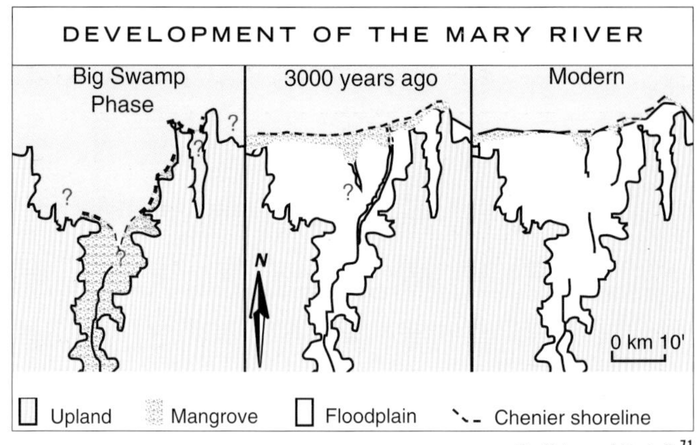

After Mulrennan & Woodroffe[71]

which only the Sampan and Tommycut have breached. They have formed rapidly expanding dendritic networks of creeks, invading freshwater wetlands and causing dieback of large areas of paperbarks (*Melaleuca* spp.). A large tidal range, flat plains and the existence of palaeochannels has greatly facilitated the expansion of the creek networks. Over 17 000 hectares of the total area of floodplain (90 000 hectares) has been destroyed by saltwater intrusion, including 6000 hectares of paperbark forest and 11 000 hectares of grazing land. Another 30 to 40 per cent is immediately threatened.[71, 72, 73, 74]

The plains include several pastoral properties and a number of conservation reserves. The area contains some of the most productive pasture in the Northern Territory, supporting cattle and buffalo on native grasses such as *Hymenachne* and introduced para grass (*Brachiara mutica*). The wetlands are teeming with wildlife, most notably large breeding populations of magpie geese, egrets and other waterbirds. Barramundi fishing spots are popular with tourists. Because of the economic importance of the region, remedial work was undertaken on the plains in 1987. A barrage has been constructed at Shady Camp Billabong, limiting the penetration of salt water into the freshwater systems upstream. Blocks have been placed on major palaeochannels and in breaches in the cheniers near the coast.

Changes to the shoreline have resulted in coastal retreat of up to 400 metres to the west of Tommycut Creek. The mangrove fringe has retracted, evidenced by a zone of dead *Avicennia* on the landward side of the zone and stumps on the mudflats on the seaward side.

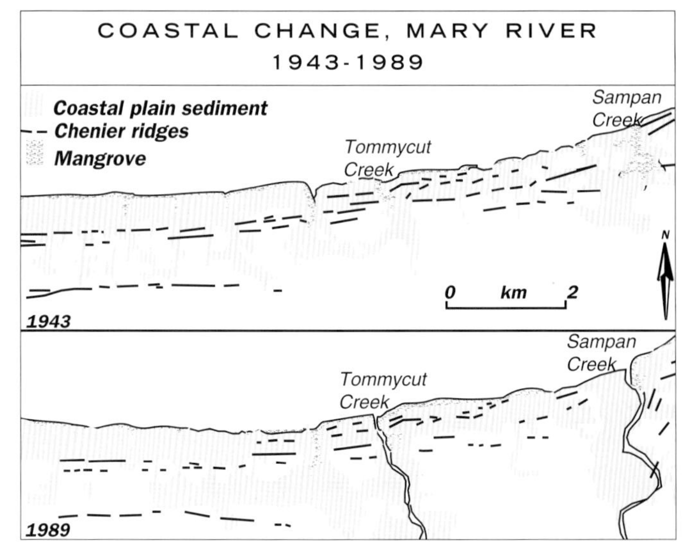

After Nanson *et al*[65]

THE IMPORTANCE OF ESTUARINE AND COASTAL RESOURCES TO ABORIGINAL PEOPLES

A shell midden on the McArthur River floodplain. COLIN WOODROFFE

Abundant shell middens along Australia's northern shoreline indicate the importance of shellfish in the diet of Aboriginal peoples in the mid and late Holocene. Eustatic sea-level changes have been a major constraint on distribution and preservation of the coastal archaeological record, and sites proliferate only after about 6000 years ago, when sea levels stabilised at about their present height. Before 6000 years ago, the sea rose from a low at the last glacial maximum, displacing the shoreline landward and progressively submerging archaeological evidence.

Impressive shell mounds are found at Milingimbi, Weipa and Princess Charlotte Bay, while smaller shell middens are scattered along the southern shore of Van Diemen Gulf, and on and near the coast in Anbarra territory in Arnhem Land. The middens on the coastal and estuarine plains of the South Alligator River fall into four classes:[70]

- *Coastal middens*, mainly of shell, on the chenier ridges, usually less than 1 metre high, comprising intertidal and shallow marine molluscs. These probably result from ephemeral wet season camps on the chenier ridges.
- *Surface mounds* 15–20 metres in diameter and less than 50 centimetres high on the plains, composed of shelly silt or clay with rock and bone fragments. Mangrove bivalve shells dominate in these mounds, and among the bones a human mandible has been found. Worked stone in these sites includes grindstones, mortars and pestles and chipped stone artefacts.
- *Palaeochannel middens*: at several locations on the estuarine plains former positions of the river channel can be identified. Several middens occur in these locations, and the characteristic shells are of bivalves which occur towards the rear of mangrove forests.
- *Surface scatters*: a few shells of landward-fringe mangrove bivalves represent the debris from one or a few meals.

PALAEOCHANNELS AND MIDDEN DISTRIBUTION IN THE BULLOCKY POINT AREA

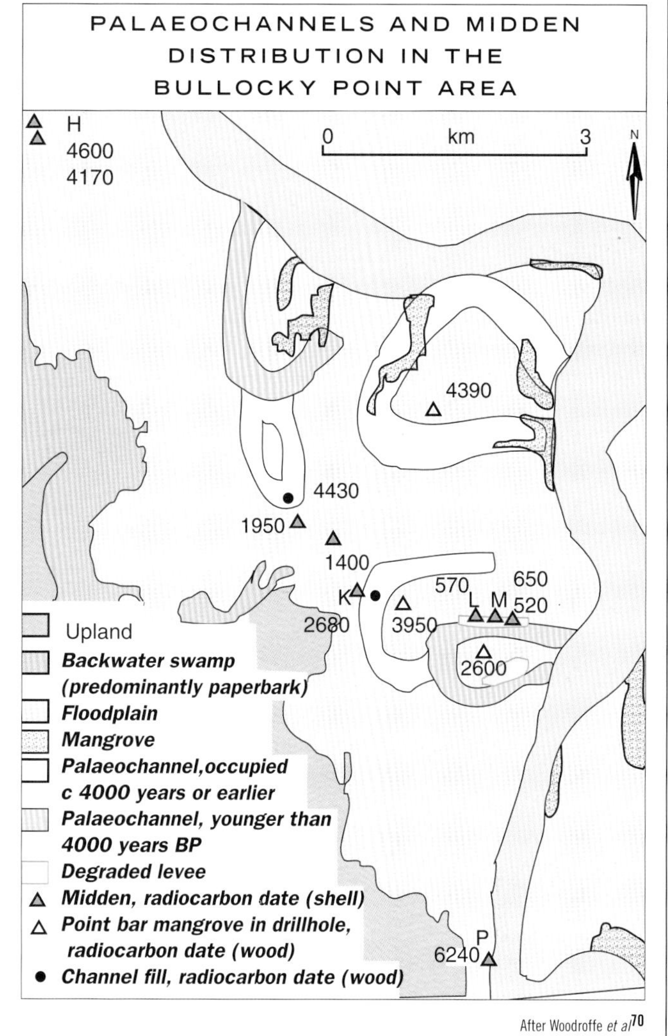

After Woodroffe *et al*[70]

Sea level relative to northern Australia was approximately 150 metres below present at the last glacial maximum, about 18 000 years ago, and the shoreline was several hundreds of kilometres to the north of its present location. From 18 000 to 6500–6000 years ago the sea rose rapidly. At 8000 years ago, sea level was approximately 12 metres below present, rising rapidly until 6800, when it was 5 metres lower than today. The rate of rise then declined until about 6000 years ago. Middens less than 6000 years old on the plains are only shallowly buried by freshwater clays and there is no evidence of sea level higher than today.

The overall area of mangroves has not changed greatly—the zone has moved to accommodate the changed patterns of erosion, accretion and saltwater intrusion.

Neither sea-level changes, which have been minimal, nor rainfall changes are thought to account for the changes in the last 40 to 50 years. Anecdotal evidence of dynamiting chenier ridges in the 1970s to make better access to barramundi fishing grounds; of artificially cutting channels like Tommycut (cut by a mythical Tommy); and of boat wakes causing bank destruction, does not account for the grave situation which threatens so much of the floodplains. However, the direct impact of feral buffalo had a considerable effect. Numbers are believed to have peaked around 1937–38, coinciding with the onset of the tidal creek network expansion, and have declined in recent years as a result of the Brucellosis and Tuberculosis Eradication Campaign (BTEC), during which buffalo numbers were decimated, but an estimated 35 buffalo per square kilometre were still present on the Mary River plains in May 1981. The modifications which have occurred on the Mary and Alligator floodplains have not occurred on the Adelaide, Finniss and Reynolds Rivers, where buffalo numbers were also high, so local conditions must have exacerbated

Saline intrusion into the Mary River plains is killing the paperbarks in originally freshwater swamps. COLIN WOODROFFE

A palaeochannel of the Mary River is a clearly visible feature of the coastal plains. COLIN WOODROFFE

An expanding creek and saltwater intrusion on the Mary River coastal plain. D. KNIGHTON

the effects in the Mary and Alligator Rivers. The more erodable natures of soils and landscapes in the region are believed to be contributing factors.

Buffalo compact the muds with their hooves, they wallow, slither on banks, break down levees and chenier ridges and create 'swim channels' upstream, in addition to their heavy grazing pressure which decreases ground cover and increases erosion. There is no doubt that feral buffalo have contributed to the changes, but overall grazing pressures on the plains as a result of pastoral practices has apparently led to compression of soils and erosion all over the plains, not only in the wet areas favoured by buffalo. Changing ecosystems by introducing para grass, which locally forms monocultures and is becoming 'feral' over large areas, enables much more continuous grazing than was the case when grazing was good during, and for a while after, the wet season, but virtually stopped when the plains dried.[68]

No single reason for saltwater intrusion can be identified but a combination of several factors has tipped the balance, reverting the system from a predominantly freshwater wetland environment to one dominated by saltwater conditions. The situation is interesting because, with globally rising sea levels as a result of the Greenhouse Effect, the extensive coastal floodplains of northern Australia are all potentially at risk. Not only are they so flat that small rises will involve big changes—they are actually slightly concave, such that the parts (often with paperbarks in them) furthest from the coast or the tidal rivers are lowest, often below the level that the high tide reaches at the coast. Even under present sea level, these areas will be inundated if the higher levees are breached.

Already it can be seen that current land-use practices are not sustainable in more vulnerable areas.

MAGELA CREEK

Magela Creek, a major tributary of the East Alligator River, is another northern river which has left a detailed sedimentary record of a fluvial landscape dominated by climatic and sea-level changes associated with the glacial–interglacial fluctuations of the Quaternary.[65] Uranium-series dating of floodplain deposits has shown that ferricrete formation is presently ongoing (as in the Gilbert River) and that the floodplain history can been traced to early Pleistocene or Tertiary times, possibly nearly 3 million years ago, when the old terraces in the lower valley were laid down.

Inset within this older alluvium is a valley fill which, by TL dating, was initiated 300 000 years ago. With each glacial climate change and associated fall in sea level, distinct palaeochannels have been eroded into these floodplains, infilling later with alluvium when climate and base-level conditions were conducive to deposition of river sediments. Channels were deeply eroded during low sea level. The most recent palaeochannel beneath Magela Creek, cut at the last glacial maximum, was a deep trench. It started to fill about 8000 years ago and fill now occupies a 30-kilometre stretch upstream from the coastal floodplain, rising to 30 metres above present sea level. It filled at an accelerating rate, probably as a result of declining stream competence associated with the drier conditions in the late Holocene, and augmented by the backwater effects of sea-level rise. The continued deposition of sediment blocked the mouths of tributary valleys along the course of Magela Creek, forming alluvial-dammed tributary lakes (billabongs) and diverting the streams to junctions further down stream. The bases of the billabongs now lie

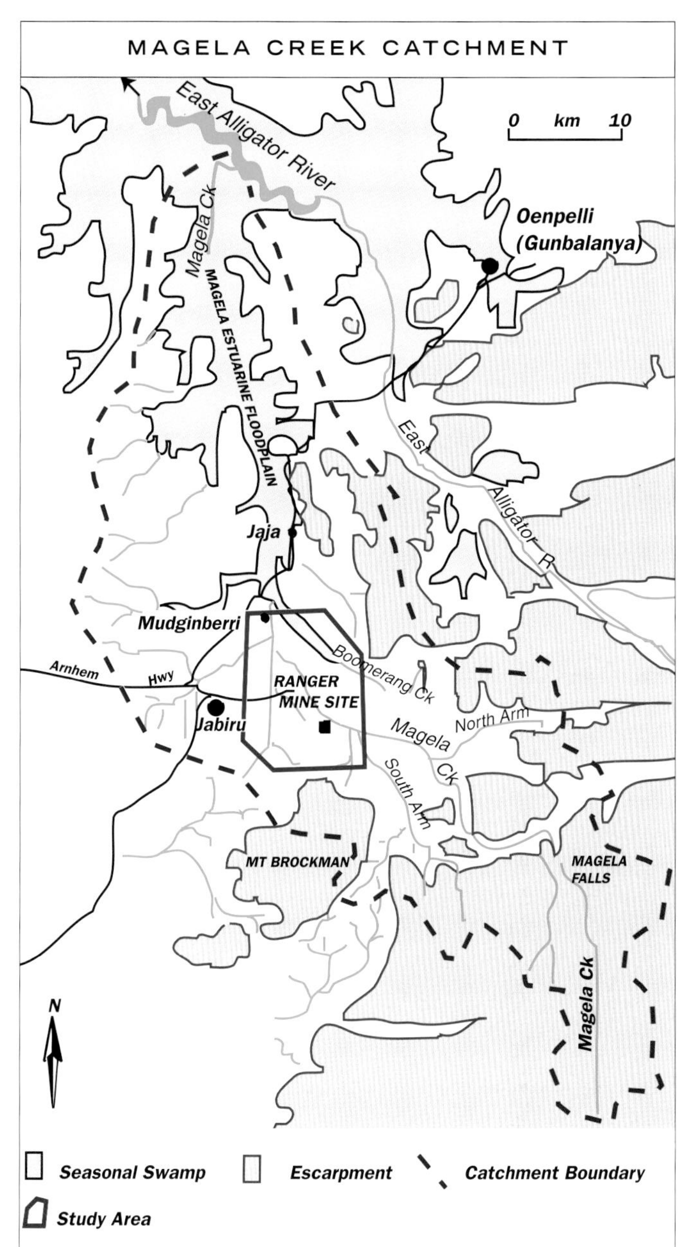

After Nanson *et al* [65]

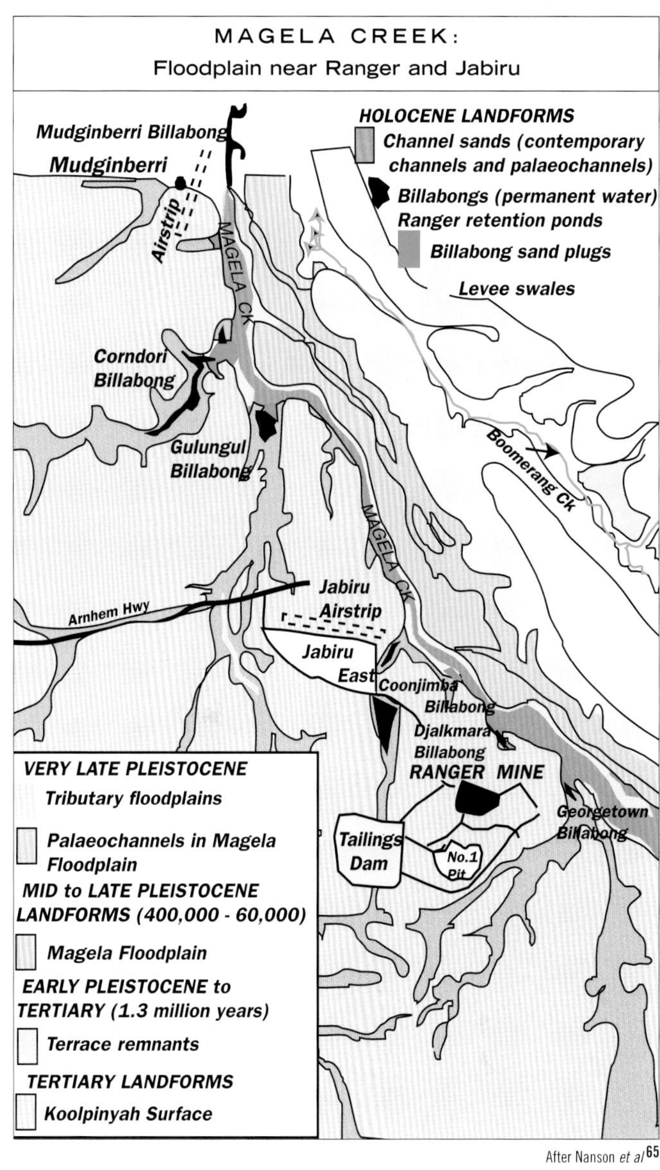

After Nanson *et al* [65]

Magela Creek, Northern Territory. COLIN WOODROFFE

Magela Creek, draining through the Arnhem Land escarpment, Kakadu National Park. The accumulating alluvium in the creek from the surrounding steep slopes is a source of evidence of the progressive drying of the Australian climate in the tropics over the last 400 000 years. GERALD NANSON

The Arnhem Land Plateau terminates in a dramatic escarpment. Magela Creek, arising on the Plateau, cuts down through the resistant sandstone and then traverses the slopes and the coastal plain on its way to the sea. GERALD NANSON

about 2 metres below the bed of the creek. Clearly, the Magela with its extensive sandstone source area has accreted much more rapidly than the tributaries which flow from the low-gradient, clay-rich old land surface below the escarpment.

Infilling of the palaeochannel underlying the present Magela Creek is now largely complete. Further alluviation will result in sand spilling onto adjacent floodplains and islands, possibly infilling the tributary billabongs.

Magela Creek flows across the Arnhem Land Plateau surface in a shallow, mostly bedrock channel, flanked by isolated pockets of alluvium. At Magela Falls the creek plunges into an 11-kilometre long gorge littered with big boulders and lined with monsoon forest. The escarpment in the Magela catchment is deeply fretted with small tributary gorges, joints, towers, tumbled sandstone blocks and caves. This highly dissected landscape supplies sand to Magela Creek from a large surface area. Immediately downstream of the Bowerbird gauge, the creek leaves the escarpment and flows in a broad shallow valley stripped of sandstone and cut into the old land surface. In these reaches it is an anastomosing-channel, sand-transporting system. In the vicinity of Ranger Mine, steep-sided canal-like channels separate islands and floodplains that have well-defined levee banks, many protected by dense mats of tree roots. Many islands support large monsoon forest trees. (See the map of Ranger Mine on page 245 of *Listen … Our Land is Crying*.)

In profile, the creek is seen to have an abrupt change in slope at Mudginberri Billabong, which separates the steeper gradient upper section from the almost flat, virtually unchannelled infilled estuaries of the coastal floodplain. A channel reforms close to the junction with the estuary of the East Alligator.

Magela Creek has been intensively studied because of concerns about its possible contamination by the Ranger uranium mine. To

Magela Creek, with an enormous load of sediment. GERALD NANSON

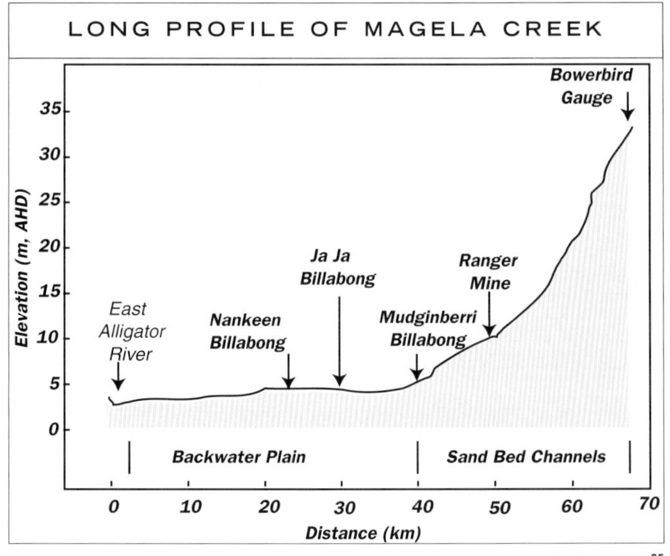

After Nanson *et al*[65]

date, activities there have led to no such contamination and in fact to a lessening of the detected radioactivity in the river, which has natural background levels as a result of the natural weathering of the radioactive rocks in the region. A full account of the mine and how it manages the water within its boundaries safely is given in *Listen ... Our Land is Crying*, on pages 244–5. In the monsoonal tropics, where in the wet season there are many occasions when the floodplain of the river is inundated and the run-off from the whole catchment is rapid, such management is both difficult and vital. All water falling on the Ranger property is kept on-site and passes through dams and purifying wetlands before entering the surrounding environment.

Magela Creek has accumulated alluvium as the climate dried progressively over the last 400 000 years. GERALD NANSON

THE FLOOD-PULSE CONCEPT

Understanding the principle that every river is a complex ecosystem and not just a channel that conveys water is basic to understanding how it functions and what it requires to remain healthy. The floodplain rivers of semi-arid and arid Australia are so far removed from 'typical' rivers that understanding their requirements assumes another dimension.[85] The periodic connections made between different parts of the river–floodplain system are integral to their overall functioning. There are three important elements to this connectivity: up and down the river channel; laterally where connection with the floodplain allows transfer of materials and resetting of ecological systems; and a vertical connection where water passes from the channel or floodplain to the groundwater, providing a sediment–water interface which is an important refuge for organisms during dry periods.

Floods are a driving variable in the dynamics of river–floodplain systems. Periodic changes in water level allow important exchanges of biota, organic material and other nutrients in the river channel, floodplain and wetlands. In some environments, flood pulses may be annual events; in others they may be infrequent and even irregular.

Variations in duration, rates of rise and fall, frequency and predictability are important hydrological drivers of river functions. The highly variable flows in the Barwon–Darling River, for example, have resulted in complex channel morphology. Prior to the regulation of this river, cross-sections of the channel were characterised by a series of flat surfaces, or 'benches', which provided important habitat heterogeneity and performed some functions ascribed to floodplains. During high flow events they enabled the accumulation and temporary storage of organic matter and other nutrients. The more variable the flow, the greater the number of benches present. With increased physical complexity of the channel, more surface area is available to store organic matter and to provide a food source, as well as to provide more diverse habitat for aquatic organisms.

Where floods are unpredictable in terms of their timing, species with flexible life cycles are likely to have a selective advantage. Sedentary organisms such as attached algae or aquatic macrophytes will be at a disadvantage in those environments which experience rapid rises and falls in water levels.

WHAT HAPPENS WHEN FLOODPLAINS ARE INUNDATED?

When floodplain soils are wet there is an immediate release of a surge of nutrients. This surge is partly a legacy of the last flood; comes partly from the biological and physical effects of wetting dry soil; and is partly due to the rapid breakdown of organic matter which has built up between floods.[86] Wetting also awakens the many different small freshwater species that have been dormant since the last flood. When the presence of water cuts off the oxygen supply to soil bacteria they switch over and function anaerobically, releasing nutrients into the water in forms available to other living things. The biological productivity of floodwaters peaks quickly and then falls away slowly until it reaches a new plateau. Water that has been standing for a time cannot support the same rapid biological growth as fresh floodwater. Floodplain species are well adapted to take advantage of the sudden surge of vitality.

Within hours of wetting, floodplains teem with tiny aquatic species, mostly microscopic, starting the food web—and opportunistic species come from afar. New waterborne species are introduced in the floodwaters, isolated pockets of water are connected with each other and with the parent river, and breeding cycles of plants and animals are switched on. Whether floodplains act as sources or as sinks for different materials has important implications for riverine ecosystems—determining everything from the health of fish to the production of cyanobacterial blooms (which result in toxic waters). Under natural conditions, rivers supply water, nutrients and sediments to floodplains; floodplains supply carbon, living organisms and water treatment to rivers—and breeding grounds for plants and animals. As the floodplains remove excess nutrients and pollutants from the water and supply food and life to the rivers, their separation from rivers results in impoverishment and degradation.

It has been shown that floodplains are biologically rich, harbouring between 100 and 1000 times more species than do the rivers which flow across them. Billabongs, intermittent lakes, backwaters, anabranches, swamps and wetlands provide habitat diversity. It has been found that a high degree of endemism exists in billabongs spaced out across a floodplain, even though they become connected during floods. Some have been likened to islands which develop their own species in isolation. Much work still needs to be done to establish details, but it looks as though the flooding which has an effect of mixing biota does not completely alter the composition of communities in billabongs, and some endemism persists. This means, of course, that destruction and loss of just one billabong which has endemic species results in extinctions. There is intriguing research ahead for freshwater biologists.

SECTION FOUR

THE GREAT ARTESIAN BASIN

The Great Artesian Basin (GAB) is one of the world's largest groundwater basins. It occupies an area of about 1.7 million square kilometres, or 22 per cent of the Australian continent, underlying the arid and semi-arid lands which comprise most of Queensland, and parts of New South Wales, South Australia and the Northern Territory. The northern end of the Basin underlies a relatively narrow region extending to the southern margin of the Gulf of Carpentaria and up Cape York Peninsula, where a tropical monsoonal climate prevails. A submarine extension of the GAB lies under the southern margin of the Gulf. (There has been very little development of the artesian waters in the Gulf plains.)

The GAB has an estimated total water storage of 8700 million megalitres, which is roughly 17 000 Sydney Harbours. The aquifers are quartzose sandstones of continental origin and Triassic, Jurassic and Cretaceous ages, and the main confining unit comprises marine sediments of Early Cretaceous age, laid down when all Australia's major continental basins were inundated by epicontinental seas. (See map on page 5.)

The Basin is up to 3000 metres deep in parts. It has been uplifted and exposed along its eastern margin, and tilted south-west.[76] Recharge occurs mainly in the eastern marginal zone, an area of relatively high rainfall, and large-scale groundwater movement is generally towards the south-western, southern and western margins. The narrow northern sector drains to the north. Water travels through the permeable sandstones, in pores between the sand grains which are incompletely cemented together to form the sandstone, at between 1 and 5 metres a year.

Recharge also occurs in the western margin of the Basin, with groundwater flow direction towards the south-western discharge margin. Natural discharge occurs in those areas from flowing artesian springs, most of which have built up mound-shaped deposits of sediments and carbonates. Some discharge also occurs by diffuse seepage where the overlying confining beds are thin. Many springs are associated with structural features like faults, abutment of aquifers against bedrock, or thin confining beds near the discharge margins. Because of the very slow movement of groundwater, much of the water emerging in springs near Lake Eyre is up to 2 million years old.

Prior to development, the Basin was in a natural steady-state condition with an equilibrium between recharge and natural discharge from springs and vertical leakage. The potentiometric

An artesian bore near Lake Callabonna in the Strzelecki Desert pours precious water into a bore drain which runs away through the sand—evaporating, going nowhere, growing nothing. EG MORRISON

Aboriginal artefacts gathered from round mound-springs south-west of Lake Eyre in the Strzelecki Desert, S.A. Aboriginal peoples depended on the permanent waters of mound-springs for their survival in the deserts. REG MORRISON

This Mound Spring National Park landscape captures the feeling and fascination of the flat, salt-encrusted land and the amazingly wide sky of the dead heart of Australia. M.E.W.

Hamilton Hill, an extinct mound-spring which stands like a mesa in the landscape, across the salt-encrusted plains of the Mound Springs National Park, S.A., where artesian waters flow naturally to the surface. M.E.W.

Pigeon Hill, where the old Ghan Railway embankment causes a back-up of water which comes to the surface where the wide plains abut against the ancient hills. Sulphur makes yellow patches in the salt. M.E.W.

Spring-fed pools at Dalhousie Springs in the Strzelecki Desert are real oases in the desert. REG MORRISON

Sedge swamp and salt in the run-off from the Bubbler Mound-spring. M.E.W.

surfaces (the levels to which water can rise under its own pressure) of the confined aquifers in the Jurassic to Early Cretaceous sequence were above ground-surface over the whole Basin before exploitation began in the 1880s, which meant that water emerged under pressure whenever bores intersected aquifers.

Most of the GAB is characterised by extremely variable rainfall. Periods of drought, often prolonged, are punctuated by pulses of high-intensity rainfall which may generate plentiful herbage, which may in turn lead to wildfires. The natural biota had adjusted through time to these uncertainties orchestrated by ENSO, and possibly also to changes brought by human occupation of the country during the last 60 000 years. Many native animals do not need access to reliable

The Bubbler, in the Mound Spring National Park. M.E.W.

drinking water, or depend on the availability of refuge areas like the mound-springs, permanent wetlands and waterholes during dry times. The native plants and animals of the arid and semi-arid GAB (and elsewhere in this driest of all continents) have evolved to cope with extended drought and uncertain rainfall.

At the time of European settlement, mound-springs provided the only permanent source of water in isolated areas throughout the rangelands. Aboriginal peoples and early explorers knew their value, moving from one to another in their journeys across the country, but mostly the springs were too scattered and restricted to be of any real pastoral value. Ephemeral rivers characterise the arid and semi-arid rangelands, and the natural waterholes on creeks and rivers were too unreliable for stock, particularly in droughts. So it was the discovery of the GAB aquifers that fed the mound-springs and supplied the widespread artesian water below the rangelands that really opened them up to grazing. Without the artesian water, most of the rangelands of the GAB could not support a livestock industry.

Today, more than 200 000 people live and work within the GAB, dependent on artesian water to a great extent. Apart from the pastoralists, about 60 towns in outback Queensland, New South Wales and South Australia are almost completely dependent on artesian water; mining and development of oil fields, and agriculture, also depend on the underground water resource. Sixty per cent of the GAB is used for grazing; 15 per cent for Aboriginal homeland; 4 per cent for conservation reserves; 21 per cent is technically unoccupied but most of this portion is currently under claim by Aboriginal people.

HISTORY OF DEVELOPMENT OF THE GREAT ARTESIAN BASIN

In 1864, a well was dug 30 metres into the 'Wee Wattah' mound-spring on Kallara Station in western New South Wales. It was replaced by a bore in 1878, the first artesian bore into the GAB.[77] The bore located near the well was initially 10.6 metres deep but was later deepened to 53.3 metres and flowed at 7.1 litres a second (about 600 000 litres or 0.6 megalitres a day). It stopped flowing in 1939 and became sub-artesian because of the general drop in pressure all over the Basin which has resulted from excessive water extractions.

In 1885, a Queensland Government geologist decided to drill a deep bore at Blackall to see if artesian water could be located there. Before that bore was completed, private bores struck artesian flows at Back Creek, east of Barcaldine, in 1886, and at the Ungamilla Bore on Thurulgoona Station near Cunnamulla in 1887. By 1908 the whole Thurulgoona Station was fully watered with bore drains, earth tanks and natural waters, and all over the GAB exploitation of the

At the turn of the century, new bores were a source of civic pride. These ladies in their Sunday best were enjoying an outing at the Noorama Bore, south-east of Cunnamulla. USED WITH THE PERMISSION OF THE DNR QLD. (ARCHIVES BRANCH)

artesian waters was taking place. By 1915, bores in the Basin were producing 2050 megalitres a day, and declining flows in some of the older bores were warning that all was not well even then.

Archival photographs in Queensland's Department of Natural Resources record the celebrations that accompanied the successful drilling of bores at the turn of the century. Water was seen as the only thing missing to make the region fertile and productive. (It is ironic that it is now evident that it has been the availability of enormous quantities of water which has been the desertifying agent in the grazing lands of the GAB.)

No thought was given to the future, and most of the water ran away from the bores in open drains across the landscape, evaporating as it went. In 1994, there were 28 000 kilometres of open bore drains in Queensland and 6000 kilometres in New South Wales.[78]

The Murweh Bore in the Warrego District, between Wyandra and Charleville, drilled in 1899. This bore still functions, having been rehabilitated, but still runs to waste in a bore drain. USED WITH THE PERMISSION OF THE DNR QLD. (ARCHIVES BRANCH)

PROBLEMS WITH EARTH TANKS

Earth tanks provide supplementary waters to bore drains throughout the country that can be watered from artesian sources. Some are constructed adjacent to bore drains and filled from high winter flows to provide water during summer when the drains either stop flowing, or their water is too mineralised for stock use. Others are constructed on minor flow lines well removed from bore drains. Earth tanks have problems with evaporation—water quality deteriorates in summer; siltation is rapid in borewater-fed tanks; the tanks contain high levels of nutrients—stock foul the water and may become bogged; disease can be spread by water from earth tanks.

PROBLEMS WITH BORE DRAINS

According to Ross Blick, who has made a detailed study of artesian water use in Queensland,[81A] the major problems associated with bore drains are:

- **Water loss** Less than 2 per cent of water in bore drains is used by stock or wildlife; 98 per cent is lost to evaporation and seepage.
- **Pressure loss** Virtually all bores have lost pressure since the first tapping of the Basin, and many have become sub-artesian (needing pumping to the surface) over time. Many mound-springs have dried up as a consequence of the pressure loss, and the extinction of some of their endemic species is a concern.
- **Salting** All artesian waters contain salts of different types, and these have accumulated in and on the soil.
- **Summer dryback** Water flows the full length of drains on cool winter nights but dries back on hot summer days, so earth dams are constructed to provide continuity of supply, and bring their own problems.
- **Water quality** Increased concentrations of salt and algae occur progressively along the drain's length and can reach levels poisonous to stock.
- **High maintenance costs** are involved in keeping the drain flowing, and in repairing damage by pigs and cattle which can allow water to flood out locally, causing scalding.
- **Uncontrolled access to water** It is virtually impossible to control access by feral grazers as well as stock and kangaroos, so heavy, widespread grazing pressures result.

A bore drain runs through the landscape, evaporating. ROSS BLICK

- **Limited vehicle access**, restricting movement to defined crossings.
- **Impeded crossflows of storm run-off**, depriving pastures of natural watering by rainfall.

Bore drains have also led to the spread of noxious woody weeds such as the prickly acacia, which has formed dense thickets along some drains. Stock spread the seed from one infested area to another, eating the pods, voiding seeds wherever they come to drink along the bore drains.

THE GREAT ARTESIAN BASIN BORE REHABILITATION SCHEME

With declining pressures, one-third of the bores in the GAB had become sub-artesian by the 1980s, and in 1989 a scheme was instituted by government to help with bore capping and rehabilitation. The program is jointly funded by Federal and State governments. Financial assistance was provided and those farmers who wished to cooperate had to contribute 20 per cent of the costs. The scheme did not then extend to piping of the bore drains. Times were hard and few were able to take part in the scheme, and most of the available government funds were not spent.

However, the failure of bores on the eastern margin of the GAB demanded immediate attention. The Milroy Bore, 60 kilometres north of Moree in New South Wales, was one of the affected bores; a cost–benefit analysis showed that capping the bore and piping the bore drains would be economically beneficial—so the project was undertaken and completed in 1991, and showed the anticipated economic advantages and a huge saving in water as well as improved pressures. Control of feral grazers became easier so grazing pressures could be controlled. At the same time the Kayanna Bore Trust in Queensland investigated and implemented the largest bore-piping scheme completed to date, delivering piped artesian water through a 236-kilometre network of polypipes to 35 properties and servicing an area of nearly 68 000 hectares.

By 1997, eight bores had been capped and piped in south-west Queensland to replace 338 kilometres of bore drains, with 535 kilometres of pipeline to 263 water points and 10 homesteads. A further four bores were piped in the 1996–97 financial year. The advantages of being able to turn off water when it is not needed for use at a specific water point, and to use the rainfall and conserve artesian water during favourable seasons, are obvious.

By 1997, about 256 of the 900 uncontrolled bores in Queensland had been rehabilitated under the program, with a saving of water of 30 000 megalitres annually. A change to piped distribution throughout the GAB could eliminate all waste. Much of the rehabilitation to date has been uncoordinated; an overall strategic plan involving the rehabilitating, capping and piping of strategic bores, shutting down the rest and providing water to regions around the strategic bores by a reticulation system of pipes and water points would be the most economic and profitable from all points of view.

ADDED WATER *What the availability of added water in a naturally arid or semi-arid region has done to the natural environment*

Prior to settlement, most of the GAB rangelands were effectively waterless, except in good seasons, and therefore protected from the grazing pressures of large native herbivores. The provision of water, coupled with high stocking rates of introduced sheep, cattle and horses up to the droughts of 1895 to 1902, brought a fundamental change to the country.

Charles Sturt observed during drought in 1828 near Mt Oxley, in the Bourke district of New South Wales, 'I never saw anything like the luxuriance of the grass … waving as it did higher than our horses' middles as we rode through it.' Contrast that mental picture with the scalds round Mt Oxley which characterise the region today.

When the first settlers came to the semi-arid rangelands they thought they had found limitless feed for their animals and that the only restriction was the shortage of water. They set about to augment the few permanent waterholes with dams and wells and they steadily increased stock numbers, not realising that the natural vegetation with its high biomass was not a sign of high productivity. High stocking rates simply devastated the vegetation, leading to erosion and degradation of the ecosystems. The 1901 Royal Commission into the plight of farmers in the Western Division of New South Wales recorded the degradation which had occurred in that region, and the inability of farmers to make a living in consequence. Woody weed invasion of pastures and predation by dingoes received special mention.

In the well-documented case of Thurulgoona Station in Queensland, the 'achievement' of the watering of the whole property did nothing to slow vegetation changes and decline in carrying capacity—the grassy plains lost their good pasture species; mulga and native pines died and were replaced by brigalow; saltbush and bluebush replaced palatable grasses; and by 1915 the land was 'much eaten out'.[77] The wider use of the whole property for grazing, made possible by the provision of artesian water, had caused the first real losses in the vegetation.

A scientific investigation of the implications for biodiversity of water management in the Great Artesian Basin has been conducted by the CSIRO.[57, 82] This is an important study for it shows how Australia must manage its arid regions (more than 70 per cent of the

A scald on Eucumbene Station, Bollon, where a bore drain had flooded regularly, 30 years ago, killing trees. (The drain has since gone.) ROSS BLICK

The impact of stock concentrations on degraded country surrounding watering points is clearly seen. JIM GASTEEN

Deep erosion along a cattle pad. JIM GASTEEN

continent) if native biodiversity is to be conserved. While we all know that water is necessary for life, it does not follow, in arid lands at least, that the artificial provision of abundant water is beneficial. Nowhere has this been more clearly demonstrated than in the arid and semi-arid grazing lands of the Great Artesian Basin.

The provision of artificial water sources has allowed more widespread grazing (and over-grazing) by livestock and by larger native and feral herbivores, posing threats to native plants and animals that do not use the water. Grazing-sensitive species are now confined to tiny patches in the landscape. Known as *decreaser* species, these grazing-sensitive plants disappear from areas regularly grazed, while *increasers*, often unpalatable but resilient, proliferate, benefiting from decreased competition. Some of the nature reserves declared within the GAB in order to preserve biodiversity (wisely selected to include examples of all the different bioregions) retain many artificial water points. Many of these should be closed over time to reduce negative impacts on grazing-sensitive plants and animals, especially those species which are inadequately protected elsewhere. In those regions where the ratio of artificial to natural waters is still low, consideration should be given to balancing provision of water for livestock with conservation of biological diversity by maintaining a patchwork of areas remote from water. In regions where the density of artificial waters is high, conservation of biodiversity on freehold and leasehold lands might be enhanced with a mixture of approaches accommodating the needs of the biota and the aspirations of landholders, tailored according to land type and condition. This requirement for dryland refuges to maintain biodiversity has been described by CSIRO scientists as providing 'ungrazed "islands" in a "sea" of herbivory'.[57]

A study was conducted recently to map the distribution of artificial water points and to determine the position and extent of areas lying beyond the influence of water-dependent grazing animals in Australia's rangelands.[79, 80, 82] The results showed that water points are now so widespread in the more productive rangelands that reference areas for determining the pre-grazing spatial patterns of biodiversity are now extremely rare. Most large mammals require regular access to drinking water and in arid environments its availability determines where they graze. The water sources become foci for grazing activity, resulting in a zone of attenuated impact (a 'piosphere') around each water point. The main grazing impact probably occurs within 5 kilometres of water for sheep and 10 kilometres for cattle, although sheep can walk 10 kilometres to preferred plant communities and cattle 20

BIOREGIONS OF THE GREAT ARTESIAN BASIN

CYP
GUP
MII
MGD
DEU
SSD
CHC
ML
BBS
FIN
STP
GVD
DRP
FOR
BHC
0 km 400
N

CHC - Channel Country; ML - Mulga Lands; SSD - Simpson-Strzelecki Dunefields; GUP - Gulf Plains; MGD - Mitchell Grass Downs; BBS - South Brigalow; STP - Stony Plains; DRP - Darling Riverine Plains; CYP - Cape York Peninsula; DEU - Desert Uplands; GVD - Great Victoria Desert; Fin - Finke; BHC - Broken Hill Complex; FOR - Flinders & Olary Ranges; EIU - Einasleigh Uplands; MII- Mount Isa Inlier

After Noble *et al*[57]

THE GREAT ARTESIAN BASIN

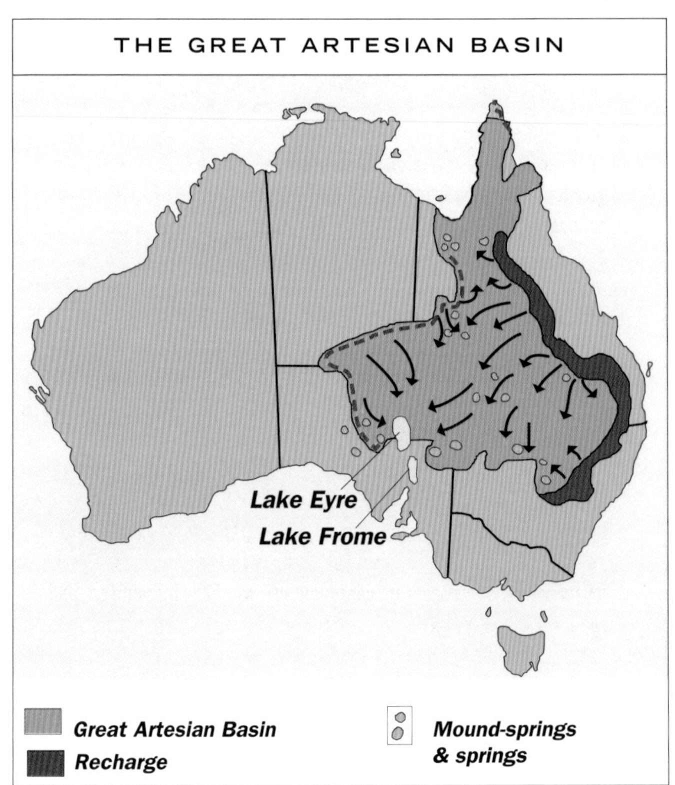

IT'S TIME FOR A VISION FOR THE GREAT ARTESIAN BASIN!

The Great Artesian Basin Environment Advocacy Group (environment representatives of the Great Artesian Basin Consultative Council and advisory committees) envisages a future where the GAB is managed in a holistic way. In this vision, water pressure in the Basin has been significantly restored (an achievable objective), and as a result a significant number of mound-springs has been restored. The vision has two concepts—the protection of the land, water and associated natural biota; and ongoing human occupation of the arid and semi-arid lands of the GAB.[81A]

To conserve the region's biodiversity, a system of Representative Reserves is established, comprising 15 per cent of every bioregion in order to conserve the full range of native species in the GAB landscape.

PROBLEMS WITH BORE DRAINS

IN THE RESERVES

- All artificial water has been removed, restoring the ephemeral status of surface water. All bores have been capped.
- Natural billabongs and waterholes have been de-silted, feral fish removed.
- Feral animals and weeds have been controlled.

ON GRAZING LANDS, PROPERTIES ARE MANAGED TO RESTORE ARIDITY

- All free-flowing bores have been capped or shut down. Artesian water from reconditioned bores has been piped to strategically placed troughs; bore drains are no longer used.
- Properties adjacent to reserves have no artificial water within 5 kilometres (sheep grazing) or 10 kilometres (cattle) of the reserve boundary.
- Earth tanks and dams have been replaced by piped water, and the number of water points increased so that graziers can move stock and rotate pasture. Water points not in use are turned off, and the amount of water saved has resulted in the re-pressurising of the artesian aquifers.
- Self-mustering traps have been installed at all water points. The properties are managed to enable the re-establishment of palatable grasses and forbs, using rotational grazing and spelling pastures for two or more years. Total grazing pressure has been reduced by more effective control of feral animals and native herbivores. In highly significant *aquifer recharge areas*, land use has changed from grazing to other uses which do not damage the infiltration qualities of the soils. (The Desert Upland Bioregion where a great deal of the recharge occurs has been subjected to the greatest amount of land-clearing over the last few decades. There is considerable concern that the infiltration of water into the aquifers may be impeded as soil structure declines and erosion and other problems follow. Immediately following clearing, measurements by soil experts suggest that infiltration rates increased locally. However, this bioregion is fragile, marginal country with light, sandy soils. That it is being cleared at all is irresponsible. After a few years its nutritional levels will be so low that its continued use will be compromised, and its soils will be compacted and no longer as receptive to the rainfall needed to keep up recharge levels to the GAB.)
- **Mound-springs** that had dried up have been restored with the re-pressurising of the aquifers that feed them. A national data base on their biota has been established and their unique biology understood and protected. On grazed land, mound-springs have been fenced off, their water piped to troughs.

HUMAN OCCUPATION OF THE ARID AND SEMI-ARID LANDS OF THE GAB

Human occupation of these areas involves a range of uses—towns, mining, tourism, grazing.

- **Towns** continue to use water for domestic purposes. Where possible, surface water storage has been removed to allow surface flow from rain to spread across the land, spreading the sediments which were trapped in the dams. Adjacent streams have been rehabilitated. New industries have been established throughout the GAB, increasing employment, while the infrastructure renewal programs and the biological and agricultural research have also provided jobs. Energy has been produced from the pressures of the artesian waters.
- **Mining** has continued within the GAB—the highest value use of artesian waters. Regulations provide long-term protection of the landscape and the water, and protocols have been established for the rehabilitation of the mined areas when mining is finished. Given the high value from mining, joint venture projects to improve knowledge about the hydrogeology and geochemistry of aquifers are undertaken by government and mining companies.
- **Grazing** in the GAB is opportunistic, relying on weather cycles to determine stocking rates—recognising the concept of great climatic variability determined by the Southern Oscillation–El Niño phenomenon (ENSO). Productivity of the land has improved through the rationalisation of artificial waters (with a resulting reduction in weeds and feral animals), restoration of ephemeral water in the landscape and rotation of paddocks. Scalded country has been restored. Graziers can now control total grazing pressure, adjusting stocking rates according to availability of feed. After rainfall, there are times when groundwater is not needed. Kangaroos are harvested for meat production and value-added industries are established. Fire is used as a tool to re-establish grassland and control weeds.
- **Tourism** brings prosperity to a region which offers unique landscapes: Aboriginal history—with Aboriginal rangers informing visitors about land and water, bush foods and medicines; natural spas have been developed and yabby farms established in their tailwater dams.
- **Infrastructure renewal** involving a strategic approach to capping, piping and rehabilitating bores has restored pressure to more than 50 per cent of the original. In the four regions of Queensland (Goondiwindi, South West, Longreach and Julia Creek) the program focussed on the most wasteful bore or group of bores. These were capped and piping laid to minimise the number of bores required to supply water to each region. Funding to achieve this strategic approach has been continuous, as State and Federal governments recognised the benefits of improved productivity of the land. The infrastructure renewal program has extended to areas where dams and earth tanks were decommissioned. Regionalised planning of property water points has given improved landscape ecology and function, and improved profitability. Previous programs, which did not address this holistic approach to GAB management, were defunded.
- **GAB management** An interstate organisational structure, the Great Artesian Basin Commission, has been established, and is responsible for

managing the waters of the GAB and its associated impacts on the land which overlies the water. The Commission has worked with State governments to ensure that a comprehensive, adequate and representative reserve system is in place to preserve native species of the arid and semi-arid landscapes and that infrastructure renewal programmes are linked with an overall land management strategy to ensure the protection of the arid and semi-arid ecosystems.

CONCLUSION

The water and land of the GAB is recognised by all Australians as a national treasure and is managed to ensure the protection of its natural processes and native species. Humans living and working in the GAB recognise the nature and cycles of the area and have developed an economic system aimed at the long-term occupation of the land which is respected throughout the country and the world.

IS THE VISION AN UNATTAINABLE DREAM?

NO! It is attainable, based on good science, and requires only our joint will and determination to achieve it.

kilometres. The main feral grazers of northern Australia, the horses, donkeys and camels, can also range as far as cattle. Less is known about the distances covered by kangaroos and goats, which are responsible for more than half of the total grazing pressure in some southern rangeland regions.

This study of piospheres showed that there is little of the rangelands in the GAB that is untouched by grazing because of the proliferation of artesian water points—a disturbing situation for conservation of biodiversity. In the Cunnamulla region, for instance, less than 3 per cent of the 1:250 000 map sheet area is now more than 5 kilometres from a water point. However, more than 30 per cent is further than 5 kilometres from any natural source of water, and 90 per cent is further than 5 kilometres from any permanent natural water. This shows just how much change has occurred. Few areas of the GAB are now further than 10 kilometres from water—a major transformation of a vast area that was once essentially waterless. As Ross Blick said to me: 'By providing so much water we have made the landscape more arid.' I would have gone further and said that in many cases the provision of so much water had *desertified* the landscape.

POSTSCRIPT

Arising out of the study described above—on the eve of sending the completed manuscript to the publisher—comes a project which lifts the spirits and gives hope for the future—the BIOGRAZE: Waterpoints and Wildlife Project has just come into being. It is a joint project of the CSIRO's Centre for Arid Zone Research at Alice Springs; the Biodiversity Branch of the South Australian Department of Environment, Heritage and Aboriginal Affairs; and the Parks and Wildlife Commission of the Northern Territory. It also involves land managers in the regions where research is to be centred—the Barkly Tableland in the Northern Territory and the Kingoonya region of South Australia.

The aim of the project is to discover the response of plants to different grazing pressures and establish general principles for managing pastoral land without affecting the species composition of natural vegetation in the rangelands. The information already available on decreaser species highlights the need to find ways of maintaining a variety of habitats in a variety of grazing states. The project will examine what would be the costs involved to landowners if 10 per cent of the land area were to be set aside for conservation of decreaser species. (It is quite likely that the better overall management of the land involved would lead to higher productivity from smaller area, and the land, the biota and landowners would all be winners.)

FOOD WEBS IN STREAMS AND RIVERS

Carbon is the essential building block of all plant and animal tissue. Understanding the sources and fate of organic carbon in river ecosystems is an essential requirement for the ecosystems' sustainable management and health. Many aspects of stream and river degradation which result from changes in land use are a direct result of fluxes in organic carbon. Increases in nutrients lead to imbalances and proliferation of unwanted species; decreases in nutrient as a result of catchment clearing impoverish the food chains; changes to light regimes through removal of riverbank shading affects the biota.[48]

Food webs in *forested* streams and rivers are strongly dependent on inputs of nutrients and energy from the surrounding catchment. Leaf litter, and other organic matter from insects and other animals, enters the river directly from riparian vegetation or is blown or washed in; dissolved organic matter leaches from decomposing vegetation on the forest floor. All this terrestrial carbon provides a major source of energy and nutrients. Some small forms of algae in the stream can be food sources for some aquatic creatures, but the dense riparian canopy restricts the production of aquatic plant life by providing too much shade.

Much of the research on food webs in rivers has been conducted in Europe and North America on forested streams. Little research has been done on any Australian rivers, although this deficit is gradually being addressed. It appears that findings for our forested headwater streams have shown much the same patterns of organic carbon and energy flow as Northern Hemisphere rivers, but so few of our rivers fall into this category that they are the exception.

In the wet–dry tropics and the arid and semi-arid interior where streams with a relatively open riparian canopy drain sparsely vegetated catchments (and many of these rivers are ephemeral), the situation is completely different. Some experts favour the 'river continuum' concept, which emphasises the importance of terrestrial carbon and nutrients which enter the network of tributary forest (or at least, upland wooded) streams and eventually 'leak' downstream. Filter-feeding and browsing aquatic invertebrates are an important functional group in downstream reaches, capturing the organic matter and forming the basis of the food chains. Direct inputs of leaf litter from adjacent riparian vegetation are considered to be minor contributors to organic carbon in the wider, more open channels. Middle-order reaches, where shading is diminished, can show an increased dependence on in-stream algal production. (See page 69 for information on algae in very turbid Channel Country rivers.) Rivers according to this 'continuum' concept are 'receiving' rivers, dependent on upstream sources of carbon and nutrients, and to protect their health the first priority would be protection of their headwaters.

More recent research has emphasised the importance **of organic matter derived from floodplain sources [83, 84] rather than from upstream processes**, and the 'flood-pulse' concept has been developed; this applies, in particular, to our arid-land rivers. River—floodplain interactions are vital in this model, and reductions in the magnitude and frequency of flood flows by various control mechanisms will effectively isolate the river from the floodplains which feed it and have a profound effect on river and floodplain health.

The truth of what is needed for river health probably lies in accepting that to be healthy and function sustainably a river needs input from its headwaters, its lower reaches and their in-stream and riparian ecosystems, and its floodplain and associated wetlands.

EXPLORERS AND THE RIVERS OF EASTERN AUSTRALIA

In 1813 the Blue Mountains were penetrated by Lawson, Blaxland and Wentworth, and some months later the Deputy Surveyor General, George Evans, crossed to the Bathurst Plains and there discovered the first of the rivers which rise on the inland side of the Great Divide. He named it the Macquarie, after the governor of the day. In 1815, on a second expedition, Evans discovered a second river flowing west—the Lachlan, again named for the governor at that time. The idea of an inland sea, an 'Australian Mediterranean' as Sturt later named it, was central to the thinking of those times and was to lead many explorers westward on incredible, and sometimes fatal, journeys to find it.

The Surveyor General, Oxley, in 1817 and 1818, led expeditions to trace the two rivers, believing that they would lead him to the inland sea. Both rivers ended in impenetrable marshes, which Oxley believed bordered that sea; that the Macquarie reformed below the marshes and ran on to flow into the not-yet-discovered Darling, and the Lachlan reformed below the Cumbung Swamp to join the Murrumbidgee, was not known.

The mystery of where the rivers went became more intriguing when in 1824 Hamilton Hume (walking from Sydney to Port Phillip on Bass Strait with William Hovell) crossed the west-flowing Murrumbidgee. Continuing south to near present-day Albury, the explorers 'arrive suddenly on the banks of a fine river'. They had discovered the upper reaches of the Murray—the Hume, flowing westward again, no one knew where—recognising it as a major snow-fed river.

The botanist Allan Cunningham, who had accompanied Oxley on the expeditions to the Lachlan and the Macquarie, was also a believer in the inland sea. When in 1827 he led an expedition northward from Sydney in which he discovered the Darling Downs, his belief was strengthened. He discovered four more westward-flowing rivers—the Namoi, Gwydir, Macintyre and Dumaresq.

All these explorations set the scene for Charles Napier Sturt, who arrived in Sydney, became private secretary to the Governor, Sir Ralph Darling, and undertook his first exploration into the interior in 1828. He was charged by Darling to continue the exploration of the Macquarie where Oxley had left off, and to find out where the westward-flowing rivers went. This expedition coincided with a severe two-year drought. The marshes which had stopped Oxley were largely dry and could be traversed, but mosquitos, mud, leeches and kangaroo flies made the journey a nightmare. Sturt was sure that the marshes were not a barrier between the Macquarie and the inland sea as Oxley had thought, but were an 'ultimate evaporation and absorption of its waters, instead of their contributing to the permanence of an inland sea, as Mr Oxley had supposed'. Proceeding downstream from the Marshes:

> *As the path we had observed was heading northerly, we took that course and had not proceeded more than a mile upon it, when we suddenly found ourselves on the bank of a noble river ... The party drew up upon a bank that was from forty to forty-five feet from the level of the stream. The channel of the river was from seventy to eighty yards broad, and enclosed an unbroken sheet of water, evidently very deep, and literally covered with pelicans and other wild fowl. Our surprise and delight may better be imagined than described... Coming from the N.E. and flowing to the S.W., it had a capacity of channel that proved that we were as far from its source as from its termination... Its banks were too precipitous to allow of watering the cattle, but the men eagerly descended to quench their thirst ... nor shall I ever forget the cry of amazement that followed their doing so, or the looks of terror and disappointment with which they called to inform the water was so salty as to be unfit to drink.*

Sturt's party had discovered the Darling, but at a time during drought when salt springs were discharging highly saline groundwater into the stream.

On the return journey from that point at which they had discovered the Darling, the party discovered the junction of the Castlereagh and the Darling.

Having shown that the marshes were finite, and the Darling was 'the chief drain for carrying off the waters falling westerly from the eastern coast', Sturt arrived back in Sydney after four and a half months, having lost neither 'man nor beast'.

Sturt's subsequent journey to trace the Darling in 1829 is described in the Murray–Darling section of this book (page 214).

SECTION FIVE

EASTERN AUSTRALIA

THE GEOLOGICAL HISTORY OF EASTERN AUSTRALIA

Australia's Cretaceous 'Congo-style' rivers: a river system of Amazonian proportions

Can you imagine Australia with a river system so active that it could carry enough sediment from eroding mountains along the ancestral Great Dividing Range, straight across the continent, to deposit a dump 9 kilometres deep in the Ceduna Depocentre, in the sinking rift valley off South Australia's coast? Such a river system did exist about 100 million years ago.[81] It is paralleled in modern times by the Congo drainage in tropical Africa, which has its headwaters in the eroding mountains of the African Rift, and high rainfall consistent with an equatorial latitude. The Congo carries sediment from the eroding mountains straight to the sea, shedding little on its way and accessing no significant amounts from the terrain across which it flows.[88] The headwaters of the Nile were once headwaters of the Congo. They were lost to the Congo when downwarping occurred along the Lake Victoria axis when the African Rift Valley formed.

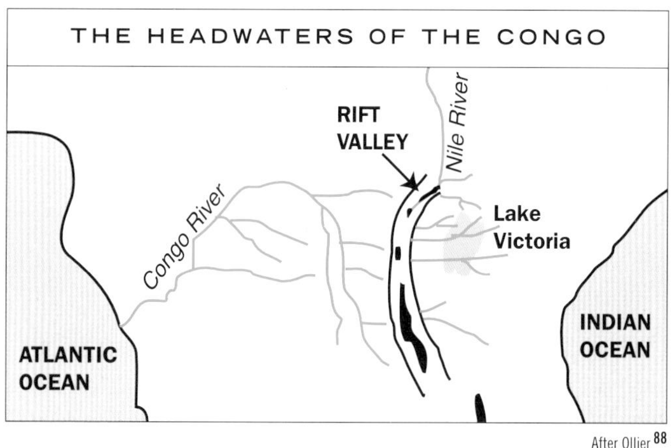

Around 100 to 96 million years ago, eastern Australia had high mountains on the continental edge, 'rift' mountains which resulted from Chilean-style volcanic activities during the Early and Middle Cretaceous. When this regime changed and the volcanic arc moved away from Australia's margin, the mountains were eroded and the river energy and consequent erosion diminished. The climate of the Middle to Late Cretaceous was warm temperate to subtropical—this being one of the 'greenhouse' phases of the prehistoric past when

Narooma, New South Wales. The east coast, south of Sydney, has many beautiful estuaries and inlets. REG MORRISON

AUSTRALIA'S CONGO-STYLE DRAINAGE
100 to 96 million years ago

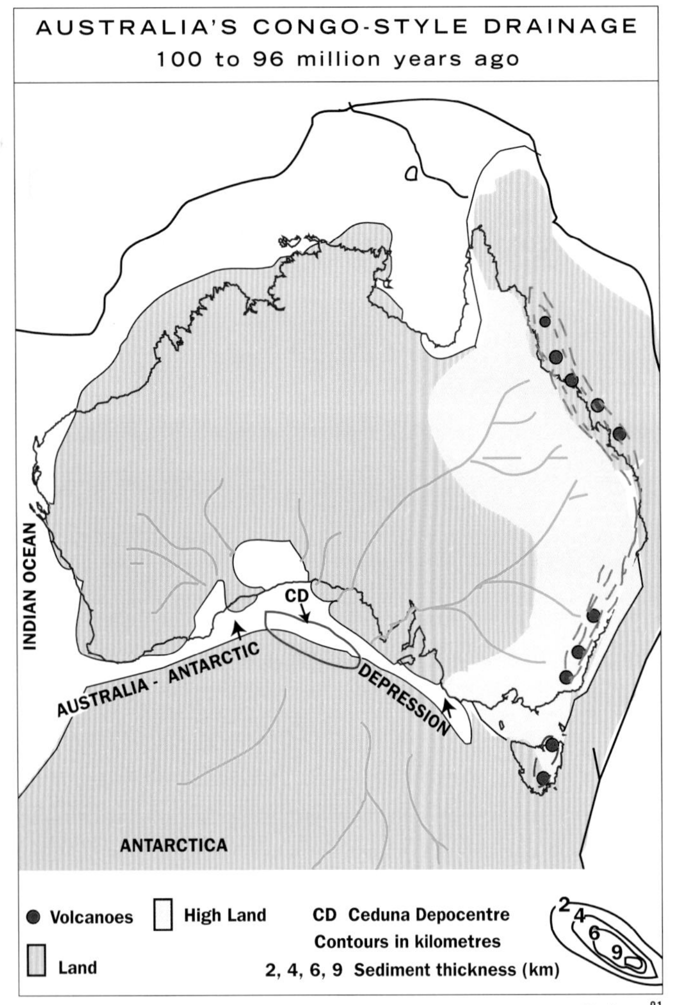

After Veevers[81]

MAJOR DIVIDES AND BASINS OF SOUTH-EAST AUSTRALIA

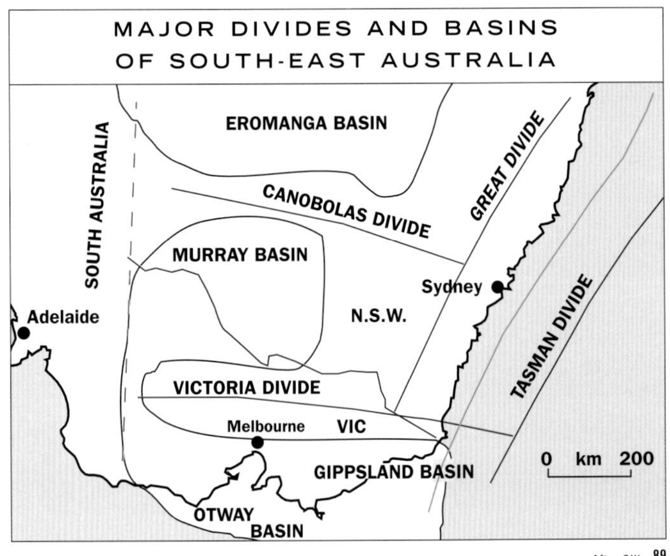

After Ollier[89]

equitable warmth prevailed from the Equator to the Poles without steep temperature gradients. Our continent lay in high latitudes between 75° and 50° South, a well-watered, well-vegetated land, still attached to Antarctica through Tasmania and the South Tasman Rise but with the proto-Southern Ocean invading the widening, lengthening and sinking rift between the two continents.

LANDSCAPE EVOLUTION IN EASTERN AUSTRALIA:
Mesozoic to the Early Tertiary

River development, basin sedimentation and the evolution of major divides in eastern Australia are correlated.[89] According to Professor Cliff Ollier, a leading geomorphologist, the sequence of events up to 60 million years ago created basins and divides in eastern Australia in the following order.

The region has a basement of Palaeozoic rocks, eroded to a palaeoplain, on which lie two sedimentary basins separated by a system of divides. The Great Artesian Basin is Mesozoic, sinking and receiving sediment during the Jurassic and Cretaceous from 170 to 65 million years ago; the Murray Basin is Cainozoic, starting to subside in the Early Tertiary about 60 million years ago. The Cretaceous–Early Tertiary Gippsland–Otway Basin lies to the south.

In the Jurassic, before the break-up of Gondwana, Australia was connected to Antarctica to the south, and the New Zealand Subcontinent (also known as Pacifica) was connected along the eastern edge of Antarctica and Australia. Rivers from the south, and from the Tasman Divide on the New Zealand Subcontinent several hundreds of kilometres to the east, supplied coarse sediment to the Great Artesian Basin. This drainage is older than the formation of the continental margin, the Eastern Highlands and the Murray Basin. Traces of these north-flowing rivers can still be found today—old channels with gravels, crossing the Victoria, Great and Canobolas Divides.

In the Cretaceous, with rift valley formation between Antarctica and Australia and development of the proto-Southern Ocean progressively from the west, the southern part of the palaeoplain tilted, creating the Victoria Divide. This watershed, approximately where the Victorian Highlands are today, separated the Great Artesian Basin from the Gippsland–Otway Basin. Rivers ran from the Victoria Divide northwards to the Great Artesian Basin and southwards to the Gippsland–Otway. (In brief, in the Mesozoic a simple system of rivers drained north and north-west over most of south-eastern Australia, but the pattern has since been modified by tectonism and vulcanism.)

The Otway Basin originated as a rift associated with the opening of the Southern Ocean; the Gippsland Basin formed as a result of incomplete rifting of an arm of the Tasman Sea. The west to east progressive rifting which was creating Australia's southern margin beheaded rivers which had run from elevated landscapes in Antarctica into southern parts of the already very flat Australian continent—like the palaeodrainages near Esperance in Western Australia described on pages 14-17. Rifting and opening of the Tasman Sea from 80 million years ago led to the establishment of the modern Great Divide along the eastern margin of the continent. Rivers which had run west from the New Zealand Subcontinent were also beheaded and new drainage patterns created. Some reversed rivers occur on the seaward sides of the Victoria and Great Divides. (The evidence for reversal and the processes involved are described in detail for the Clarence River, page 158.)

In the south-east, drainage from the Victoria Divide and the inland side of the new Great Divide continued to flow northwards to the Great Artesian Basin, until the Murray Basin started to subside from 60 million years ago. Then a new warp axis, the Canobolas Divide,

EASTERN GONDWANA

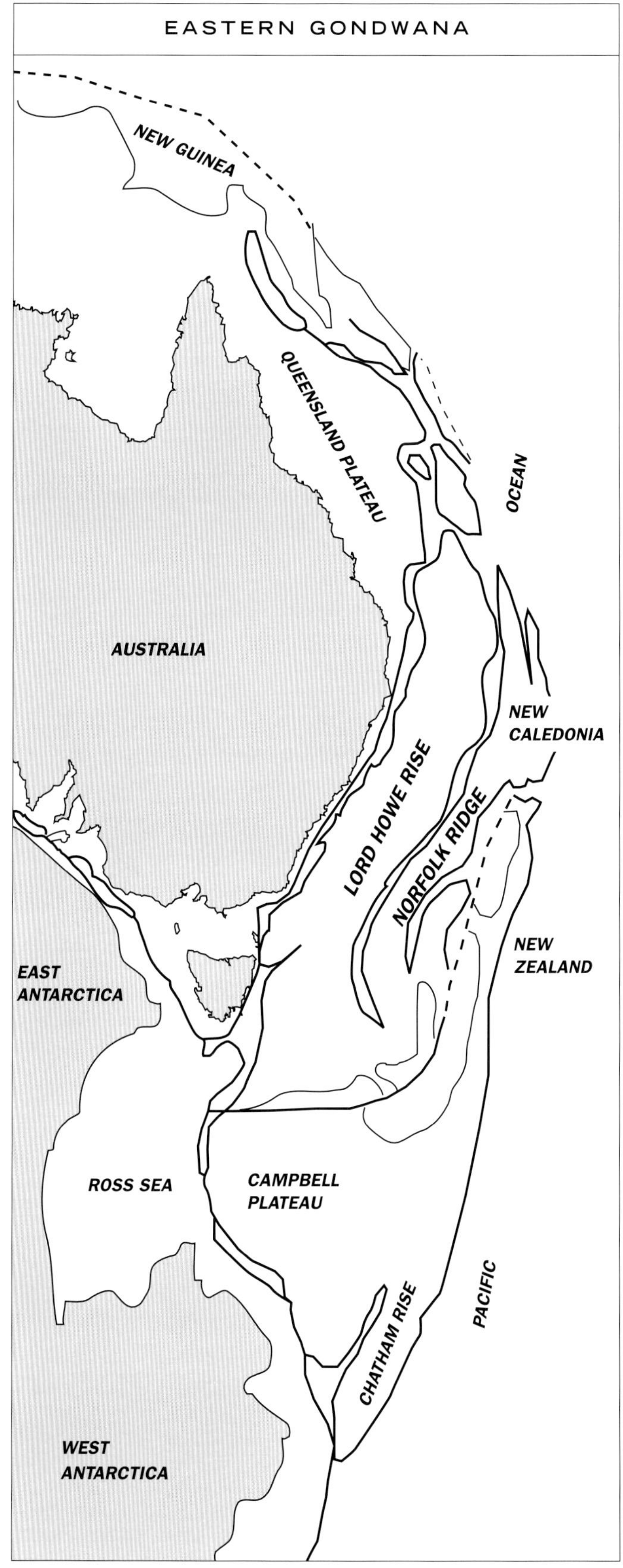

After Ollier[89]

developed between the two basins, formed by the downwarping of the Murray Basin rather than the uplifting of a divide.[89] Predominantly west-flowing drainage developed in the Murray Basin and the sediment supply to the Great Artesian Basin ceased.

PALAEODRAINAGE AND MODERN DRAINAGE OF SOUTH-EAST AUSTRALIA

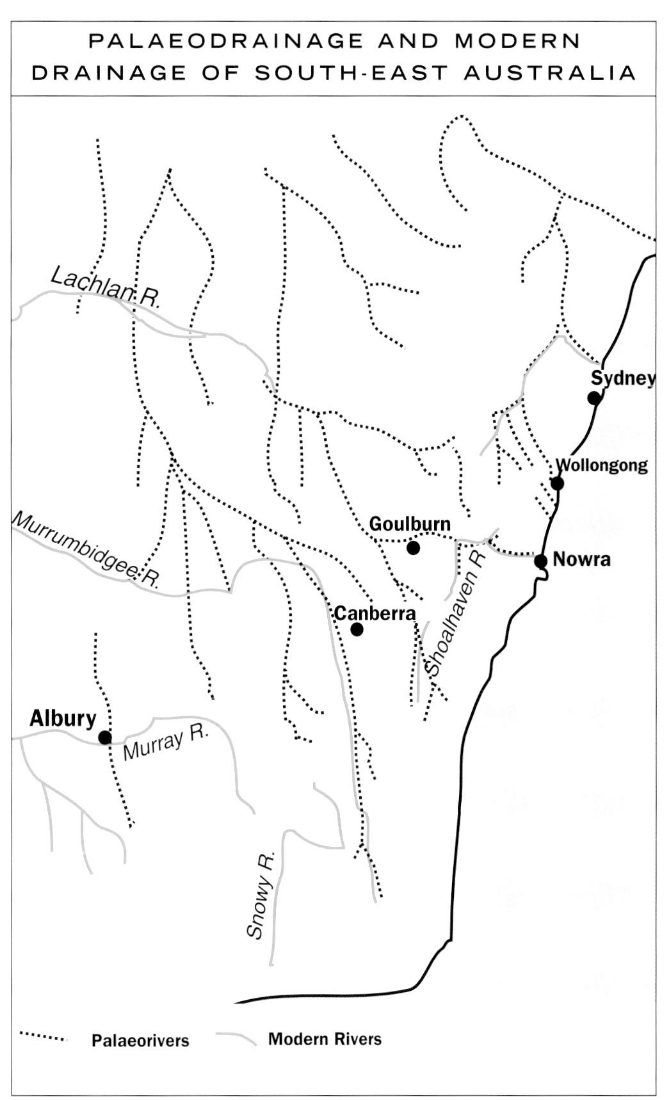

After Ollier[89]

The warp axes which form the divides separating basins are lines of crustal weakness where volcanic activity has occurred. Volcanoes erupted throughout the Cainozoic—for example, the Ebor, Barrington, Warning and Monaro volcanoes on the Great Divide; Mt Canobolas on the Canobolas Divide. Lesser lava flows with small or unknown sources were widespread on the Great Divide and eastern Victoria Divide in the Oligocene and Miocene, finishing in the Late Miocene.

The palaeodrainage of south-eastern Australia according to Professor Ollier is seen in the map on this page. The model of highland development and erosion he used emphasises a substantial westward movement of the divide as a result of tectonism and stream capture. The subject has become highly controversial. New research, in which basalt flows down river valleys have been dated and the on-ground evidence re-assessed, has produced a different story—one of great antiquity of river valleys, rivers maintaining their present courses since very early Tertiary times, stability of the highlands since the Mid Tertiary, and little movement of the divide through the Tertiary.

As is always the case in natural science, nothing is ever simple and a single explanation never satisfies all the facts everywhere, so although one interpretation satisfies most situations, there are others to which another explanation could equally well apply.

LANDSCAPE EVOLUTION OF EASTERN AUSTRALIA DURING THE TERTIARY

The widely distributed basalt flows in eastern New South Wales in the early Cainozoic, especially in the Hunter Valley and adjacent areas, the Shoalhaven region and on the Monaro, which have been recently mapped and dated, have supplied much evidence supporting the revised view on drainage patterns and their evolution. Case studies following in this section will describe this evidence, and the results of tectonism and regional geological structures and processes.

We will see that if this reading of the geological evidence is right, then the following situations apply:[21]

- The pre-basaltic topography of the Monaro differed little from the present day, and the present drainage is comparable with that developed before the Early Eocene to Oligocene volcanism.
- Headwater reaches of the Lachlan and Wollondilly Rivers were close to their present beds by the beginning of the Miocene.
- Drainage disruption was caused locally wherever volcanic eruptions occurred and radial drainage patterns often developed around volcanic areas.
- Very little volcanic activity occurred in the basins except for the Late Tertiary and Quaternary lava flows in western Victoria.
- Major uplift of the Gippsland–Snowy Mountains block took place prior to the Miocene when blocks of terrain bounded by faults were elevated.
- Several major faults affect the Gippsland and the Kosciusko areas, with throws of several hundred metres, possibly even a kilometre, in some areas.

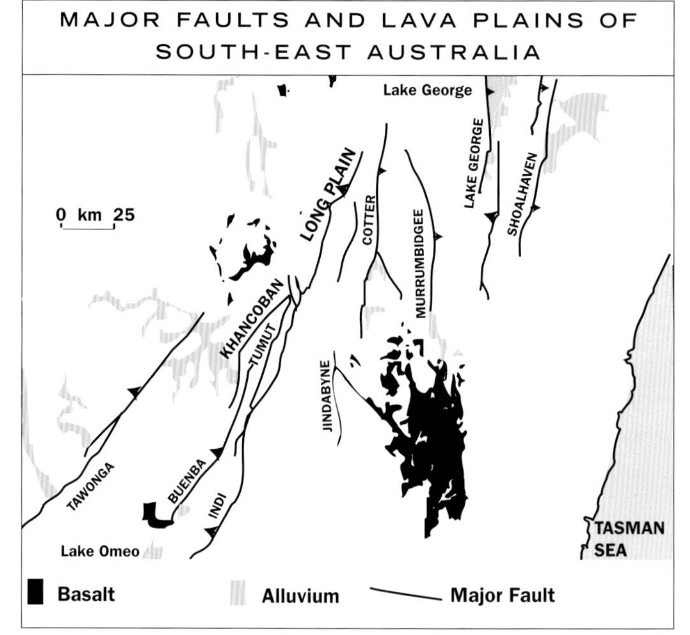

After Ollier[89]

- On the Southern Highlands, a localised tilt block movement on a fault created Lake George and caused diversion of northern tributaries of the local river system.

Sea level changes through time have affected the lower reaches of rivers in south-eastern Australia. A Late Miocene eustatic sea level fall led to valley downcutting, especially in Gippsland. When sea level rose again in the Early Pliocene, the Murray Basin was inundated and tidal conditions penetrated far inland.

LAKE GEORGE, NEW SOUTH WALES

Lake George, in the Southern Tablelands of New South Wales, presents a record of historical water level changes covering the last 100 years.[91] It has responded sensitively to seasonal variations in rainfall intensity, remaining filled for the early years of the century, but drying up during the 1940s and during the drought of 1982–83. Evidence from previous strand-lines round the lake and from the lake floor sediments has been studied, making this one of the best documented semi-permanent lakes of Australia. It is of particular value as its situation is intermediate between the south-eastern mountain catchments and the inland semi-arid plains.

Lake George is a fluctuating closed lake. Groundwater in its 932-square kilometre catchment is mainly of low salinity, but high salinity groundwater is evident beneath the lake bed, confined by a clay layer. As the lake dries, salt is concentrated in its waters by evaporation. Complicated interactions take place between the pressurised saline underground aquifer and the lake water. In effect, salt has been stored away below a level where a hydrostatic balance exists 10 to 12 metres below the surface, during the last 1 to 2 million years.

The lake basin was originated in the Mid Miocene, or earlier, by faulting. It is a normal fault-angle basin and the sediments deposited in the lake reveal the history of the region through time. A series of river and lake sediments 150 metres in depth lying below the lake bed have been discovered by drilling. Three sedimentary layers have been identified:

- The oldest, the **Gearys Gap Formation**, consists of deeply weathered sand and gravels. These Early Tertiary river sediments were deposited on the Palaeozoic bedrock and are associated with a prior drainage system which was incised into the bedrock, flowing north-west. Palaeochannels can be traced under the lake bed. There is a break in sedimentation between this oldest formation and the two which follow.
- The Ondyong Point Formation consists of river sands and lake clays. A laterally persistent layer of sand and silt separated it from the uppermost formation, but there is no break in sedimentation.

Lake George, with a reed island during one of its lake-full stages. M.E.W

- The **Bungendore Formation** comprises lake clays and silts and, with the Ondyong, represents deposition in a closed drainage basin. The differences in sediment type are largely a function of climatic changes.

Margins of the Basin are the low watershed of the Great Divide in the east and the Lake George Range in the west. Its southern and northern ends have subdued topography with low saddles and a complex of natural and artificial drainage lines and swampy and lagoonal areas.

Streams in the Basin occupy wide, open valleys in their upper reaches. They converge towards Lake George (which only constitutes about 16 per cent of the total area of the drainage basin)—Collector Creek from the north, Allianoyonyiga, Taylor and Butmaroo Creeks from the east, and Turallo Creek from the south. Ancient strand lines, up to 37 metres above the lake bed, testify to very large fluctuations in water depth in late Quaternary time. Above this level the lake overflowed via Gearys Gap into the Yass River drainage system.This last happened between 27000 and 21 000 years ago.

LAKE GEORGE, NEW SOUTH WALES

After Jacobson *et al*[90]

LAKE GEORGE

After Ollier[92]

Lake George and its associated features presents a simple and elegant example of drainage strongly modified by faulting. The Lake George Fault cuts across drainage lines, which must therefore be older than the fault. The old Taylors Creek used to flow right across the area from east to west, but is now dammed back to make a lake in the fault angle depression. The fault responsible for creating the basin and lake disrupted drainage, and headwater streams of the Yass River were beheaded. To the north, a number of creeks that formerly flowed north-west were diverted by the fault, and after barbed junctions they flow along the base of the fault scarp into Lake George. To the south of Lake George, the fault crosses the Molonglo River, which has a much bigger catchment than the northern rivers and was able to maintain its course across the rising fault block, cutting the Molonglo Gorge.

Prior to European settlement, the landscape was eucalypt woodland. The pollen record in sediments shows that fire and proliferation of eucalypts has increased over the last 130 000 years at the expense of a less sclerophyllous vegetation with vine thickets and abundant casuarinas. The largely erosional landscape today is the result of the impact of European clearing, grazing and agriculture, and gullying is widespread. Sand and gravel extraction from strand line and aeolian deposits around the southern and northern margins of the lake have contributed to land degradation, as have the massive roadworks which have occurred in the last few years.

The hydrological regime of Lake George has been one of cyclical wet and dry conditions. Salt is transported into the lake from the catchment in wet periods. The lake salinity is inverse to water volume. In dry periods the lake acts as a playa, with capillary zone evaporation. It has been described as a natural analogue for a wastewater disposal basin.

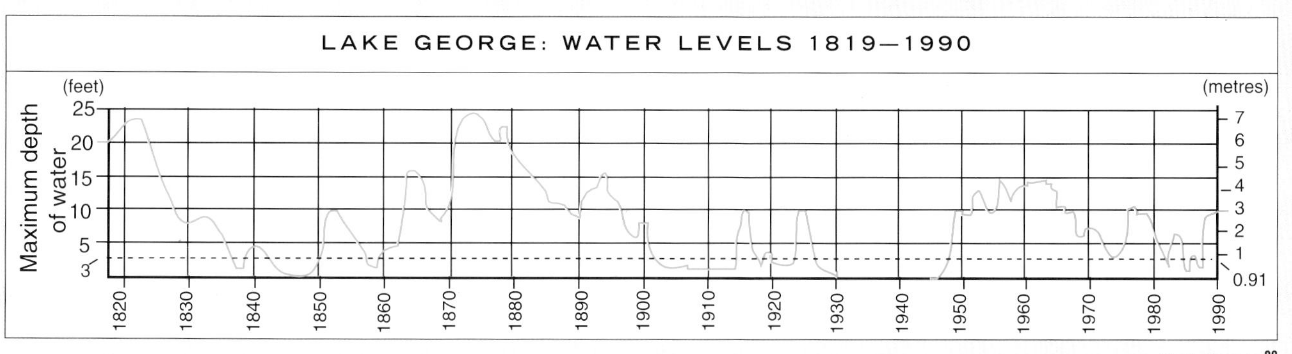

After Jacobson *et al*[90]

South-eastern Australia

In spite of the essentially stable setting for eastern Australian rivers, European land-use practices have resulted in widespread channel degradation in less than a century. It is increasingly recognised that effective rehabilitation of these rivers and floodplains requires an understanding of the rivers' histories, which had determined their form and function before they were altered by our activities.

It has become very clear as research into river function in Australia has expanded in recent years that Australian rivers, their function and geomorphology, are unique, and that trying to classify them or manage them like those of Northern Hemisphere lands simply does not work.

The characteristics of the rivers in the uplands and eastern slopes of south-eastern Australia differ greatly from those further to the west; the former are usually shorter and steeper and are confined to narrow valleys in relatively rugged terrain.

The valleys of south-eastern Australia are very old, resulting from landscape uplift and river incision certainly over more than 30 million years, and probably up to 100 million years.[94] There seems little doubt that much of the eastern Australian landscape has remained essentially unaltered since at least Middle Tertiary times—about 40 million years. During all that time the rivers eroded their valleys under regimes of rainfall and sediment distribution which were not very different from those which prevail today. Quaternary fluctuations in climate have produced terraces, floodplains and changed channel styles, but did not create the valleys themselves. The Pleistocene ice age resulted in fluctuation between drier glacial stages with decreased river activity, and wetter interglacials with more active rivers. The major but gradual modifications which resulted from the climate changes of the last 2 million years were surface changes imposed on a landscape that has been tectonically inactive for tens of millions of years. The fluctuating sea levels, from low during glacial stages, to high during interglacials, would have had a profound effect on the lower reaches of coastal rivers, however.

Research to date has shown that not only the coastal rivers, but the rivers of eastern Australia generally, have erosional and depositional records that correspond to the Late Quaternary climatic fluctuations of the last glacial cycle. This will be demonstrated in accounts of rivers whose history is described in this book. The long-term trend in the last 130 000 years, which encompass the glacial cycle, shows a declining flow regime.

The Pleistocene ice age and south-eastern Australia

The area affected by glaciation on the continental mainland, centred on Mt Kosciusko, contains 13 abandoned cirques of which Blue Lake and Twynam have been closely studied.[91] Early estimates were that about 1000 square kilometres were glaciated. It is now known that only about 50 square kilometres were under a permanent ice cap, and the other features which had been recognised as glacial were in fact periglacial. Block streams, scree slopes and solifluction terraces are widespread over the summits in the Snowy Mountains and Victorian Highlands. Gelifluction deposits are particularly widespread, occurring down to 900 metres in the south-east, rising to nearly 1400 metres in the New England district. Some doubt remains about the dates at which the icy conditions started, but they were in place by 30 000 years ago; deglaciation was rapid after 14 000 years ago and complete by 11 000 years ago, with close to modern conditions by 10 000 years ago.

This timetable fits with the most recent (1997) research on dating climate changes, studies of emu egg shell in Central Australia (discussed on page 74).[53] The most probable reasons for the small glaciated area and the large periglacial one are simply the absence of very high ground and the lack of sufficient water vapour for massive ice production in that extremely dry and windy phase.

Declining cyclical trends of flow regime in south-eastern Australia

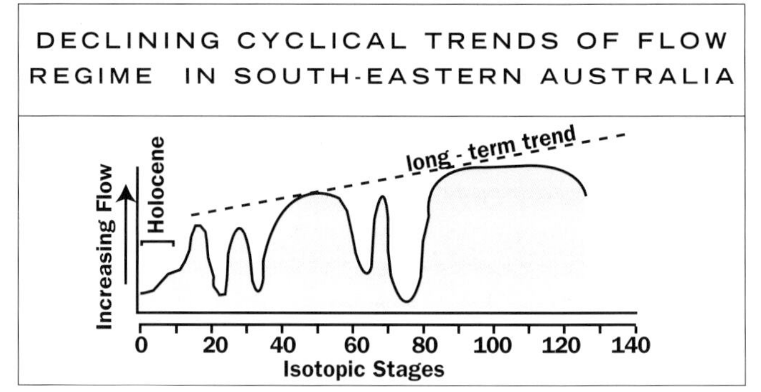

After Nanson & Doyle[94]

Earlier in the last glacial cycle, between 125 000 and 25 000 years ago, the coastal rivers were relatively high energy, laterally active systems transporting coarse sediment. In low-gradient large valleys where these older, coarser deposits remain as terraces, the rivers have not been especially responsive to European impact. In contrast, those in the smaller, steeper-gradient valleys have been particularly responsive to European land-use practices and catchment clearance, and due to local stream power conditions, the middle reaches of their valleys have proved the most vulnerable. These river systems had formed during the last 5000 years—low energy, fine grained, small channelled systems, and had been stabilised by riparian vegetation in the last 2000 years.[94]

Streams in the smaller coastal catchments are today laterally unstable and have relatively mobile gravel or sand beds. Because they are in ancient bedrock valleys whose sediments have been reworked several times in the last glacial cycle, and because the rate of production of sediment by erosion of their headwaters is very slow, the sediment in their beds and on their floodplains and terraces is a non-renewable resource. Some sediments in coastal rivers have been dated at more than 100 000 years, notably in the Hawkesbury–Nepean and Shoalhaven catchments, providing evidence of long-term stability. It would take tens or hundreds of thousands of years to replace these existing valley sediments. This fact has not been considered where commercial sand and gravel extraction has occurred on a large scale. With the increased passage of sediments out to sea as a result of changes to the rivers which have made them fast-flowing conduits during floods, it is possible to visualise contemporary valleys stripped of their fill.

The pattern of river change in south-eastern Australia which results from European disturbance is strikingly similar in all the rivers so far studied. Originally discontinuous watercourses have become incised, continuous channels; 'chains of ponds' have largely disappeared and the few that remain are considered heritage items. Systems which were previously in a state of dynamic equilibrium have become disequilibrium systems characterised by alternating erosional and depositional reaches. The character, pattern and timing of changes to river morphology have varied from catchment to catchment, dictated by the nature of human impacts, and the relative sensitivity of differing river reaches to change. Ultimately, the history of river changes is dictated by the history of flood events and the relative condition of the river and catchment at the time of major floods.

Today in south-eastern Australia annual rainfall is usually over 800 millimetres, and can be up to 1800 millimetres in some areas, leading to some catchments originally having wet sclerophyll or rainforest vegetation. Clearing of such forest in regions with frequent coastal storms, steep channel gradients, confined valleys and unconsolidated sediment has resulted in many highly unstable rivers.

Statistically significant increases in mean annual rainfall have been recorded in south-eastern Australia since the 1940s.[97] A corresponding increase is seen in the number and magnitude of floods in streams near Sydney. The increased flooding is related to increases in channel erosion and bedload transport.

A study of the Nepean River, west of Sydney, in 1987 described periods of higher-than-average and lower-than-average rainfall and flooding as flood-dominated regimes (FDR) and drought-dominated regimes (DDR) respectively.[98] The FDR and DDR characterise the flood record of New South Wales coastal rivers, with mean annual flood discharges two to four times higher in FDR. The study concluded that various kinds of change to river form and function were linked to the alternating pattern, and that each of the regimes lasted for three to five decades. The changes were considered as distinct from those orchestrated by human activities in catchments.

The impacts of the regime shifts were described as being widespread, affecting long lengths of channel, but the timing of the responses depended on the individual sensitivities of different parts of the river systems. In certain erodable reaches, the presence of two distinct channels, an inner, smaller one related to DDR, and a larger one related to FDR, was used as evidence of the cyclic pattern. A recent (1998) reassessment of the evidence for alternating FDR–DDR regimes orchestrating channel changes has concluded that there is no convincing evidence of these alternating regimes as an inherent part of the New South Wales climate, nor is there evidence that our rivers evolved over centuries under those alternating conditions.[99] There is no evidence that channel adjustments following European settlement of south-eastern Australia were controlled by natural climate cycles rather than by disturbance from vegetation changes and other human impacts. This is not to say, however, that extreme climatic variability is not a major feature of river flow in south-eastern Australia. Equally, this does not mean that river morphology has not responded to changes in flow since European settlement of the continent.[99A, 99B]

It is interesting to note that although much research into river history has been undertaken over recent years, there is no evidence of changes to flow patterns and river behaviour induced by Aboriginal activities over the period that Aboriginal people were in the south-east before the arrival of Europeans. Their arrival in these parts is variously assumed to have been anywhere between 40 000 and 20 000 years ago, and if their burning practices had the profound effect on vegetation that is suggested by some writers, evidence of that would be recorded in river deposits. In contrast, there is no doubt at all that in many New South Wales catchments bank erosion and channel degradation are directly attributable to introduction of foreign animals and over-grazing; deforestation, and in particular the removal of vegetation from stream banks; destruction of reed beds; swamp draining; agriculture; and other changes.

The nature of river response to floods is dependent on channel (and catchment) boundary conditions. Localised, catastrophic channel and floodplain changes are responses to factors within the systems themselves—where reaches and floodplains which have evolved to a point where they have become erosionally vulnerable and are primed and ready to be transformed by the next significant event. (In other words, the changes result from the crossing of internal thresholds rather than from climatic forcing.)

THE WINGECARRIBEE SWAMP: *a catastrophic event terminating a long period of destabilisation*

Dr David Tranter of the Robertson Environment Protection Society began his description of an unusually dramatic environmental catastrophe: 'On the night of August 8, 1998, Wingecarribee Swamp, a wetland of national and international significance, which had taken 10 000 years to form, blew out and collapsed …'.[100] The event made news headlines and captured the imagination and the media carried stories of 'exploding swamps', conjuring up a range of strange mental images. Seldom is there such a nice example of cause and effect, or such a visible demonstration of how human activities can destabilise a natural system so that it reaches a critical threshold and ecological collapse is suddenly triggered by a not-unusual following event.

Wingecarribee Swamp is situated in the Southern Highlands between Robertson and Moss Vale, at the headwaters of the Wollondilly River, a major tributary of the Hawkesbury. It is fed by the Wingecarribee River, which drains a catchment of only 40 square kilometres. The swamp originally occupied an area of 600 hectares, and clear, fresh water of excellent quality emerged from its lower end.

Evolution of the swamp

The evolution of the swamp took place over a period of about 10 000 years.[100] Originally, a waterway ran down the shallow bedrock valley and became progressively choked with reeds and sedges. Anaerobic conditions and lower pH developed in the plant litter which accumulated below the swamp vegetation as it expanded across the valley floor. The incomplete breakdown of the vegetable matter resulted in the formation of peat, and over time the depth of the peat deposit increased and living vegetation took root in it. Year by year this vegetation rose higher until it formed a continuous heath-like meadow of some 100 wetland species suspended 4 to 5 metres above the bottom on a column of peat of its own making. This organic sponge-like structure was strong enough to hold in dynamic equilibrium against the force of gravity 1 million tonnes (1000 ML) of water, which filtered through the peat to deliver clear, pure drinking water.

The Wingecarribee Swamp is home to at least five rare and/or endangered species—an orchid, a gentian, a yellow loosestrife, a giant dragonfly and a rare bird, the Australasian bittern. The heritage value of the swamp was recognised by the Fauna Panel as long ago as 1967 and this was drawn to the attention of other government agencies by the National Parks and Wildlife Service, its successor during the 1970s. In 1990, the swamp was listed on the Register of

CHAINS OF PONDS

Explorers in the early nineteenth century repeatedly described river channels on the Southern Tablelands and highlands as running across fine-grained sediments and characterised by 'chains of ponds' with intervening swampy areas vegetated by native reeds. The naturalist J. Lhotsky, on a journey in 1834 from Sydney to the Australian Alps, described 'round or oval basins of from 20 to 200 feet in diameter or length, excavated or sunk in the superficies of an alluvial soil which is commonly of a rich kind'.[95] These sluggish streams led Lhotsky to ask, 'why with such apparently equal materials to those which other countries possess, our continent could not develop itself, so far as to the more perfect formation of creeks and rivers'. (Poor old Australia! Its native animals were 'vermin' and even its rivers did not know how to behave!)

A few examples of chains of ponds still exist on the Southern Tablelands, consisting of reed-lined small channels flowing between ponds up to 11 metres deep, set in a swampy floodplain.[96] The floodplains of these streams consist of organic-rich loam and clay loam overlying a variety of alluvial and colluvial deposits of Late Quaternary age. There are very small quantities of sand and gravel in the floodplains adjacent to the chains of ponds, a characteristic feature that, along with the organic content and sedimentary fabric, allows recognition of this depositional environment in stratigraphic sections.

Between 1850 and 1950 almost all these chains-of-ponds reaches of rivers were transformed into continuous incised channels in which there is significant bedload transport. In fact, many of the rivers were transformed within a couple of decades of European settlement, and frequently the cause was the erosion caused by over-grazing by introduced stock, often exacerbated by rabbit plagues, which decreased ground cover and increased run-off. Much of the damage was done in this way prior to large-scale land or river-bank clearing. Droughts, which then as now were inevitable, compounded the problems of over-grazing and subsequent erosion.

The swampy valley, headwaters of the Bogong River, in the Namadgi National Park, on the fringes of the High Country. M.E.W.

A chain of ponds in the swamp headwaters of the Bogong River. M.E.W.

the National Estate by the Australian Heritage Commission; in 1992, as a Landscape Conservation Area by the National Trust; and in 1993 as a wetland of national significance by Environment Australia. In 1996, the New South Wales Heritage Council recommended that the swamp be protected by an Interim Conservation Order, but this was not implemented until March 1998.

The swamp as water supply

In the early 1900s, the New South Wales Government resumed the swamp as a water supply for Bowral. In 1974, a dam and water treatment plant was built across its mouth, inundating the lower half and forming the Wingecarribee Reservoir. In 1993, the catchment, swamp and dam was acquired by Sydney Water, and a statutory plan was prepared in collaboration with the National Parks and Wildlife Service to manage and protect the ecosystem and the quality of water it delivered.

Peat mining in the swamp

Peat mining began in 1967 under a pendency grant from the Mines Department, followed by a 20-year lease in 1972. When the lease expired in 1992, mining was allowed to continue. In the light of the number of protection orders and recognitions of the heritage value of the swamp listed above, this was indeed surprising (and irresponsible?). The peat was dredged from the swamp with an orange-peel grab mounted on a pontoon, floating in a 'dredge pool' of its own making. In late 1996, the Minister set up a Mining Warden's Inquiry, by which time the excavation had tripled in size since the leases expired. The Inquiry continued through 1997, during which time mining was allowed to proceed. The mining company, Emerald Peat, argued for renewal of the leases and the Department of Mineral Resources supported its application, while Wingecarribee Shire Council, five State Government agencies and many semi-government bodies and environmental organisations opposed the application. In 1998, *after* the collapse of the swamp, the Minister determined that the leases would not be renewed.

Collapse of the swamp

The collapse of the swamp took place on the night of 8 August 1998 after heavy rains. The pent-up head of water contained within the peat blew out, the water table dropped by several metres and the sides of the swamp fell in. The suspended mass of peatland vegetation was churned up, the dredge was washed into the Wingecarribee Reservoir immediately downstream, together with an archipelago of peat some 5 million cubic metres in volume. A broad channel opened up right along the length of the swamp, allowing agricultural and urban run-off to flow directly into the dam.

The swamp is draining, the peat is drying out (and will present a fire hazard—peat fires can burn underground for hundreds of years). We are witnessing the destruction of an entire ecosystem and irreparable damage to a valuable urban water supply. The two as yet intact bays of peat at the southern side of the swamp appear also to be drying out.

The cause of the collapse, as determined by two independent consultancy firms commissioned by Sydney Water, was the peat mining operations.

Sydney Water declared the collapse of the swamp to be an 'incident', and set up an inter-departmental Steering Committee to advise on how to contain the damage. The peat that had washed into the reservoir was kept clear of the water treatment plant by means of a strong rope, which is now being replaced by a permanent fence. Physical and chemical changes are being monitored. For six months, the water was too turbid to be handled by the treatment works (late 1998) so silt curtains were placed round the intake and the water so enclosed was dosed with alum to precipitate organic debris. The treatment works was upgraded at great expense and is now able to handle the water without alum dosing. The cost is being passed on to Wingecarribee Shire Council in water charges. Nutrient levels are very high and there is danger that cyanobacterial blooms will break out when the turbidity declines.

The Wingecarribee Reservoir, through the Glenquarry Cut, is used as a back-up for Sydney and Wollongong's water supply—so more than Bowral's water quality is at stake.

Wingecarribee Swamp, June 1998—a more or less continuous meadow. HELEN TRANTER

Peat dredging during the 1997 Mining Warden's Enquiry. DAVID TRANTER

What makes the whole affair even sadder is that peat need not be mined—many other products can replace it in the urban horticulture for which it is exploited. In fact, peat mining anywhere in Australia is to be deplored and should cease.

DRAINAGES WEST OF THE GREAT DIVIDE

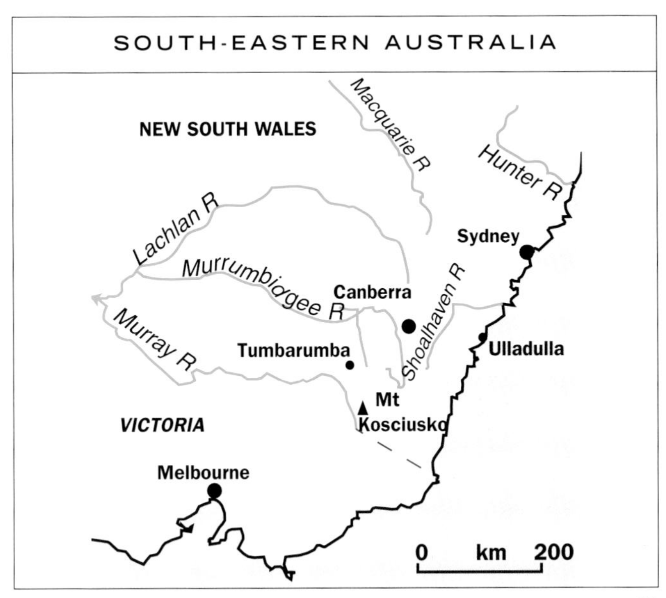

After Young & McDougall[101]

The highlands of south-eastern Australia have recently been proved to comprise ancient landscapes in which the drainage patterns have changed little in possibly as much as 40 million years.[101] Headwater streams of the Murray, Murrumbidgee and Lachlan Rivers rise in these highlands. (Studies of these major rivers are presented in the Murray Basin section, pages 210-232.)

Mean annual rainfall over the region exceeds 1000 millimetres, and in the higher areas near Cabramurra is 1400–1500 millimetres annually. At times during the Pleistocene, slopes above 1000 metres elevation were subject to periglacial activity, while an area of about 50 square kilometres around Mt Kosciusko was subject to glaciation.

Basaltic lavas that flowed into river valleys draining westwards from the Tumbarumba region near Mt Kosciusko have been dated and mapped, revealing a history different from that originally envisaged for the south-eastern highlands.[101] (The original ideas were based on the traditional cyclical models involving planation and disruption of drainages with stream capture and scarp retreat as major factors.)[89]

The basalt flows have been dated at between 19 and 23 million years, and they covered and preserved Early Miocene landscapes to a remarkable degree. A geographer from the University of Wollongong and a geologist from the Australian National University, Canberra, have reconstructed the landscapes and have shown that the major topographic features of the highest part of the Australian continent had assumed essentially their present form by the Mid Tertiary, and that some of the features had been formed in the Cretaceous. They also showed that plateau surfaces had been lowered only by about 2 to 5 metres per million years, and that most major streams have incised by about 5 to 18 metres per million years, with a maximum incision of 30 metres per million years, since the Early Miocene. Even more surprisingly, despite the passage of 20 million years, stream gradients have not waned and major breaks in stream profiles have not been eliminated. The modern stream profiles are strikingly similar to their Miocene counterparts, suggesting that the evolution of the profiles has been determined mainly by variations in stream power, rather than by headward retreat of knick-points through the drainage system. These findings imply a rate and pattern of denudation very much at odds with widely accepted models derived from areas in other parts of the world, once again emphasising the unique nature of Australia.

The area studied is centred on Tumbarumba. The terrain consists of dissected plateaux, with elevations between 1700 metres and about 1100 metres, split by the 700-metre deep gorge of the Tumut River, which flows northwards. West of this gorge, isolated peaks (notably Granite Mountain) rise up to 300 metres above the general level of the Bago Plateau. The western edge of the plateau is bounded by a major escarpment that drops steeply 500 to 600 metres into the valley of Tumbarumba Creek. West of Tumbarumba Creek the terrain is hilly with some summits rising to 1000 metres, but with most ridges between 600 and 700 metres above sea level. The Murray River has

The Glenquarry Cut which transfers water to Sydney and Wollongong. M.E.W.

Northern (Kangaloon) shore of the Swamp immediately after its collapse, showing slumping of the banks and creation of an open channel. 12 August 1998. SYDNEY WATER

The archipelago of peat in the Wingecarribee Reservoir (approximately 5 million cubic metres) that was washed out of the Wingecarribee Swamp (upstream to the left) four days earlier. 12 August 1998. DAVID TRANTER

An aerial view of the lower part of the Wingecarribee Swamp, 12 August 1998, four days after the swamp collapsed. Caalang Creek, which drains the unsewered village of Robertson and the dairy farms downstream, now feeds directly into Wingecarribee Reservoir via the new channel that has opened up, instead of being filtered by the swamp. DAVID TRANTER

An aerial view of the dredge pool during the Mining Warden's Enquiry before the swamp collapsed. A line of buoys across the pool indicated the pipe through which the peat slurry was pumped to the onshore processing plant. DR JUDY MESSER

BASALT FLOWS IN THE SOUTH-EASTERN HIGHLANDS

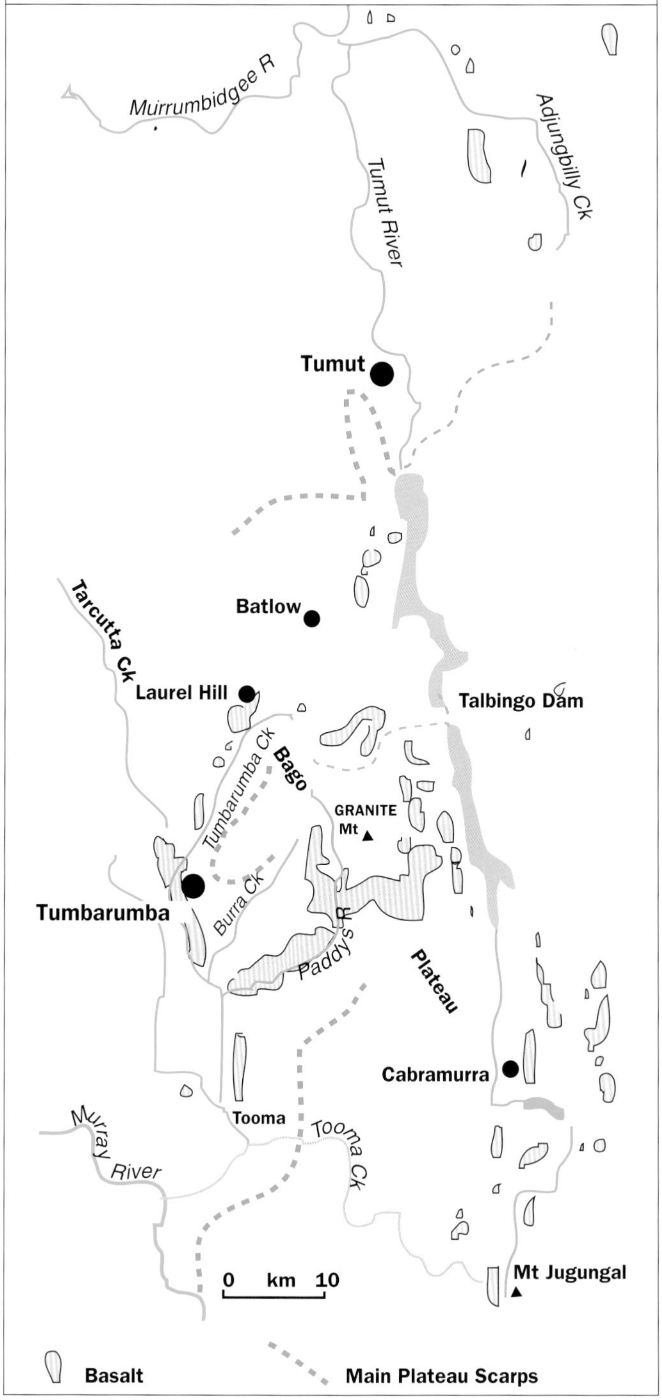

After Young & McDougall[101]

cut a narrow trough, with an alluvial floor generally less than a kilometre wide, through the southern parts of these hills. The upper reaches of the trough of the Murray River and of its tributary, the Tooma River, lie at only 250 to 300 metres above sea level, but apart from well-developed benches at about 800 metres, the land to the east rises steeply to the major plateau surface at about 1700 metres.

A discontinuous sheet of basalt, with a maximum thickness of about 80 metres, covers an area of about 30 square kilometres on the central part of the Bago Plateau east of Tumbarumba. Elsewhere, the basalt consists of dissected remnants of long, narrow tongues that flowed down the main valleys draining to the south-east and north-west of the plateau. Originally, one tongue probably flowed 50 kilometres down Tumbarumba Creek. Some basalt also spilled over the low divide between the Tumbarumba Creek and Tarcutta Creek valleys, and remnants are found there today. The basalt tongues covered valley sediments, and because some of these 'deep lead' gravels are gold-bearing, much information on the depth of valleys, their fill and the depth of basalt is available. From all the evidence it has been possible to reconstruct the Miocene landscape onto which the basalt was extruded and to show that:

- The escarpment forming the western edge of the Bago Plateau was present in the Miocene before the lava erupted.
- Relief in the region was not very different from today before eruption of basalt.
- The highest section of the uplands of south-eastern New South Wales was already rugged terrain at approximately its present elevation long before the eruption.
- Valley incision began in the Early Tertiary or Late Cretaceous, so the origins of the upland surfaces must date from the Mesozoic. The depth of stream incision has been remarkably small.
- The upper Murray valley dates from the Early Tertiary, as do the valleys of the Tumut, the Tarcutta Creek and tributary streams of the Murrumbidgee and the Lachlan.

The clear evidence of 1200 to 1600 metres of Mid Tertiary relief revealed by this work at Tumbarumba shows that all the uplands in New South Wales south of Sydney had been lifted to their present altitude prior to 30 million years ago.

TARCUTTA AND UMBANGO CREEKS

Tarcutta Creek is a south bank tributary of the Murrumbidgee River and contributes suspended sediment to it. Its history and its present situation are similar to that of the other tributaries, the Hillas, Jugiong, Umbango, and Yaven Yaven Creeks.[102] The Tarcutta is characterised by alternating unstable reaches of predominantly erosional and depositional character, the instability being the result of alterations to the environments in the creek's catchment since European settlement. In the erosional reaches, bed and bank erosion has increased channel capacity, and has destroyed significant areas of fertile floodplain land and reduced the frequency of overbank flooding. Coarse-grained sediment from the erosional reaches has accumulated downstream, where it has caused bed aggradation, reduced channel capacity and increased incidence of overbank flooding. Adjustment in the system has continued throughout this century, with incision in the upper alluvial reaches sending slugs of sandy sediment into the downstream reaches. The conversion into a continuous bedload system appears to date from about 1950, when the last of the deep swimming holes filled with sand.

This pattern of changed behaviour is typical of countless creeks and rivers in Australia, so the Tarcutta Creek situation, which has been scientifically monitored, serves as a description of the processes which have widespread occurrence.

Tarcutta Creek rises in rugged hills 1000 metres above sea level near Tumbarumba in the western foothills of the south-eastern

TARCUTTA CREEK AND OTHER MURRUMBIDGEE RIVER TRIBUTARIES, SOUTH-EASTERN NEW SOUTH WALES

Murrumbidgee R
Nangus
Gundagai
Wagga Wagga
Hillas Ck
Yaven Yaven Ck
Forest Hill
Borambola
Tarcutta
Adelong
Kyeamba Ck
Tarcutta Creek
Janey Harvey Bridge
Humula
Umbango Ck
Batlow
Westbrook
Carabost
Holbrook
Tumbarumba
0 km 10

After Page & Cardin[102]

Tarcutta Creek—a dry stream bed in May 1999 where the highway crosses the river. M.E.W.

A section of Tarcutta Creek near Borambola, showing 1 metre of Post Settlement Alluvium over old floodplain clays. YALE CARDIN

The deeply incised reach of Tarcutta Creek at Janey Harvey Bridge where much bank erosion has occurred in the last 15 years. YALE CARDIN

highlands and flows northwards in a bedrock-confined valley before swinging north-west near Tarcutta. It joins the Murrumbidgee near Wagga Wagga. Before settlement, eucalypt woodlands, modified by Aboriginal fire into grassy woodlands, were the general vegetation type. The region has been cleared heavily and has problems now with introduced plants, particularly willows and blackberries in the riparian zone. It contains some large conifer plantations.

The early explorers Oxley, Hume, Hovell and Sturt travelled through the region between 1817 and 1830. They recorded the creeks as chains of ponds, swamps and reedbeds, like those described by Eyles in the Yass, Molonglo and Wollondilly Rivers.[96] Today, Tarcutta Creek's swamps are reduced to remnants at Courabyra, 10 kilometres north of Tumbarumba. Clearing and grazing have depleted ground cover, leading to erosion in storms, when sediment from gullies and channel incision is deposited over alluvial flats. Such eroded sediment today forms a metre-deep layer overlying the original floodplain.

The changes to the creek's catchment since European settlement have destroyed the natural form of the creek, which had been established over a long period and was in equilibrium with the topography, climate and vegetation cover. From a sinuous channel with shallow reaches, deep pools and reed swamps, the creek became a deeply incised drain and the water table in the valley was lowered, drying the swamps. Dramatically increased run-off with higher flood peaks during rainfall events caused widespread erosion and the incision of the creek bed, and the deep pools were filled with sediment. All these effects were the inevitable results of tree clearing and depletion of ground cover (aggravated by a rabbit plague between 1910 and 1920). The deliberate destruction of reedbeds by burning, and channelling of the creek, which was recommended to prevent flooding, hastened the changes to the system.

Progressive changes have been documented in Tarcutta Creek since the early days of settlement. In the 1930s, an extensive swamp near the junction with Umbango Creek was drained. Local channel incision followed and the deeper channel exposed large logs in deposits under the swamp, indicating changes in the drainage patterns in the valley in the past. Some swamp habitats, and pools with platypus living in them, remained until the 1940s. Near Tarcutta a deep pool capable of being dived into safely from the Hume Highway bridge in the 1890s was shallow by the 1930s; a deep pool near the Umbango Creek junction was a swimming and fishing hole until it silted up in the late 1940s. In Umbango Creek, upstream of Humula, the channel was already narrow and deeply incised before 1950, and the big floods of the 1950s widened the channel by up to 30 metres.

More recently, deeper incision of Tarcutta Creek followed 'channel improvement works' aimed at reducing local flooding. In 1985, the Wagga Wagga Council removed trees and logs from the channel at the Janey Harvey Bridge and straightened the channel. This stopped flooding, but 3 to 4 metres of incision followed and the bridge collapsed! Downstream, severe bank erosion and silting of a long stretch resulted from these 'improvements'.

The deep floodplain exposures revealed by the incision at the bridge site showed indurated basal gravels overlain by beds of sand, silt and clay, some of which show old soil layers. It is probable that most of this revealed section was deposited in the Late Quaternary when the Upper Murrumbidgee River flow regime was characterised by higher flood peaks and a greater discharge of coarse bedload than occur today.

Concerns about the channel instability by local landowners has prompted the Oberne–Tarcutta Landcare group to exclude stock from the riparian zone and start replanting it with native trees.

DRAINAGES EAST AND SOUTH-EAST OF THE GREAT DIVIDE

European disturbance of catchments has profoundly altered the sediment regime of coastal rivers in south-eastern Australia, just as it has altered those on the other side of the Great Divide, such as the headwater streams of the Murrumbidgee. Channel incision, followed by channel expansion and rapid downstream transfer of sediments, has characterised the changes. Granite catchments have been particularly prone to destabilisation, having large stores of reworked saprolitic material (decomposed rock) in their headwater valleys.[230] Incision into upland valley fills has released these vast volumes of material, often cutting down through them to bedrock and then scouring out the valleys. The sediments have washed downstream, completely altering the character of channels in the middle reaches of the rivers and in the distal lowland or coastal plain regions where much of the sediment ends up. Bank erosion, floodplain damage with channel widening, and changes to regional hydrology have followed.

The same scenario has applied in all the streams arising in the Southern Highlands and it is interesting to note that it has been established in many instances that the first destabilisation was not through anything as drastic as land-clearing. The introduction of herds of introduced grazing animals, over-grazing in which ground cover was damaged, particularly where kangaroo grass or other palatable species disappeared under constant grazing pressure, with the addition of the inevitable drought conditions, followed by flooding, was enough to start the processes. Gully erosion, which was widespread within a couple of decades of settlement and grazing of native vegetation, provided the sediment, and large gullies appeared in headwater stream valleys. Many of these early gullies are still features of some upper catchments because a more or less stable state was reached after the initial release of sediments and the gullies have not remained as highly active, continuously, since. In some catchments, however, no quasi-stable state has been reached.

The following series of case studies demonstrates the processes and changes induced by the impact of European settlement.

RIVERS OF EAST GIPPSLAND, VICTORIA

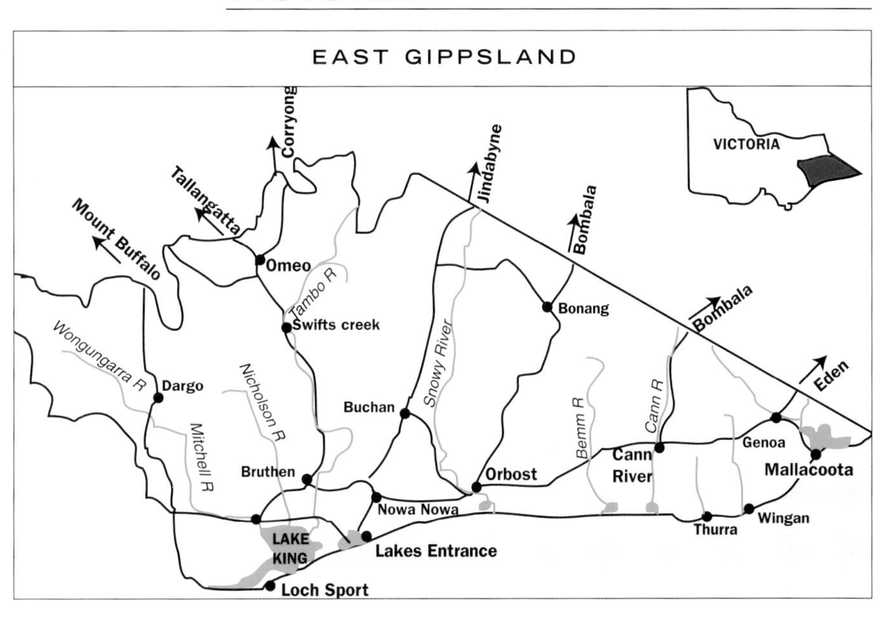

The La Trobe River

The lower La Trobe River is one of the most disturbed river systems in Australia, having been long subjected to 'river improvement', including 66 artificial meander neck cut-offs which reduced the length of the lower part of the river by 25 per cent.[103] With the exception of the seven lowest, the cut-offs were planned to be shallow floodways, but most of them progressively eroded to full river depth. In the 1970s the erosion was dealt with by using the expensive techniques of bank battering and rock protection. By the 1980s it was hoped that the alternative engineering approach of fencing and planting would provide a better long-term solution, but the work was often destroyed by continuing erosion.

In an attempt to reduce velocity and therefore alleviate erosion of the lower river, some of the cut-off meanders are now being reinstated (at a cost of about $25 000 per cut-off) with the expectation that this physical restoration will improve habitat conditions. However, the sediment stored in the cut-offs is now entering the river, and there are questions about its toxicity as well as about the probable consequences of the sediment input on the form and function of the river.

Archival records of the lower La Trobe River have enabled a study to be made of the impact of clearing of riparian vegetation, de-snagging, re-channelling to remove meanders, and other man-made changes over a century and a half of European management.[104] The same impacts have affected the alluvial reaches of almost all large rivers in eastern Australia, but few have such detailed records.

The La Trobe valley was settled by Europeans in the 1840s, and by the 1890s nearly all the trees had been cleared from the La Trobe River's banks, to the extent that when attempts were made at de-snagging the channel there were no large lift trees remaining. At the time of settlement, the lower La Trobe was an example of a low gradient, low capacity, low energy suspended-load river. The mean annual flood (recurrence interval RI, 2.33 years) was approximately 200 cubic metres per second, but bank-full channel capacity at Rosedale was about 60 cubic metres per second. (The catchment area at Rosedale is about 4000 square kilometres.) The river flowed at bank-full or more for at least 50 days a year at the time of settlement, and the width of the river was 25–30 metres, according to early records.

A major river survey, conducted in 1925, provided a useful baseline for subsequent measurements. Some 283 cross-sections were surveyed along a 100-kilometre stretch of the lower La Trobe, and comparisons with the early records showed that no measurable change had occurred in bank-full width of the river, in spite of many meander cut-offs and other European changes having been made in the intervening years.

Between 1927 and 1937 many large floods occurred, one even being a 350-year event with 15 times the mean annual flow. The high magnitude floods caused only a small increase in bank-full width over a short length of the river, despite the fact that riparian vegetation had been largely cleared.

In response to the monster flood of 1934 a program of de-snagging was undertaken between 1936 and 1940. Once again, records at the Rosedale gauge show only a small response in channel capacity as a result of the de-snagging, and two other large floods in 1952 and 1953 caused little change in bank-full width.

The greatest changes in dimensions, hydraulic and sedimentological characteristics of the lower La Trobe River were caused by the relatively minor floods of the 1970s, after the installation and abandonment of the artificial meander cut-offs during the 1950–70 period. These cut-offs provided upstream-migrating knick zones (steeper, eroding fronts) as they eroded to reach river channel depths. This affected the main channel gradient, and led to the undercutting and collapse of banks. (Numerous artificial meander cut-offs installed in the Macalister and Cann Rivers, also in south-eastern Victoria, have most likely been responsible for similar effects.)

The Snowy River

The headwaters of the Snowy River are located in the Australian Alps in New South Wales. Nearly two-thirds of the catchment of 15 800 square kilometres lies in that State, with the rest in Victoria. The majority of the Victorian catchment is in the Snowy River National Park, uninhabited and relatively inaccessible. Below Jarrahmond gauging station, where the river leaves the foothills, it runs across the alluvial Snowy Flats, which have been largely cleared for agriculture.

The Snowy Mountains Scheme (SMS) is a complex system of diversions, impoundments and hydro-power stations that transfers water from the Snowy to the Murray and Murrumbidgee Rivers. The SMS started to affect flows in the Snowy River in 1955 and by 1967 regulation was complete with the closure of the Jindabyne Dam. This dam resulted in a release of only 1 per cent of mean annual flow to the river below it. It is now widely recognised that the reduction of flow has been killing the Snowy (and other rivers); recently, a Federal Government Inquiry (Snowy Water Inquiry 1998) recommended that the release from the dam should be 15 per cent of mean annual flow, a figure which would not affect the functioning of the SMS but which falls far short of the 28 per cent demanded by interest groups located below the dam in New South Wales and Victoria.[106]

The river in the Snowy Flats area was subject to massive flooding in historical times. The largest flood is thought to have occurred in 1870, not long after settlement, and may have been twice the magnitude of the big floods of 1934 and 1971. Floods with a peak discharge of between 40 000 to 170 000 ML per day have occurred regularly since 1922, with their frequency (for a given discharge) being unaltered after regulation. Here, as in other regulated rivers, the peak floods get through as they did before regulation, or they come from unregulated parts of the catchment, and it is the pattern of lesser floods which is affected. Regulation has resulted in a much-reduced frequency of flows between 2000 and 30 000 ML per day, due to the storage of low magnitude spring snow-melt floods in Lake Jindabyne.

It is not only regulation of the river for the SMS that has changed the lower Snowy. Land use on the flats has contributed. The Snowy on the flats is shown in an early map to have been from 50 to 100 metres wide in 1865, while 1934 surveys show it to have been 375 metres wide at Lynn's Gulch and up to 150 metres wide at Gilbert's Gulch. Available data suggests that the river widened catastrophically in the very early period of European settlement. The big flood of 1870 may have been the trigger for, or cause of, the widening. The clearing of riparian vegetation and the removal of woody debris by early de-snagging parties contributed to the changes. Early de-snagging was to improve the river for navigation; the de-snagging program between 1932 and 1941 was to control the build up of sediment; in the 1950s the Snowy River Improvement Trust spent a large proportion of its funds on de-snagging. The present bedform of the river—flat, featureless, shifting sand as a result of siltation and altered flows—is similar to conditions found in other eastern Australian rivers which have been destabilised by human activities. Loss of deep pools and variety of in-stream habitats results in decrease in riverine biodiversity.

Another consequence of the regulation of the Snowy has been the increased distance to which tidal effects have penetrated upstream from the river mouth. The salt wedge has migrated several kilometres

The Snowy River at McKillops Bridge, flowing after snow-melt in the headwaters. REG MORRISON

The Snowy River—a 'river' of sandbanks, starved and strangled by lack of water.
TOTAL ENVIRONMENT CENTRE, SYDNEY

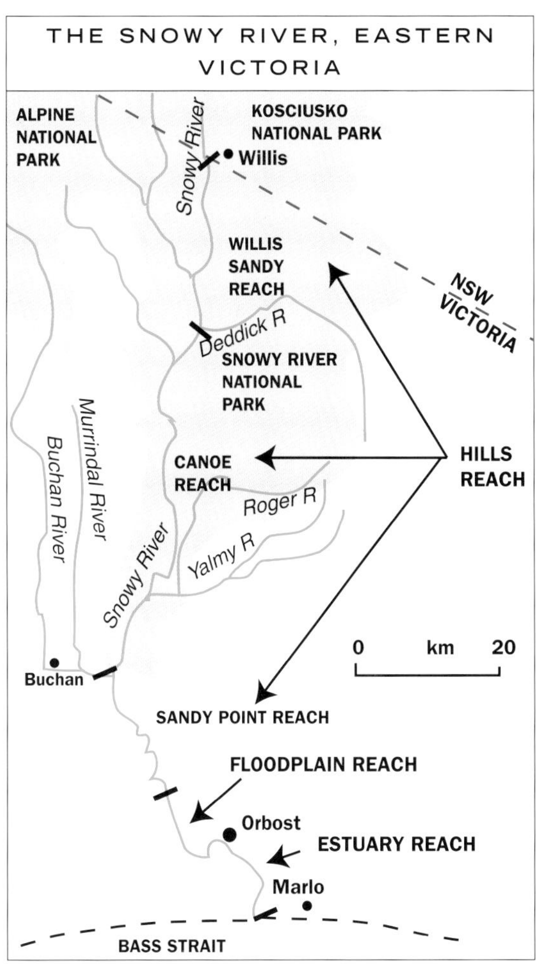

After Bain & Tillyeard[105]

upstream of its pre-SMS location, and some wetlands have become permanently salt. The estuarine lakes Corringle, Curlip and Wat Wat are also affected by the upstream migration of the salt wedge.

Prior to European settlement the Snowy Flats were a vast wetland system. Apart from the lower estuarine Lakes, which remain intact, there are only about 2 to 3 square kilometres of wetland left on the floodplain—representing less than 4 per cent of the 1865 area. Clearing and draining occurred in the first couple of decades after settlement; the river channel became larger; construction of levees and channel widening prevented flooding. The loss of ephemeral wetlands and riparian vegetation are the greatest man-made alterations to the river environment on the flats.

Features of the lower Snowy River and proposals for rehabilitation

A multi-disciplinary team of biologists, engineers and a geomorphologist, from the Department of Civil and Environmental Engineering, University of Melbourne, and consulting firms Fluvial Systems Pty Ltd and ID & A Pty Ltd, have studied the river and made recommendations for its rehabilitation.[105,106] In this recent survey, four sections of the river with different characteristics and problems have been identified:

- **Hills Reach** The Willis Sand Reach and Canoe Reach within the Snowy National Park have been colonised by willows and blackberries, in response to the contraction of the channel under the reduced flow regime.
- **Floodplain Reach** Change in flow regime has resulted in reduction of flow and in changes of bedform, which is now characterised by longitudinal bars. Deep pools and alternating bars, which existed prior to the SMS and provided favourable aquatic habitat, are no longer characteristic of this reach. The loss of riparian and verge vegetation and replacement in some cases by willows has also had effects.
- **Estuarine Reach** Lower flows have allowed the salt wedge to penetrate further upstream more frequently. Wetlands which were only brackish at times of natural low flow are now essentially saline. Loss of streambed vegetation and agricultural activities have increased erosion of banks.
- **Wetlands** Wetlands are of three types: floodplain wetlands fed by rain, not by the river; wetlands fed by the river; and estuarine wetlands fed by both tidal and river flows. All wetlands have been affected by conversion into farmland; regulation of flooding by levee banks; and regulation of flows by headwater dams. Despite varying degrees of degradation, all of the large permanent wetlands below Orbost, except Cabbage Tree Lagoon, are listed on the Register of the National Estate.

Rehabilitation of the river requires the establishment of sufficient quantity environmental flows and physical intervention in the river and its environs to compensate for man-made changes. The physical and biological state of the bed, banks and verge of the river have to be strategically managed.

The environmental flows should not only be sufficient for the needs of the river but should also mimic the pre-regulation flow patterns. Increased flows alone will not bring back the in-stream morphology which is most favourable for the aquatic biota. In the lower Snowy River, the original pool and riffle patterns were associated with regular alternating side-attached bars, which are practically non-existent today. Installation of timber pile fields in the form of retards at the historical location of the bank-attached side-bars is proposed to re-create pools. The retards increase flow resistance and lead to deposition of sediment around the pile fields. This will eventually increase the velocity of the channel opposite the toe of the bar and a scour pool will develop near the river bank. Alternating bars can be installed from Jarrahmond to Orbost, and this will help to bring back the variability and range of riverine habitats.

The re-creation of bars and pools will provide an opportunity for reinstating woody debris in the pools, where it should be aligned to improve habitat and to have a stabilising effect. The woody debris should be anchored so that it remains in position during major flow events.

Wetland rehabilitation will depend on the cooperation of landholders, who should be encouraged to fence sensitive areas, remove grazing and replant where necessary. The high value wetlands below Orbost are highest priority and Crown land frontages must be fenced and replanted and measures taken to limit saltwater intrusion.

Riparian vegetation has to be replaced, verges fenced so that stock do not have access, and a program of willow removal has been advocated. The willows issue is a contentious one and by no means clear-cut. (See below for further discussion.) Weed control is an essential part of the rehabilitation program.

THE CANN AND THURRA RIVERS

Comparison of a river affected by settlement with a pristine river

These two rivers were selected as the subject of a doctoral thesis by Andrew Brooks of Macquarie University because the Thurra River is the last temperate lowland alluvial river on the eastern seaboard which is in virtually pristine condition from its source to the sea, whereas the adjacent Cann River is highly disturbed.[109] The Cann was originally analogous, pre-settlement, and it is now a much altered, sand-filled, wide channel. Because of the direct comparability of the two catchments, these two rivers offer a unique opportunity to evaluate the impact of human activities on river morphology. Not only the nature of changes to the river system but also their magnitude can be demonstrated, and it can be clearly shown that the changes are attributable to human activities—largely de-snagging and riverine vegetation clearance. The Thurra provides firm evidence of what the original Cann was like—its floodplain vegetation as well as its physical form and function—so the element of speculation involved in reconstructing the river's original conditions from limited information is removed, in contrast to the situation of most other rivers which have suffered change.

The catchments of the Cann and Thurra Rivers lie side by side, with their rivers running parallel and about 15 kilometres apart, in east Gippsland. The geology of the two catchments is equivalent, both being dominated by Devonian granites and metamorphosed Ordovician sandstones. Average rainfall in both ranges between 1000 and 1200 millimetres, and their hydrology is similar. Vegetation along the undisturbed Thurra currently comprises two community associations—warm temperate rainforest and wet sclerophyll forest. The vegetation is believed to have been the same for the Cann River, making the two rivers as closely analogous as any pair of rivers could be. In addition, both rivers have been subject to logging activities in their headwaters (more intensive in the Thurra than in the Cann), and both have suffered wildfires—particularly the Ash Wednesday fires of 1983, when the whole of the Thurra catchment and 60 per cent of the Cann catchment were burnt out. Thus the factors which have resulted in the dramatic changes to the Cann can be attributed to the de-snagging and riparian vegetation clearance and other severe human-induced impacts which have occurred only on the Cann.

Settlement on the Cann River floodplain started in the 1860s and the whole floodplain had been largely cleared by the early years of the twentieth century, with the exception of the immediate riparian zone. In order to control overbank flooding an extensive de-snagging program was carried out along the alluvial reaches between 1940 and the late 1960s. De-snagging and the riparian vegetation clearing which accompanied it resulted in major channel expansion, which progressed downstream. The large flood of 1971 contributed further to channel expansion, which has continued and escalated in

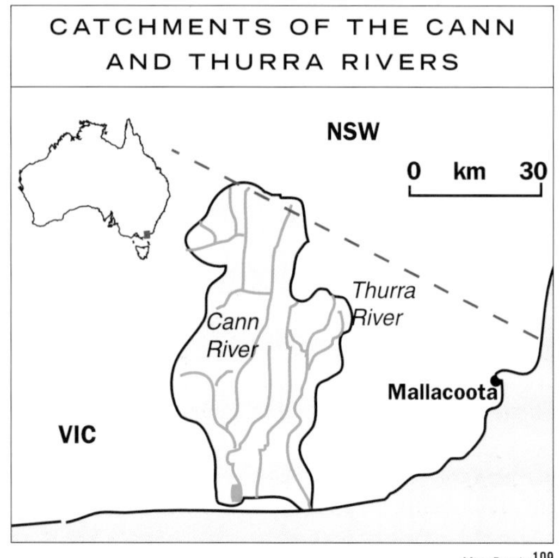

After Brooks[109]

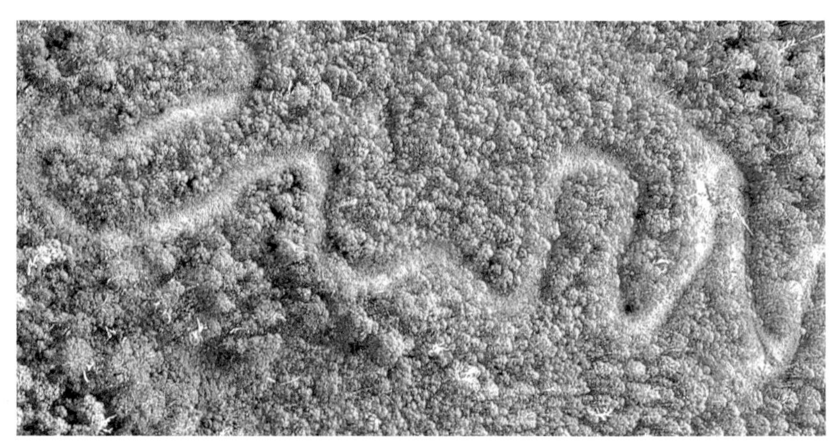

A sinuous palaeochannel is clearly seen in the densely vegetated floodplain of the Thurra River. The palaeochannel was abandoned 5600 years ago. Compare this picture with the palaeochannel on the Cann River floodplain. The contemporary Thurra River channel is obscured by forest in the upper left-hand corner of the picture. ANDREW BROOKS

Two studies, mainly involving two different species of willow, have looked at the advantage or disadvantage of planting willows on river banks.

WEEPING WILLOWS (SALIX BABYLONICA) IN THE UPPER MURRUMBIDGEE CATCHMENT[107]

This study gives an insight into the willows' widespread distribution and into the arguments in favour of their use as riverbank stabilisers, or against their continued presence in the landscape.

Willows have been established in the upper Murrumbidgee catchment of New South Wales since the 1820s. They may have been imported originally as garden trees, to soften the harsh Australian landscape and remind early settlers of 'home'. They took readily to local conditions and grew so obligingly fast that they were obvious candidates for gully and river bank stabilising when the effects of over-grazing, land clearing and other European activities caused the massive erosion and change in river form and function, which has been described in this book for streams of eastern Australia. The period between 1850 and 1870, in particular, was one when 'perfectly astounding' erosion was recorded by W. B. Clarke (1860) in the southern goldfields of New South Wales. Up until the 1920s, willow planting was a regular activity.

Willows provided stability, and reaches of rivers that have had their protection have been less prone to erosion than those without them. Main channel erosion was largely overlooked while soil conservation and other agencies were worrying about hillslope stabilisation. In the upper Murrumbidgee catchment, a change in emphasis occurred in the 1990s with attention being given to channel erosion. (Research had shown that sediment and phosphorus sources were dominated by channel erosion, prompting the change in emphasis, and Landcare groups are now actively working to reduce flood nuisance and riverbank erosion.) The use of weeping willows as a pioneering species in actively eroding sites became part of the programs.

Willows are well suited to stabilising eroding banks. Their dense root systems of fine roots provide a surface mat as well as an efficient bond between a sand and gravel substrate and the soil of banks. The speed of willow growth is a great advantage. While native vegetation may take a couple of decades to establish sufficiently to supply protection, willows do it in five years. To establish a willow, all that is required is to set a branch in a hole. If the branch is tall enough, no protection is needed from grazing stock. In addition, willows have few pests; they provide valuable stock feed in droughts; they grow in frost hollows where no native trees will; they provide welcome shade in summer; and they have aesthetic appeal. There is simply no Australian candidate with all these advantages, or even most of them, to replace the willow. (This applies in many man-made degradation situations, where only exotics with the vigour of weeds are able to cope with the changed, pioneering conditions which we have created.)

The disadvantages of willows are numerous: Their vigour and ease of growth has led to domination of many riparian zone ecologies; they exclude understorey; new strains are now seeding where once only vegetative growth was involved in their spread; they create dense shade and they do not provide hollows or food for native fauna; they decrease channel capacity and deflect erosive flow against banks; and they cause avulsions where they have blocked channels. (Most of these disadvantages only present themselves where there has been no control, and perhaps we need to apply 'farm management' to our rivers, and their banks, along the reaches which we have altered by our activities.)

The study concluded that total removal of willows from the Australian scene would be impossible even if it were to be proved to be desirable. Removal of those growing within channels is obviously essential, but even this has to be staged and managed so that there is no sudden release of slugs of trapped sediment and no rapid change in flow patterns. Because of the variability of flow in most of our rivers, any sudden and large-scale removal of willows may endanger bank stability and cause more problems than existed before.

WILLOWS IN TASMANIA[108]

This study, where the species involved is mainly crack willow (*Salix fragilis*) presents a more clear-cut picture of the disadvantages of willows in the riparian zone.

Crack willow is now present on every lowland river and stream where land clearing has taken place. It was used at first to control riverbank erosion when fertile river flats were cleared for agriculture. Where it has become the only vegetation on small streams, its vigorous growth has meant that no other vegetation can become established. In the majority of these cases it invades the stream bed to some degree, and many streams now have a bed consisting mainly of willow roots, fine sediment, and organic matter resulting from willow leaf fall. When old willows fall over they frequently re-sprout, collecting debris and creating log jams, which increases flood levels.

The effects of the willow invasion have been increased flooding, reduced summer flows, loss of habitat for flora and fauna, and low levels of dissolved oxygen—which limits aquatic life. Invertebrate fauna is depleted in numbers and in species in willow-infested reaches of the rivers monitored. Willows are deciduous and contribute large amounts of organic material over a short period, compared with native species with more continuous leaf fall. The bacterial activity which is involved in breakdown of the leaf fall material results in low oxygen levels in the water. Nutrient levels also increase because the willow roots impede throughflow. The shading of the river has an effect, with less photosynthesis by in-stream plant life, while water temperature is also affected by the shading. The graph shows a comparison of dissolved oxygen and water temperature between an area infested with willows and an area which has been cleared in Quamby Brook in northern Tasmania.

Through the advent of the National Landcare Program and, more recently, the National Heritage Trust, the Tasmanian community has become increasingly interested in rehabilitating the river corridors. The availability of funds has made willow removal a priority. Landcare Tasmania produced a booklet entitled

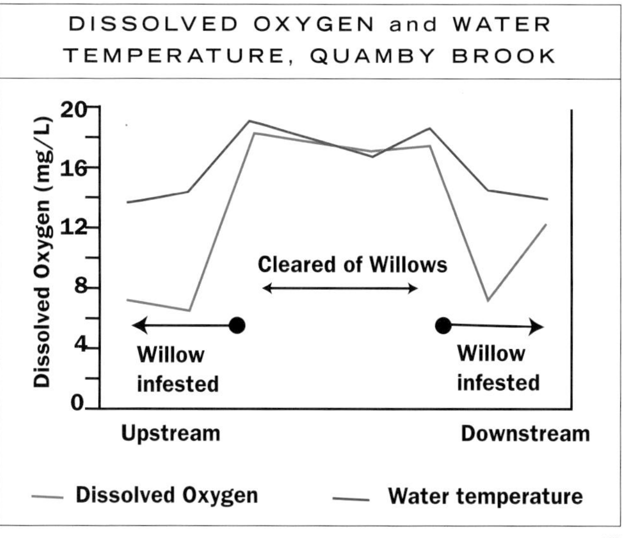

After Bobbi[108]

Willow Management Guidelines, which provided invaluable advice for the small groups of people operating in their local areas. With availability of greater funding under the National Heritage Trust's Rivercare Program more ambitious clearing programs on larger rivers have become possible—but, ironically, there is a greater risk of environmental damage if the clearing is too rapid and its consequences are not understood. The Department of Primary Industry, Water and Environment has joined with Landcare to develop *Rivercare Guidelines* to help planning and managing river rehabilitation.

The hazards of willow removal are several:

- The creation of pastured stream banks with no other form of vegetation is believed to present the greatest hazard. The increased light levels cause blooms of aquatic plants and algae which can result in supersaturation of dissolved oxygen during the day; removal of shading also increases water temperature. Invasion by native and exotic weeds such as cumbungi (*Typha* spp.), parrot's feather (*Myriophyllum* spp.) and Canadian pond weed (*Elodea* spp.) creates new management problems.
- Between the time when willows are removed and bank stabilising vegetation is established, movement of sediment and nutrients which had been trapped by the willows presents the greatest threat. Algal blooms can result, particularly in agricultural areas. Erosion, siltation of downstream reaches and general instability are threats. Leaving stumps on banks is preferred to complete removal, but the ability of the willow to regenerate from any remaining living fragment makes follow-up management essential.
- Management of heavy machinery required to undertake willow removal is an additional challenge, as there is a temptation to remove all in-stream obstacles, and large woody debris is important as habitat for aquatic organisms. There has also been a tendency for some working groups to try to modify the stream channel. Straightening results in greater velocities and erosive power during high flows.
- During heavy machinery operations there is a risk that surviving native vegetation will be damaged or destroyed. It is much better to retain what is there than to start from scratch with replantings.

The contemporary Cann River channel at West Cann Bridge. Note the excessive sedimentation and the lack of LWD in the channel. ANDREW BROOKS

The contemporary Thurra River channel. Note the low capacity of the channel, the extensive amounts of LWD in the channel and the deep pool in front of the observer. ANDREW BROOKS

the last three decades.

Channel capacity, as compared with pre-European river conditions, has increased by 700 per cent; bank-full volume has increased 45-fold; sediment transport capacity 1000-fold. By studying the floodplain sedimentology it has been possible to show that these changes are unprecedented over at least the last 27 000 years. The rate of channel erosion and the enormous changes to systems dynamics are well outside the bounds of any that might have been caused by natural changes to climate, flood regimes or extreme fires. The Cann River is still adjusting to the changes today, and it can be demonstrated that restoring the channel to its original dimensions is an unattainable goal.

Disturbance to the Thurra River and its floodplain has been minimal in comparison. Very high LWD (large woody debris) loading is a feature of the river. The sandbed Thurra channel exhibits the pool and riffle morphology characteristic of gravel streams. This is largely a function of the large volume of LWD in the channel. The fact that the bed is reinforced with wood over considerable stretches, combined with the hydraulic roughness associated with the logs, has led to a situation where the transport of sediment is so low that pools and riffles have been able to develop. The de-snagged Cann, on the other hand, has a relatively uniform bed with minimal roughness.

Even the Thurra, however, with its well-forested floodplain, is subject to channel change. The aerial photograph shows a meandering palaeochannel reach of the river. Apparently the river, hidden among the trees in the top left part of the frame, abandoned that reach with its meanders and now runs in what would have been

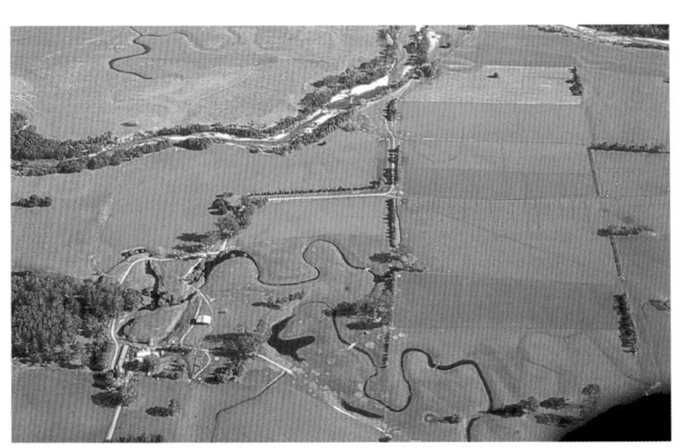

Oblique aerial photo showing the cleared floodplain of the Cann River with low capacity, highly sinuous palaeochannels clearly preserved on the floodplain, and the relatively straight, high capacity contemporary channel. ANDREW BROOKS

an overflow channel during floods, the whole new meandering loop rejoining the main channel downstream. Cut-offs, anabranching and avulsions are normal parts of river behaviour even in forested catchments, but they occur slowly over considerable lengths of time. The palaeochannel section in the photograph was abandoned 5600 years ago. Channel migration rates in the Thurra and pre-disturbance Cann are in the order of 10 to 30 millimetres a year. (Recent Cann River migration rates are up to 4.5 metres per year.)

The pre-disturbance channel of the Cann River is well preserved in the floodplain and is clearly of comparable dimensions to the contemporary Thurra. Modelling from the information supplied by the original Cann channel gives a good correspondence with the observed situation of flow, sediment transfer and other functions in the Thurra. Thus in this case the conclusions about the amount of change experienced by the river drawn from analysis of floodplain evidence can be validated by comparison with a very similar river which is in pristine condition. And, in particular, it is shown that the agent for the vast changes in the Cann has been human activity, particularly damage to riparian vegetation and removal of large woody debris from channels.[111]

The widespread practice of de-snagging, only recently falling out of favour, has been exacerbating the very processes which it was intended to combat. If it were feasible to restore woody debris to degrading streams they would have more chance of repairing themselves over time, but where gross channel widening has occurred it is not feasible—it would take many times the vast amount of wood which has been extracted from the rivers and there is no source of such material. By restoring riverbank vegetation, in time rivers will again have access to the LWD they need for health and stability.

The importance of LWD in controlling river dynamics in pre-disturbance rivers, particularly perennial alluvial rivers, is shown by the study of the Cann and the Thurra Rivers. When a river channel is narrow, a falling tree can span it and become fixed by both banks; when channel widening occurs, snags no longer have the same effect and the LWD factor in stream stability is lost or impaired.

THE GENOA RIVER AND JONES CREEK

The **Genoa River** has its headwaters in the mountainous country in the south-eastern corner of New South Wales and runs parallel to the New South Wales–Victoria border, terminating in Mallacoota Lakes.[110] The Croajingalong National Park, which includes the Mallacoota Lakes system, is part of an internationally recognised Biosphere Reserve. Rainfall in the catchment is about 1000 millimetres a year, evenly distributed but with frequent intense rainfall events which often cause flooding. Jones Creek is a short south bank tributary with a forested subcatchment largely within the Coopracambra National Park.

The Genoa River lies within the East Gippsland Basin, of which 97 per cent retains its native vegetation, and most of the streams are in good to excellent condition. However, in the mid 1980s the lower 10 kilometres of the river through the agricultural floodplain reach was assessed as being in poor to very poor condition in a State-wide survey.[112]

Clearing of the river margins and grazing predisposed the banks to erosion and channel widening. The severe floods of 1971 and 1978 resulted in a straightened, shallow, sand-filled riverbed up to 65

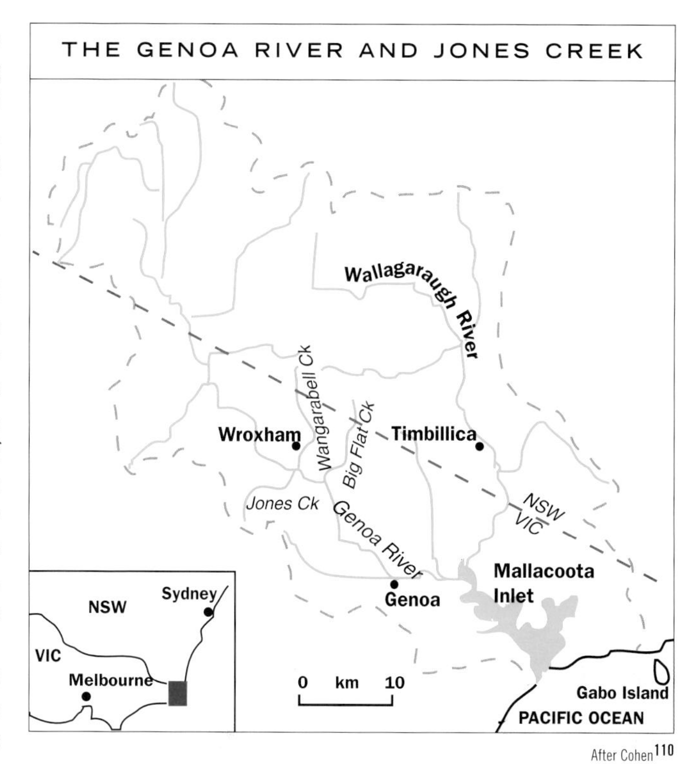

After Cohen[110]

metres wide covering over 60 hectares, with 4-metre high vertical banks largely devoid of vegetation.

The East Gippsland Catchment Management Authority is continuing a program of river rehabilitation commenced in 1990 by its predecessor river management organisation. The program has created conditions that have allowed hundreds of thousands of indigenous plants to establish by regeneration on the bed and banks. It relies principally on river flow and the resulting natural disturbance to create and stock seedbeds with indigenous species. Creating the conditions for germination and managing the successional establishment of the indigenous species is termed 'assisted regeneration'.[113]

The main management actions were the stabilisation of the bed and banks through rock beaching and retardation fencing at key locations, cattle exclusion through 20 kilometres of fencing (1990– 93), weed control and strategic planting (1994 and ongoing). The critical elements are the creation of stable river conditions to allow protection of the seedbed and plants from browsing and weed competition.

Assisted regeneration recognises succession, and uses the ongoing changes in the composition and structure of vegetation communities as an aid in the rehabilitation of the disturbed river. Part of assisted regeneration is planning management actions to influence succession by controlling some species (weeds) and promoting others, for example, by planting the slow-to-regenerate kanooka (*Tristaniopsis laurina*).

Success on the Genoa River has depended on planning, vegetation knowledge, communication, the support of adjacent landholders, and persistence. A geomorphological study determined the system conditions, and a strategic plan (Community Works 1994) set priorities, management objectives and directions. An investigation of vegetation strategies by Riparian Australia determined that regeneration was more appropriate and cost-effective than planting, and that good farm management on adjacent properties would allow successful regeneration by ensuring a low adjacent weed population. Practical techniques for maximising regeneration were developed with local expertise, and demonstration sites were used to communicate the new approaches.

Persistence over years is the essential prerequisite of assisted regeneration. On the Genoa River, an annual weed and fence maintenance program helped to produce an environment conducive to successful germination. The example following illustrates the need to sustain such preconditions to take advantage of the right circumstances when they arise in the unpredictable river environment. Ten kilometres of river (even when severely degraded) is highly heterogeneous.

On the lower Genoa River, while regeneration occurs across the sandbed in suitable microhabitats, floods usually wash away most of the plants, leaving only those in protected areas towards the bank. The thousands of remnants provide more protection for each other as they grow. The vegetated width increases by annual increments, and in places is now 20 to 30 metres wide; acacias are dominant. In June 1990, a moderate flood of 4.8 metres followed three dry years in which there had been no regeneration beneath the acacia trees. The flood filled the channel and deposited fine silt and seed beneath the acacias. A massive germination of shrubs followed, and tens of thousands of plants, principally tea-tree and paperbark (*Leptospermum* spp. and *Melaleuca* spp.) are now establishing. Without sustained weed control, blackberry and other environmental weeds would have dominated and prevented the germination and without further control this year the blackberries, which also germinated, will grow and smother the shrubs. This is just one of many such plant and environment interactions which management aims to promote. (Ed Thexton of Riparian Australia kindly provided the updated information on the Genoa River project.)

Management by assisted regeneration is protecting banks and immobilising sediment movement downstream into the Mallacoota Lakes, achieving the central aim of rehabilitating the lower river. The diverse range of indigenous plants is binding the sediments and reducing channel width.

Assisted regeneration has application for the rehabilitation of upper catchment river reaches, in particular at or downstream of diverse remnant vegetation populations. As demonstrated on the lower Genoa River, it is a cost-effective model that results in the establishment of thousands of plants of an array of species and forms for minimal cost in the volatile and unpredictable river environment while maximising biodiversity.

The Genoa River, downstream of Genoa. TIM COHEN

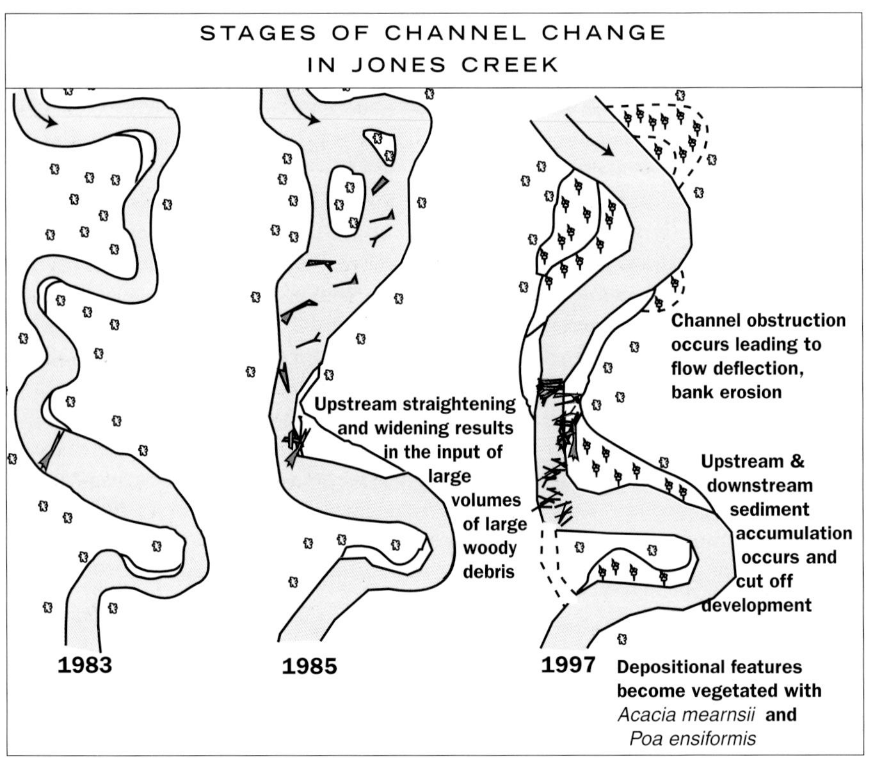

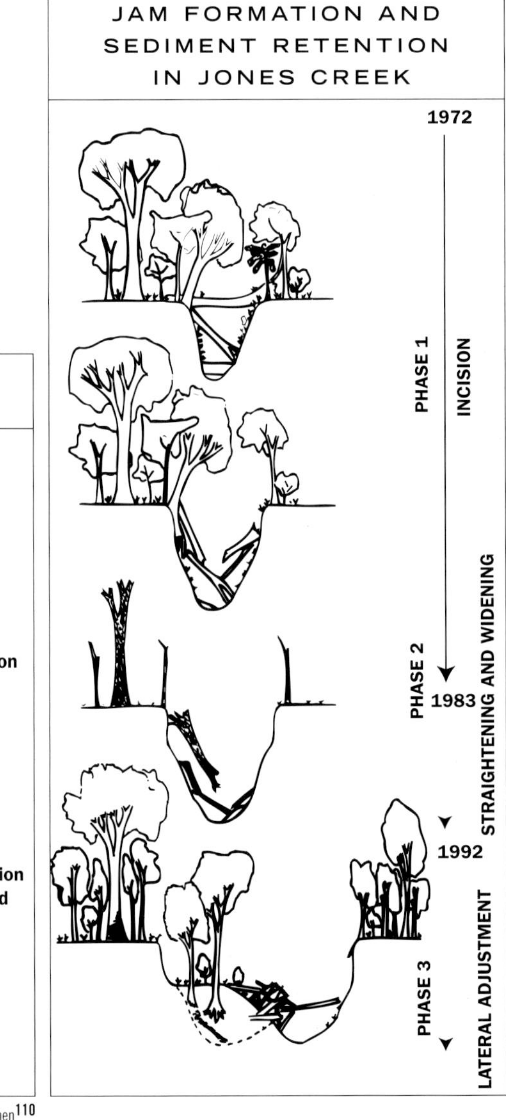

After Cohen[110]

Tree-fall in Jones Creek where the stream is narrow enough for the fallen tree to form a barrier across it. TIM COHEN

A megajam, Scaly Reach, Jones Creek. TIM COHEN

Jones Creek is a 30-square kilometre subcatchment of the Genoa River. This ephemeral stream transports silt, sand and gravel and has undergone extensive change in channel pattern and dimensions over the last 30 years. It traverses a forested floodplain which increases in width from 50 metres to 500 metres towards the confluence of the Genoa River. Much can be learned about natural mechanisms for channel recovery from a recent study by a scientist in the Geoscience Department of the University of Wollongong.[110]

The lower reaches of Jones Creek above its junction with the Genoa River were affected by the channel widening of the Genoa which followed the floods of 1971 and 1978. With the twofold widening of the Genoa, a lagged tributary response resulted in significant changes to channel form in the 3.5 kilometre stretch of the river studied. Channel widening and bed incision and steepening in Jones Creek resulted from the destabilising of the system, but now the creek displays evidence of natural mechanisms which may enable it to recover, and which illustrate how other degraded streams can be managed to enable their recovery.

Three phases of channel change have been identified along Jones Creek since 1972:

- **Phase 1 (1972–83)**: channel depth increased by 1.5 metres and bed slope increased, resulting in an increase in stream power. There was some channel widening during this phase but the river maintained its sinuosity.
- **Phase 2 (1983–92)**: large increases in width and reduction in sinuosity occurred as the river straightened. Bed aggradation and alternate bar formation also occurred. By 1992, the bed was 32.3 metres wide, compared with 16 metres in 1983. The Ash Wednesday wildfire burnt out the whole area in 1983, and may have had an effect on the already destabilised river. It is estimated that 171 000 cubic metres of material was eroded from the study area during the expansion stage, including 99 200 cubic metres of sand. One-third of the sand fraction is presently stored within the channel in realignment features, cut-offs, point bars and bed storage. The storage of this material is part of the recovery process, with 65 per cent of it 'locked up' in vegetated realignment features (and not available to be reworked as it is in so many destabilised rivers where the rate of colonisation of the sand features is not sufficient).
- **Phase 3 (1992 to present)**: a period of further localised channel expansion and lateral adjustment, but there has been only minor change to sinuosity.

Further straightening occurred in the recent 1998 flood with continuing aggradation and some degradation in areas that have recently straightened. High stream power has been maintained.

System recovery in Jones Creek depends on within-channel aggradation and subsequent stabilisation of depositional surfaces. The deposition and colonisation of channel marginal features or incipient floodplains is part of this process. Long-term channel recovery is dependent on the reduction of stream power, the increase of sinuosity (within a wider channel), the decrease of channel width, and subsequent reconnection of the channel to its floodplains so that floodplain inundation results.

Rates of colonisation of depositional surfaces are extremely high in Jones Creek. Black wattle (*Acacia mearnsii*) is the dominant early successional species, which quickly locks up sediment, reducing material for reworking and reducing channel cross-sectional area. Secondary canopy species with similar growth rates, such as river peppermint (*Eucalyptus elata*) and manna gum (*E. viminalis*), provide for longer-term stabilisation by establishing a forested surface with a range of longer-lived species. Hazel pomaderris (*Pomaderris aspera*), spiny-headed mat-rush (*Lomandra longifolia*) and sword tussock grass (*Poa ensiformis*) further enhance the stability of within-channel features. Rush and sedge species which grow within the river bed itself are valuable stabilisers. (*Juncus usitatus, Cyperus eragrostis, Phragmites australis* and *Lomatia longifolia* are the most prominent of these.)

The large volume of coarse woody debris and the log jams within Jones Creek reduce flow velocity and increase channel roughness—both factors which help to stabilise the system. The log jams are reworked during floods but, given time, they locally induce channel marginal deposition and subsequent colonisation and stabilisation.

The recovery mechanisms seen in action in Jones Creek suggest that river management policies for other degraded coastal rivers should encourage natural recovery by retaining (and assisting in the regeneration of) in-channel vegetation; and by retaining or reintroducing woody debris.

SOUTH COAST RIVERS OF NEW SOUTH WALES

The Bega River Catchment

As occurred in so many east coastal rivers, stream courses and river behaviour in the Bega River catchment were completely altered within a few decades of European disturbance.

The Bega catchment drains an area of 1040 square kilometres. The catchment has a steep escarpment, rising to around 1200 metres, on three sides, descending abruptly to foothills through which most of the river system flows. About 70 per cent of the original eucalypt forests and woodlands in the catchment were cleared within a few decades of European arrival.[117] This is a sensitive granite catchment where clearing rapidly changed flood regimes. Devonian granites and granodiorites dominate the local geology, giving rise to deep, well-drained and erodable sandy soils. Rainfall varies between an average annual 750 millimetres and 1200 millimetres, depending on topography, but is highly variable from year to year and throughout the year.

At the time of first European settlement of the region between 1830 and 1840, the middle and upper courses of the Bega River's tributaries were discontinuous, and characterised by extensive swamps. A continuous, low capacity channel flowed across the lowland plain. Upland sections of the subcatchments had continuous valley fills characterised by chains of ponds. Drainage of upland swamps, over-grazing, and a range of indirect responses to early agricultural pursuits triggered head-cut incision into these large sediment sources. Incision was quickly followed by extensive channel expansion, supplying massive volumes of sediment to the lower catchment—up to 10 million cubic metres.[118] Incised channels are locally more than 10 metres deep and 100 metres wide. As a result, channel floors were disconnected from their perched valley fills, and swamps ceased to exist where the incision had occurred.

Downstream, the lower Bega River was also completely transformed.[119] Originally, it had a deep, narrow channel lined by river oaks, with deep pools for platypus habitat. It was a suspended or mixed-load system. Banks were fine-grained and relatively cohesive silt and clay, as was the floodplain, which graded to a series of valley-marginal swamps and lakes. Clearance of riparian and floodplain vegetation, drainage of swamps and wetlands and the critical disturbance of tributary headwaters completely changed the lower river. Channel change began after 1850, following European settlement.[102] The deep, tree-lined, suspended mixed-load channel widened by up to 340 per cent and shallowed significantly. The sediment accumulation on the floodplain surface was transformed to fine to medium sands. The channel widened from about 40 metres to 140 metres between 1851 and 1896; pools were filled with sediment; up to 2 metres of sand accumulated on floodplains which had previously accumulated fine silt.

By 1926, the channel in the reach which winds round the Bega township was 250 metres wide and dominated by broad sand sheets. It had become a sandy bedload system transporting coarse granular sand in its channel and depositing fine to medium sands on its floodplain. It has been estimated that up to 4.5 million cubic metres of sand have been added to the floodplain, while 8 million cubic metres of material resides in the channel on the lowland plain. Remarkably, less than 20 per cent of the sediment released from the catchment has reached the estuary.

It seems that relatively little change to river structure occurred between 1900 and 1960. (The first half of this century was comparatively dry.) Since then, the lower Bega has become choked with willows and other exotic vegetation and a complex pattern of bars and islands has developed within the braided river channel.

The changes to the form and function of rivers in the Bega

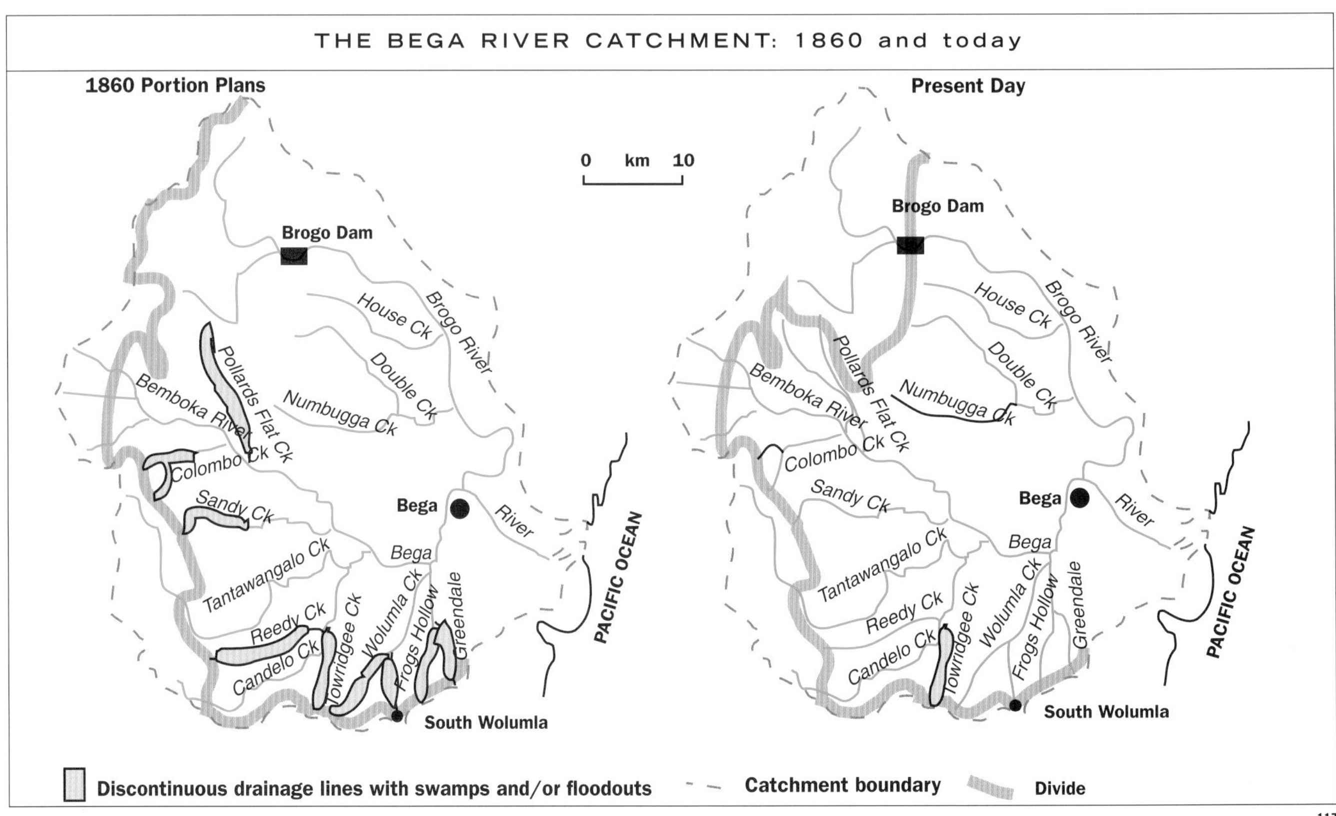

After Brierley[117]

LAKES CREATED BY GEOLOGICALLY RECENT VULCANISM IN VICTORIA

MAAR LAKES

A maar is a broad, roughly circular, flat-floored volcanic crater with steep inner walls and a low surrounding rim built from fragments of rock material blown out in the eruption. Thirty such craters occur in the region west of Geelong, concentrated near Colac and Camperdown, and Tower Hill west of Warrnambool.

A maar develops as a result of violent volcanic activity. Superheated steam and other gases emerge from the ground ahead of a rising column of lava. The steam is the product of boiling groundwater, and the combined pressure of steam and the ascending lava and its gases blasts a pipe, or vent, through the overlying surface rock. Continual eruptions widen the vent into a broad crater, which forms the floor of the maar. Blown-out fragments of tephra or pyroclastic rock (tuff) forms the rim of the crater.

Lake Keilambete, in particular, has acted as a rain gauge. It is one of three maars near Terang which have been extensively investigated in order to establish a climatic record for the last 16 000 years.[114] Lake Keilambete and Lake Gnotuk have a history of about 10 500 years, Lake Bullenmerri of 16 000 years. Gnotuk and Bullenmerri occupy twin craters of high relief, with rims about 130 metres above lake level, separated by a saddle comprised of ejected tuffs. A variety of volcanic eruptions in the vicinity preceded the eruption of the maars, whose craters were also formed in multiple volcanic events, unlike most maars within the region. All three maars have erupted through the Port Campbell Limestone.

Lake Bullenmerri is a twinned crater and is clover-leaf shaped in consequence. Lake Gnotuk is single and almost circular. The overflow point for Bullenmerri is into Lake Gnotuk over the saddle between the two, and the last recorded overflow was in 1841. There are two sources of groundwater in the strata surrounding the lakes. The lower source is a confined aquifer of low yield in the Port Campbell limestone, the upper an unconfined aquifer at the base of the tuff, forming a perched water table that surrounds the crater. The two aquifers merge towards the lake, discharging as springs round the water's edge. The basal sediments in all three lakes overlie aquicludes, preventing any significant downward seepage. The soils and tuffs within all three craters are highly permeable, absorbing most rainfall which then infiltrates down to the water table or is transpired upwards. Run-off is practically non-existent and infiltrated soil moisture ends up as baseflow. Thus the lakes act as giant evaporimeters, being dominated by surface processes, with some input from surrounding groundwater. The record of water level is contained in organic muds and shoreline sands deposited over time.

Today the climate in the region of the three lakes is sub-humid with rainfall

Lake Gnotuk, separated from Lake Bullenmerri to the left of the picture by a ridge, is twice as salty as seawater. It is effectively an evaporation basin. M.E.W.

Lake Bullenmerri is brackish to fresh, receiving more local run-off than Gnotuk. M.E.W.

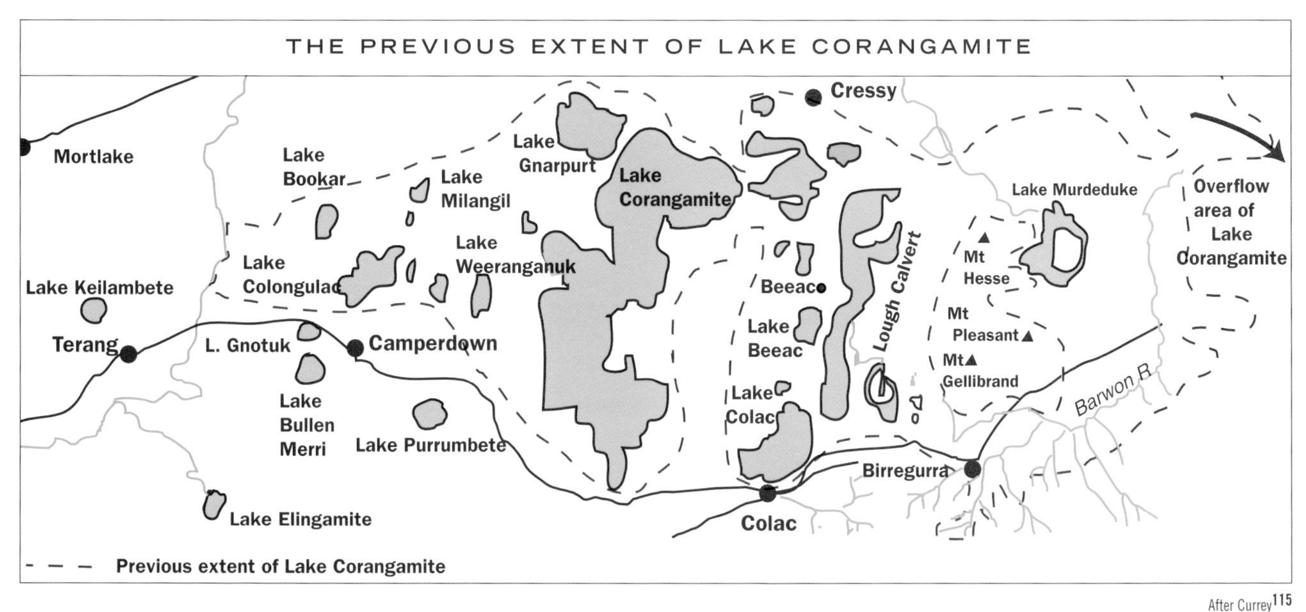

After Currey[115]

averaging between 815 and 825 millimetres a year. Before European occupation, the lakes were surrounded by grassy woodlands to open forest with river red gums (*Eucalyptus camaldulensis*), *E. viminalis*, *Allocasuarina stricta* and several species of *Acacia*. The grassland was dominated by the perennial tussock grasses *Themeda triandra* and *Poa labillardiera*. Densely vegetated swamps and large shallow lakes were widespread in the vicinity. Grassy woodlands tending to grasslands occurred to the north on basaltic soils; open forests on calcareous soils to the south. The current vegetation is exotic grassland with some tree belts dominated by cypress (*Cupressus* spp.) and sugar gums (*E. cladocalyx*). Almost all the swamplands have now been drained and areas which used to be flooded annually are farmed year-round. Land use is mainly grazing. Lakes Gnotuk and Bullenmerri are listed on the Register of the National Estate.

Climate history recorded in the lake sediments of Bullenmerri and Gnotuk reveals the following information:

- The Late Pleistocene was a period of increasing temperature and rainfall following the last glacial maximum. Bullenmerri was the only lake holding water during this period. From 16 000 to 14 500 years ago it was shallow, ranging from 10 metres to a few metres deep (compared with today's depth of 60 metres). Swamp conditions may have existed in Keilambete and Gnotuk at 14 500 years ago.
- At about 14 500 years ago Lake Bullenmerri shallowed, producing carbonate platelets at its deepest point, indicative of dry times.
- Between 10 500 and 10 000 years ago the region was wetter and all three lakes held water.
- A sustained rise started at about 8500 years ago and resulted in all three lakes overflowing by 7000 years ago.
- Between 7000 and 5500 years ago the three lakes were overflowing with brackish water.
- Progressively drier conditions resulted in a long-term decline in lake levels and by 3500 years ago the lakes approached their minimum depths and highest salinity levels.
- Between 3100 and 2000 years ago lake levels were low and salinity high, but several recoveries in lake level took place within this interval.
- From 2000 years ago, the water levels were rising in all three lakes, and Bullenmerri remained at or higher than current levels from 1700 years ago. Trees were drowned round the water's edge in this filling; in the 1960s, recreational divers reported trees at depths up to 27.4 metres below the surface in Bullenmerri.

Today Bullenmerri is a brackish to freshwater lake, and Gnotuk has twice the salinity of sea water. Bullenmerri, and Purrumbete—which lies further east towards Lake Corangamite—are the two main lakes in Victoria where salmon are liberated. An aerator in Bullenmerri helps to prevent algal blooms by mixing the lake's warm surface water with the cold water below.

The Tower Hill Lake lies within the crater rim, a large expanse of water with a central ash island. The volcanic area is a refuge for native animals in the cleared surrounding landscape and a great deal of re-vegetation and maintenance has been carried out by local Landcare groups. Now koala numbers in this 'island' situation have become a problem—too many for the limited food source. M.E.W.

The cliff-like rim of the Tower Hill crater shows the horizontal layers deposited while the volcano was active 12 000 years ago. M.E.W.

LAKES FORMED WHEN VOLCANIC ACTIVITY DAMMED RIVERS

Lake Corangamite, Victoria's largest lake, is one of many which were formed when volcanic activity resulted in basalt blocking streams and diverting them into hollows which were also partly a product of the vulcanism, created by the overlap of different lava flows.[115] As time went on, the major rivers which had filled Corangamite were themselves diverted by the formation of divides, and now this lake and other lakes, like the maar lakes, depend mainly on groundwater and rainfall run-off (mainly infiltration on the rims of the maars) from their surrounding catchments for their water.

Lake Corangamite today has an area of about 160 square kilometres. It receives the Woady Yallock creek from the north and the Pirron Yallock from the south. Mt Emu Creek drains the western flats of the area to the sea. The lake area was originally seven times larger, as shown by the distribution of lacustrine sediments and lunettes, and extended from Inverleigh and Winchelsea in the east to Mt Emu Creek in the west; to Foxhow and Cressy in the north; and Camperdown, Colac and Birregurra in the south.

Maximum extent of the lake was between 7000 and 6000 years ago.

CHANNEL CHANGES, LOWER BEGA RIVER

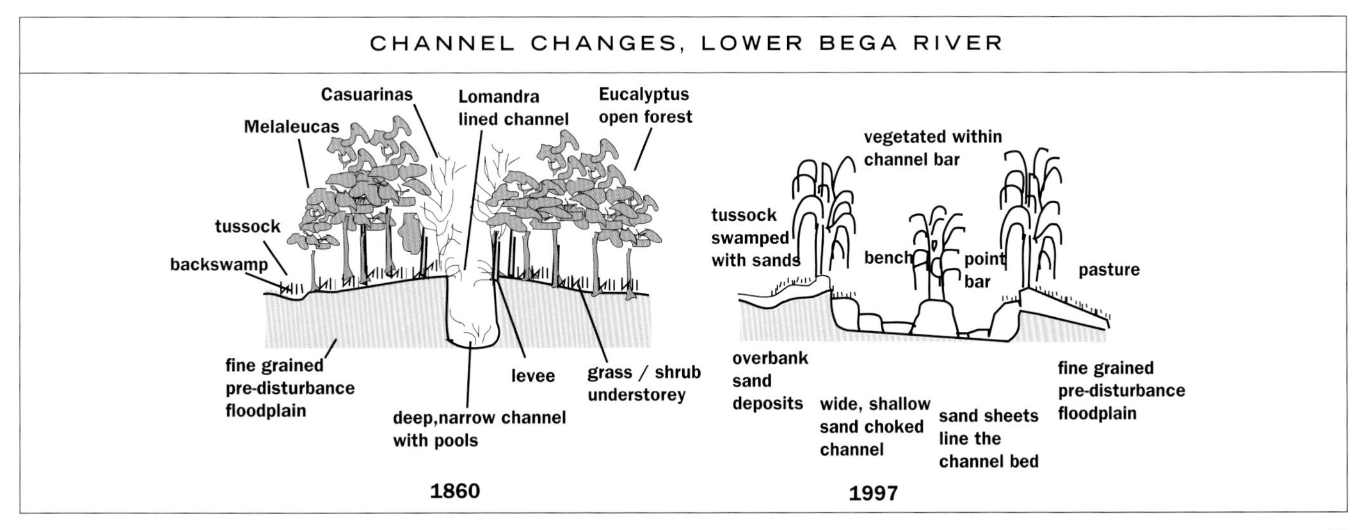

After Brierley *et al*[117]

DISTRIBUTION OF RIVER STYLES IN THE BEGA—BROGO CATCHMENT

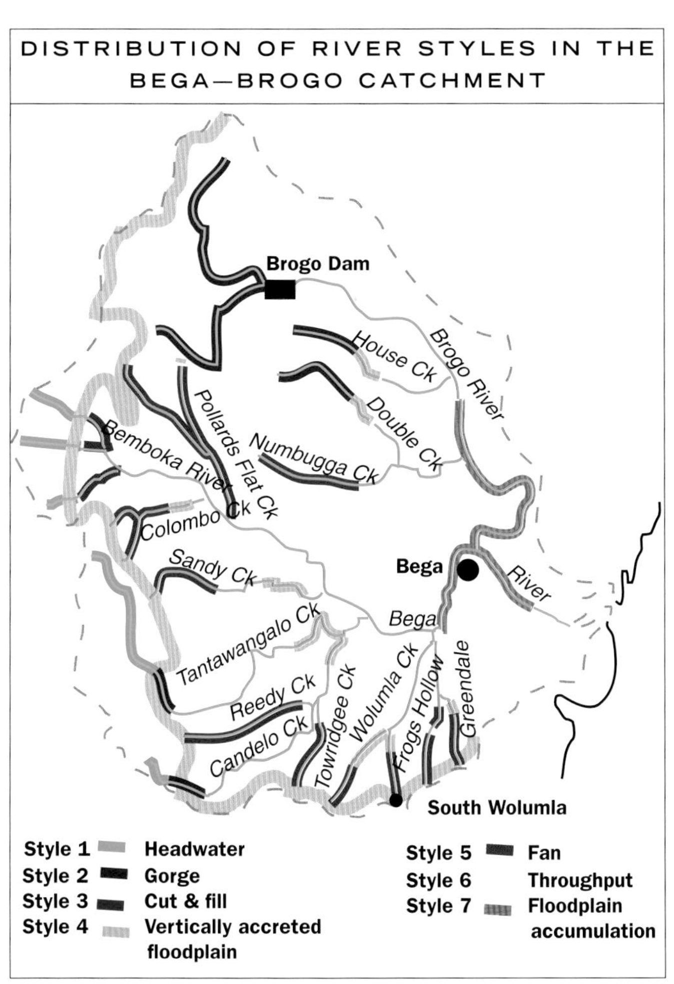

After Brierley[120, 121]

PRIORITISATION OF RIVER MANAGEMENT SITES IN THE BEGA—BROGO CATCHMENT

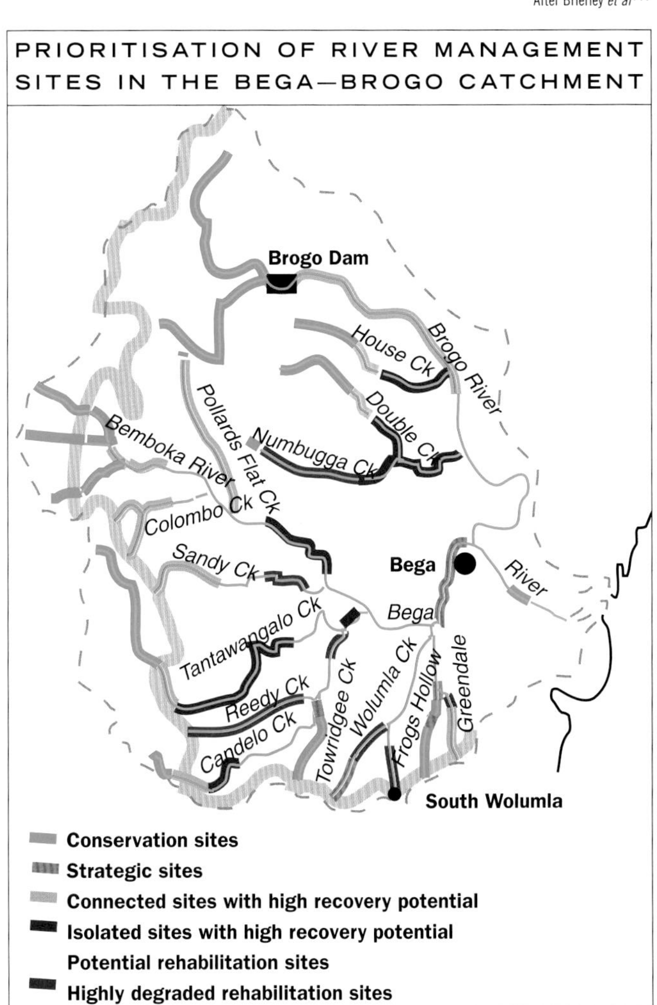

After Brierley[120]

catchment over the last 150 years have no equivalent analogue over the last 6000 years. The adjustments have been catchment-wide and the magnitude of changes such that the prospects for system recovery are significantly constrained.[120] To reduce the dimensions of the over-widened channels is virtually impossible—only very limited materials are available along river courses. This is a serious dilemma, and one that is faced in the management of so many of our rivers. The biophysical consequences of such changes cannot be overlooked. Unless stable physical processes are restored along river courses, efforts at river rehabilitation and attempts to maintain aquatic ecosystems will be compromised.

In an attempt to address these issues, scientists at Macquarie University have developed a generic and open-ended catchment-based approach to classification of river styles.[120] ('River style' summarises river character and behaviour within a reach.) This framework can be used to prioritise management efforts at river rehabilitation. Conservation is the basis for the strategies. As a first priority, those sections of the catchment which retain inherent values in geo-ecological terms must be protected and maintained. Should

A wide, sand-filled reach of the Bega River. GARY BRIERLEY

Incised cut-and-fill river style in Upper Wolumla Creek. The site of a former swamp, the channel is now 10 metres deep and 60 metres wide. Bands of sand and mud materials can be seen in the banks, indicating recurrent phases of cut-and-fill river processes in the past. KIRSTIE FRYIRS

The wide, willow-crowded Bega River winds round Bega. GARY BRIERLEY

Mid Frogs Hollow Creek, an example of a floodout river style. The wide valley on moderate slope has complete valley fill extending over the whole valley floor. Sand sheets splay over the swamp surface (foreground) and muds accumulate in the seepage zones downstream (background). Discontinuous watercourses and ponds characterise the surface. Tussock grasses dominate vegetation associations. KIRSTIE FRYIRS

Oblique aerial picture of Upper Wolumla Creek showing the escarpment in the background and the incised cut-and-fill river style in the foreground. Note the distinct break in slope, demarcating the escarpment from the valley floor, and the extensive incised valley fills at the base of the escarpment. Rolling foothills also extend from the base of the escarpment. CAPTION KIRSTIE FRYIRS, PHOTO ANDREW BROOKS

these reaches become degraded, it may be impossible, or impractical, to re-establish natural conditions, and the potential off-site implications of changes to the remnants may be severe. For example, if the remaining swamps in the Bega catchment become incised, large volumes of material would be released, potentially modifying river character and behaviour downstream.

Second priority is given to strategic sites which protect the high conservation reaches. Head-cuts which threaten swamps would fall into this category. Upland swamps, associated flood-outs, and threatening head-cuts are considered to be the highest priority river rehabilitation sites in the Bega catchment.

Rehabilitation is then a case of working with the inherent recovery potential of the river—with its in-built ability to adjust. In the natural state, form–process associations are in balance. So the Bega catchment

THE WOLUMLA CREEK SUBCATCHMENT

Bega River
Wolumla Creek
Frogs Hollow Creek
Greendale Creek
Creek
South Wolumla
Frogs Hollow Swamp
0 km 5
Escarpment zone
Catchment boundary

After Fryirs & Brierley[122]

rehabilitation strategies must work from the headwaters, protecting the first priority conservation sites, and then proceed downstream into those with recovery potential, and finally deal with the most degraded reaches. The degraded reaches cannot be effectively managed until the upstream sections are stabilised. The sediment-choked lowland plain of the Bega River can only be rehabilitated when upstream sediment sources are managed effectively.

With Landcare and Rivercare movements, Catchment Management Committees and River Trusts in Australia showing the way for the rest of the world, the aim is to develop sustainable river management practices. However, in the Bega catchment the geomorphic changes have destroyed the pre-disturbance character and pattern of habitat availability and the potential for restoration to original state is virtually non-existent. Irreversible changes have occurred. In particular, the nature of ecosystem functioning has been fundamentally altered. Remnant aquatic ecosystems in the upland areas beyond the escarpment are functionally disconnected from the remainder of the catchment. In such a catchment, we have to work with the altered system to achieve maximum rehabilitation success under altered conditions. This requires a change in vision and the acceptance of the new parameters.

The assessment of biophysical character and behaviour of differing river styles, and how they fit together within a catchment framework, may provide a firm basis for a wide range of management issues, including water reforms. Particular water allocation strategies may have differing implications for different river styles, impacting, for instance, on the management of ecological refugia along river courses. A catchment-based river styles framework may also provide a basis for design and implementation of appropriate environmental monitoring strategies that record the effectiveness of management strategies.

WOLUMLA CREEK, A TRIBUTARY OF THE BEGA RIVER

Wolumla Creek is a north–south aligned subcatchment in the south-east of the Bega catchment, draining 90 square kilometres.[123, 122] *Wolumla* is an Aboriginal word meaning a waterhole. First settlers came to the area in 1851, creek lines were claimed by first landowners and hill slopes were divided among selectors. Water was 'always available' in the well-grassed swamps. Soon after settlement the valley flats and adjacent hill slopes were cleared, swamps were drained and cultivated, and access roads were constructed. Portion Plans dated from 1865 refer to the study area as the 'Wolumla Big Flat' and show large areas of swampy terrain.

The valley fills of Wolumla Creek's upper catchment have been studied in detail and give an insight into the evolution of these 'cut and fill' rivers which are found in wide, steep-sided valleys at the base of the escarpment.

The valley fills are formed from a combination of cut and fill processes. The deeply weathered granites of the Bega Batholith provided a generous supply of sediment. The valley fill at the base of the escarpment consists of alternating horizontal layers of sand and mud. Sand units were deposited as sand sheets or splays on floodplain surfaces or in floodouts that form on top of intact valley-

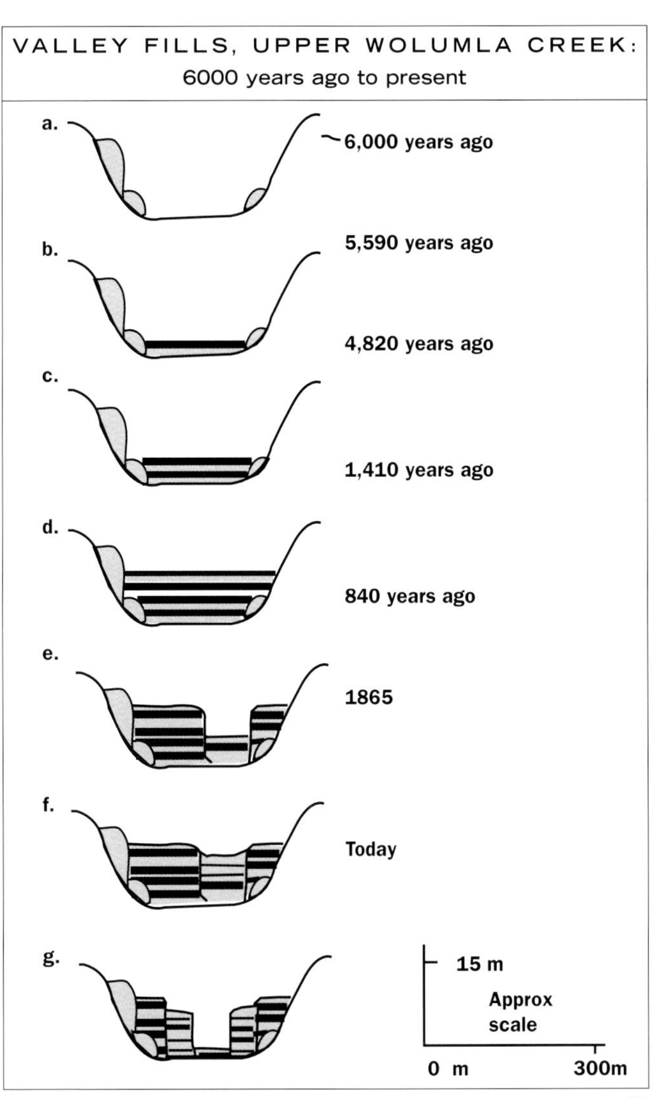

After Fryirs & Brierley[122]

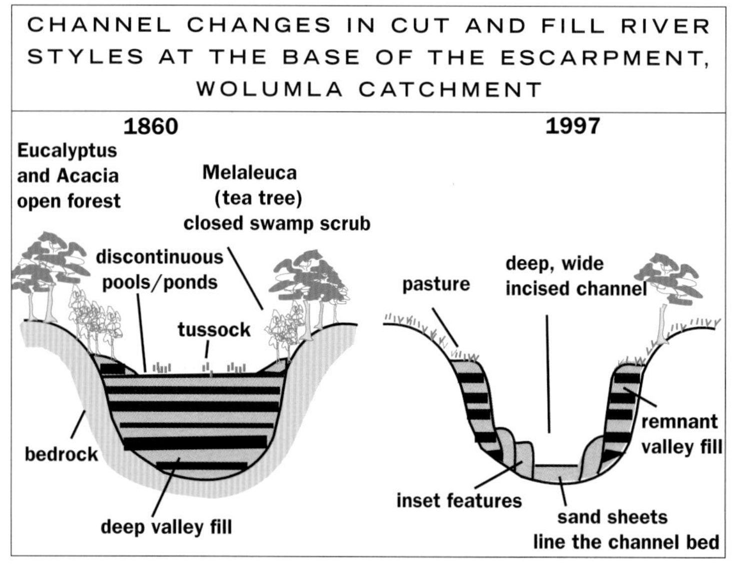

After Fryirs &Brierley[122]

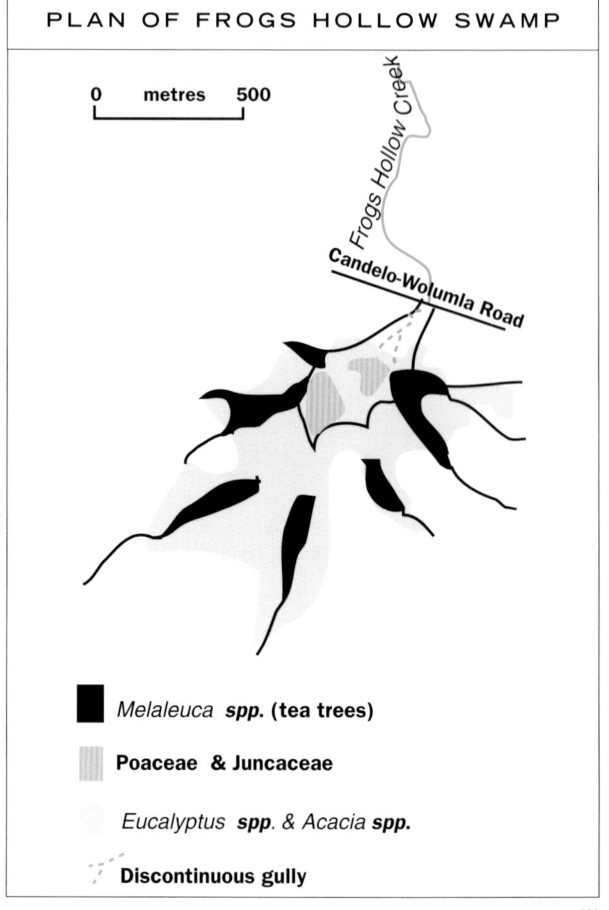

After Fryirs & Brierley[122]

fill surfaces downstream of discontinuous gullies. Organic-rich muds were deposited from suspension in swamps or in seepage zones at the distal margin of flood-outs. Within 5 kilometres of the escarpment, valley deposits grade from sand to mud, indicating the distribution of these processes across the landscape.

Radiocarbon dates indicate that virtually the entire valley fill of upper Wolumla Creek was excavated prior to 6000 years ago. Only remnant terraces are evident in valley margins. The valley subsequently filled between 6000 and 1000 years ago, producing valley fills about 12 metres deep but no greater than 300 metres wide. Re-incision into the valley fill, on a scale smaller than the present incision, is indicated at about 1000 years ago, after which phase the valley refilled. Within a few decades of European settlement the valley was incised again and it is now characterised by a channel over 10 metres deep and 100 metres wide in places—for a catchment of less than 20 square kilometres!

Headwaters of the Frogs Hollow Creek provide a rare example of a remnant swamp of the sort which comprised the Wolumla Big Flat before 1900 (when incision is believed to have started there). A distinct pattern of vegetation associations, dominated by melaleucas, still exists in upper Frogs Hollow, despite vegetation clearance and cattle grazing.

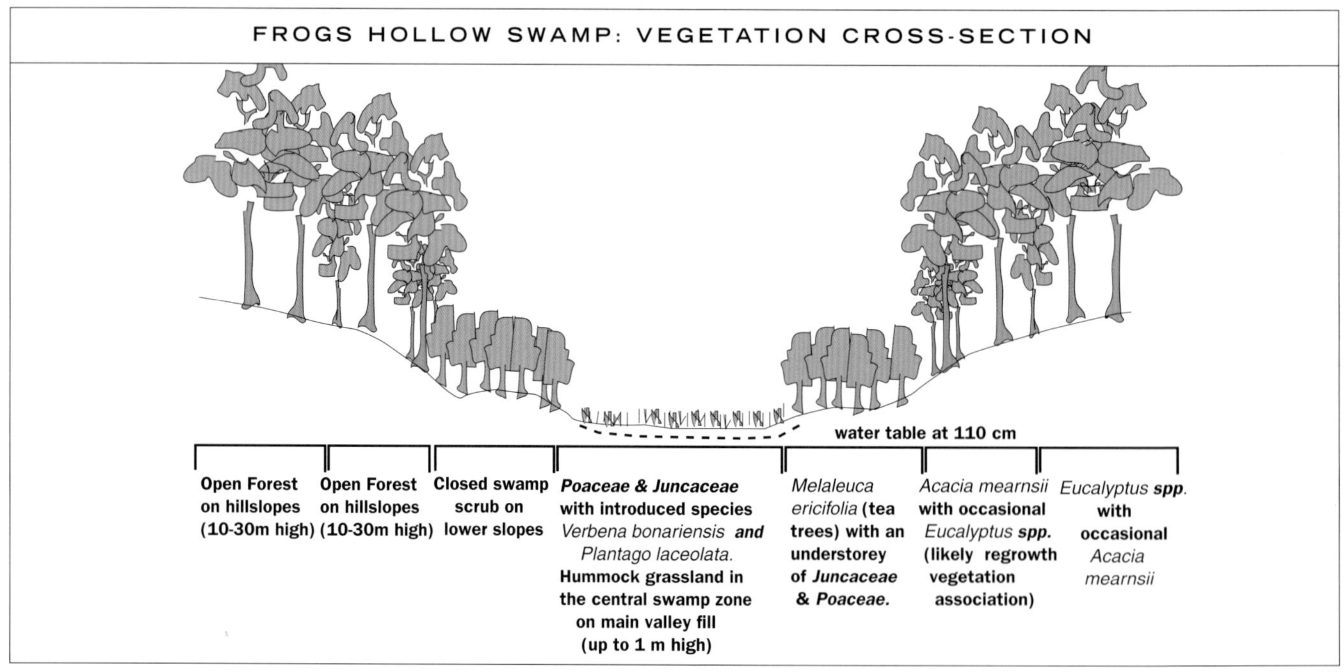

After Fryirs & Brierley[122]

THE COBARGO CATCHMENT

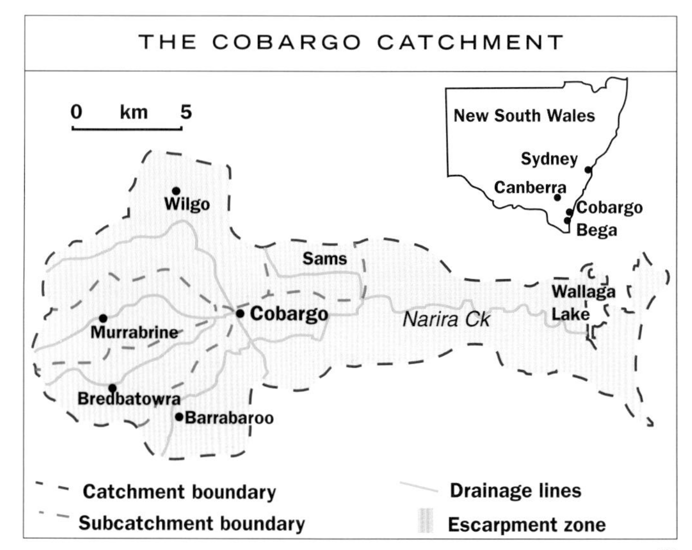

After Brierley & Murn[124]

ADJUSTMENT OF CHANNELS TO REMOVAL OF VALLEY-FILL SEDIMENTS AND TRANSFER TO LOWER REACHES

PRE-DISTURBANCE (1850) | TODAY

SEDIMENT SOURCE ZONE — A | (i)

SEDIMENT TRANSFER ZONE — B | (ii) erosion of colluvial footslope

SEDIMENT ACCUMULATION ZONE — C ????? | (iii) point bar deposition; sand accumulation on floodplain ?????

After Brierley & Murn[124]

This small catchment, with Narira Creek running down its centre, drains a funnel-shaped area of only 137 square kilometres in the Wallaga Lake National Park just north of Bermagui.[124] The ranges of the Great Escarpment in which three tributaries of Narira Creek rise are close to the coast in this region. Early Devonian granites comprise this section of the escarpment. The granites are the source of large volumes of sandy sediment for the headwater valley fills. The town of Cobargo lies near the junction of the three tributaries—the Wilgo, the Murrabrine and the Bredbatowra. Two other tributaries—the Barrabaroo to the south and the Sams to the north—also flow into the Narira. They are not sourced by the escarpment granite country and, coming from Ordovician meta-sedimentary rocks, do not have significant valley fills in their upper reaches, and thus their sediment input into Narira Creek is minimal.

Today, the Cobargo catchment supports dairy and beef production almost exclusively. Average annual rainfall at Cobargo since 1930 is 975 millimetres, but ranges from 362 millimetres in 1982 to 2171 millimetres in 1956. (Highest recorded monthly rainfall was in February 1971 when 790 millimetres fell, with 487 millimetres in one 24-hour period, resulting in the biggest flood on record for Narira Creek.) Flow along Narira Creek is ephemeral.

Upland areas of the catchment are densely forested with eucalypts. Prior to European disturbance, the upland valley fills of the catchment had intact, vegetated surfaces and were sediment accumulation zones with sands and organic-rich muds deposited in horizontally interbedded sequences. The unincised valley fills were transitional downstream into swamps with occasional ponds which extended downstream to Cobargo. Long flats were vegetated mainly by grasses (*Phragmites* spp.) and sedges; margins supported acacias. River courses upstream of Cobargo were largely unvegetated, but river oaks (*Casuarina cunninghamiana*) are the primary channel marginal vegetation in the middle and lower catchment.

European settlement commenced in the 1830s; by the 1860s ringbarking had begun; the whole catchment, apart from the steep upper slopes, was cleared by the turn of the century. No creek was evident along the Murrabrine Valley, and Wilgo Creek was a chain of ponds as late as the early 1890s. (The Aboriginal word *wilgo* means 'deep waterhole'.) Erosion was noted following the droughts in the 1880s, when cattle tracks became pronounced between waterholes, leading to incision and draining of swampy flats. A continuous channel developed along Murrabrine and Wilgo Creeks, but the channel remained discontinuous along Bredbatowra Creek.

Headwater reaches of the streams are bedrock-dominated, with shallow steps in steeper sections. Extensive valley fills extend roughly 3 kilometres downstream from the base of the escarpment. These constitute contemporary sediment source zones, and virtually all banks are eroding in this zone.

The Murrabrine subcatchment shows the erosive changes which have resulted from destabilising the headwater regions. Upland incised channels are now up to 50 metres wide and 8 metres deep—for a catchment area of less than 10 square kilometres. Contemporary rates of lateral expansion of upland channels are relatively trivial, implying that rates of headwater retreat and widening of incised channels occurred mainly, and at much higher rates, in the decades immediately following disturbance.

Mid-catchment reaches have narrower channels, between 10 and 15 metres wide, and 1.5 to 3 metres deep, and because they are confined by their valleys there is little space for floodplain development. Most of the middle and lower catchment acts as a sediment transfer zone. Bed aggradation (silting up of the stream bed) and formation of sand bars in the stream are a primary response to the sediment supply from the upper catchment. Sand bars in the stream, particularly point bars which develop against bedrock spurs, deflect the flow against the opposite bank, which is undermined, adding more sediment to the stream. The sediment has spread out on a discontinuous floodplain along the 10 kilometres of the Narira Creek approaching Wallaga Lake.

Bredbatowra Creek behaves somewhat differently, having a steeper headwater zone but shallower sediment transfer zone. Incised valley fills grade into an intact, grassed swamp in which occasional pools, up to 30 metres long and 2 metres deep, are separated by saturated ground. Up to 2 metres of sedimentation has occurred on this swamp in the last 40 years, with shallow sand sheets spreading as floodouts downstream of the continuous channel. Aerial photography shows that this accumulation zone has extended several hundred metres upstream since 1957, infilling the lower reaches of the former channel, which is now marked by willows growing on the swamp.

The downstream margin of the swamp in Bredbatowra

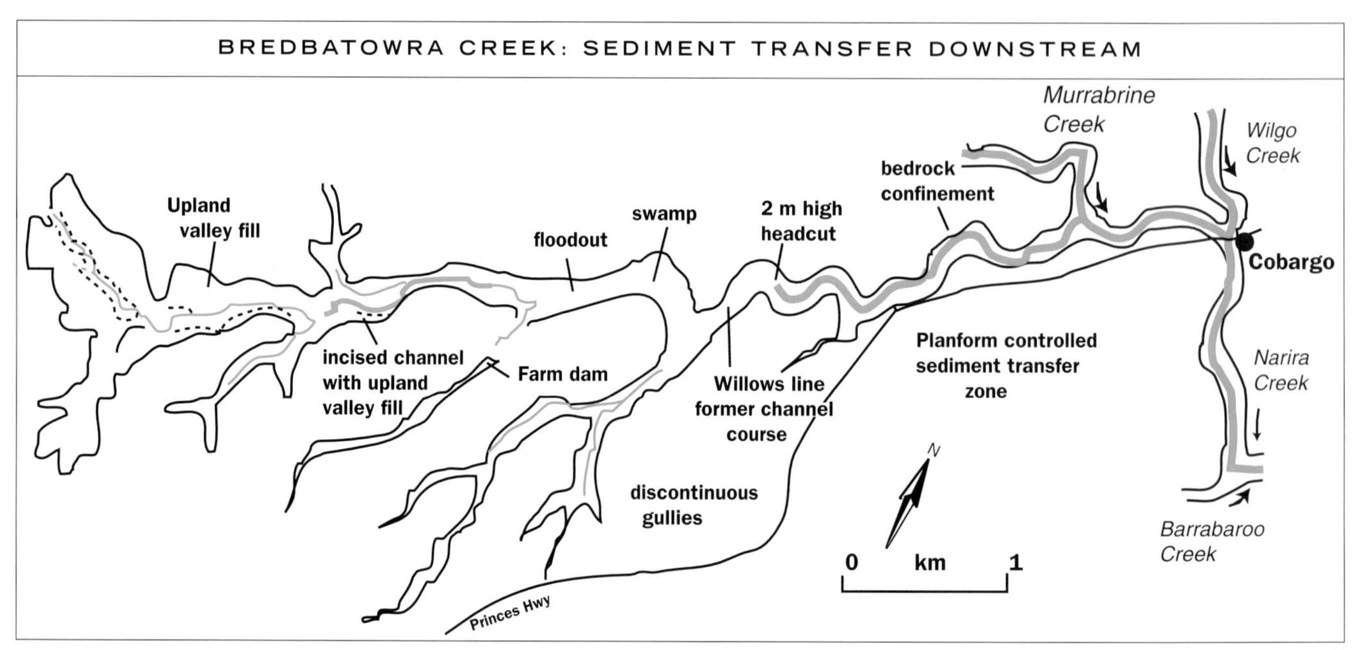

After Brierley & Murn[124]

subcatchment is defined by a 'head-cut' 2 metres high (an erosional front where the drainage below is cutting back into the alluvium in the swamp). Aerial photographs, and anecdotal evidence, suggest that the head-cut started in the 1971 flood and has been migrating upstream, and has migrated upstream by more than 1 kilometre, stripping the floodout deposits from the floodplain surface and revealing the black muds of the underlying swamp deposits. The sands that it has released have added to those moving downstream. A deep swimming hole at Cobargo, in use until the 1940s, has disappeared without trace. (The 1971 flood is also marked in the floodplain of the Murrah River, midway between the Narira and Bega Rivers on the New South Wales South Coast.[3])

The present tidal limit in Narira Creek extends upstream approximately 2–3 kilometres from Wallaga Lake. Aerial photos indicate that between 1944 and 1969 a narrow lobe of the Narira Creek bird's-foot delta extended 150 metres into the lake. A second lobe, 225 metres long, developed between 1969 and 1994 (in response to an enhanced sediment supply), essentially maintaining the same form of delta. (Approximately 100 000 cubic metres of sand have been added to the delta in the last 50 years; the delta is expected to continue growing at a similar rate over the next 50 years because of the amount of sediment stored in the system's transfer zones.)

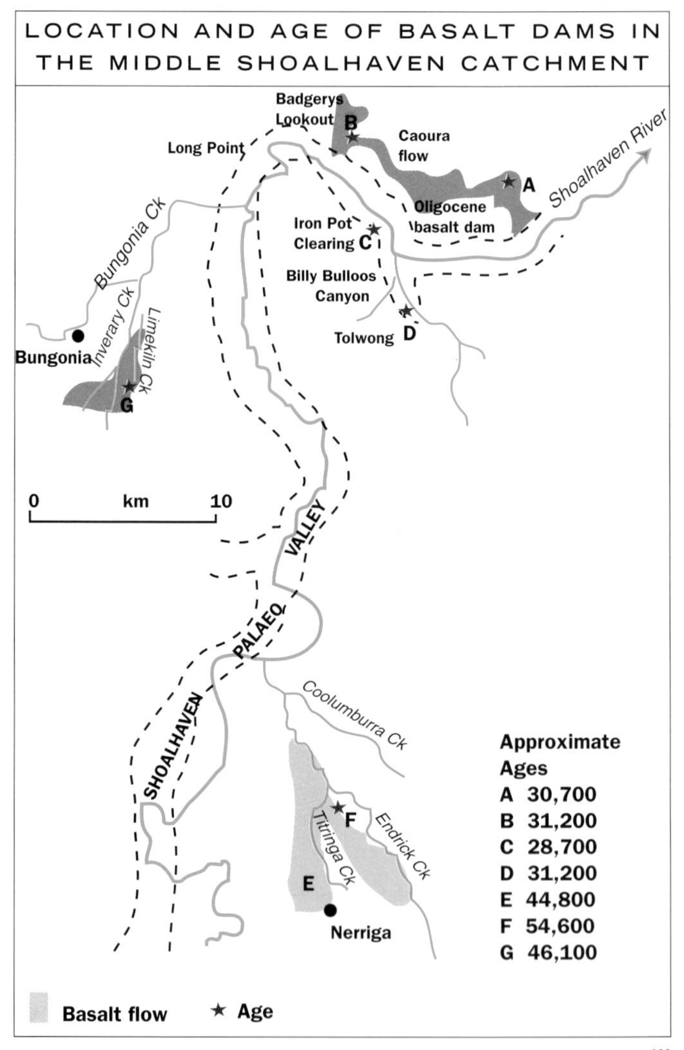

After Nott[126]

THE SHOALHAVEN RIVER

The Shoalhaven River catchment of 7300 square kilometres traverses two major geological provinces. From its headwaters to the gorges near the Tallong bend it runs through Lachlan Foldbelt terrain; downstream, it lies within the southernmost extension of the Sydney Basin. The Triassic sandstone cliffs which characterise the escarpment south of Sydney end on the northern edge of the Shoalhaven valley near Nowra. To the south, the escarpment is cut into the older Permian sandstone. The Triassic sandstones once extended about 20 kilometres further south, according to geologists, and have been eroded off the Permian layers.[125]

The long-term evolution of streams in the Shoalhaven catchment of south-eastern New South Wales has been a contentious issue for decades. Several geomorphologists believed that the Shoalhaven was captured at the sharp eastward bend near Tallong, and that this was evidence for the westward migration of the Great Divide in this

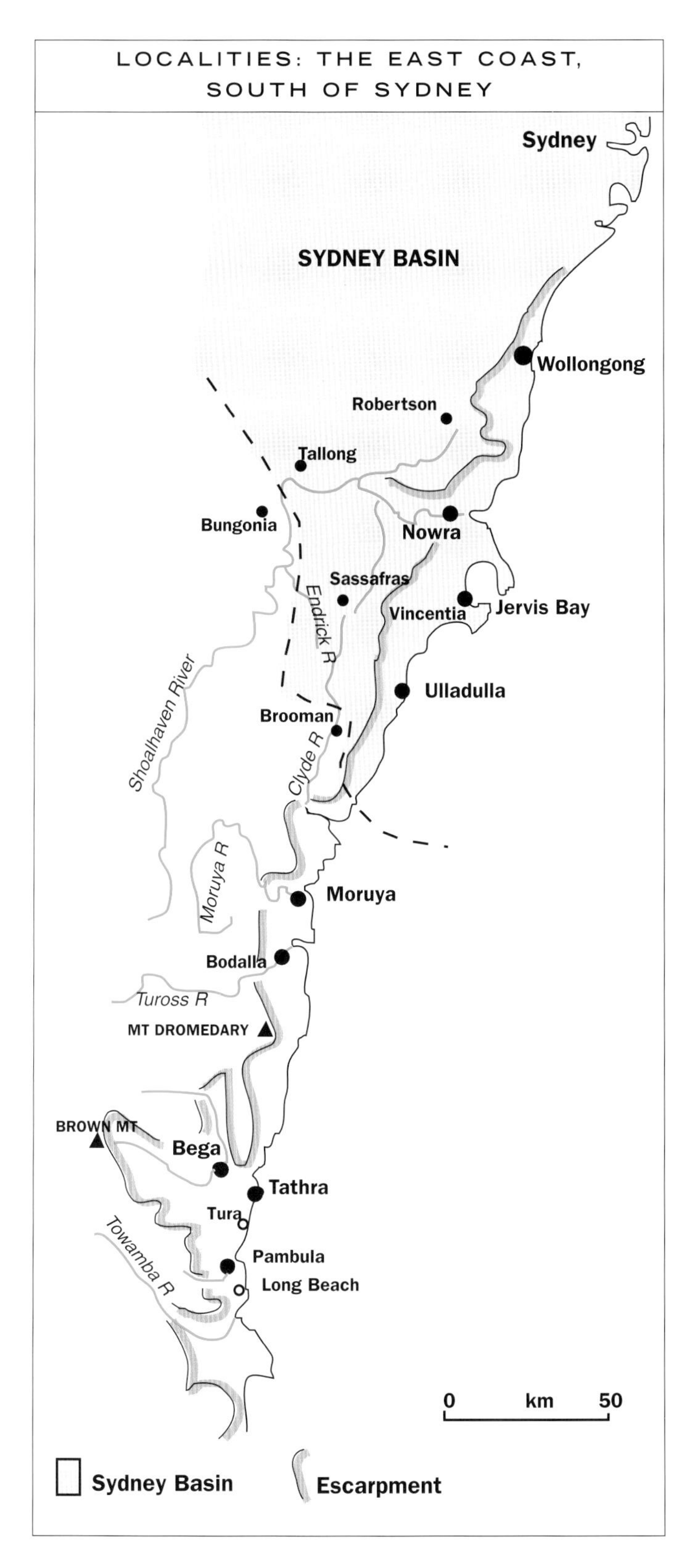

region. Other workers have argued that capture did not occur and the location of the divide had been stable throughout the Tertiary. Studies of sediments in palaeovalley networks, and of basalt flows in valleys, have now shown that the Shoalhaven and many of its tributaries have maintained almost the same course since at least the very early Tertiary.[17, 101, 125, 126, 127] This provides strong evidence against the capture hypothesis, and the information obtained also suggests that elevation of the Eastern Highlands had occurred prior to the Miocene, and probably even pre-Oligocene. (Studies of rivers flowing off the Southern Highlands' western and southern flanks, such as the Lachlan, the upper Murrumbidgee, the Snowy and the Towamba, concur with findings in the Shoalhaven catchment.)

The headwater streams of the Shoalhaven lie in the steep, narrow valleys which characterise the hilly to mountainous country of the south-eastern highlands, between 1200 and 1400 metres above sea level. The upper and middle reaches lie within the Palaeozoic Lachlan Foldbelt; the lower reaches extend into the southern extremity of the Sydney Basin; the middle catchment is dominated by the undulating Shoalhaven Plain. A network of palaeovalleys underlies the plain, the proto-Shoalhaven and proto-Mongarlowe Rivers, and the proto-Nadgigomar and proto-Budjong Creeks. The buried channels of the proto-Shoalhaven (deep leads) are clearly exposed in former alluvial gold quarries at Spa Creek, Black Springs Creek, Old Timberlight, Oallen Ford and Spring Creek. The Shoalhaven palaeovalley runs very close to, and in places crosses, the present-day river over the 120 kilometres that can be traced.

The middle Shoalhaven catchment contains a vast sheet of lacustrine and alluvial sediments covering about 2000 square kilometres, providing plenty of material for study. It is clear that a number of the catchment's streams were dammed by basalt flows, resulting in two phases of lake development, the first in tributary valleys in the Mid Eocene (40 million years ago) and the second within the palaeo-Shoalhaven valley during the Mid Oligocene (30 million years ago).

Near Tallong, the Shoalhaven River flows through a spectacular cliff-lined canyon 550 metres deep. Oligocene basalts (between 28 and 31 million years old) flowed from shallow upland valleys into the main canyon and also into a tributary canyon. These tongues of basalt, most of which cap silicified fluvial sediment, extend 50 to 70 metres below the rim of the canyon, which lies at 600 metres above sea level. Although this section of the canyon has since been incised another 380 metres, the upper part of the canyon is known to be at least 30 million years old. An even greater age is indicated for the narrow canyon section of the Towamba Valley at the southern end of the region where basalt flowed about 800 metres from the adjacent upland down to an elevation of just 300 metres above sea level. The dating of the basalt is uncertain but ranges from at least 34 to 54 million years old, with field evidence suggesting it may be Palaeocene (60 million years old), a suggestion backed up by the fact that plant fossils in the canyon just upstream of the basalt have been dated at that age.

The effects of basalt entering the Shoalhaven Gorge can be traced far upstream.[127] The palaeovalley which runs through the gorge was dammed by the Caoura, Billy Bulloos and Iron Pot Clearing basalt flows, producing a 100-metre high basalt dam which resulted in an equivalent-depth lake that extended upstream for over 100 kilometres. This lake has been named Lake Tolwong.[126] Another lake, Lake Titringo, of much more limited extent, was formed when basalt blocked the palaeo-Titringo Creek, a small western tributary of the palaeo-Endrick River.

A western tributary system of the Shoalhaven River near Bungonia is partly filled with a 35-metre thick unit of Eocene age, the Limekiln Creek Siltstone, which was deposited in a lake. When the palaeo-Limekiln Creek was blocked by basalt, Lake Bungonia was formed, its extent determined by the presence of the lake sediments.

Studies of the basalt flows into the Shoalhaven also enable estimation of the rate of headward erosion since the time when the flows ran down the valleys, and have shown it to be in the order of 100 kilometres in 60 million years.

JERVIS BAY: HOW CHANGES IN SEA LEVEL AFFECT LANDSCAPES

The Biodiversity Group of Environment Australia, formerly the Australian Nature Conservation Agency, or ANCA, is a body dedicated to the protection of biodiversity. In 1995, it produced a major publication on Jervis Bay, edited by one of its officers and three scientists from the Applied Ecology Research Group at the University of Canberra, and containing specialist chapters on all aspects of the region contributed to by about 60 experts. A beautifully produced volume in the KOWARI series, it is the sort of publication which will do much for conservation and protection, bringing into focus as it does the value, the beauty and the uniqueness of the area, which is a favourite one with tourists, holiday-makers and residents.[128]

Jervis Bay is part of the City of Shoalhaven, a major urban centre of the Illawarra Region of the south coast of New South Wales, about 200 kilometres south of Sydney. A fair proportion of the Jervis Bay region is comprised of the Commonwealth Territory of Jervis Bay—Boonderee National Park of about 6300 hectares, including Bowen Island and Commonwealth waters of the bay (900 hectares); Defence land at HMAS *Creswell*; and the Wreck Bay Aboriginal Community (400 hectares). Beecroft Weapons Range forms the northern peninsula to Jervis Bay and is under the control of the Department of Defence. The rest of the region is administered by the New South Wales Government.

This beautiful piece of the coast shows so many of the features created by changing sea level that it has been chosen here to represent the drowned valleys around our continental margin (which include Sydney Harbour). The coastline has received its sculpturing and achieved its present-day form from the rises and falls in sea level during the last 2 million years—the Pleistocene ice age. The last glacial cycle of that ice age has had the final control. (We are living in the interglacial following the last glacial maximum, which had its peak 18 000 years ago.)

A series of palaeogeographic maps illustrates the changes which have occurred in the last 140 000 years. (These maps are based on those in the KOWARI 5 publication, which were prepared by AGSO, with kind permission of the Biodiversity Group of Environment Australia.)

(The scenario described for Jervis Bay is repeated again and again around the Australian coastline. The evolution of Broken Bay, Sydney, described briefly on page 144, shows the same sequence, including the attachment of an island (Barrenjoey) to the mainland by a sandspit or tombola.)

THE SANDSTONE COUNTRY OF THE SYDNEY BASIN

The headwater streams of rivers dammed to supply water to the Sydney region rise in rugged sandstone country of dissected plateaux. The rough and infertile terrain precluded agriculture and pastoral land-use and, as the land was considered useless from the time of European settlement, large areas were set aside as the Royal National Park and the Ku-ring-gai Chase in 1879 and 1894 respectively. Today, about one-third of the sandstone plateaux is

JERVIS BAY, NEW SOUTH WALES

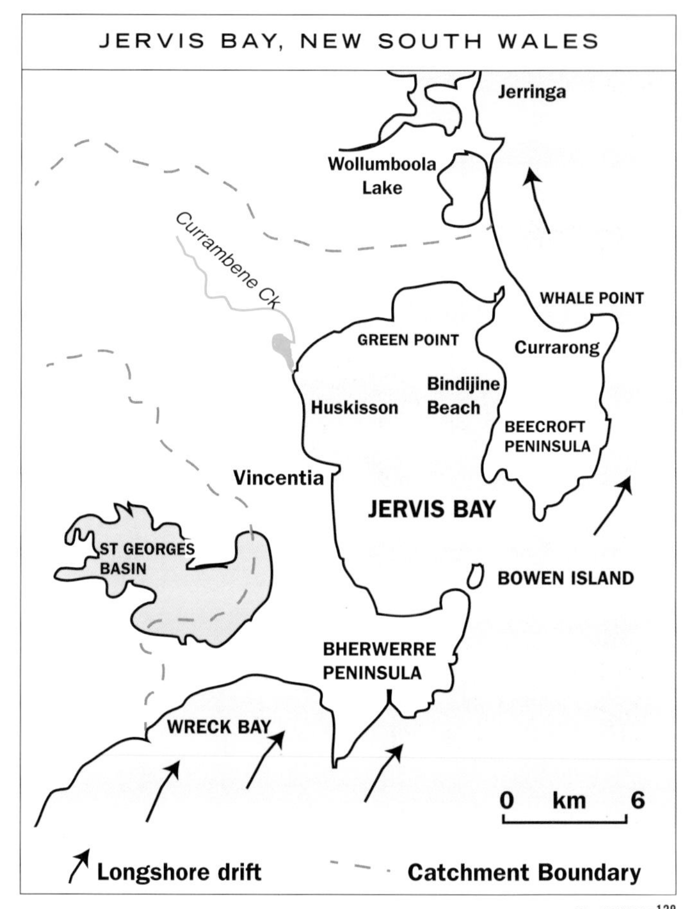

After KOWARI 5[128]

FLUCTUATIONS IN SEA LEVEL IN THE LAST 200 000 YEARS

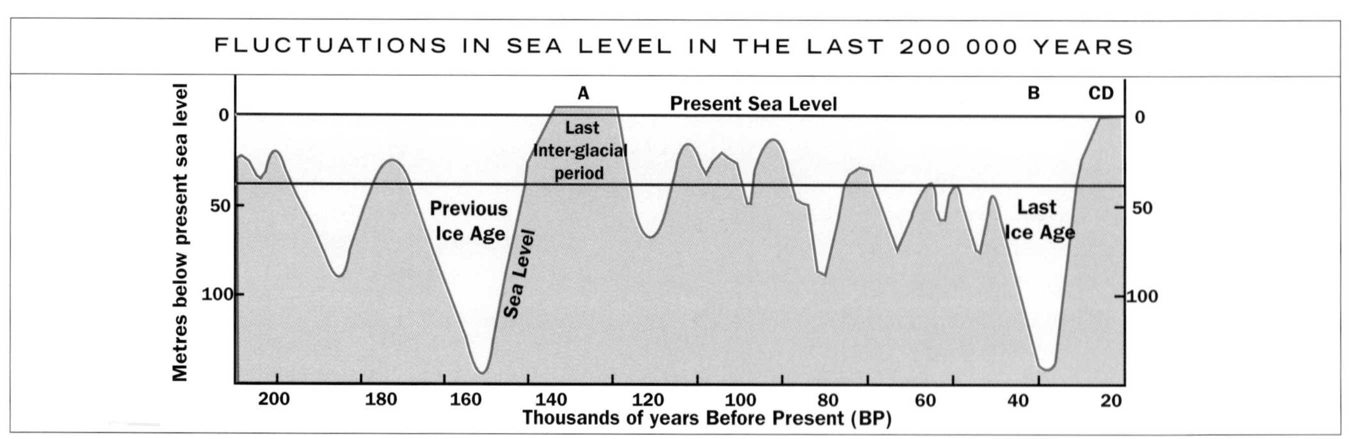

After KOWARI 5[128]

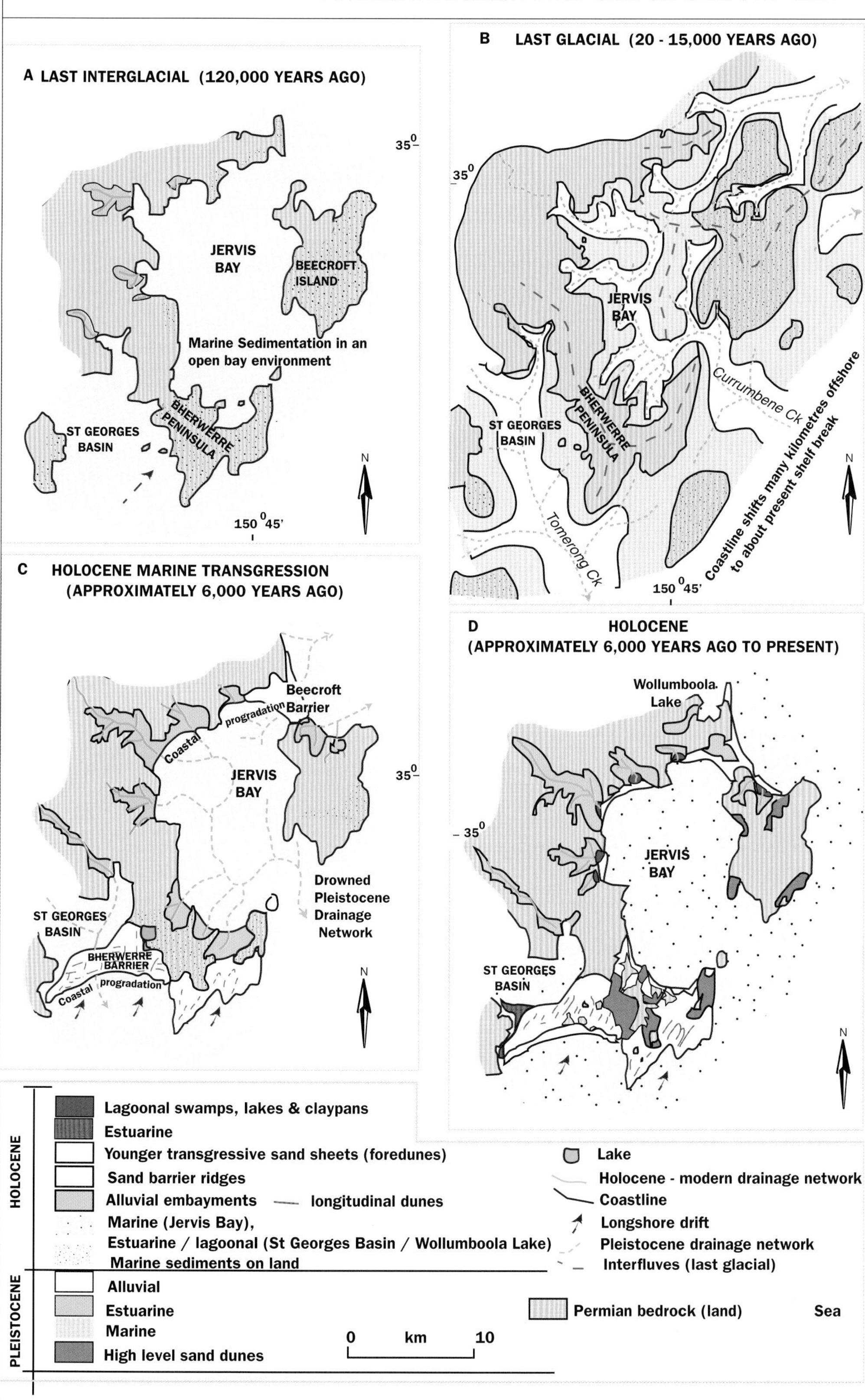

Map 1: During the last interglacial 140 000 to 120 000 years ago, global sea level was higher than it is today, 'St Georges Basin' was open to the sea and the 'Beecroft Peninsula' was an island.

Map 2: During the last glacial maximum, between 20 000 and 15 000 years ago, sea level was almost 150 metres lower than it is today. A wide expanse of the continental shelf was exposed and 'Jervis Bay' was a valley with a river system—-the palaeo-Currambene Creek and its tributaries. A small divide in its northern sector gave rise to a north-east flowing stream which joined another flowing from the 'Beecroft Peninsula'. 'St Georges Basin' was the valley of the palaeo-Tomerong Creek. All the streams flowed out to sea, cutting down deep across the exposed continental shelf to reach the coastline many kilometres to the east of today's margin.

Map 3: Sea level rose steadily after the glacial maximum, drowning the valley systems and their palaeodrainages. When sea level stabilised at about 6000 years ago, Jervis Bay was established. The longshore drift had created a wide sand barrier at the mouth of St Georges Basin and built dunes on Bherwerre Peninsula. A sandspit had joined the Beecroft Peninsula to the mainland and it was no longer an island.

Map 4: In the last 6000 years of more or less stable sea level, beaches have formed between headlands around the bay and estuarine conditions have developed at the entrance to St Georges Basin.

After KOWARI 5[128]

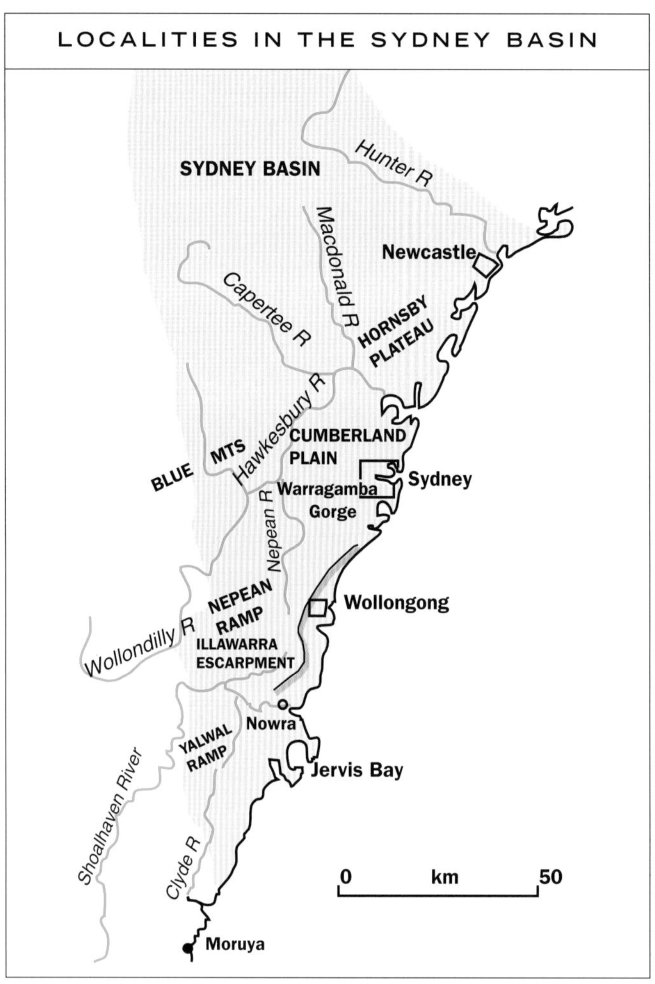

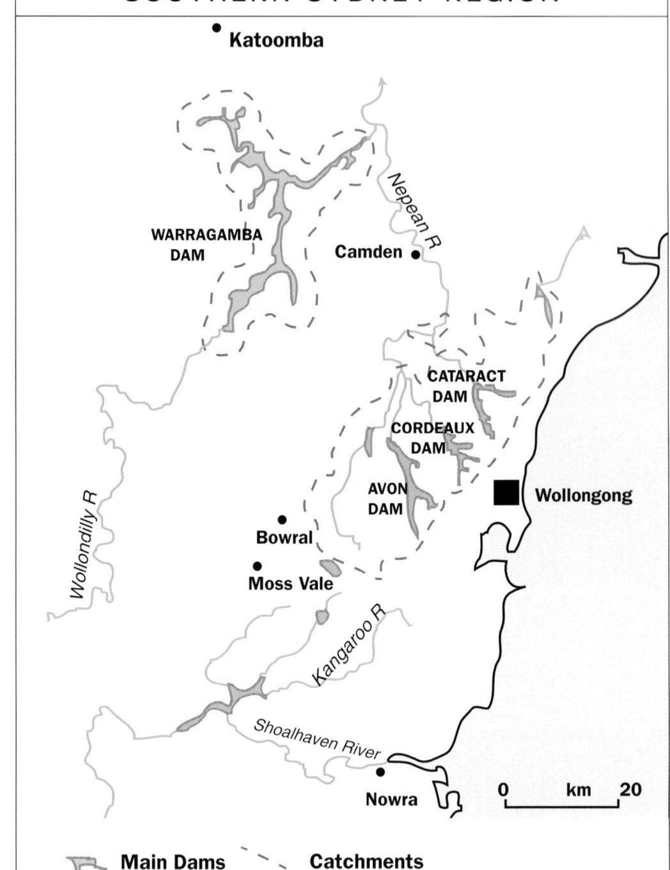

After Young & Young[129A]

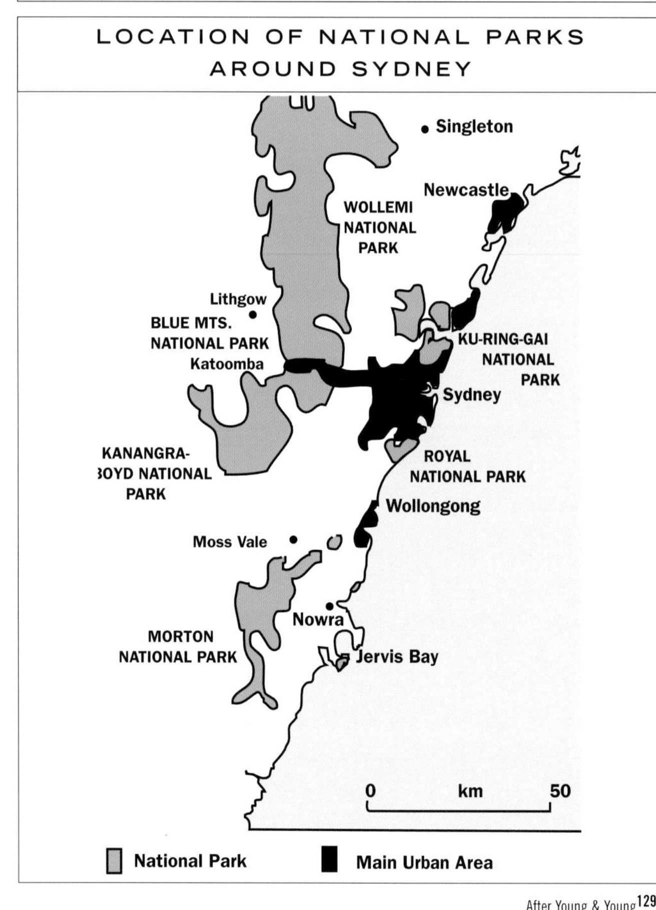

After Young & Young[129A]

incorporated in national parks (963 430 hectares in total). A further 130 000 hectares are set aside as protected catchment for the Sydney region's water supplies.

As people in Sydney know to their cost and inconvenience, protection of catchments has not been total—urban encroachment, rural activities and general man-made changes to river catchments, and to river form and function, have made water management and the supply of quality, germ- and contaminant-free water a nightmare for Sydney Water.

THE WORONORA PLATEAU AND THE ILLAWARRA REGION

The Woronora Plateau, west of the Illawarra Escarpment, lies at low elevation, falling from 300 to 500 metres above sea level along the escarpment to about 100 metres at the Nepean River. The Plateau is drained by headwater streams of the Georges, Cataract, Woronora, Cordeaux and Avon Rivers.[129] Their upland valleys are commonly broad, gently sloping and trough-like, swampy and occupied by virtually treeless sedgeland and heath. Such valleys are termed 'dells' and are characterised by low discharges, well-vegetated surfaces, and sediment accumulation—their competence is too low to move the sandy load washed into them from the surrounding ridges because of the gentle slopes of the valleys. There is a high water table within the trapped sediments because the dells lack open channels and

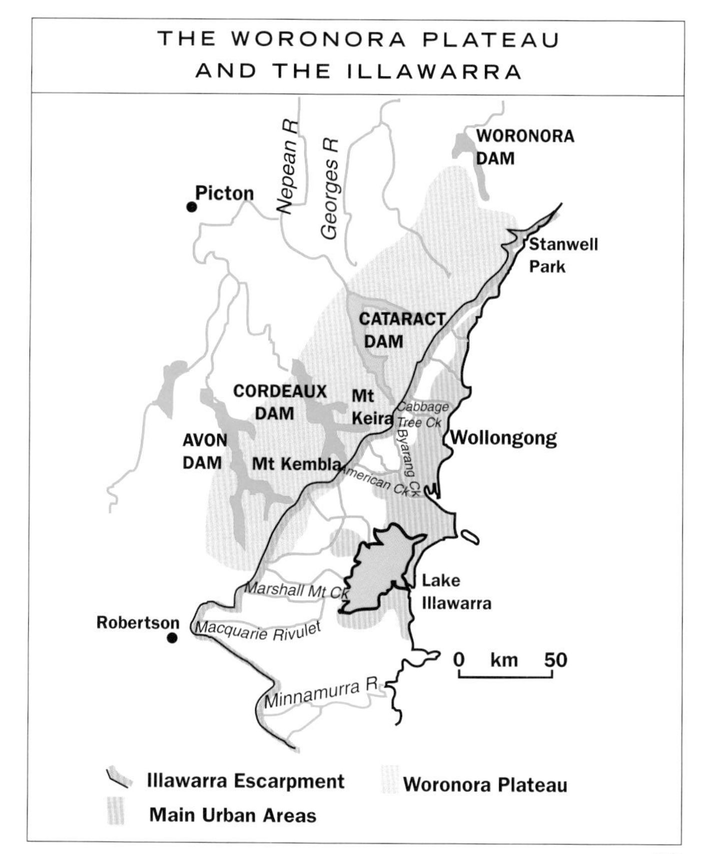

After Young & Young[129A]

because rainfall exceeds evaporation in the area in which the dells are found. Dells are mires, in which high water tables are due to terrestrial inflow (not merely from rainfall directly onto them) and in which, therefore, both detrital and organic accumulations contribute to the sedimentary mass.

The sediments deposited within the broad bedrock floors of the dells are largely organic-rich sands. Ages of the different sediment layers have been dated as between 11 500 and 6500 years, with a pollen and plant macrofossil record showing that there has been virtually no change in the vegetation over the last 11 500 years.

The dells, with their swamps and pools and rich biodiversity, are important purifiers and protectors of the quality of water which flows into the rivers which they feed, which flow into our dams. River water in natural environments which is slowed in its passage by reedbeds, swamps and deep billabongs, is cleansed. When river behaviour is changed and rivers become deeply incised drains carrying water rapidly from source to dams, as has become the case in some of the major tributaries running into city water-supply dams, no natural filtering and purifying occur. The water brings its pollutants, acquired from altered landscapes around it, and from deliberate addition to it, straight to our distribution points, and man-made purifying systems do not always work.

Swamps of a very different nature to the dells occur on some near-vertical sandstone cliffs in the Blue Mountains. Where permeable sandstone layers overlie clay-rich layers in the cliff faces, water percolating through joints in the sandstone meets the less permeable claystone and emerges from the cliff face as 'seeps'. Hanging swamps develop in these seepage zones. Root mats bind a few centimetres of organic-rich soil and a curtain of vegetation hangs down the cliff face. Hanging swamps are largely confined to the Blue Mountains (and do not occur on the Woronora Plateau) because the extensive claystone layers which occur there are not present to the same extent elsewhere in cliffs in the Triassic sequences.

On the Hawkesbury sandstone plateaux within and around the northern suburbs of Sydney, 'peat swamps', like the mires, were a feature of the landscape prior to development. Balgowlah Heights was originally considered unsuitable for building on because of shallow swamps feeding the small streams which ran down to North Harbour. The streams now are piped drains, the swamps' original presence no longer visible and forgotten.

SMALL STREAMS OF THE ILLAWARRA

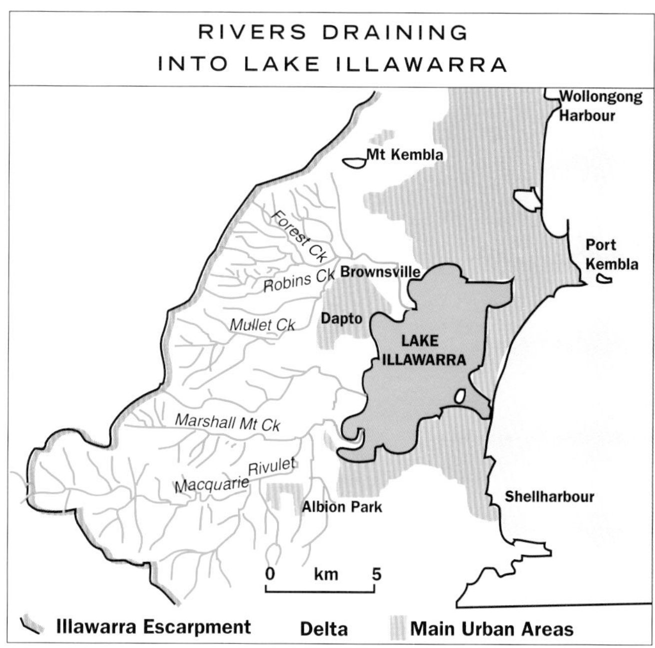

After Nanson & Young[130]

To the east of the Illawarra Escarpment, small streams that flow down to Lake Illawarra and the coast show an unusual planform. Because the size of a river channel normally relates to the volume of water it must carry, most streams show a downstream increase in channel size corresponding to the downstream increase in flood discharges.

Streams flowing off the Illawarra Escarpment, however, show a marked reduction of channel size, accompanied by a downstream increase in flood frequency, in their lower reaches.[130] Each stream can be divided into three segments. Steep boulder-filled channels with numerous rapids and small falls drain the escarpment. Beyond the base of the escarpment, channel slope declines dramatically but the streams and their narrow floodplains are confined by bedrock spurs and old alluvial terraces. Here, the channel beds are formed of cobbles. On the gently sloping coastal plain the channels are unconfined, flanked by extensive floodplains formed of fine sand, silt and clay, and flow predominantly over sandy beds.

Within the confined and deeply sloping valleys of the escarpment foothills, bed and bank sediments are relatively coarse and uncohesive, and channels increase in size downstream. However, once these streams emerge into the more open valleys at lower slopes, where floodplains are formed from fine, cohesive sediments, there is a dramatic reduction in channel size. This decrease in channel size can be explained by the sudden decrease in slope and therefore in

A downstream reach of American Creek, a coastal stream of the Illawarra, where the channel has almost disappeared and even minor flows are largely carried on the floodplain. GERALD NANSON

A downstream reach of American Creek, where channel capacity has reduced to a point where the channel cannot carry even minor floods without spilling overbank. GERALD NANSON

An upstream reach of the Minnamurra River, a coastal stream of the Illawarra, where the channel is capable of carrying regular flood discharges without spilling overbank. GERALD NANSON

THE WEST DAPTO FLOOD OF FEBRUARY 1984

In this land of drought and flooding rains, localised floods are part and parcel of daily living. In 1998 those at Wollongong made news headlines, because landslips followed heavy rain in new suburbs which had expanded onto steep slopes of the Illawarra Escarpment. It was not just water that flowed through houses and gardens and down roads, but half a hillside, carrying all before it. The major flood of February 1984 should perhaps have been a warning, because this is a region where localised storms with unusually high rainfall are frequent, and the catchment drainage is such that experts have said that a major flood, in excess of the 100-year event, might be expected within the urban boundaries every 25 to 50 years.

The Illawarra coastal region is physically dominated to the west by a high, cliff-faced escarpment which reaches a maximum height of 770 metres and is the headwater source of all major local streams.[131] The lower slopes consist of steep hill country, partly mantled with colluvium. Between the foothills and the sea lies a coastal lowland formed of alluvial valleys, low alluvial terraces and gentle dividing ridges which are outliers of the escarpment. The scarp is vegetated with wet sclerophyll forest, interspersed with patches of rainforest. The higher parts of foothills are still forested, but cleared grazing land extends well up into them.

The small streams draining the escarpment have very steep upper reaches with boulders and numerous small falls; moderately steep, well-defined and relatively large channels with narrow floodplains extend through the foothills; on the gently sloping coastal plain the channels are small and flanked by extensive floodplains. The upper channels have reworked their margins, widening their beds, and have formed shallow floodplains; downstream reaches are low-sinuosity, non-migrating channels which have aggraded with the post-glacial rise in sea level. They have deposited a deep layer of sand and gravel below the channel, and a deep cover of overbank sediments on the floodplain.

The storm in the West Dapto area of Wollongong in February 1984 produced a 24-hour total rainfall of 796 millimetres at Wongawilli—the largest 24-hour fall ever recorded in temperate Australia. Damage, excluding that to private property, was in excess of $5 million (which equates to a much larger sum today) and in addition there was massive flooding in the Mullet, Marshall Mount and Robins Creeks which flow into Lake Illawarra. The three creeks are relatively undisturbed rural streams. The effect of the flood on the creeks could be accurately determined because observations had been made on the channel widths, sediment distribution and other features the year before.

It was found that the upper reaches, where a large proportion of the flood flow was confined to the actual channel, suffered severe change with bank erosion and widening; the downstream reaches were little affected because low gradients and wide, well-grassed floodplains ensured little erosive capacity. On the floodplains, the areas showing damage were those where roads or rail causeways were built, interrupting the flow of floodwaters.

stream power; the cohesive nature of the alluvium and its retention on the channel banks by a dense cover of pasture grasses; and the availability of an extensive floodplain to carry displaced floodwater. The floodplains are thus an important part of the channel system. Studies have shown that where urban development has increased peak run-off and reduced the effective floodplain, stream channels formed in this fine alluvium rapidly entrench and increase in cross-sectional area by two to three times. Minor man-induced channel alteration and maintenance appear to trigger this enlargement.

The Minnamurra River, 30 kilometres south of Wollongong, has been little changed by human activity except for straightening of several reaches near Jamberoo at the turn of the century (and stream flow has since adjusted to these changes). The river tumbles over spectacular falls across the cliffs of the scarp. From an elevation of only 60 metres above sea level at the foot of the escarpment, the lower 80 per cent of the river's length flows in an alluvial channel flanked by floodplain. The downstream increase in width is seen in the upper half of the basin, channel-width reduction in the lower half. Marshall Mount Creek and American Creek, both mainly rural, show a similar pattern of channel geometry.

In contrast, streams through urbanised areas show cross-sectional areas two or three times larger than rural channels of the same drainage area. Channel maintenance has usually involved straightening and regular clearance of channel vegetation. Floods have scoured away the fine sediments forming the banks and have entrained the coarser material under the floodplains. In the memory of elderly residents along Byarong Creek, the creek has changed from a small rural stream with vegetation growing down to the water's edge into a chasm 2 to 3 metres deep and 20 metres wide. In the 20 years since urbanisation commenced, Byarong Creek has eroded 25 000 cubic metres (47 500 tonnes) of sediment per kilometre in the downstream 2.6 kilometres of trunk channel—in a 12 square kilometre basin. Similar changes have occurred in the urbanised Cabbage Tree Creek.

THE HAWKESBURY–NEPEAN RIVER SYSTEM

The large coastal catchment of the Hawkesbury–Nepean is 22 000 square kilometres in area.[132] Relatively empty sandstone country lies north, west and south of Sydney, providing a green zone and limiting spatial growth, while providing an ideal water collection area.

Sydney was established in 1788 as Australia's first permanent European settlement. The Hawkesbury–Nepean River formed the boundary of the County of Cumberland, Sydney's administrative district, in the early days. The river valley's fertile alluvial soil was discovered by explorers in 1789 and settlement, clearing and intensive farming began soon afterwards.

The river has two names because it was mapped by explorers in two directions in 1789. As the Nepean, it rises on the sandstone Woronora Plateau to the south of Sydney, flows north across the extensive floodplains of the Cumberland Plain (on shale) and through gorges as it twice flows in and out of sandstone topography. Below its junction with the Grose it becomes the Hawkesbury and turns to run east across the fertile alluvial floodplain between Richmond and Windsor, before it again heads north and north-east to Wisemans Ferry. The lower Hawkesbury then runs south-east into a wide estuary which is an arm of Broken Bay. The river is tidal all the way up to Windsor. The different characteristics of the landscapes through which the river runs have determined different land uses and as a consequence the condition of the river and the adjacent land is different in reaches where it cuts through sandstone or floodplains. The sandstone areas remain little modified, mostly retaining native vegetation, but floodplains are severely altered.

Floodplains were characterised by high alluvial levee banks, generally of Recent geological age, separating depressions or freshwater 'back-swamps'. Levee banks were originally covered in tall open forest of varied composition, collectively named River-flat Forest. Small pockets of mesic or rainforest vegetation occupied sheltered positions in gullies and at the base of cliffs where run-on water and fertility conditions were favourable.

THE HAWKESBURY-NEPEAN RIVER SYSTEM

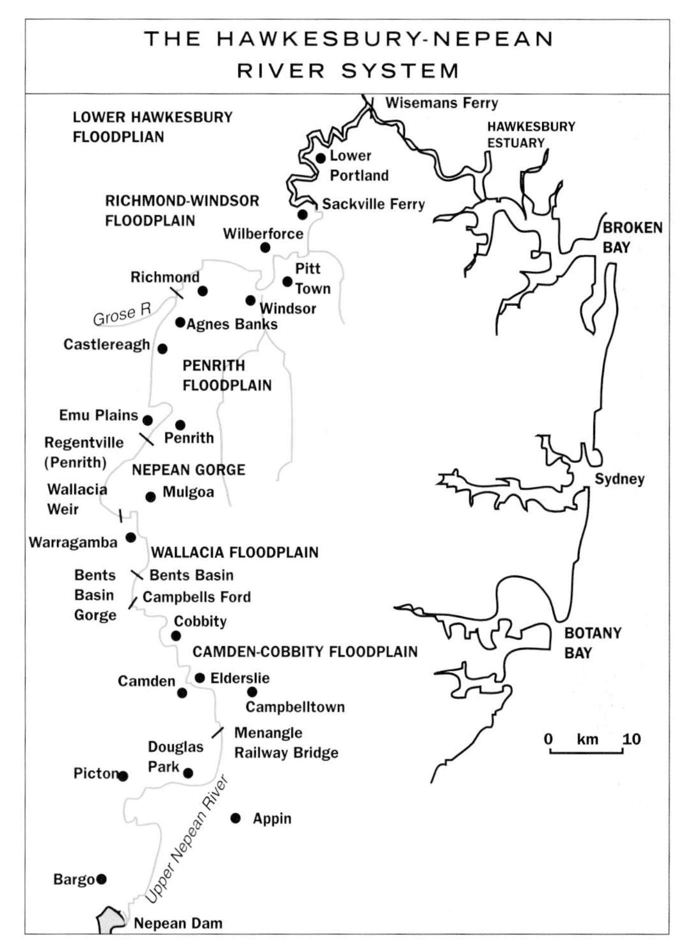

Early settlers immediately cleared the forested alluvial lands, and subsequently they have been farmed for up to 200 years. In 1803, Governor King issued an edict prohibiting clearing within 10 metres of the water, so severe was tree loss and river bank collapse. By 1826, the greater part of alluvial land was under cultivation, and today on the Cumberland Plain only 3 per cent of the original River-flat Forest remains. (A major strategic plan is being implemented to restore riparian vegetation and protect remnants.)

Although most of Sydney's population today lives in the smaller catchments of the Parramatta and Georges Rivers, increasing numbers are occupying the shale areas flanking the lower Nepean and the upper Hawkesbury. This growth, and anticipated further expansion, is creating environmental problems for parts of the catchment and the lower river.

The main threats are:

- Sewage disposals from at least 56 sewerage treatment plants, and algal and bacterial blooms from over-nutrified water. Unsewered areas present additional risks: for example, some of the subcatchments of the Wollondilly drain into Monkey Creek, which is a shale-dominated right-bank tributary flowing into Warragamba Dam. The area is largely without conventional sewerage. Stormwater carries concentrations of pollutants at two to three times background quality even in sewered areas; aerobic waste-water treatments give five times background values; and septic tanks even higher concentrations.
- Sand and gravel extraction, now causing bank erosion, loss of real estate and soil loss.
- Leakages from a toxic waste dump at Castlereagh; and leakages

THE EVOLUTION OF BROKEN BAY

Broken Bay is the estuary of the Hawkesbury River. Pittwater, a tidal waterway, opens into Broken Bay between Barrenjoey and Commodore Heights. The Hawkesbury River system has existed for about 40 million years, with much the same drainage.

At the last glacial maximum, 18 000 years ago, sea level was about 130 metres below its present level. The continental shelf all around Australia was exposed, and the Hawkesbury River had to cut a deep channel to the sea, which was then about 25 kilometres further east than the present river mouth. Pittwater was a valley with a major stream running through it and joining the lower Hawkesbury drainage on the continental shelf across the rock bar that joins Palm Beach to Barrenjoey. (The old drainage patterns across the continental shelf and below Pittwater are etched into bedrock and can be detected by seismic methods, inspite of now being filled with silt.)

As sea level rose, the coastline crept back towards its present position. When it reached its highest level, before falling back and stabilising at about 6000 years ago, it breached the rock connection (a volcanic dyke) that connects Barrenjoey to West Head, and the sea entered the Pittwater Valley and gradually scoured it out, creating the tidal waterway which exists today. Sand built up on the rock ledge between Palm Beach and Barrenjoey as a consequence of the longshore drift, creating the modern tombola.

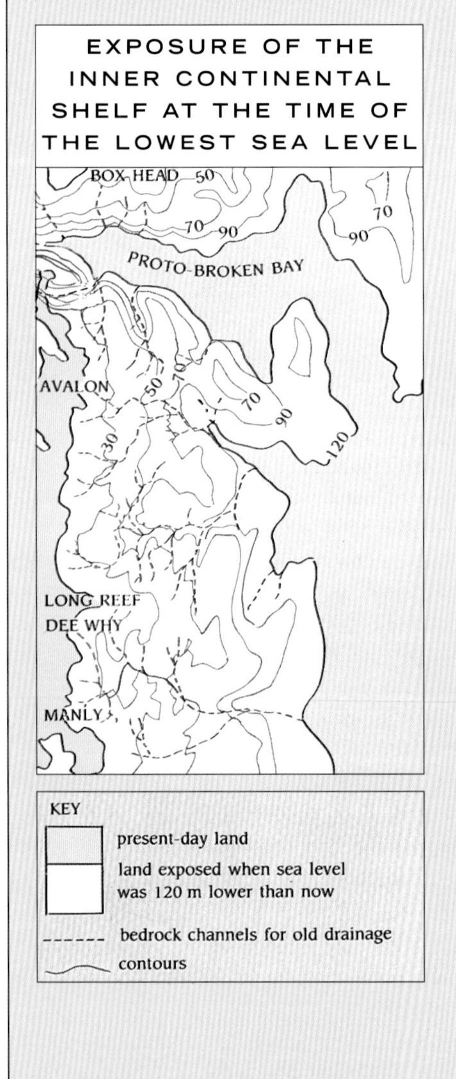

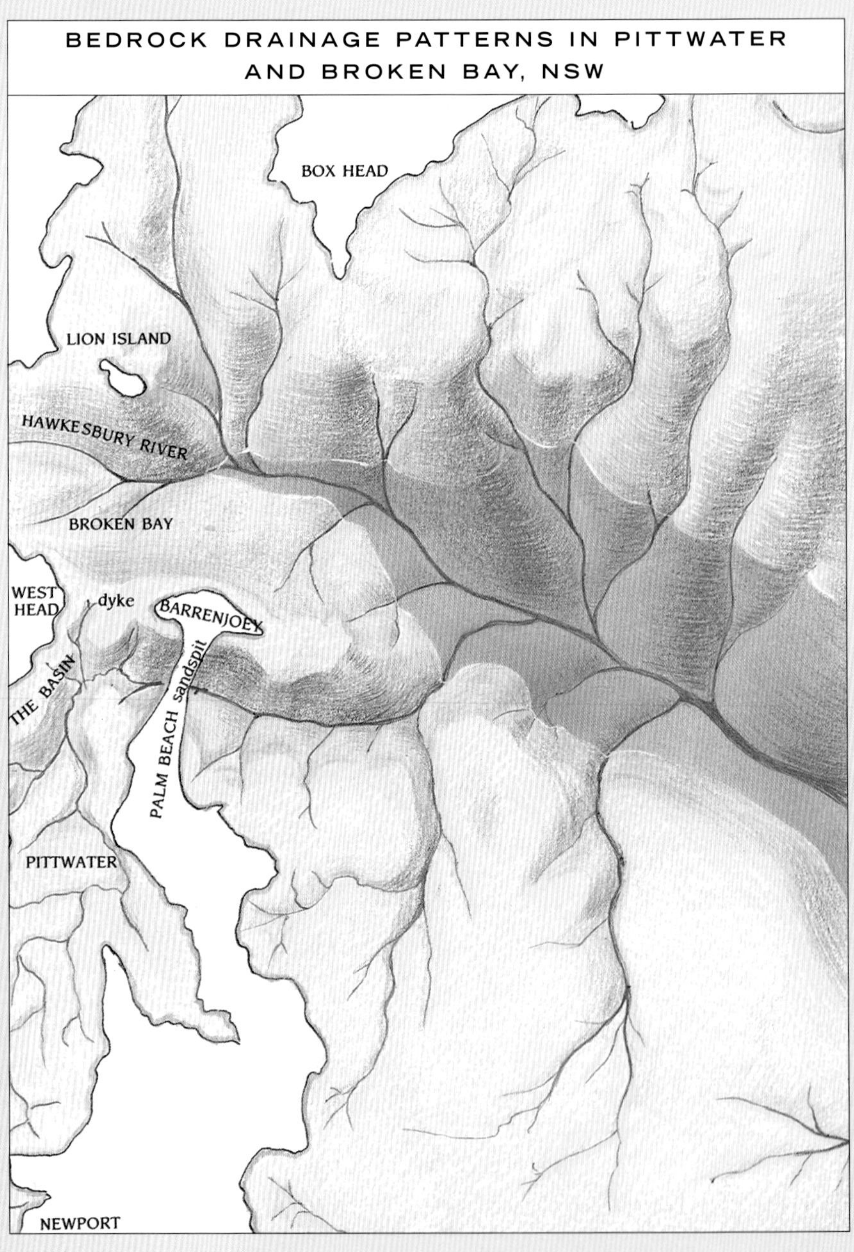

Mining gravels from the Cranebrook Terrace on the Nepean River. These gravels are indicative of a powerful flow regime about 70 000 to 100 000 years ago. GERALD NANSON

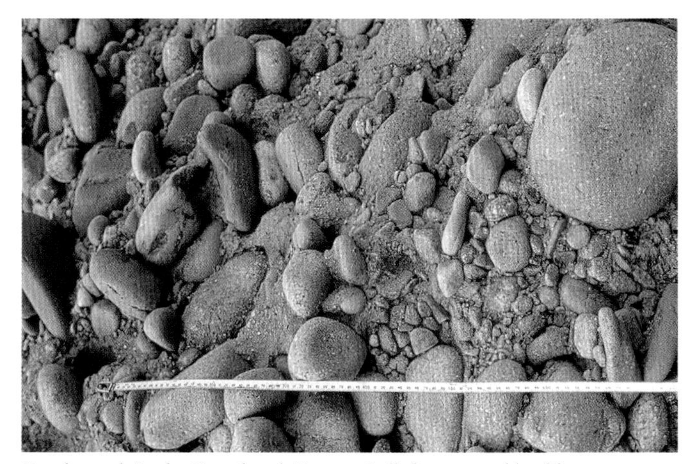

Basal gravels in the Cranebrook Terrace. Rolled stones and boulders were transported during an active phase of the river between 70 000 and 100 000 years ago. GERALD NANSON

from numerous waste dumps with material not registered as 'toxic' but still highly suspect.

- Urban growth, with increases in polluted stormwater run-off.

Dams and weirs constructed on the river cause sediment starvation below their structures, and siltation above. The downstream alluvial channel erodes. In the gorges few changes are likely, but below the Warragamba Dam the channel has incised up to 5 metres into rock. The weirs across the river upstream of Wallacia caused serious sedimentation of weir ponds, and a sand extraction program was approved. Downstream of the dam, the river channel has not adjusted its width to the decreased flow by contracting—because the flow regime has remained high and millions of cubic metres of sand and gravel have been extracted from the bed, the floodplains and the terraces.

The four Nepean dams mainly cut off sands; the Warragamba has stopped the movement of sand and gravel, starving lower reaches of the river of sediment. Sand and gravel extraction at Penrith, Richmond and Windsor has further exacerbated the situation and has led to bank and bed erosion and increase in channel capacity for much of the river below Windsor, with the effects decreasing towards Sackville.

The catchments of the Nepean and Wollondilly Rivers comprise much of the southern part of the Sydney Basin. Evidence of alluvial deposition along the Nepean near Penrith extends back into Tertiary times. Where the river emerges from the Blue Mountains, a widespread lateritised alluvial sheet is found—an ancient analogue of the Quaternary Cranebrook alluvial deposits which today occur in the Nepean Valley between Penrith and Windsor. The ancient course of the Nepean River is marked by the Rickabys Creek Gravels which occur in small pockets on the crest of the Lapstone Monocline. The river ran there before the uplift of the Blue Mountains, which occurred no later than Mid Miocene (about 20 million years ago).

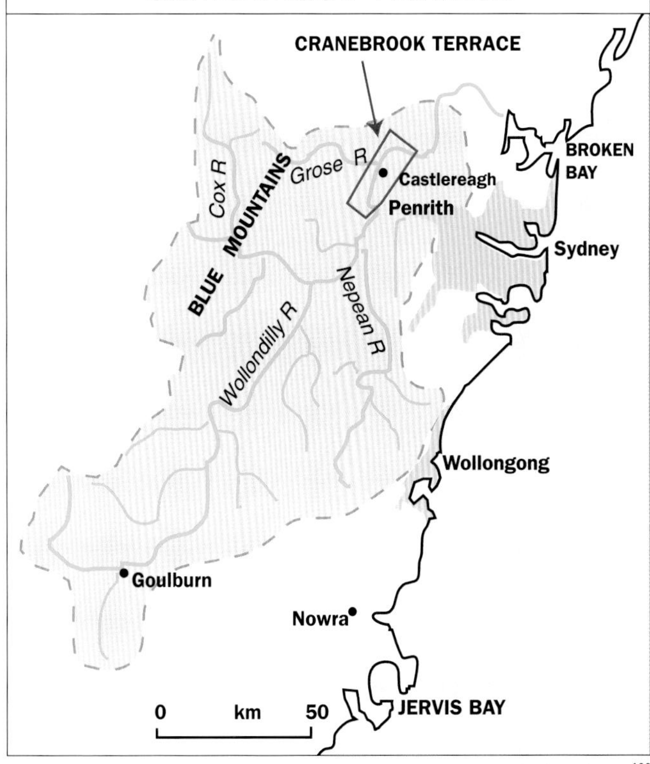

After Nanson & Young[133]

Below its confluence with the Wollondilly, the Nepean flows out from a Triassic sandstone gorge onto lowlands underlain by shale where it has laid down alluvial deposits 4 to 12 kilometres wide. Parts of these deposits comprise the **Cranebrook Terrace.**

The Cranebrook Terrace contains the largest area of Quaternary alluvium in the Sydney area, and detailed studies have been undertaken to elucidate the history of the region. By determining the ages of sediments, and assessing how they were deposited, and in what sort of environments, an interesting picture has emerged of a very different palaeo-Nepean River in a very different landscape over the last 110 000 years.[133, 134]

A considerable extent of the terrace is currently exposed by quarrying for building aggregate and shows the sequence of depositional layers. Gravels lie on the weathered bedrock of the Ashfield Shale and some large silcrete boulders several metres in diameter are sporadically located across this contact. Carbonised remains of large horizontally placed tree trunks occur within the basal gravels. The general character of the gravel unit is one of a proximal braid plain. The gravels were deposited by flows which, during flood events, would barely have been channelised. The upper gravel surface is either level or gently undulating and forms an abrupt contact with the fine overburden. This overburden was partly

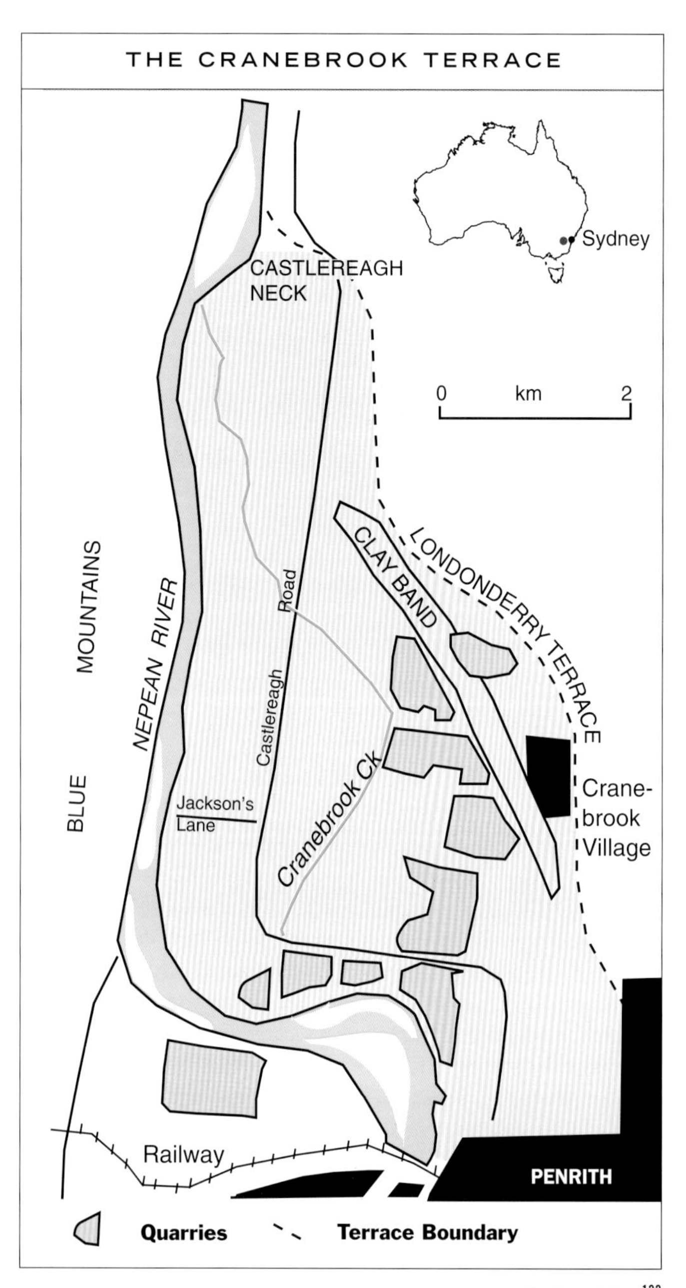

After Nanson & Young[133]

laid down at the time of the braidplain activities and is also partly the result of ongoing processes of stripping and accumulation afterwards. A palaeochannel runs along the eastern margin of the terrace and is known locally as the Clay Band.

The gravels have been dated as having accumulated during a pluvial period which ended between 40 000 and 45 000 years ago. This Cranebrook Pluvial period, with very active rivers, was followed by a stable period when the river became confined to two laterally stable channels. Additional overbank deposition probably accompanied the stable phase. The easternmost channel (the Clay Band) was abandoned between 37 000 and 34 000 years ago and the Nepean has occupied the western channel since. Some stripping of the western side of the terrace appears to have occurred about 14 000 years ago. The stability of the system during the glacial maximum reflects the absence of any glaciation in the headwaters of the Nepean–Wollondilly catchment.

The Hunter River

The Hunter Valley was the springboard for the pastoral invasion of lands further north and west, including the brigalow lands.[135] Livestock was overlanded from the Sydney region in 1821 and the whole of the valley was occupied by 1825 and severely overgrazed by 1826. The need for grass and water drove squatters around the Liverpool Range and onto the Liverpool Plains in 1826. (The Liverpool Plains are the subject of a major study in *Listen … Our Land is Crying*, pages 192–201.) The major drought of 1827–29 halted expansion temporarily. The grazing potential of the plains had been assessed by Oxley on his journeys in 1818 and 1825, and by Cunningham in 1827. Mitchell, in his explorations of 1831–32, travelled as far north as the Darling Downs and west to the Barwon by 1832, reporting on the excellent conditions and preparing the way for settlement.

The Hunter was a major artery for riverboat transport in the early days. Its siltation, resulting from over-grazing of the catchment, progressively reduced the distance to which vessels of any size could travel. (The Hunter is the subject of an account relating to management of waters discharged from coal mines into the river in *Listen … Our Land is Crying*, pages 285–7.)

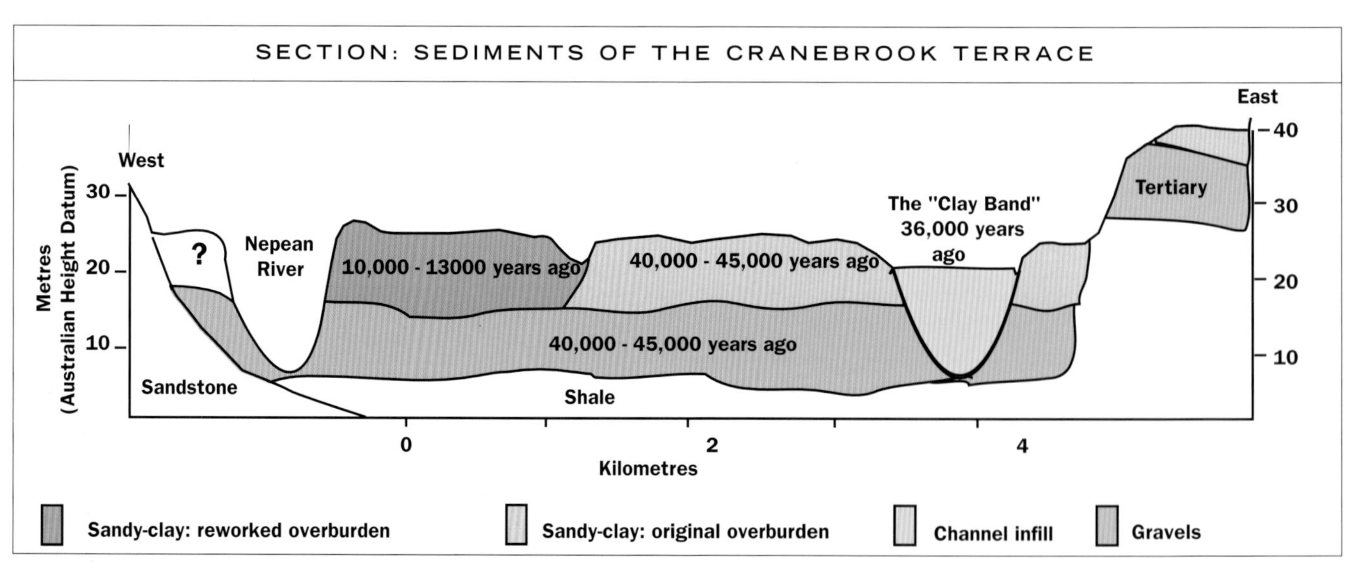

After Nanson *et al*[134]

North-eastern Australia

NEW SOUTH WALES NORTH COAST CATCHMENTS

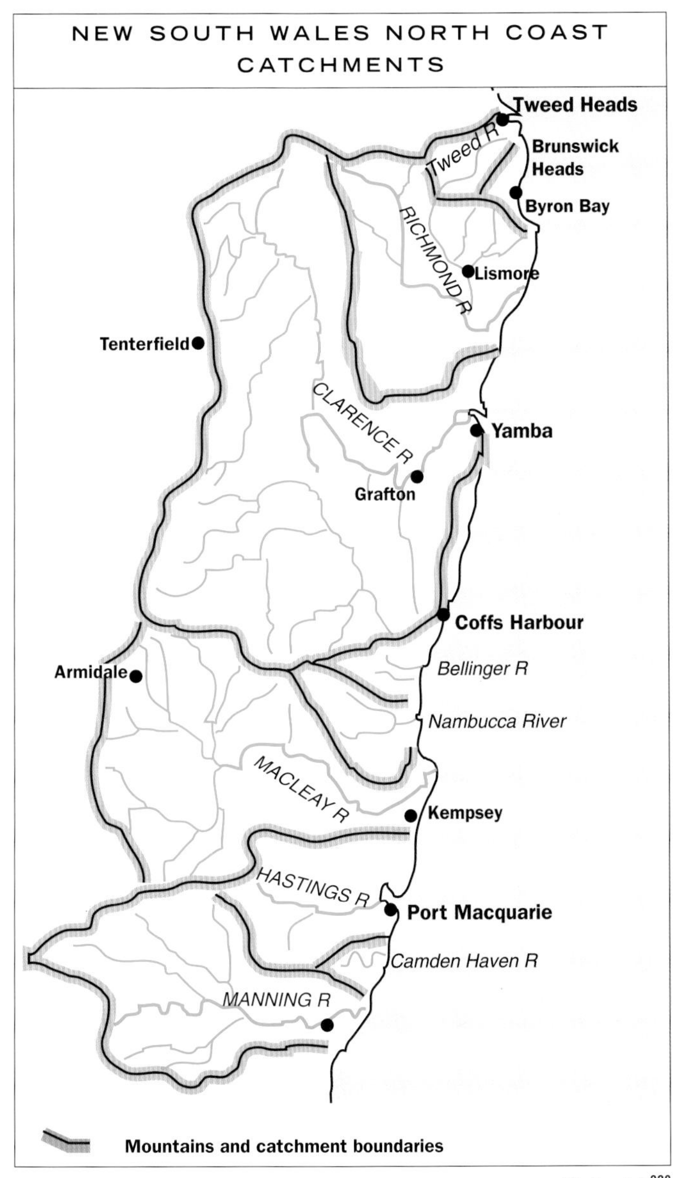

After Lines-Kelly[228]

The Manning River catchment

The Manning River catchment is bordered by the Barrington Tops in the south, the ranges of the Great Divide to the west and north-west, and the Bulga and Comboyne Plateaux in the north-east. The climate is warm humid with rainfall from 860 millimetres near the coast to 1300 millimetres in the western highlands. Tropical storms can produce high magnitude floods over the basin, which is also drought prone.

The Manning River and its major northern tributary systems

Floodplain stripping during some large floods in the 1970s on the Manning River near Charity Creek. GERALD NANSON

(Barnard, Nowendoc, Mummel and Rowleys) rise in the ranges of the Great Divide. Its southern tributary systems, the Barrington and Gloucester, rise on the Barrington Tops, while a number of small streams feed into the lower Manning and its estuary from the plateaux.

The Manning River consists of confined channels and floodplains in irregularly meandering gorges.[102] At Charity Creek near Mt George the Manning Basin is 6500 square kilometres of dissected, forested plateaux, rising to 1500 metres in elevation; and steep, pastured and forested hill country. On certain non-migrating reaches of the river, like that at Charity Flat which was a study area for recent research, floodplains were found to have been formed gradually by vertical accretion, and then destroyed catastrophically by large floods.[136] (This also happens on the Clyde floodplain in southern New South Wales where it runs to its estuary approaching Batemans Bay.) The channel at the study site is well-defined and about 70 metres wide and 5 metres deep, with a pebbly bed with gravel and bedrock riffles and short shallow pools. The floodplain has well-defined levees and floodplain backchannels, all confined within a bedrock valley.

The vertical growth of these floodplains brings about their own destruction. As steep levees gain height they confine larger and larger flows to the main channel and to steep-gradient backchannels on the narrow floodplains. Eventually, an erosional threshold is exceeded and the floodplain is destroyed, to build again until the unstable point is again reached.

At Charity Creek, a series of major floods between 1968 and 1978 stripped a large part of the upper unit of sandy loam from the basal gravels at the downstream end of the floodplain. The floods have

THE MANNING RIVER CATCHMENT

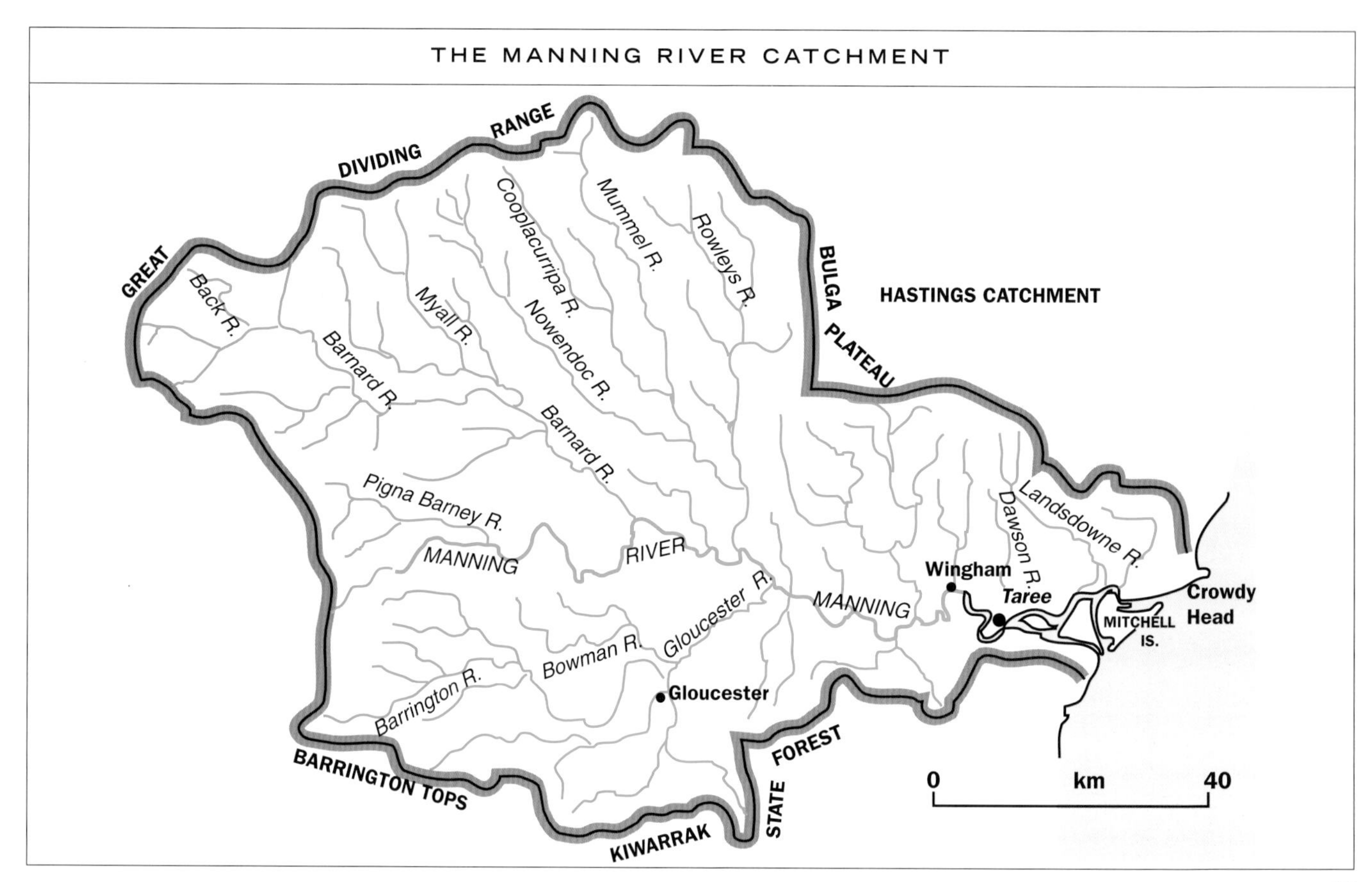

After Lines-Kelly[228]

THE CYCLE OF STRIPPING AND ACCRETION FOR FLOODPLAINS ALONG THE MANNING AND CLYDE RIVERS

1. STRIPPED
2. BUILDING
3. MATURE
4. ERODING

Bedrock · Gravel · Sand & silt

After Nanson[136]

been categorised as a 14-year flood in January 1968; 4.9- and 3.8-year floods in January and June 1974; 4.4- and 9-year floods in January and March 1976. This period of intense flooding culminated in a 52-year event in March 1978. Most of the severe erosion occurred in the 1976 floods; the final big one did not have the same large erosional consequences. Altogether, 2 to 4 metres were stripped from the floodplain over a 100 to 140 metre wide area along nearly 600 metres of floodplain. More than 200 000 cubic metres of material was removed, leaving less than half of the floodplain uneroded. Floodplains immediately upstream and downstream of Charity Flat were not eroded in these events, but severe stripping took place at several other sites in the basin, suggesting that floodplain stripping results in a mosaic of erosional activity along these rivers.

The Hastings and Camden Haven River catchments

The small Camden Haven catchment is separated from the Hastings catchment by the Broken Bago Range. It contains three spectacular volcanic piles—North, Middle and South Brother, ignambrite plugs which tell of explosive volcanic activity 16 million years ago. The Watson Taylor Lake, and the Queens Lake at Laurieton, are part of extensive coastal wetlands and mangrove swamps.

THE HASTINGS AND CAMDEN HAVEN CATCHMENTS

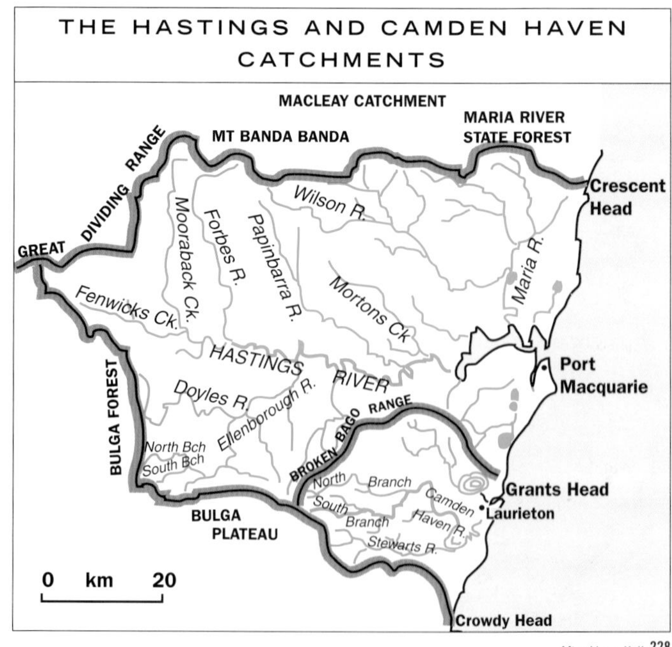

After Lines-Kelly[228]

The Hastings catchment is bordered to the north by the Banda Banda Plateau where the Forbes, Pappenbarra and Wilson Rivers rise; and to the south by the Bulga Plateau where the Doyles and Ellenborough Rivers rise. The Wilson River joins the Maria River, running parallel to the coast through extensive wetlands and into the Hastings estuary.

The upper Hastings River at Mt Seaview. M.E.W.

The Macleay River catchment

The Macleay River rises west of Armidale on the New England Plateau and flows out to the sea at South West Rocks, north of Kempsey. Ranges of the Great Divide form the western and north-western boundaries of the catchment; the north-eastern boundary is formed by the Snowy Ranges and the Macleay Hills; and the Banda Banda Plateau and hills of the Maria River State Forest are boundaries to the south.

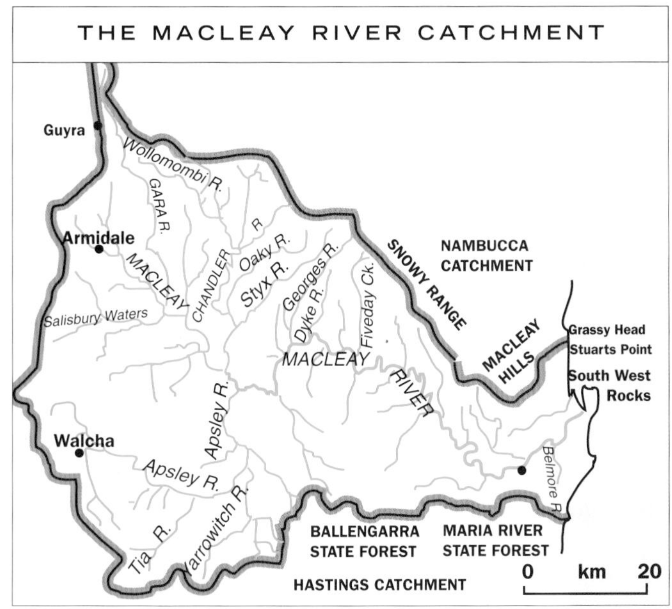

After Lines-Kelly[228]

The Bellinger and Nambucca River catchments

The Bellinger and Nambucca Rivers rise in adjacent valleys in the mountainous fringes of the New England Plateau. The Bellinger catchment is dominated by the Dorrigo Plateau. The Bellinger River has two arms—the southern is the Kalang, the northern the Bellinger. Some headwater tributaries, such as the Never Never, are high energy gravelbed streams which plunge over the 100-metre escarpment formed by the Tertiary Ebor volcanics and then make their way down alluvial valleys to join the Bellinger. The Nambucca is fed by several creeks, most of which rise in the north-west of the catchment.

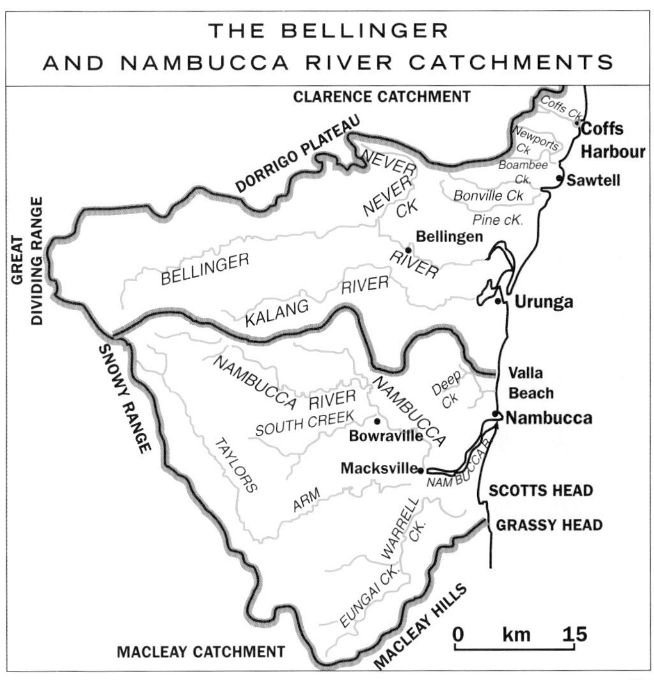

After Lines-Kelly[228]

EVOLUTION OF THE DORRIGO PLATEAU: *drainage patterns in the Bellinger and Nambucca catchments*

The Dorrigo Plateau is covered in basalt which was produced by the Ebor volcano 18 million years ago.[137] The volcano erupted on a palaeoplain of moderate relief. Subsequent uplift and tilting led to erosion of the Nambucca Beds, together with much of the volcano, and creation of a major escarpment—part of the Great Escarpment. In this area the Great Escarpment is therefore younger than 18 million years. The neck or plug of the Ebor volcano is an intrusion in the middle of the Bellinger Valley, known as the Crescent. The reason for the uneven erosion of the basalt, where the south-eastern sector has gone while the rest forms the crescent of the Dorrigo Plateau, lies in the nature of the geology below the basalt. The plateau lies on erosion-resistant rocks of the Coffs Harbour structural block, while the rest was on the easily erodable Nambucca Beds and was not only eroded but also undermined.

The drainage pattern, developed since the volcano erupted, is a classic example of radial drainage.

Ebor Falls, one of many spectacular waterfalls which descend from the volcanic escarpment left by the Ebor volcano. M.E.W.

THE DORRIGO PLATEAU— A REMNANT OF THE EBOR VOLCANO

Ebor
Dorrigo
GREAT ESCARPMENT
Neck

After Ollier[137]

DISTRIBUTION OF VOLCANIC ROCK FROM EBOR VOLCANO, AND RADIAL DRAINAGE

Blicks R.
Nymboida R.
Guy Fawkes R.
Dorrigo
Ebor
0 km 10
Bellinger River
Kalang River
Nambucca River
Styx River
Georges River
Dyke River
Five Day Creek
Nulla Nulla Creek
Taylors Arm
MACLEAY RIVER
Volcanic Rocks

After Ollier[137]

THE NAMBUCCA RIVER

Studies on the Nambucca River catchment have presented a nice picture of the changes to the river which have occurred as a direct result of European activities.[94] The same sequence of changes is seen in many of the smaller coastal catchments in south-eastern Australia.

At the time of European settlement the river channels of the Nambucca catchment were narrow, sinuous and relatively deep, with well-forested banks and floodplains. The floodplains were fine gravel and silty sand laid down during the last laterally active phase, a low energy period from about 3000 years ago. The only coarse sediments were terrace remnants. Following the arrival of Europeans there have been four recognisable phases of catchment destabilisation. **The primary cause of degradation is entirely attributable to human impact—but the timing of each phase is partly the product of the irregular nature of flooding.** Large floods, clustered into two distinct episodes, precipitated the changes.

- **1830–70** Cedar-getters arrived in the 1830s and selectively logged until the 1850s, probably without causing great disturbance because gaps in the forest recovered. The first pastoralists who came in the 1840s did not stay long and they, too, probably had limited impact—perhaps a bit more suspended sediment in the rivers.

- **1870–96** Serious land clearance took place for farming above the tidal limit. By the 1890s this would have had an appreciable effect on the rivers of the middle and upper catchment. Three major floods in 1890, including the largest on record, were followed by large floods in 1893 and 1894. Cleared banks were undermined and collapsed, the rivers widened, eating into the floodplains, gravels entered the streams as bed material and the fine silts were either deposited on floodplains as overbank deposition, or flushed downstream to the tidal channels and the estuary. The coarse gravels in the river bed contributed to the stream bank erosion and channel widening because of their erosive power. Flow speed increased because with bank clearing there was less woody debris in the stream to slow down the flow, and big snags were washed away during the floods. Thus the last decade of the nineteenth century became a flood-induced period of severe channel erosion.

- **1897–1947** The first half of the twentieth century was a comparatively dry period with reduced river activity for much of coastal New South Wales. During that period the Nambucca experienced only four floods larger than 8.9 metres at the Bowraville gauge. During this fairly quiet phase some of the changes wrought by the earlier active flood regime were slowed, and although some bank collapses continued, channel widening and further changes were virtually on hold, waiting for the next active flood session to complete the rivers' modifications. There were still deep pools for fishing and swimming and the channels, though severely degraded, still had some of the characteristics of the pre-disturbance channels. During this phase many of the channels were primed, ready for the greater disturbance that would come with the next flood-dominated phase.

- **1948–present day** This phase contrasts dramatically with the previous half century. Between 1948 and 1965 there were 11 major floods of over 8.9 metres (Bowraville gauge). Those in 1949 and early 1950 were widely reported as channel-modifying events in other New South Wales coastal rivers. These floods greatly destabilised the system, particularly in the northern part of the catchment—deep pools disappeared, channels widened and straightened. The availability of modern mechanical equipment after World War II influenced the extent to which channel modification works and rates of extraction of sand and gravel were achieved, adding to the rate of change. More recent floods have continued the process, and the Nambucca catchment has now made an almost complete transition to an expanded rural system modified by European land clearance.

In the Nambucca River, the erosion of stream banks and adjacent areas of floodplain caused gravel to appear along most of the channels. One profitable 'solution' was seen to be gravel extraction directly from the river bed, for use on roads and for other construction works. This mining process simply caused more gravel to be eroded, at an even faster rate, from the flanking alluvium. The situation was not solved, it was exacerbated. The solution for rehabilitating these severely degraded streams is to stabilise the most severely eroded reaches of channel, stop gravel removal, and allow the system to repair itself naturally over subsequent decades.

The aerial photograph shows the present situation at the mouth of the Nambucca River, where it is joined by the Warrell River. This is a river which follows a northward course from Scotts Head to the Nambucca estuary, running behind the dunes like so many of the coastal rivers do because of the northerly sand drift which has progressively moved their mouths further north as it has built the coastal dune systems. Massive sand banks are seen in the inner harbour and in the joint mouth of the two rivers.

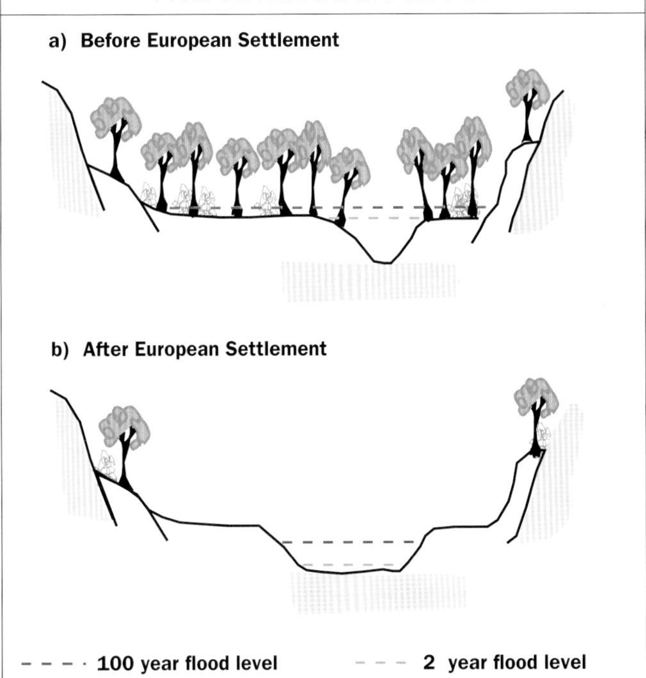

After Nanson & Doyle[94]

A 'training wall' was built in the Nambucca estuary in 1907 with the intention of creating a safe, reasonably deep inner harbour. But the planners did not take into consideration the amount of siltation of the estuary which was occurring as a result of catchment erosion, or the powerful activities of the northwards sand drift. Since the

Aerial photograph of Nambucca Heads showing siltation of the estuary. The V in the training wall and the breach which allows the tide to pour into the inner harbour at the eastern end of the training wall are clearly visible. NAMBUCCA SHIRE COUNCIL

PLANFORM OF PRE-EUROPEAN AND POST-EUROPEAN STREAM CHANNELS

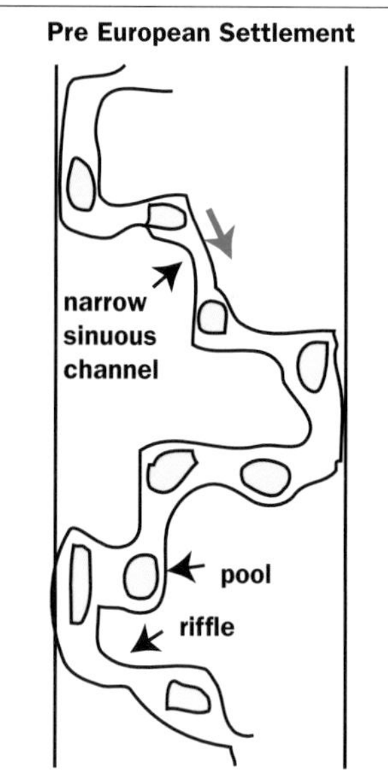

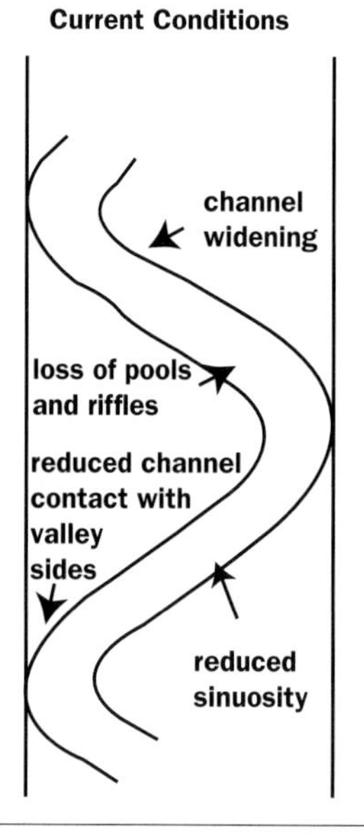

After Nanson & Doyle[94]

Well-vegetated and stable reach of river in the Nambucca valley. GERALD NANSON

Nambucca Heads. The sand island has developed in the centre of the inner harbour. FR TOBY KLEIN

Nambucca valley—severe bank erosion as a consequence of channel incision in response to European land clearance. GERALD NANSON

South Arm, in the Nambucca valley. A knick-point migrating up a tributary as a consequence of channel erosion and incision of the trunk stream in response to land clearance. GERALD NANSON

building of the wall, the inner harbour has half filled with sand, and the sand island which resulted has been colonised by river oaks. A flood in 1972 breached the training wall a little westward of the V-wall, leaving a channel through which the tides pour daily.

A study of the lower Nambucca River was commissioned recently to investigate options for flood management. Nuisance flooding at Macksville and in Nambucca Heads occurs frequently and major floods caused economic and social losses as recently as 1963, 1972 and 1989. A previous study, the Macksville Flood Protection Study, had proposed the construction of levees to protect the residential and commercial areas at Central Macksville from periodic flooding, but although the study concluded that the project would result in 'large overall benefit to the community' there was community opposition and the idea was abandoned.

A number of options were considered in the 1995 Floodplain Management Study, including: dredging the river from the entrance to Stuart Island (1.5 kilometres upstream) or from Stuart Island to Macksville; increasing the hydraulic capacity of the railway embankment at Kings Point; replacing the causeway at Stuart Island with a structure which allows throughflow; closing, or widening, the breach in the V-wall; or removing the railway embankment and floodgate north of Macksville to improve flows upstream of Macksville to Newee Creek. Two further options were raised by the Floodplain Management Committee but were not considered feasible—the construction of a dam on Taylors Arm and/or the

COFFS HARBOUR

The making of the harbour at Coffs Harbour involved constructions which joined South Coffs Island and Muttonbird Island to the coast, as well as the construction of the long eastern breakwater protecting the entrance. The location was a prominent headland, diverting the sand drift even before the constructions, similar to Smoky Cape, 100 kilometres to the south, and Cape Byron on the far north New South Wales coast, where the volume of sand in the drift is of the same magnitude.

Fraser Island, off the Queensland coast 200 kilometres north of Brisbane, and the Great Sandy Region to its west and south-west on the mainland, are also products of the longshore drift. Fraser Island is the largest sand island in the world, with the narrow Great Sandy Strait lying between it and the coast. At low sea level times in the Pleistocene ice age, the island was a promontory on the wide, exposed continental shelf. The accumulation of sand to make the island results from the presence of a rocky bar on the sea bed which is emergent at the north of the island. This obstruction caused the current to drop its sand, and the immense volume which has accumulated southwards has Seventy-five Mile Beach on its eastern side as evidence of the amount of sand which has been transported over time. The whole sand island acts as an unconfined aquifer, and streams of pure, fresh water run year-round. Perched lakes, naturally clay-lined, which means that their waters do not seep away in the sand, are situated in the southern half of the Island and, surprisingly, rainforest is present in pockets where it has built up its own nutrients over time.

The effect of the northward sand drift up the eastern Australian coast is everywhere evidenced by the dune systems on the continental margin and the sandy shoals which make navigation so difficult for small craft. How it has affected river mouths is seen when one looks at maps which show west to east flowing streams which approach the coast and then turn abruptly and run northwards behind the dunes until they can find an exit to the sea. Maps of the Richmond, Tweed and Brunswick catchments (pages 159-160) show this feature particularly well, as does the northward migration of the mouth of the Herbert River, documented on page 171.

The problems of sand accumulation in estuaries are exacerbated by the increased sediment transported downstream by rivers which have suffered catchment disturbance as a result of land-use practices; and where river regulation has occurred, and normal flow has been restricted.[149]

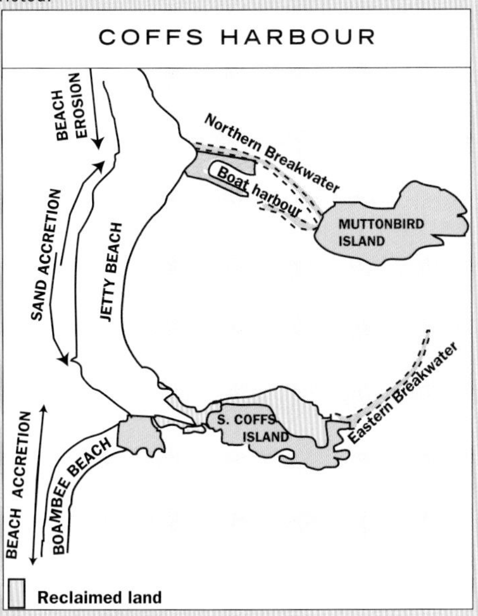

SAND, GLORIOUS SAND: THE LONGSHORE DRIFT AND FRASER ISLAND

Along the eastern coast of Australia vast quantities of sand are carried from south to north by ocean currents. The average longshore transport rate has been estimated at 75 000 cubic metres a year. The sand accumulation which results in harbours and estuaries when construction of breakwaters and other structures impede the flow is a major problem, and the effects of the diversion of the drift round such obstructions leads to changed sand conditions on the beaches on either side. For example, the Tweed River breakwaters decrease the sand on Queensland's Gold Coast Beaches; the changes made at Coffs Harbour with the construction of the harbour have led to accretion on Boambee Beach to the south, and on beaches within the harbour (which is constantly silting up), as well as to erosion of beaches to the north.

A permanent stream on Fraser Island, fed by the unconfined aquifers in the sand dunes which comprise the island. M.E.W.

Nambucca River, and the provision of a second entrance through the dunes to the ocean. The study modelled the effects of the various options for different flood intensities and concluded that the only option which would have any real benefit was dredging from the entrance to Stuart Island, but that even that option was hardly viable because the rate of sand transport (the longshore drift of sand northwards along the coast) would replace the sand as rapidly as it could be removed. The costs and the engineering problems involved would seem to be prohibitive.

The problems experienced on the Nambucca floodplain are outlined here because they are common to so many coastal communities which have built their towns and infrastructure on floodplains or coastal wetlands, relying on engineering expertise to solve the problems which are inherent in such areas and which have been greatly exacerbated by the destabilising of the rivers by human activity. Even now little thought is being given to how much more serious the problems will be, and how much harder to solve, as sea level rises under Greenhouse warming—or, more immediately, as rainfall events become more intense (one of the expected, and arguably already evident, consequences). Even small changes will make a difference—consider higher than normal king tides coinciding with a flooding river and increasing back-up in already choked river systems ...

WETLANDS OF THE NORTHERN TABLELANDS

Little Llangothlin Lagoon from the air. STEPHEN GALE

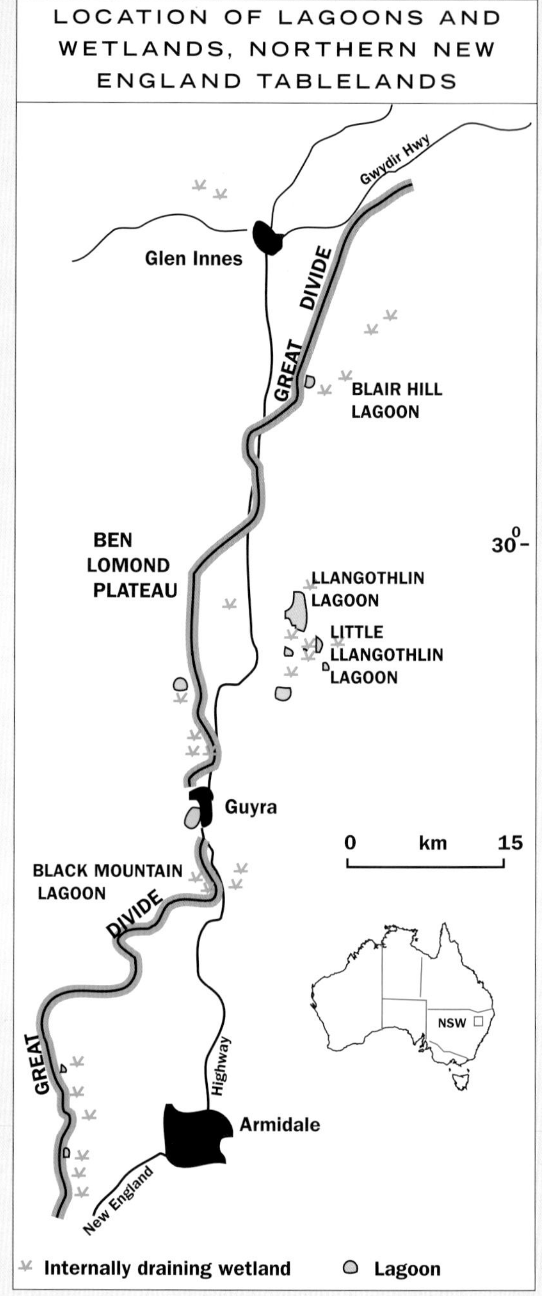

After Haworth *et al*[138]

The Northern Tablelands, or New England Region, of New South Wales represents the second highest land region in Australia. Its landscapes tend to be flat or of low relief, forming an undulating surface ranging in altitude from 800 to 1500 metres. An ancient Mesozoic surface of relatively low relief was blanketed by successive accumulations of Tertiary basalts which have largely been stripped away to again reveal the Mesozoic landsurface, resulting in a situation similar to that described for the south-eastern highlands. Rainfall on the tablelands is relatively high, while evaporation rates are low.

Many streams, among them some of Australia's major rivers, have their headwaters on the tablelands, and the low relief of the terrain results in an abundance of extensive and important wetlands. The state of environmental health of these wetlands varies greatly, depending on the degree of disturbance, which in turn depends on whether the wetlands lie within the belt of good agricultural land that runs generally north–south along the western edge of the tablelands. This is also the zone of fertile basalt-derived soils, as opposed to the granite-dominated land to the east, which is largely within National Parks or State Forests, apart from a pocket of richer volcanic soil in the Ebor–Dorrigo district.[139]

The parallel basalt–granite zoning also divides the two major types of wetland in the region: shallow lakes (lagoons) are located on or near major divides in the heavily farmed basalt belt to the west; and heath bogs fill the broad upland valley heads in the eastern, wetter granite belt. The Ebor district is again anomalous, having no wetlands of the lagoon type although it is a basalt region with high rainfall. It has extensive swamps in the flat upper valleys of the former Ebor–Dorrigo volcano.

Llangothlin Lagoon—a toe of granite enters the reed-filled lagoon. BOB HAWORTH

Both basaltic lagoons and granitic heath bogs act as high altitude hanging swamps intermittently feeding the headwaters of major coastal and inland streams, like the chains of ponds in the headwaters of streams in the south-eastern highlands, and the dells or mires on the sandstone plateaux in the Sydney Basin.

The other types of wetland in the New England region are riparian wetlands on the floodplains of some of the tableland streams; upland basaltic swamps of the Ebor type (which are also found in the southernmost portion of the region as the tablelands approach the Liverpool Range south of Walcha); and seepage springs and bogs at the base of basalt flows and the steeper and more extensive granitic outcrops. One small but unusual category of wetland is the sphagnum (peat) bogs of Round Mountain, found only above 1300 metres on southerly aspects and representing the northernmost extent of this type. (Round Mountain is in the Cathedral Rock National Park just west of Ebor.)

River regulation, damming and extraction have had far-reaching and probably irreversible effects on riparian wetlands. The rapid increase in numbers of farm dams in the last 50 years has undoubtedly reduced overland flow, base flow and groundwater reaching rivers and their associated wetlands. Changes to river form and function in the New England catchments have been similar to changes elsewhere since settlement. Swampy, discontinuous watercourses have been replaced by enlarged, incised continuous channels; siltation has made channels more homogeneous and lessened the diversity of ecosystems, reducing the range of habitats; linkages of biophysical processes between floodplains and channels have been reduced or eliminated.

The other great change to tableland hydrology has been the extensive draining of upland seepage sites, which commenced with settlement 150 years ago. The widespread boggy zones which made dray traffic all but impossible over many areas have disappeared except in very wet years. The range of the aquatic biota has been drastically reduced, particularly in the basalt areas, and the remaining and increasingly isolated wetlands are vitally important for retaining biodiversity. Much of what is left of the aquatic ecology is now concentrated in the lagoons, which have themselves suffered a great reduction in numbers and extent.

DRAINAGE DIVIDE LAGOONS

Fifty-four shallow upland lakes, all occurring above 930 metres, on or close to the Great Divide or adjacent ridges, and inland of the 1000-millimetre rainfall isohyet, have been identified in the New England region. More existed in the recent past but have been destroyed since the beginning of European settlement. Almost all of the remainder have been modified in some way, usually by partial drainage, sometimes by damming and raising the outlet. They are closed drainage systems most of the time so most of the products of their catchment erosion are retained in their lake-floor sediments. After flooding rains, water spills over into the adjoining river catchment. All but six of the lagoons spill into either the Clarence, Macleay or Gwydir river systems.

Under natural conditions, water levels fluctuated. The original ecological richness of the lagoon habitat depended on the fluctuation in width between wet and dry seasons of the surrounding vegetation zones. Few lagoon basins dried out completely before they were artificially drained; the practice of raising the lip to maintain deep water has been as damaging to ecological richness as has draining. The richness of these wetlands (as with other wetlands and also with rivers in Australia) depends on their progression between extremes, allowing a wide variety of intermittently dormant species to flourish at least some of the time.

Some of the lagoons have lunettes on their eastern (lee) side, probably formed under different local climatic conditions in the Pleistocene ice age. Some Tertiary coal deposits have been found to exist at depth below some of the lagoon localities, indicating that similar lakes have existed in high and exposed positions on the New England Tableland for at least 20 million years. Radiocarbon dating of peaty material in today's lagoons indicates that continuous sedimentation only started 15 000 years ago, after the last glacial maximum. (During glacial stages the lake basins would have been deflated, their sediments scoured out by wind erosion.)

Studies of sediment in several lagoons have been carried out by researchers from the Geography and Planning Department of the University of New England, Armidale,[140, 141] and co-researchers from the University of Sydney, ANSTO and a consulting company,[138] using lead isotope dating techniques. Results show the same patterns of sedimentation and changes to the aquatic environment in the lagoons as that which has been documented for the rivers of eastern Australia. There is also an interesting record of pollutant accumulation in the sediments.

POLLEN DIAGRAM: LITTLE LLANGOTHLIN LAGOON, GUYRA

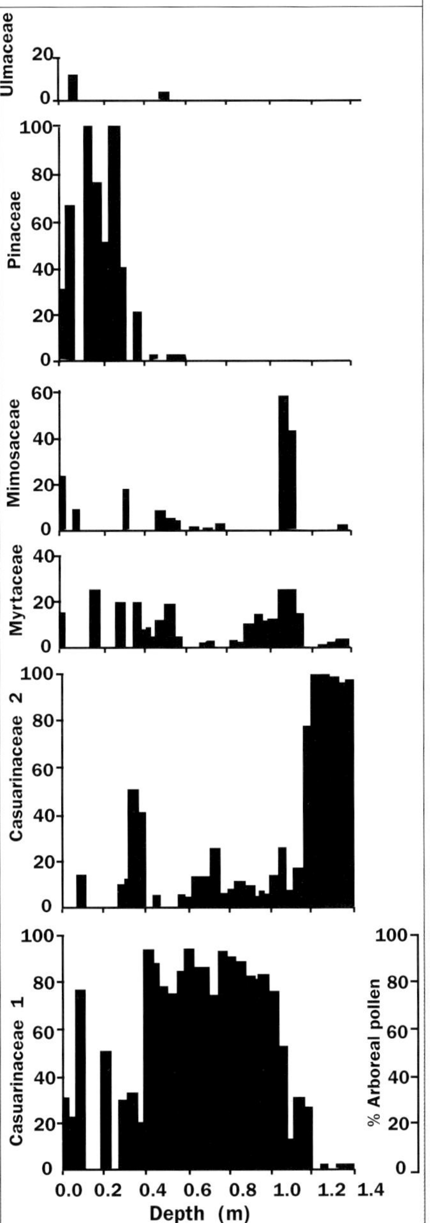

After Gale *et al*[140]

LITTLE LLANGOTHLIN LAGOON

The lake sediments of the Little Llangothlin Lagoon, 18 kilometres north-east of Guyra, revealed a number of interesting facts. An initial period of extreme erosion followed the arrival of Europeans and their cattle and sheep. Above this rapid zone of sedimentation in the lagoon, rates have returned to approximately their pre-settlement levels and have remained steady. The end of rapid and voluminous sedimentation is believed to have been due to most of the available material having been freed and passed downstream into the lake in that first flush. This corresponds to the rapid phase of erosion and sedimentation which is seen in rivers of south-eastern Australia and which in many cases could be shown to have been the result of over-grazing even before large-scale tree clearing occurred.

A pollen record in the sediment layers shows vegetation change from lowest sediments upwards:

- Lowest sediments contain an assemblage dominated by Casuarinaceae 2 (with small pollen grains), with a low incidence of Myrtaceae and Mimosaceae, and of Casuarinaceae 1 (with large pollen grains)
- At 1.2 metres depth, abundance of Casuarinaceae 1 rises
- At 1.16 metres, Casuarinaceae 2 decreases significantly and Myrtaceae and Mimosaceae increase (briefly)
- Above 1.08 metres, pollen of Mimosaceae

occurs only rarely and sporadically though Myrtaceae is not as comprehensively reduced

- Casuarinaceae 1 decreases dramatically above 0.4 metres
- From about 0.6 metres, pollen of exotic trees like Pinaceae and Ulmaceae appear for the first time in the record and increase dramatically.

Casuarinaceae decreased as land use increased—the trees were chopped down for firewood and other uses. Massive clearing for agriculture occurred in all the suitable areas, and dieback has since killed most of the remaining eucalypts.

Phosphorus and lead levels in sediments were shown to start to increase between 1–0.5 metre in sediment cores, and above that level to increase dramatically. In the case of phosphorus, the rise corresponds to fertiliser use, which escalated dramatically after World War II. The increment in lead levels is believed to be related to airborne lead which has increased worldwide in aerosol because of industry and automobile emissions.

BLACK MOUNTAIN LAGOON

Black Mountain Lagoon is situated adjacent to the Great Divide, seven kilometres south of Guyra. It is a shallow basin less than 0.5 square kilometre in area in a basalt catchment of 1.3 square kilometres. When full it overflows to the south-west into Boorolong Creek, a tributary of the Gwydir River and the Darling. A stream drains into the south-east corner of the lagoon, and a usually dry gully into the north-east corner, but the lagoon's water is mostly derived from shoreline seepages and springs. The New England Highway runs along the eastern side of the catchment, more or less along the watershed of the Great Divide.

Once again sediments in the lagoon have recorded the effects of European settlement and land use. They show that before European influence the lake was open water, which used to dry partially, retaining an eastern pool. From the time of settlement it has rapidly evolved into a vegetation-choked swamp. The same dramatic period of sedimentation as in Little Llangothlin Lagoon followed the arrival of grazing herds; after that initial burst of erosion, sedimentation continued at a more or less constant lower rate approximating the rate before settlement. This constant rate has been maintained in spite of changing and accelerating land use, the occurrence of significant floods and droughts, and the absence until recent times of the application of conservation practices. (An increased rate in very recent times relates to the disturbances caused by the building of the highway.) It contrasts, however, with sedimentation rates in the period from 12 300 to 5600 years ago, which were an order of magnitude lower than those of the last 100 years.

The lake sediments record the use of agricultural chemicals on the catchment during the twentieth century, with phosphorus, zinc, arsenic and lead being stored in them. A big increase in phosphorus corresponds to the 1950s, when a new generation of owners began draining the catchment and applying large amounts of 'super' (superphosphate) to their pastures. Zinc levels rose then too, and probably came from the superphosphate fertilisers; high lead levels, higher than the base level which results from the general airborne fallout, probably result from the emissions of vehicles on the highway on the edge of the catchment; and the arsenic detected is believed to have accumulated as a result of pesticide usage.

The concentration of toxic elements in sediments in such a small basin with such a small catchment should start alarm bells ringing in general about accumulations of pollutants in soils, river and lake sediments, and in waters in large catchments where intensive agriculture is practised. Though not as easily measured as in a closed basin, such levels of toxins probably exist much more widely than has been thought. The most likely places for their accumulation are obviously in the low parts of landscapes where swamps and lakes occur. Perhaps consideration should be given to testing lakebed sediments in areas downstream from intensive agriculture *before* they are cropped, as often happens where irrigation has dried wetlands, enabling them to be used in this way. Some plants are known concentrators of certain elements and they can presumably get into human food chains (cadmium from superphosphate is a recognised hazard of this sort).

Dieback has killed eucalypts on the New England tablelands. M.E.W.

PHOSPHORUS AND LEAD CONCENTRATIONS, LITTLE LLANGOTHLIN LAGOON

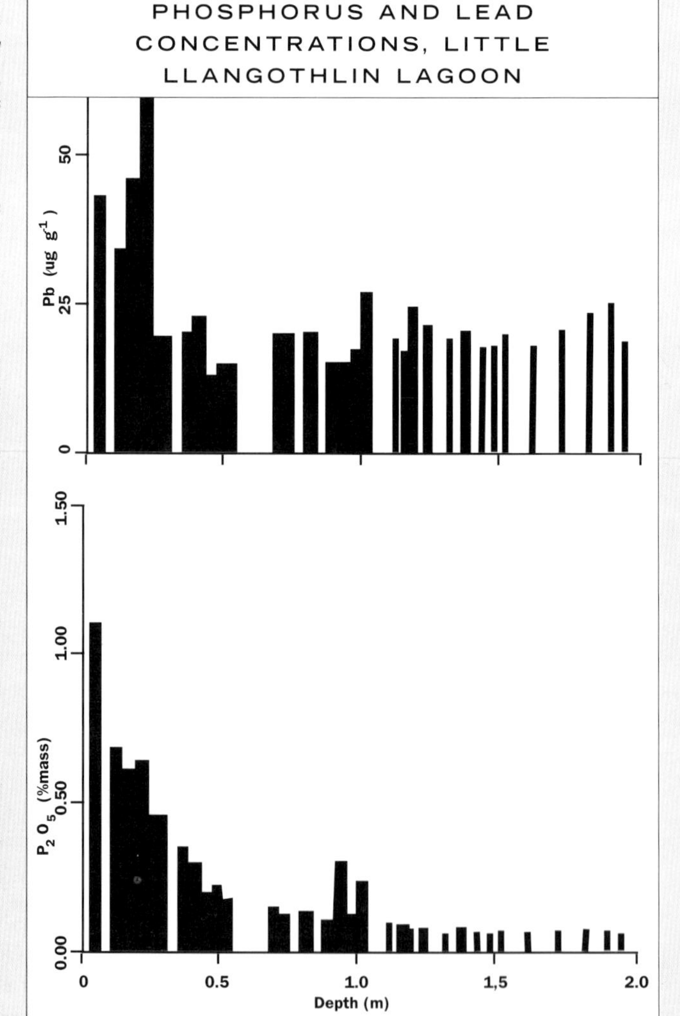

After Gale *et al*[140]

THE CLARENCE RIVER CATCHMENT

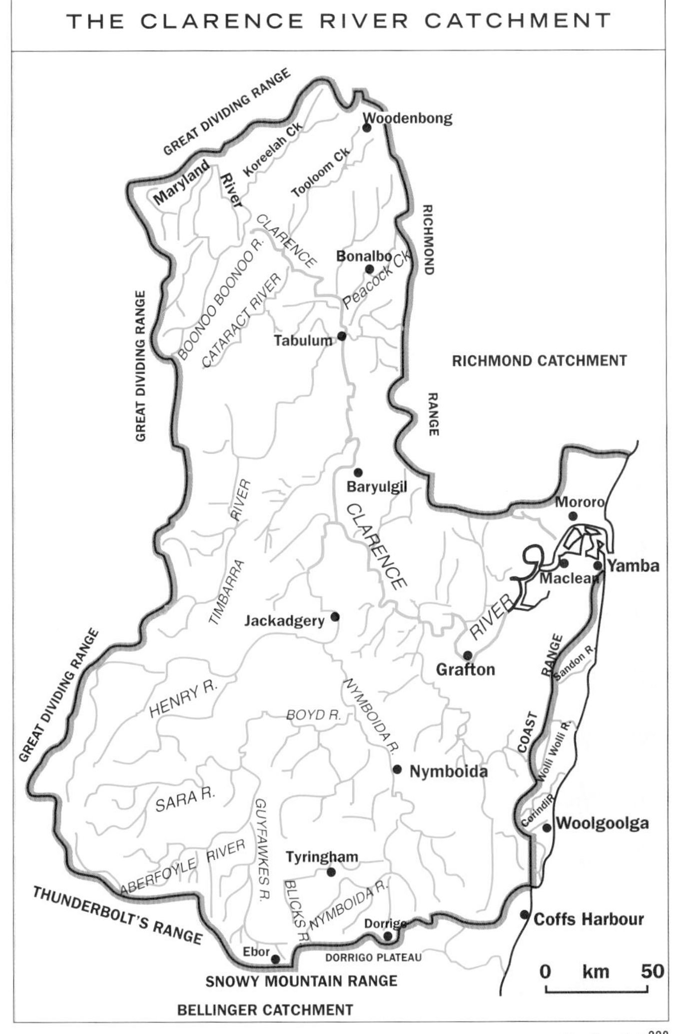

After Lines-Kelly[228]

The Clarence River catchment is the largest on the New South Wales north coast (over 20 000 square kilometres), extending from the ranges of the Great Divide along the Queensland border to the Dorrigo Plateau in the south. It is separated from the Richmond catchment, which lies between its northern sector and the coast, by the Richmond Range. The ranges of the Great Divide form its western boundary, and the Coast Range forms a divide along its eastern edge. The main valley containing the trunk stream is almost 200 kilometres long. Most of its major tributaries originate on the New England Block and cross the Great Escarpment in a series of impressive falls.

The Nymboida River and its tributary system form the main southern branch of the Clarence, rising in the high Thunderbolts, Snowy and Dorrigo Ranges and separated from the Orara River by a low divide. Since 1924, water has been extracted from the Nymboida and piped over a divide into a hydro-electric power station (North Power) on a small stream at a much lower elevation called Golang Creek. It flows down this creek into Blaxland Creek and eventually into the Orara River, which is a major tributary of the Clarence. An Olympic-style canoe course is created when North Power releases water into the creek below the power station, turning a boulder-filled dry channel into wild white-water.

Piles were driven into the bed of the Orara River at Upper Orara in an attempt to stabilise the river bank. A heavy rainstorm in July 1999 caused the river to erode the bank, stranding the piles in the stream channel and rendering the attempts at containing erosion ineffective.
PHOTO COURTESY *THE ADVOCATE*, COFFS HARBOUR

The power station is licensed by the Department of Land and Water Conservation, and North Power is obliged to maintain channel stability on Goolang and Blaxland Creeks. The Clarence River County Council has been delegated to undertake erosion control for North Power since 1967. Remedial work has not been effective, however, and accelerated bed and bank erosion has occurred—deeply incised, greatly widened channels; loss of pool and riffle morphology; land lost to erosion; sediment deposited in the Orara River filling its pools—all classic responses to human disturbance of the system. During periods when no water is discharged by the power station, the Orara often flows underneath its sand accumulations.

According to the Commissioner for Healthy Rivers, Dr Peter Crawford, in an interview in July 1999, all water in the Nymboida and Orara catchment areas of the Clarence River system was fully allocated and any future water needs must be met by a trade process—foreshadowing a licensing and metering process for users. Expanding water needs of the Clarence Valley and Coffs Harbour could not be met unless management of the resources of the whole catchment was improved.

The Orara, the source of water for Coffs Harbour, was reported to be in a state of serious degradation.[142] Parts of the river had incised by 2 metres since 1975; banks were collapsing; floodplain real estate was disappearing and other problems had become so visible that public meetings were called to enable residents and stakeholders to see the plans being drawn up for rehabilitating the river. Orara Rivercare groups have been ramming rows of poles into the river bed

THE CLARENCE AND CONDAMINE RIVER SYSTEMS: once a single river system flowing north-west

The south-east flowing Clarence River is closely aligned with the north-west flowing Condamine River just across the Continental Divide.[143] This alignment is continued by a large southern tributary, the Orara River, which flows north-west, away from the sea, to meet the south-east-flowing Clarence. Below the junction of the Clarence and the Orara, the character of the Clarence changes completely. The aligned streams—the Condamine, Clarence and Orara—represent the remains of an earlier north-west flowing palaeoriver that extended the full length of the Clarence–Moreton Basin. The lower, meandering reach from the Orara junction to the ocean is essentially an overflow channel.

The Clarence and most of the north-west-flowing part of the Condamine River (before it turns west and runs on to become the Balonne at Surat) lie within the Clarence–Moreton Sedimentary Basin. This basin is a narrow eastern extension of the Great Artesian Basin, battle-axe in shape, aligned north-west to south-east and divided at its waist by the north to south trending continental watershed. It is a 'framed basin', bounded by the bedrock which underlies and surrounds it, and which has done so since it started to sink and accumulate sediment during Late Permian and Early Triassic times 240 million years ago.

During the Permian Period, the whole area which now includes the Clarence Valley had been part of a generally flat surface, part of which was to start

CATCHMENTS OF THE CLARENCE AND UPPER CONDAMINE BASINS
With approximate position of the Great Escarpment

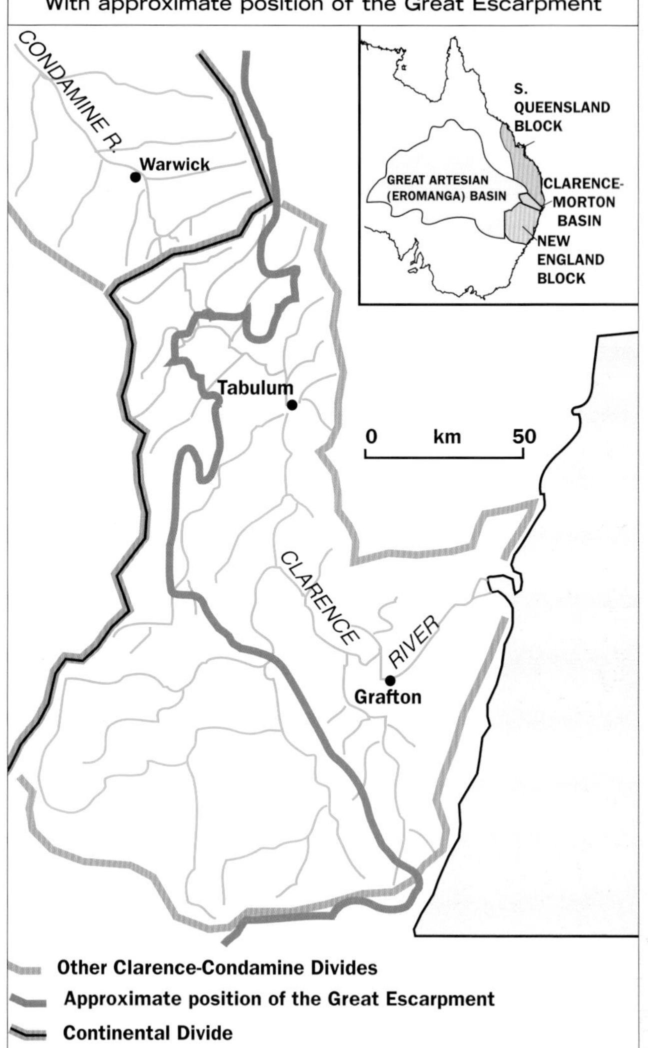

After Ollier[143]

GEOMORPHOLOGY OF THE CLARENCE—MORETON BASIN

Tertiary Volcanic Rock

Limit of Clarence-Morton Basin

After Ollier[143]

sinking to form a basin, while surrounding areas remained as blocks of elevated terrain. Terrestrial sediments accumulated as the Basin subsided from the Late Triassic, through the Jurassic and probably through the Cretaceous (although most of the Cretaceous sediments have subsequently been eroded away). The eastern section of the Clarence–Moreton Basin now lies between the New England and South Queensland blocks, and is open to the coast between Brooms Head and Schnapper Point. The remains of Miocene shield volcanoes and basalt now cover much of the South Queensland Block and extend into the central part of the Basin, where they would have disrupted some palaeodrainage patterns 23 to 22 million years ago.

At the time when the Basin was being filled with Mesozoic sediments, the present eastern continental margin had not come into existence. Major drainage originated at a watershed several hundred kilometres east of the present coast on the Tasman Divide.

A palaeoriver system drained the sinking basin, carrying sediments mainly from high land beyond the present continental edge into the Clarence–Moreton

Basin and onwards across the Great Artesian Basin, contributing to the sediment load carried by Australia's 'Congo-like drainage' (see page 107) which drained the whole eastern half of Australia at the time.

Large-scale sedimentation of the Basin ended some time before 80 million years ago, after reaching a maximum depth of about two kilometres, when rifting and opening of the Tasman Sea started. (The Tasman Sea was opened by seafloor spreading between 80 and 60 million years ago.) The palaeoriver system, north-west flowing, which used to drain the whole of the Clarence–Moreton Basin, was effectively beheaded. This system was, in part, the palaeo-Orara, the palaeo-Clarence west of the Orara junction, and the palaeo-Condamine. After rifting, an axis of uplift developed parallel to, and inland of, the new continental margin, resulting in a change in drainage patterns. The development of the Great Divide resulted in the initiation of coastward drainage and headwater erosion in the coastal zone—and led to the Clarence running backwards in its bed. The barbed pattern of its tributaries is evidence of its reversal—tributaries normally run downstream to join rivers; here they travel upstream to join.

The headwater erosion of its tributaries created the Great Escarpment locally. On the western side of the Divide, the Condamine continued its north-westerly journey, its new headwaters situated along the elevated axis of the Divide.

All the geological evidence thus suggests that a palaeo-Clarence—Condamine river system existed in the Jurassic, and that the Clarence was reversed in post-Cretaceous and probably pre-Miocene times. Drainage would have been increasingly confined to the eastern part of the basin as the new Continental Divide developed across the mid-section and would have been further constricted to the south-east in the Miocene when volcanic activity was occurring. The volcanic landforms were centred on Mount Warning, Focal Peak and Main Range shields (23 to 22 million years old) and on a series of smaller vents and flows. The Richmond River, to the east and the Logan and other streams to the north, developed as radial drainage around raised volcanic areas which also prevented most of the remaining northern basin drainage from reaching the Clarence. Another effect of the Miocene vulcanism was to push the main trunk of the Clarence further towards the western margin of the basin. This situation enabled it to capture the north-trending streams flowing from the New England Block.

The features which followed rifting and establishment of the eastern margin of Australia—the Great Divide, uplifted margin and the Great Escarpment—are characteristic of passive margins created by supercontinent fragmentation. The Drakensburg of South Africa, the Western Ghats of India and the east Brazil highlands all show similar features of structure and palaeodrainage.

Believe it or not, this is an Olympic-style canoe course! When the power station releases water at the point from which this photo was taken, the creek turns into wild white water. People in the township of Nymboida had hoped that this reach would be chosen as a training site for canoeists for the 2000 Olympics, which would have brought activity to this quiet part of the world, but no such luck! M.E.W.

to change flow patterns and have been replanting the river banks. The Coffs Harbour City Council has a three-year plan for river restoration, using environmental levy funds, grants from the National Heritage Trust and other sources, and support from the Department of Conservation and Land Management and the Clarence Catchment Management Committee.

The Richmond River catchment

The Richmond River catchment is bordered to the north by the McPherson, Tweed, Nightcap and Coonyum Ranges; to the west and south by the Richmond Range, and to the east by the coast from south of Evans Head to Broken Head. It comprises a large number of small streams. Lismore lies at the junction between the Wilsons River and its tributaries, including Terania Creek, which flow from the

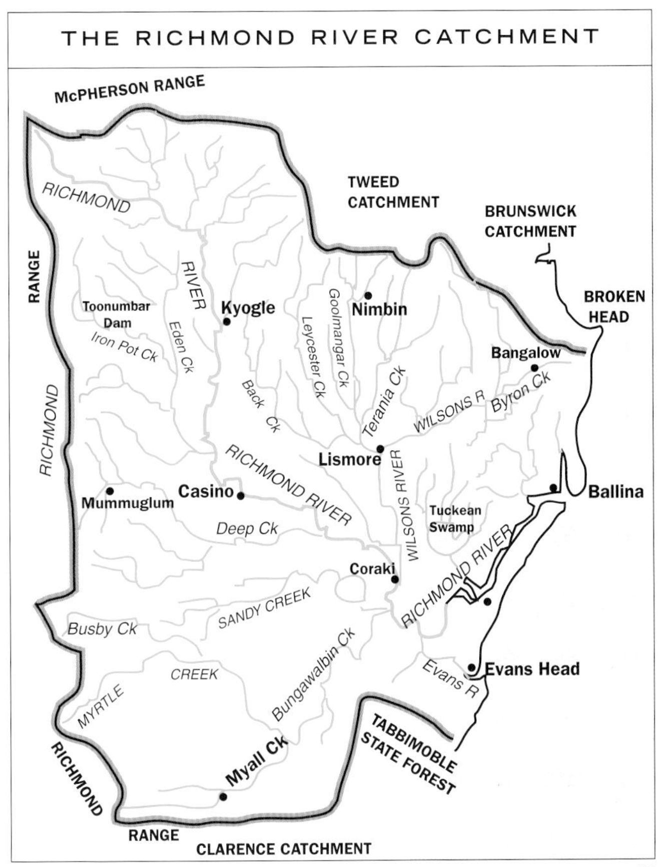

After Lines-Kelly[228]

The Nymboida River. M.E.W.

Nightcap Range. Ballina lies in the estuary near the river mouth.

The Richmond River developed as a radial drainage on Miocene volcanics.

THE BRUNSWICK AND TWEED RIVER CATCHMENTS

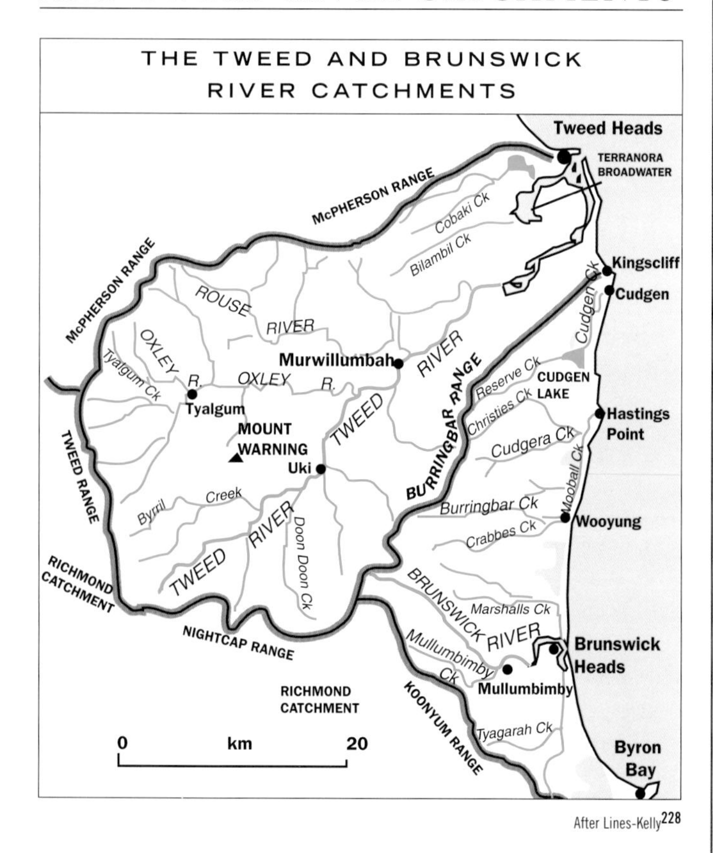

THE TWEED AND BRUNSWICK RIVER CATCHMENTS

After Lines-Kelly[228]

The Tweed and Brunswick River catchments are two of the smallest catchments on the north coast. They are separated from each other by the Burringbar Range and have similar climate and topography.

The Brunswick River has a very small catchment. It rises in the Burringbar Range and has a southern tributary, Mullumbimby Creek, which rises in the Koonyum Range and joins it in Mullumbimby town. A minor northern tributary stream, the confluence of the Lacks and Marshalls Creeks, runs almost to the coast and then, unlike all the streams along the coastal fringe to its north, swings abruptly south to enter the estuary near Brunswick Heads. The Mooball, Cudgera and Cudgen Creeks all make abrupt northwards turns to run along for considerable distances behind the coastal dunes.

The Tweed River catchment lies at the northern limit of New South Wales—a fertile valley almost completely encircled by rugged hills. Mt Warning, a shield volcano which was erupting between 23 and 22

RECENT HISTORY OF THE TWEED RIVER

Mount Warning and Point Danger are mentioned in the 1770 log of Captain James Cook. In 1823 John Oxley sighted the mouth of the Tweed and mentioned the hazards encountered in navigating the entrance. In 1844, the Government Surveyor, Clement Hodgkinson, recorded that the river was navigable for over 60 kilometres but that the entrance bar was dangerous. Cedar-getters arrived in the valley that year and selective logging proceeded rapidly, followed by clearing for farming, which commenced in the early 1850s. By the 1860s, shipyards and commerce of all sorts was established at Tweed Heads. The Murwillumbah townsite was first surveyed in 1872.

The coastal shipping trade was important to Tweed Heads and valley residents, bringing their supplies from Sydney and Brisbane, and returning with cut timber and produce. Steamers and other craft unloaded inside the river entrance because shifting shoals made navigation further upstream difficult. Flat-bottomed punts transported cargo upriver. A river trading and mail service, started by the Skinner brothers with row-boats and graduating to ferries, ran from the 1880s until 1934, when the age of motor transport arrived.

'Training walls' were built between 1891 and 1904 to restrict shoals and improve the harbour, and a breakwater was authorised in 1899 and completed five years later to improve conditions for shipping, because it was felt that the Tweed Valley had great economic potential. Extensions to the breakwater were made in the 1960s to reduce the area of dangerous breaking water which was encountered at the mouth of the river. (The northwards drift of sand along the coast has resulted in a build-up of sand south of the breakwaters, estimated at 1.5 million cubic metres, and continuing problems in the river mouth, where a further 1.5 million cubic metres is stored on the Tweed Entrance Bar. Sand loss on the Queensland Gold Coast beaches has been aggravated by this accumulation, and a scheme has been devised to construct a sand bypass plant at the river mouth to remedy the sand accumulation locally, while benefiting beaches to the north. (Contracts were let in 1999 and the works are due to begin in 2000.)

The Colonial Sugar Refining Company opened the Condong Sugar Mill in 1880; dairying developed in the 1880s and a butter factory opened in Murwillumbah in 1887. The river was used as a transport medium until the 1940s. Cane boats, banana boats, cream boats and ferries plied the Terranora, Cobaki and Duroby Creeks and Broadwaters, and considerable dredging was carried out to keep the waterways open. Much of the dredged sand was used to fill up wetlands and raise low-lying areas. (An estimated 2 million cubic metres were dredged during the 'river training' procedures.) Tourist cruise boats used the Broadwaters until the 1940s. By then cane was no longer grown in adjacent areas, channel silting had become excessive and commercial river trade was over. These days a small professional fishing fleet and shallow-draft pleasure boats are the only vessels negotiating the entrance bar.

The railway was extended from Lismore to Murwillumbah in 1894.

million years ago, lies in the middle of the valley, and the surrounding ranges and hills are the rim of its caldera—the Macpherson Range in the north; the Tweed Range forming the western wall, with high ridges in the south; and some hilly ground (Burringbar Ranges) in the south-east forming a divide between the valley and the coastal region. Much of the original rainforest vegetation still survives on the steep slopes around the edges of the valley and in the National Park around Mt Warning. Elsewhere, the land has been extensively cleared. Bananas are grown on steeper slopes, sugarcane on the flat floodplains, with intensive dairying, mainly for cream production, in the middle reaches of the valley

THE TWEED RIVER

An example of the problems of a river and estuary in the subtropics

THE TWEED RIVER ESTUARY

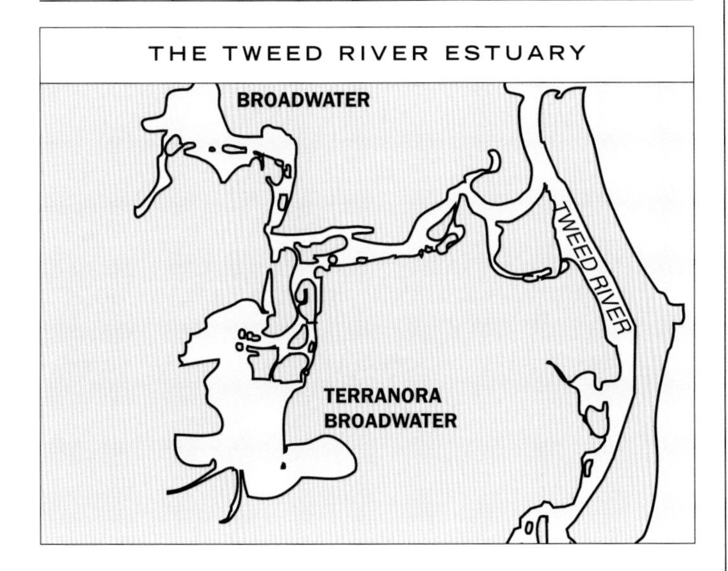

The Tweed River rises in the Tweed Range and runs essentially north-east. It has two major tributary systems, the Oxley and the Rous (also known as the Middle and North Arms of the Tweed). Tidal influence penetrates to the weir at Hindmarsh Flats, immediately west of Murwillumbah. An unusual hydrologic feature occurs just below Murwillumbah, where Mayal Creek flows across the floodplain and links the Tweed to the Rous at Dungay. The area of floodplain in the triangle between the rivers and creek is thus an island, known as Tygalgah. This feature may represent an old meander of the Tweed across the floodplain. Two lakes, Cobaki and Terranora Broadwater, connect to the lower Tweed estuary through Terranora Creek. They are shallow and are sometimes dry on extra low tides. They are fed by Cobaki, Bilambil and Duroby Creeks. The Tweed discharges into the sea at Danger Point, just south of the New South Wales–Queensland border. Clarrie Hall Dam, providing water for Murwillumbah, is situated on the southern branch of Doon Doon Creek. Its catchment is still rainforested.

The estuarine wetlands of the Tweed are of four kinds: mangrove swamps, with five mangrove species, occupy about 300 hectares of the estuary; salt marshes and intertidal mud flats and sand flats occupy about 25 hectares; and seagrass occupies about 50 hectares. Wetlands in the valley have been ravaged and largely destroyed by draining, filling, grazing, cropping, flood mitigation works and development of different sorts.[229A, B, C, D]

MOSQUITOS, CANE TOADS AND ACID SULPHATE WATERS

Acidic waters, with pH values below those that will support fish or other gilled aquatic life, can be found in many coastal areas in Australia. Ponds, drains, lakes and dams holding these acid waters become prolific mosquito breeding sites, attracting salt marsh mosquitos in particular and extending their range into non-tidal areas.

In the Tweed, two species which breed in the salt marshes—*Aedes vigilax* and *Culex sitiens*—have been found breeding in acutely acid water (with a pH of less than 3) in drains and watercourses fed by acid sulphate soils. The drainage systems in the parts of the floodplains where rivers are tidal have tidal exclusion gates with controlled leakage ports, and where these are not functioning properly the conditions favour mosquito breeding. Controlling the leakage ports can improve water quality in drains, allowing colonisation by fish larvae, copepods and other predators to control mosquito larvae.

Where wetlands are drained to improve pasture for dairy cattle, two sources of acid often become activated: the spoil from the drains spread on the surface and open to the air leaks acid; and the opening up of the acid sulphate layer by deep-cut drains allows oxidation and acid leaching. Spoil can be capped or treated with lime; drains can be made wide and shallow, removing run-off water without activating the acid layer, to decrease these problems.

Another problem which arises when acid sulphate soils are drained is of land slumping, which results from shrinkage of the deep acid-bearing clays when they are exposed. Ponding of acidic waters, and the extension of tidal inundation into areas which were previously above tidal influence, follows slumping, and new breeding areas are created for mosquitos. Even the coastal sands on which much housing and resort development occurs have problems with acid waters and mosquitos. When topsoil is removed or stripped during development, the pyrite in the sands oxidises quickly on exposure to air, producing sulphuric acid. Run-off into drains or hollows again provides breeding sites for mosquitos. Liming the sands has proved successful in controlling the acid problems associated with property development in the Tweed.

Coastal freshwater wetlands in the Tweed have been extensively drained for grazing. Even where the drainage has not activated acid production, in many cases problems have arisen because the tidal flap gates, fitted to drains to exclude river floodwaters and salt water from high tides, do not function properly. When salt water penetrates up drains into freshwater swamps, vegetation is killed or changed. The brackish water favours cane toads, which can spawn and develop into toadlets in water of moderate salinity and acidity, while native frogs are far less tolerant of the changed conditions.

(Information kindly supplied by Clive Easton, Entomologist, Tweed Shire Council.)

The Tweed Valley suffers extensive flooding and major floods, such as occurred in 1954, when about 125 square kilometres were inundated, with hundreds of houses and farms isolated. The most recent major flood was in April 1989. The floodplains of the coastal rivers occupy about 70 square kilometres, but are sparsely developed so flooding affects few people. They form a 5-kilometre wide coastal strip from Billynudgel to Kingscliff, with outlets to the sea from Mooball, Cudgera and Cudgen Creeks. All these coastal creeks turn

A MAJOR FISH KILL EVENT

An account of a major fish kill event was published in *Fishing World*, March 1989, by Clive Easton, Entomologist with the Tweed Shire Council, and is summarised here with his permission.[144]

The East Coast suffered a severe drought through 1985 and 1986, and bream, whiting and flathead became very scarce on the lower Tweed and from the beaches in the region. Apparently the fish dispersed to higher reaches of the rivers because of the generally higher estuarine salinities. When the drought broke with torrential rain in March, more than half a metre of rain fell and anglers expected a fishing bonanza which would come as the waters began to clear.

The return of clarity to the river was dramatic. One day the river was brown and turbid, loaded with sediment, the next it was 'gin clear'. But there was no fishing bonanza. Looking down through the crystal clear water along 15 kilometres of the Tweed's main branch and 8 kilometres of its north arm, hundreds of thousands of dead and dying fish could be seen on the bottom. Several days later the carcasses floated to the surface—bass, mullet, bream, flathead, luderick, gar, eels and crabs—total devastation of the upper two-thirds of the river's tidal reaches. Dead bloodworms, which had emerged from the river sediments to die, clogged fishnets in the lower estuary; Council trucks loaded tonnes of offensive smelling dead fish from river banks. The event was proved to have been due to a rush of acid entering the system, and low oxygen levels. During drought the water table drops, and the acid sulphate layer is oxidised and releases acid into soils, waiting to be flushed out, even in undisturbed areas; while this is happening, the acid drainage from cultivated sugarcane land and other agricultural activities becomes concentrated in the man-made drains. Huge amounts of sulphuric acid are thus available to be flushed into the rivers when the rain comes. The high acid concentrations dissolve aluminium and the aluminium ions are responsible for flocculating the sediment which made the water turbid, clearing it amazingly overnight. (Aluminium is often used for water clearing in domestic water supplies.)

Alhough State Fisheries officers hoped that the acute acidity would have been flushed away in the March flood, this was not the case. Two months later a second flood occurred, and four days later the same reaches again cleared magically, and dead fish, this time in numbers much reduced because of the previous episode, littered the bottom. With each subsequent rainfall over several months, kills of juvenile fish (much less visible than the major events) occurred.

Initially, sugarcane farmers were reluctant to admit that their operations were responsible for the acid problems, claiming that fish kills had occurred regularly on the Tweed for generations, and their drainage activities were only comparatively recent. However, historical records have shown that extensive drainage of what are now known to be acid sulphate soils was carried out by a local pioneer, one Joshua Bray, using an Aboriginal labour force, in the 1860s! So disturbance and acid slugs are not recent in the Tweed. (An account of acid sulphate soils is given in *Listen … Our Land is Crying*, pages 93 and 94.)

A fish kill in the Tweed River caused by the release of acid from acid sulphate soils during a rain event. CLIVE EASTON

abruptly north when they exit the coastal hills, about a kilometre from the coast, flowing parallel to the coast before running to the sea. The strong northwards sand drift along the coast and the on-shore wind pattern have caused dune formation, and the rivers run behind the dunes until they can breach them. (Since the times of low sea level when these streams ran across the continental shelf to meet the sea far from the present-day coastline, their mouths have migrated northward with the sand drift.)

The floodplains of the Tweed district are important sugar growing areas. Acid sulphate soils (ASS) are a major concern in the lower reaches of the rivers and all along the coastal floodplains. In the Tweed their potential influence extends further than the present tidal limit (just upstream of Murwillumbah), as marine penetration in the past extended a great deal further inland. (The acid sulphate soils result from the interaction of sea water and decayed vegetation. Higher than present-day sea stands in the past, when saline swamps and mangroves extended further inland, left this unfortunate legacy in greatly expanded areas.) Once the acid sulphate layer is exposed to air, acid leaches from it and enters the groundwater and river systems. The problem will then continue for literally thousands of years, making one wonder whether areas known to have active or potential problems should ever be used for cane growing or any other purpose that disturbs the soil. The short-term gains from sugar (and this southern extension of the sugar belt is not nearly as productive as the tropical sugar regions, and in some parts is barely economic) are no compensation for ongoing expenses and for the damage done to water resources. Estuaries, which are such important breeding regions for fish and other aquatic life, are severely affected and may eventually be killed by acid leaching from disturbed ASS soils.

The Tweed estuary suffers from other forms of pollution as well, as do estuaries all along the eastern coastline where population and agriculture are concentrated. In the Tweed, agricultural pollution involves enormous quantities of sediment (estimated at 230 000 cubic metres a year) being dumped into the estuary, from agricultural run-off. The siltation problem is bad enough on its own, but the silt carries high levels of nitrogen and phosphorus from fertilisers, and from dairy cattle and other livestock, nutrients which increase the risk of algal blooms and other imbalances in the system.

In addition, four sewage treatment plants discharge into the Tweed River and its tributaries—and even the treated effluent is nutrient-rich and adds to the water quality problems. A sugar mill and several large factories discharge their effluent into the Tweed estuary. Add to these the stormwater and urban run-off and pollution problems, and is it any wonder that even with the best intentions of a diligent and caring Council, the problems are barely contained now, and may be much more severe as population numbers grow inevitably in that lovely part of the coast?

The Tweed oyster scare of 1996 highlights the problems of humans, sewage effluent and estuary health. (Similar events have occurred sporadically in other estuaries and the Tweed is used merely as an example of what is inevitable wherever oyster-growing areas coincide with concentrations of population.) It is only when such events receive massive media attention that most people are made aware of the ongoing conflict between the needs of our rivers and waterways for healthy functioning, and the effects of agricultural and urban activities.

According to reports in the *Northern Star and Gold Coast Bulletin* in September 1996, 'The Tweed Shire sewage system has been blamed for the "virtual destruction" of the multi-million dollar Tweed oyster industry.' A major sewage spill coincided with the appearance of a virus which killed many oysters and was linked to about 160 cases (reported) of food poisoning. Run-off from septic tanks during rainfall events probably contributed. The oyster industry was closed down until safety of the product could be guaranteed—when the water was declared clean and the oysters had undergone purification in tanks after harvesting.

Following the oyster scare, intensive water quality and oyster-meat testing has been undertaken as part of the Tweed River Shellfish Quality Assurance Program. The testing has shown generally high water quality, and harvesting no longer takes place following periods of heavy rain. Unfortunately for the industry, in 1998 the QX (Queensland Unknown) virus appeared and decimated the crop. It seems likely from preliminary research that there is a link between acid water and the spread of the virus. New crops of oysters are being introduced as the river is now used only to 'grow out' to harvesting size, and a general down-scaling of operations has occurred.

The oyster crisis brought out in public other problems besetting the Tweed. The *Daily News* of 31 October 1996 asked, 'Just what has gone wrong with the Tweed River?'—stating that only a few short years ago fish were plentiful, pollution was unheard of and the river was easily navigable. Now sand and mud banks make river travel a nightmare, fish numbers have plummeted, pollution is serious, accounts are heard almost daily of 'sewage leaks, fish habitat destruction, oyster poisoning, rubbish dumping, acid soil run-off, diseased fish ...' Oyster farmers are only one of the groups suffering: tour boat operators, professional fishermen, even land-based tourist operators are feeling the impact.

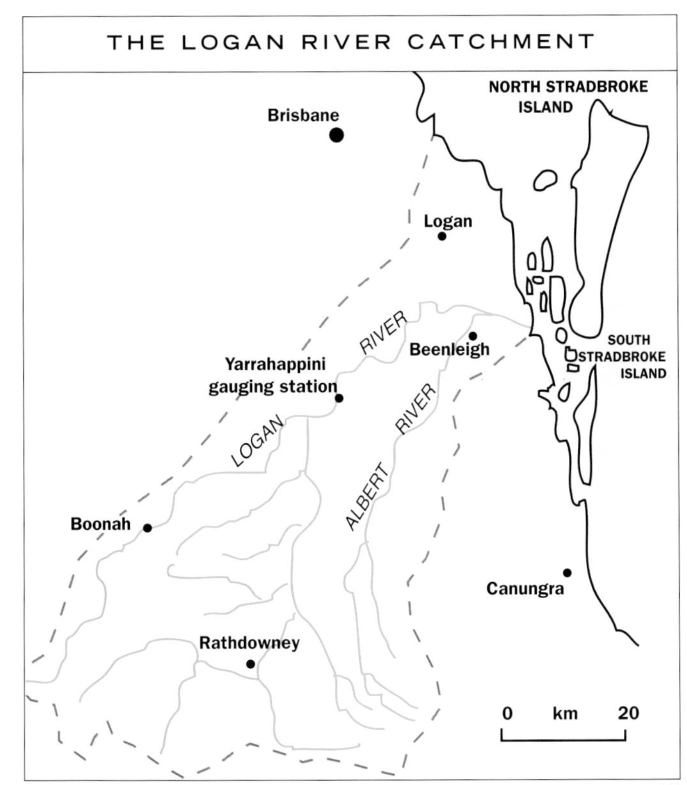

After Loneragan & Bunn[145]

RIVERS OF EASTERN QUEENSLAND

The Logan River: a study in estuarine ecology

Water that runs to the sea is NOT wasted

The Logan originated as radial drainage on a Miocene shield volcano on the South Queensland Block, north of the Clarence–Moreton Basin. It drains a catchment of about 3000 square kilometres and has a total length of about 190 kilometres, of which about 60 kilometres are under tidal influence. The Albert River is the major eastern tributary of the Logan. Annual rainfall varies considerably within the catchment, ranging from 700 millimetres in the west to 3300 millimetres in the south. About three-quarters of the stream flow is in the six months from November to April, coinciding with high summer rainfall, but timing of rainfall varies from year to year, and considerable amounts can fall in winter.

The river is largely unregulated, with only a small dam of 26 000 ML capacity in the headwaters of Burnett Creek in the southern part of the catchment supplying water for agricultural, industrial, urban, stock and domestic uses. The river in the lower estuary remains undeveloped, which makes it unique among the major rivers of southern Moreton Bay. The present flow patterns in the Logan differ little from pre-settlement flow patterns.

The Logan is one of the four river systems in south-east Queensland in which commercial fisheries are licensed to catch prawns using beam trawls. About ten fishers regularly work the navigable reaches of the Logan and Albert Rivers, during a fishing season that extends from September to April, and a further two do so on a part-time basis. Three main species of prawn are caught—greasybacks, school prawns and banana prawns. In 1983, about 180 tonnes of banana prawns were landed by beam trawls from the four river systems, of which 106 tonnes came from the Brisbane River and between 20 and 30 tonnes each from the Logan, Pine and Caboolture River systems. Most of the prawns caught in this beam trawl fishery were sold to the bait market with a total value of $2 million in 1983.

In addition to prawn trawling, commercial fishers are able to use nets and lines in the Logan River (ten full-time and six part-time operators), and to catch mud crabs (four full-time operators). The Logan is the most productive river for mud crabs in southern Queensland, and recreational fishers also target this species in the lower Logan. (Mud crabs are 12 to 24 months old when caught commercially.)

Commercial and recreational fishing are also important activities in Moreton Bay. The Bay represents about 3 per cent of the Queensland coastline but accounts for 10 to 14 per cent of the weight, and 10 per cent of the value, of commercial seafood caught in Queensland. About 400 vessels are licensed to operate in waters in and around Moreton Bay, including 200 trawlers, and most have extended licences to allow them to operate in net, line and crab fisheries. Between 1988 and 1991 about 3400 tonnes of fish, crustaceans and molluscs were caught by commercial fishers in the Bay, with about 49 per cent coming from net fishing and 41 per cent from trawling.[145]

Line fishing for commercial fish species is also an important recreational activity in the Bay, and sand, spanner and mud crabs are caught in baited pots. Worms, yabbies and small prawns are taken for bait. The recreational catch from Moreton Bay in 1985 was estimated at up to 2800 tonnes per year.

Against this background of heavy fishing pressures in estuaries and enclosed waters, research has been carried out to assess the link between stream discharge and abundance of aquatic life in the Logan estuary. It was found that high summer discharge events were correlated with high catches of fish, prawns and crabs.

The relationship between high productivity of estuaries and river discharge has long been recognised by marine biologists but, like many other bits of environmental knowledge, it is not often considered when changes in river flow and behaviour are proposed or implemented in schemes which divert water from rivers. The widespread attitude that 'water which just runs into the sea is wasted' remains, and is often the basis for justifying river regulation, irrigation, and even for proposals to turn rivers around and make them flow inland instead of running 'wastefully' to the sea.

River discharge can have an important effect on physical and biological characteristics of estuaries and near-shore waters. It affects the geomorphology, salinity and turbidity of estuaries—and these factors influence the distribution and abundance of fish and crustaceans. Seasonal and annual variation in river discharge can determine whether the mouth of an estuary is open or closed, which in turn influences the movements of marine creatures using estuaries in their breeding patterns. The number of species, and the composition of fish communities in estuarine systems, are influenced by salinity regimes, distance from the estuary mouth and the condition of the entrance channel. Marine species dominate in the lower estuary, becoming less common with increasing distance from the estuary mouth, and species capable of completing their life cycle within the estuary are increasingly abundant further inland.

It was long assumed that the terrestrial material flushed into estuaries by rivers was an important contributor to estuarine food webs. The technology available today, which can trace the source of material assimilated by an organism, has shown that the contribution of terrestrial carbon from vegetation such as salt marshes and mangroves to estuarine and coastal food webs is actually very limited. While this discovery applies to carbon, it has also been shown that the nitrogen and phosphorus exported from river systems have significant influence on estuaries and near-shore environments. Seasonal exports of nutrients stimulate phytoplankton and benthic micro-algal production, which are thought to be important primary sources in coastal food webs. (In this connection, Egypt's Aswan Dam, completed in 1965, had a marked effect on the coastal Mediterranean ecosystem. It stored all the discharge from above the dam and the water was used for hydro-electricity generation and the irrigation of land. The concentration of nitrates, phosphates and silicates in coastal waters declined markedly; phytoplankton blooms decreased; the catch of sardine, a plankton-feeding fish, declined from 15 000 tonnes in the year before the dam was built to 550 tonnes two years after its completion; and catches of prawns and shrimps in the Egyptian sector of the Mediterranean have halved.)

Excessive nutrient loads from man-made sources, such as fertiliser use in catchments and sewage discharge, can have serious deleterious effects on estuarine ecosystems. The Peel–Harvey estuary in Western Australia provides a striking example of the consequences of increases in the nutrients derived from terrestrial run-off. Increased nutrient levels resulted in massive increases in macro-algae in the 1970s and initial increases in commercial fish catches. However, by the 1980s, increase in phosphorus led to blooms of the cyanobacterium *Nodularia spumigena* in the spring and summer months, and brought changes to productivity of fisheries and the distribution patterns of some fish species within the estuary.

Increases in catches of estuarine and coastal fisheries may be related to increases in nutrients stimulating both primary and secondary production. But they may also be caused by stimulating movement into regions where the fish, crustaceans or molluscs are more readily caught by the fishery.

The high positive correlation between catches of banana prawns and rainfall in the south-eastern Gulf of Carpentaria is thought to be caused by a secondary effect of rainfall, such as lowering of salinity, or higher discharge and turbulence, which stimulates the emigration of juvenile prawns from their nursery habitats in mangrove creeks and rivers. The correlation between rainfall and catch in the Gulf has held over a 25-year period, with catches declining to 30 tonnes in a dry year (1990) but reaching the highest on record—2300 tonnes—in the following high-rainfall year. Catches of school prawns in the Hunter and Clarence Rivers have also been positively correlated with rainfall and river flow. (Prawns are short-lived species and are caught at about 6 months old.) A similar relationship between golden perch catches in the lower Murray River is also thought to be because flow stimulates the downstream migration of this species.

Whatever the relationship between river flow and seafood production, the significant positive relationship between catches of fish and crustaceans and river flow has important implications for river managers. Water flowing into the sea is manifestly not wasted. Reduction or elimination of large-flow events by dams is likely to eliminate the large seafood catches, and the economic losses must be weighed against the economic benefits of water storage. Maintaining the natural flow pattern is clearly as important, if not more important, than maintaining the magnitude of the total annual flow.

River regulation and estuarine productivity

Although river discharge is known to contribute to the productivity of estuaries, the extent of river regulation worldwide is enormous. In the mid-1980s, large dams, more than 15 metres high, were completed at the rate of more than one a day. The rate of construction has slowed since, but the trend now is for larger dams and the transfer of water from one catchment to another. At the present rate of dam construction, it has been estimated that more than two-thirds of the world's stream flow will be regulated by the year 2000.

THE BRISBANE RIVER BASIN

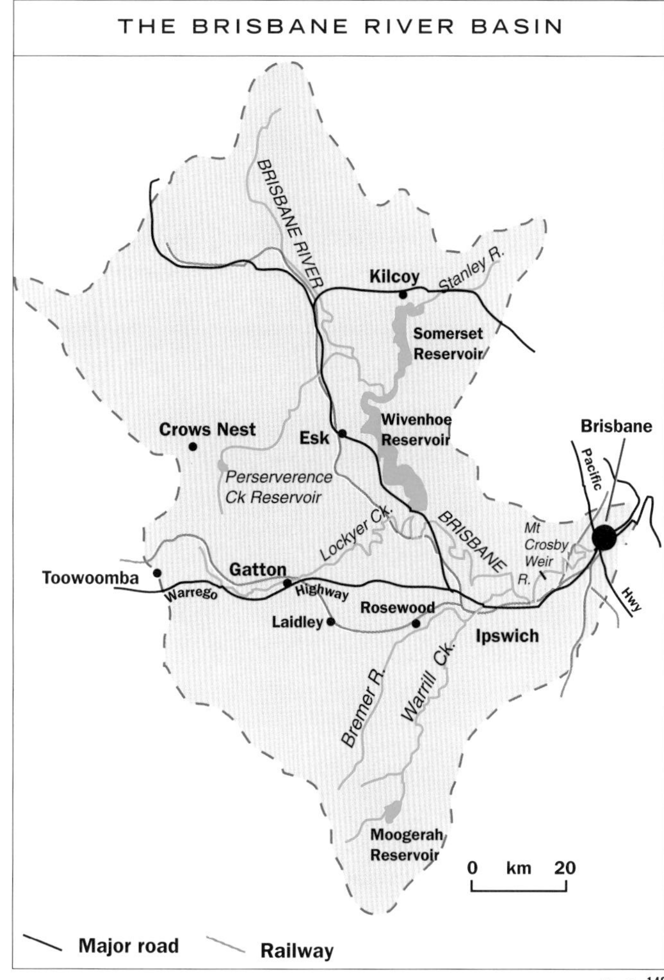

After Day[146]

THE BURDEKIN RIVER CATCHMENT

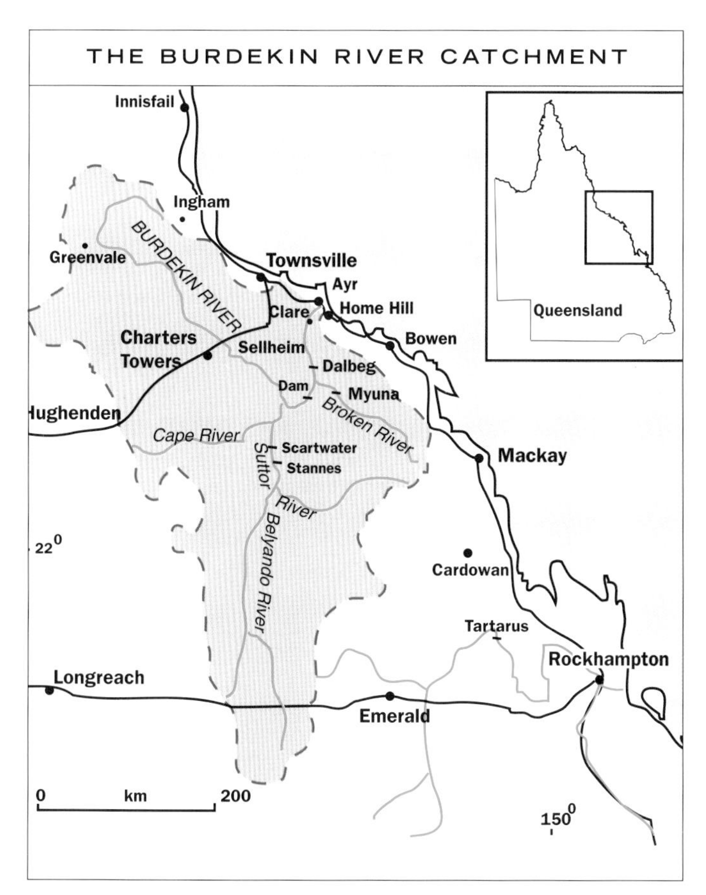

Most of Australia's 375 dams are found in the south-east of the continent. They reduce the magnitude of downstream discharges; change the seasonal pattern of flow; and reduce the frequency and magnitude of smaller flood peaks. Recently, some attempts have been made to determine the quantity of flow required to maintain downstream riparian zones, fish habitats and freshwater wetlands. The ecological needs of the estuaries and coastal zones have seldom been considered, however. (In South Africa the freshwater requirements of estuaries, which may represent up to 15 per cent of freshwater resources, are taken into account when allocating environmental flow and withdrawals. The situation in South Africa, a largely semi-arid land not unlike Australia, is so serious that by the year 2020 all freshwater resources will be fully utilised, according to recent projections.)

The fragmented nature of water management is partly at the root of many of Australia's problems. Major river projects constructed in one State often have produce their deleterious effects in another State downstream. The taking of excessive water from rivers in Queensland for cotton irrigation impacts on river and wetland health in New South Wales, and on the livelihood of graziers and others downstream; the decrease in flow of the Darling and other rivers in New South Wales into the Murray because of river regulation and excessive extraction for agriculture impoverishes the lower reaches of the Murray; while use and regulation of Murray waters has a profound effect on South Australia's water supply. When it comes to regulations governing estuaries and coastal waters there is little interaction between researchers, agencies and organisations involved in collection of information, management and regulation of marine and freshwater resources.

What Australia needs, perhaps in the best of all possible worlds, is a strong Federal Government which has control of the vital life-support systems of the nation—its water and soil resources—laying down the rules for use and management. While each State can do as it pleases there is little hope of sustainable use of these life-giving resources. Money always wins and any scheme which brings profit today (never mind disaster tomorrow) is acceptable under the current fragmented arrangements.

The Brisbane River Basin

A number of storages are found within the Brisbane River Basin, providing water for agriculture, hydro-electricity generation and urban and industrial water supply, mainly for Brisbane, and also acting in flood mitigation.[146] The Wivenhoe Dam on the Brisbane River was designed to create a 'mega' storage linking with the Somerset Reservoir and, with Lake Manchester, providing for Brisbane's urban water consumption.

The Brisbane River was reported to be 'as new' in the 1820s but its tributary the Bremer was progressively deepened by removing shingle from its basalt bed. By the 1860s there was considerable dredging in the Brisbane system, especially to remove shoals which impeded river transport. Gravel mining made considerable, long lasting changes to the rivers. The lower Brisbane River has a flood hazard rating and serious flooding can still occur.

The Lockyer Valley catchment of about 3000 square kilometres is one of the oldest irrigation regions in the State and an important vegetable-growing region. Water is derived from ground sources and weirs have been built to recharge the groundwater system. Water extraction still outstrips recharge, however, and some of the water used is saline.

THE BURDEKIN RIVER:

a river of the wet/dry tropics

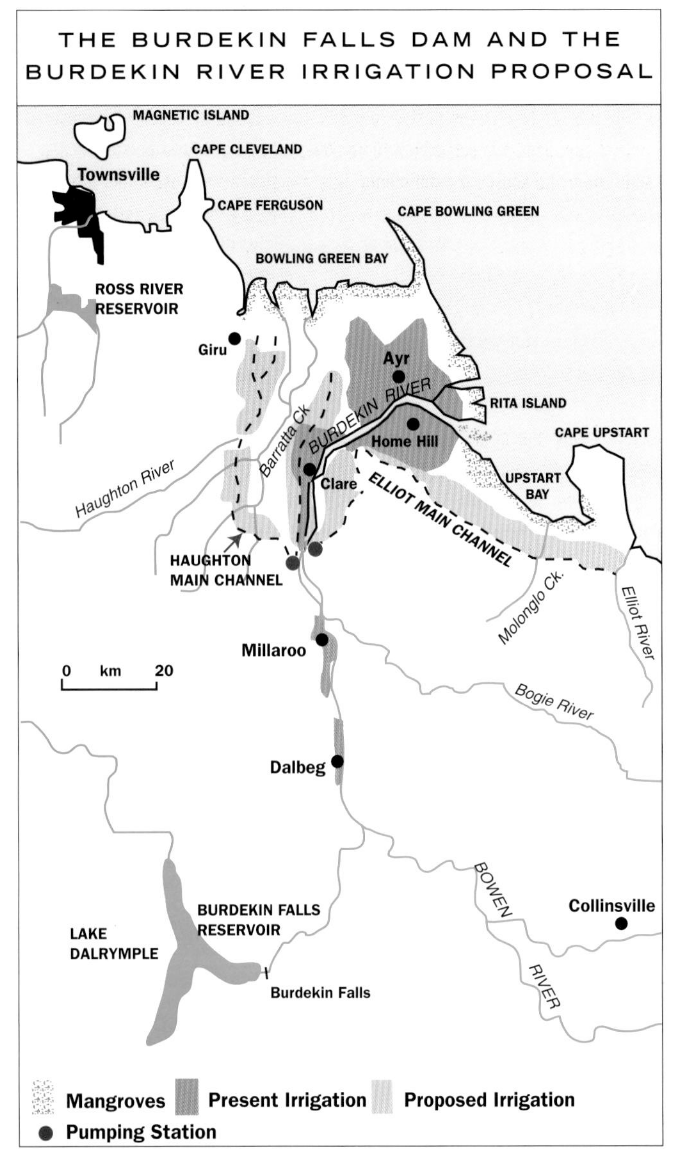

After Fleming[148]

The north coastal Burdekin Basin was settled in the 1860s, with sugar being grown on the Burdekin delta.[146] Lagoons, which are really windows in the water table, were used to irrigate the sugarcane, adjacent to the watercourses. Then a system of spear wells took over, tapping groundwater, as the area under cane expanded. A dam on the Burdekin, a dream from the 1880s onward, took 100 years to come to fruition, ultimately holding four times as much water as Sydney Harbour and intended for agricultural development and as supply for the Townsville metropolitan area.

The Burdekin delta is the largest cuspate delta in Australia, with deltaic deposits to 166 metres in depth extending over an area of about 1500 square kilometres. The system is highly dynamic with close relationships between littoral, marine and fluvial environments and rapid changes in distributory and sedimentary patterns. The Burdekin Dam traps vast volumes of coarse sediment derived from the degrading upper catchment, sediment now denied to the delta. In time this may lead to a reduction in area of the Bowling Green sector.

The Burdekin is one of Australia's great rivers, with a catchment of 134 000 square kilometres and a recorded peak discharge of 40 000 cubic metres a second—making it a significant river by world standards.147 However, almost 90 per cent of the average flow is discharged in the wet season, from January to April, and the river virtually ceases to flow naturally in October and November in 50 per cent of years. The great variability in flow is a reflection of the seasonal and highly variable rainfall which means that agriculture in the lower Burdekin is almost entirely dependent on irrigation through the dry season from May to December, and requires a wet season irrigation supplement in 50 per cent of years. (The Townsville–Bowen region is one of the driest parts of the tropical coast of Queensland, with the lower Burdekin region having an annual rainfall of between 800 and 1000 millimetres.)

The upper catchment, centred on Charters Towers, is primarily grazing country, much of it seasonally arid, and suffering from over-grazing with attendant widespread erosion and weed invasion. A description of the environments, the large sediment budget which affects the Burdekin Falls Dam, the erratic rainfall patterns and high evaporation rates is given in *Listen … Our Land is Crying* (pages 186–7). It is the Burdekin Falls Dam and the lower sections of the river which concern us in this volume.

The Burdekin Falls Dam, situated 160 kilometres upstream from the river's mouth, is Queensland's biggest impoundment.[148] Before it was built, irrigated agriculture in the lower Burdekin depended on groundwater and some direct pumping from the river. This was supported by minor storage structures. The natural recharge of groundwater was from local rainfall, supplemented by flow down former distributaries at times of significant flow, and intermittently by widespread overbank flow.

In an effort to increase recharge following over-pumping of the groundwater, pump-stations were installed to raise river water to the anabranches at lower river stages. These now operate continuously following completion of the Burdekin Falls Dam and partly act as water supply channels.

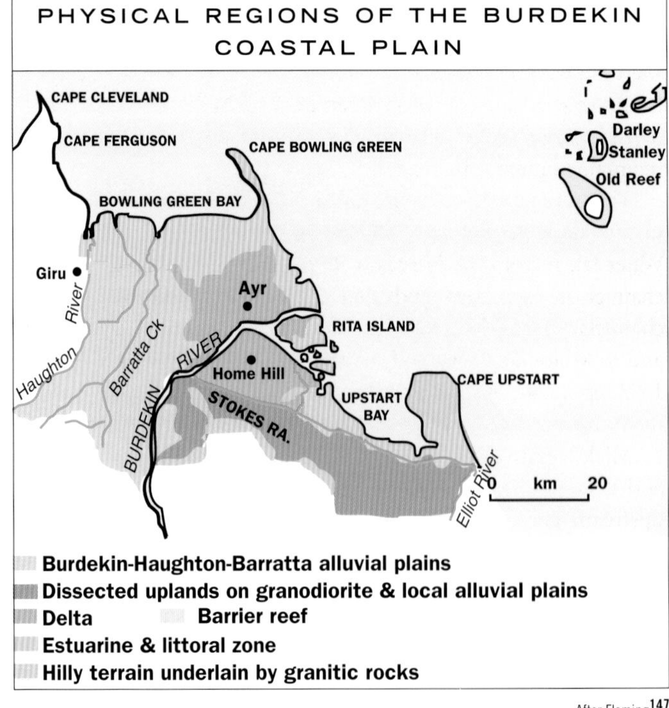

After Fleming[147]

UPDATE ON THE BURDEKIN IRRIGATION PROJECT

Kindly supplied to Mick Fleming by Peter Gilbey, Regional Infrastructure Development, Northern Region, Queensland (August 1999)

The Left Bank development is virtually complete, except for the Haughton Relift Area, which covers uncommanded land above the Haughton Main Channel. The major development has been for sugarcane, so is tied to Mill Assignments, and any increase therefore requires Cane Board approval. In limited areas, small horticultural crop development has occurred.

The Right Bank areas on the Elliot Main Channel have been the subject of continuing research and testing. Methods of controlling irrigation, reusing water and other techniques have been developed, and it is hoped that shortly eight farms will be released as a regional development trial. A proposal to extend the Elliot Main Channel beyond Molongle Creek to supply irrigation development in the lower Don River near Bowen has been costed but no decision has yet been reached.

The State Water Projects Agency, with semi-commercial status, is now responsible for all the terrestrial operations within the Irrigation Project Areas, including the major drainage works on the left bank which drain into Barratta Creek. Tailwaters from irrigation in the dry season now sustain flows and alter the water regime close to the coast. The problem of nutrients in the drainage waters and also wet season run-off are of concern, and a research proposal to develop artificial lagoons to strip nutrients is under consideration. The Queensland Environment Protection Agency is also taking an interest in the nutrient problem and its impact on the important coastal wetlands.

THE DELTA CANELANDS

The 36 000 hectares of irrigated sugarcane in the delta region is based on the use of local shallow groundwater. The North Burdekin Water Board, centred at Ayr on the left bank, and the South Burdekin Water Board at Home Hill on the right bank, levy charges to pay all operational costs. The Boards manage recharge schemes based on pumping water into former anabranches, with most recharge occurring through natural lagoons and some special recharge pits in areas without natural lagoons. There is no metering of pumps, so the efficiency of farm irrigation and recharge processes is not monitored.

Initially, water was released down the Bowen River from Eungella Dam when the Burdekin ceased to flow, but after completion of the Burdekin Falls Dam all supplies have come from there. Water stored in the Falls Dam, which comes from the southern catchments, has a high load of suspended matter, and it is thought that this turbid water is clogging the recharge process. Increasingly, the recharge channels are being used as the primary water source for irrigation of the canefields adjacent to the channels. Because there is now some permanent flow all the way to the tidal limits of both the main Burdekin River channel and its principal distributary, the Anabranch, irrigators are pumping directly from these sources.

Over-pumping of the shallow groundwater and resulting salinisation remain major concerns. The picture is becoming more complicated because areas which are supplied by surface water are producing local groundwater mounds. The Boards are currently commissioning studies directed towards developing a comprehensive groundwater model on which to base management decisions. Local groups, such as those on Rita Island, have been commissioning their own studies.

[It is obvious that management of groundwater and surface water in deltas and estuarine-littoral zones is a highly complex matter. Dam building and irrigation, which completely change hydrological cycles in an area with naturally pronounced wet and dry cycles, have predictably complex responses, and long-term effects may present many surprises.]

OTHER WATER RESOURCE PROJECTS

Investigations continue with respect to the Urranah dam site in the Bowen Basin and the Hells Gate dam site in the Upper Burdekin. These projects would permit limited irrigation of agricultural lands near Collinsville on the Bowen, and on the Burdekin upstream of Charters Towers. Earlier proposals to divert water westwards out of the catchment at Hells Gate have been abandoned.

A series of major ecological studies undertaken before the dam was built were accepted as Environmental Impact Statements for the proposed dam and irrigation schemes.[147, 148, 150] Mr Mick Fleming of the CSIRO (now retired) directed the investigations and has maintained interest in the area.

The Burdekin River Irrigation Area Project (BRIA) was proposed to provide water for up to 50 000 hectares by gravity reticulation. Water taken from the Burdekin River would be pumped to a main channel on each bank and then gravity-distributed by a series of channels as far afield as the Haughton River and Giru on the left bank and to Molongle Creek and beyond on the right bank. A review in 1993 reported that stage 1 on the right bank, involving lands between the Burdekin River and Barratta Creek, was virtually complete.[149]

Mick Fleming emphasises that the Burdekin has no true estuary, and it is only tidal for a few kilometres above its mouth—to the upstream end of Rita Island—in spite of the large tidal range. Mangroves do not extend far inland. The 'estuarine–littoral zone' is unique in its nature and its hydrological and hydraulic behaviour. The broad sandy bed of the lower course of the river, which is 500 to 1000 metres wide, carries little or no water for most of the year.

The wetland system of the Lower Burdekin floodplain has been recognised as a wetland of international importance, and is a Ramsar site (one of only two in Queensland). The adjacent Bowling Green Bay National Park, with its wetlands, makes the whole coastal region a very valuable one for wetland conservation. The Haughton River forms the northern boundary of the lower Burdekin floodplain and its estuarine–littoral zone is part of the extensive wetlands. It, like many other smaller streams of the tropical north coast, flows intermittently during the wet season and is dry for long periods in the remainder of the year. The main deltaic distributary system, below the Stokes Range–Rocks area, operates only above the moderate flood level, and so only briefly. Most of the marginal lagoons and freshwater wetlands are actually windows into the ubiquitous water table. The most active wetlands are in the outflow of Barratta Creek, which drains the back-slopes of the levees of the Haughton and the Burdekin. The Haughton frequently discharges into the Barratta. Local run-off annually floods the margins of the hypersaline flats and mangrove channels and keeps the lagoons clear and refreshes the wetlands. These wetlands now receive drainage waters year-round from the newly irrigated lands on the left or northern bank of the Burdekin. Concern has been noted for the possible environmental impacts of these drainage waters.

The floodplain is up to 50 kilometres wide and its freshwater wetlands provide significant refugia for resident, migratory and

ALLUVIAL AQUIFERS OF EASTERN QUEENSLAND

Important alluvial aquifers are developed in the Burdekin Delta, Pioneer valley, Callide valley, the Monto and Bundaberg areas, and the Lockyer and Condamine valleys.[151] As rainfall is both seasonal and variable, surface water resources are limited and artificial recharge is a common practice in the management of groundwater resources:

- In the Burdekin delta this is achieved by diverting water from the Burdekin River into a system of natural and artificial channels.
- In the Lockyer valley, where the alluvial deposits are up to 40 metres deep, weirs have been built on the river and the water which backs up as a result recharges the aquifer—but not sufficiently to balance the amount withdrawn at peak irrigation times.
- The Callide valley has a catchment area of nearly 5000 square kilometres and an alluvial plain of 450 square kilometres. Ten thousand hectares are irrigated, and water use exceeds natural recharge, which is from stream flow as well as from major rainfall events which might only occur 10 years apart. Artificial recharge is practiced by diverting the river water into the depression overlying the aquifer.
- In the Monto area, the principal agricultural industry is dairying and the major part of groundwater supply is used to irrigate lucerne and other fodder crops. In times of drought, the demand for groundwater exceeds available supplies. A scheme of conjunctive use of surface and groundwater is in operation, using a dam, a series of recharge weirs, and associated diversion and recharge channels.
- Bundaberg is a major sugar region which has relied on groundwater for irrigation and urban supplies. Natural recharge from rainfall is about 85 million cubic metres per year, with a safe yield of 55 million. Most of the groundwater is stored below sea level and average extraction is closer to 90 million. In the 1960s, intrusion of seawater into the aquifer occurred and water quality deteriorated underneath 5000 hectares. Trials showed that artificial recharge cannot make up the deficit, so additional supplies of water for irrigation have since been provided from surface storages.
- The Condamine valley contains an area of over 1000 square kilometres of water-bearing alluvium, recharged by the Condamine River. Annual recharge is only 13 million cubic metres, whereas withdrawals are 87 million. Cotton, grain crops and soya beans are irrigated, and town water supplies are dependent on the groundwater, so a massive overdraft is occurring. Stream flows are highly seasonal and the area for storage is large, so a system has been developed using surface storages, diversion weirs, channels and a system of trenches, pits and settling basins for artificial recharge.

Under Greenhouse conditions, the dry tropics should receive greater rainfall, which will possibly be less seasonal as well. This will be an advantage for recharging aquifers, but in coastal situations the rising sea level will probably negate the advantage, penetrating into the aquifers and making more of the water unsuitable for use.

nomadic waterfowl. Magpie geese, whistling and black ducks, spoonbills, ibis, brolgas and jabiru are usually present; migratory green pygmy geese and sandpipers appear seasonally; pelicans and other nomads come and go. The Burdekin duck has unfortunately become rare in the region.

The groundwater system of the Burdekin delta, where the water has been used to irrigate sugarcane for over half a century, requires careful management. The almost fresh groundwater has a water table above sea level but most of the volume is stored below mean sea level, so it is essential to maintain a positive gradient seawards. The delta groundwater system has always depended on both local rainfall and overbank floods for recharge and has always been a freshwater lens floating on saline water, in places greater than sea water in salt concentration. It is always the case in such lenses that the bulk—in excess of 90 per cent—is below the sea level equilibrium line. Any local depression of the freshwater surface by pumping will induce a rise in the salt water below and possibly cause contamination. The complex situation regarding the layering of aquifers and aquicludes below the delta complicates matters.

According to some experts the delta groundwater system is considered to be under stress, with water usage exceeding recharge since the 1980s, even though the situation has been helped by the diversion of additional recharge water from the Burdekin River.[2] They consider that an imbalance still persists and that current water use is unsustainable and, in addition, that as a result of the local lowering of the groundwater table, salt water from the sea is penetrating into aquifers below the delta. (This is a phenomenon noted also in the valleys of the Lower Don, Pioneer and Lower Burnett Rivers, where groundwater extraction has also been largely used for irrigation of sugarcane.) Mr Fleming did not believe that the groundwater was being seriously over-pumped, except on the seaward margins.

THE HERBERT RIVER

The Herbert River is the largest river system located in the sub-humid to humid tropical north-east of Australia.[152, 232] Its catchment drains an area of approximately 10 000 square kilometres to the Coral Sea. It is typical of a large number of catchments which impact on the environmentally sensitive coastal zone. Competition between alternative uses of resources poses serious threats to future economic, environmental, social and cultural potential. Balancing the demands of sugar production, other agricultural and pastoral enterprises, forestry and mining, with the need to conserve the diverse ecosystems and their unique biodiversity, requires sensitive management. The potential for unsustainable land use, which results in soil erosion and degradation of water quality, to affect the delta, the mangroves and the ocean with its marine life in this part of the Great Barrier Reef Marine Park is ever-present.

The Herbert River has its source 8 kilometres north-east of the town of Herberton, at an elevation of 1070 metres, and winds 340 kilometres to the river mouth 7 kilometres north of Halifax. Widely differing rainfall patterns characterise the catchment, with considerable differences over short distances because of changes in local topography, including the height and orientation of mountain ranges and the direction of the coastline with respect to the prevailing moist south-east air stream. Mean annual rainfall of the catchment is 1370 millimetres, ranging from 3000 millimetres on high parts of the Cardwell Range to 750 millimetres in the extreme west of the catchment. In the lower catchment, rainfall varies from the coast to the foothills. For example: Halifax 2127 millimetres to Ingham 2069 millimetres in 20 kilometres; to Upper Stone 1547 millimetres in 25 kilometres, before increasing again in the rainforest areas approximately 25 kilometres west at Mt Spec with 3169 millimetres. Three-quarters of the mean annual rainfall of the catchment occurs from December to March; minimal rainfall occurs from June to October; evaporation exceeds median rainfall from May to November.

Topographically, the catchment has three distinct zones:

- The **upper catchment** of approximately 6000 square kilometres is bounded to the north and west by the Great Dividing Range, and to the east by the Cardwell Range. Topography ranges from level alluvial plains to rugged, deeply incised ranges. In the western sector the land is gently undulating; in the east, landscapes are more irregular and much is over 1000 metres. Land use is mostly low-input grazing on native pastures with small areas of sown pastures, and agriculture on properties along the Kennedy Highway and concentrated round Ravenshoe and Herberton. Forestry activities occur in State forests (native) between Herberton and Ravenshoe and southward along the Cardwell Range. Mining has played an important role in the development of the upper catchment, with tin deposits in the north-west sector near Mt Garnet being the focus of activity. Only a small land area is involved in the many small operations but some siltation of rivers and some stream pollution results from mining activities.
- In the **intermediate catchment** of 1000 square kilometres, a large proportion of the land is National Park. Land use is grazing of native pastures and some forestry. Below Gleneagle the river flows for 70 kilometres through a deep gorge before entering the lower catchment.
- The **lower catchment** of 3000 square kilometres consists of a coastal plain bounded to the north by the Cardwell and on the south by the Seaview Ranges, which are mainly between 600 and 900 metres high, with some areas in excess of 1000 metres. Steep slopes adjoin the coastal plain and erosion of the higher ground has resulted in alluvial fans forming an apron between slopes and plains. Quaternary alluvium sediments associated with the Herbert and Stone Rivers and their major tributaries in the lower catchment comprise the coastal plain. They reach a maximum of 96 metres depth and show stratification with layers of clay, sandy clay, sand, gravel and some conglomerate and mangrove mud, reflecting sea level oscillations. Beach sands remain in strips and patches, separated in many places from the present coast by

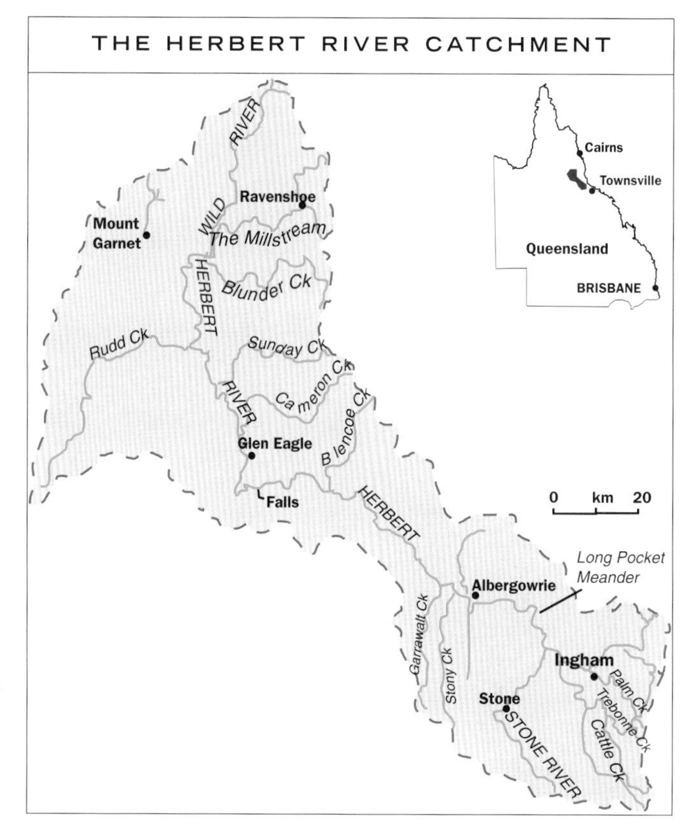

The Herbert River, looking east towards Mount Cordelia. PHOTOGRAPHS OF HERBERT RIVER CATCHMENT ALL SUPPLIED BY CSIRO TROPICAL AGRICULTURE AND HERBERT RIVER CATCHMENT COORDINATING COMMITTEE.

An abandoned tin mine, Nettle Creek, south of Innot Hot Springs. A certain amount of sediment and pollutants have entered the Herbert River from its upper catchment where a number of mines operated.

swamps. The plain forms an asymmetrical delta between the Seaview and Cardwell Ranges. This delta is a major feature of the Queensland coast, its alluvial fan covering about 500 square kilometres. Hinchinbrook Island restricts the deltaic area to the north. A strong longshore drift from south to north along the coast has been responsible for the gradual migration of the Herbert northwards. Conditions in the lower catchment are far from stable: an anabranch is threatening to develop where a distributary channel takes water from the Herbert River to the Seymour River channel and into the tidal wetlands. It is likely to continue to develop and would eventually capture the Herbert River. Several reaches of the river are prone to major flooding and channel widening, like the Washaway at Halifax and Long Pocket upstream of Lannercost, where in flood times the water takes a cut-off short cut through a very large meander loop.

The Herbert River Gorge.

The Halifax Washaway is one of the first areas of river bank overtopped in flood flow, and flows of magnitude sufficient to cause overbank flooding at this point occur as frequently as five years out of six. Bank erosion has been massive in the area, extending over 180 metres, and parts of the original township were washed away many years ago. Things have been stable since 1943 but problems continue on adjacent land. A large flood in 1991 caused extensive damage and scour-holes were formed in cane land as the water flowed to Victoria Creek. There was also scour at the river bank. Without intervention, these scours would continue to develop both upstream and downstream and the overflow course would carry increasing amounts of water until a distributory system formed.

The Lower Herbert River catchment, looking east from Trebonne.

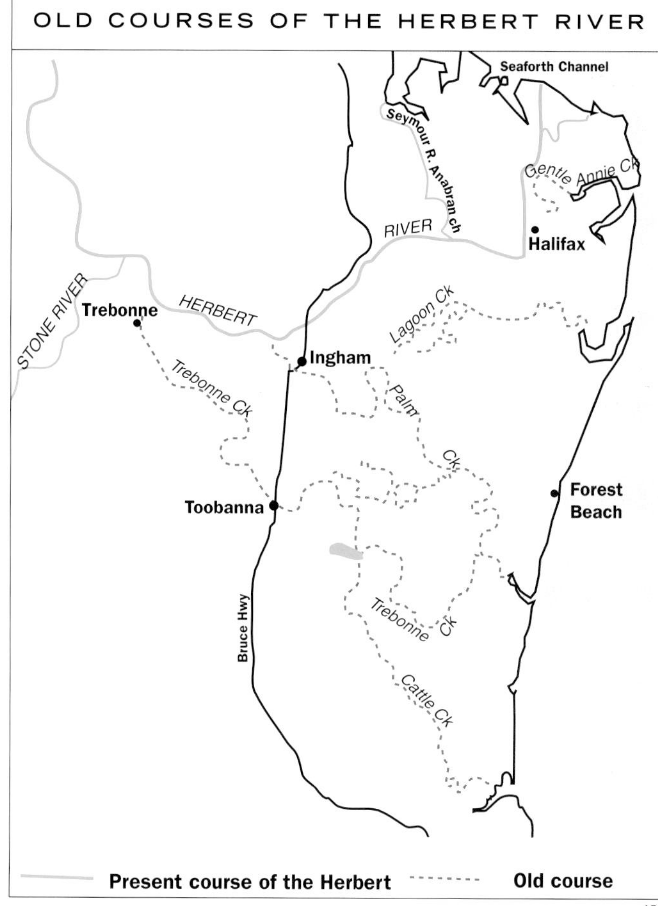

After Drummond & Associates[153]

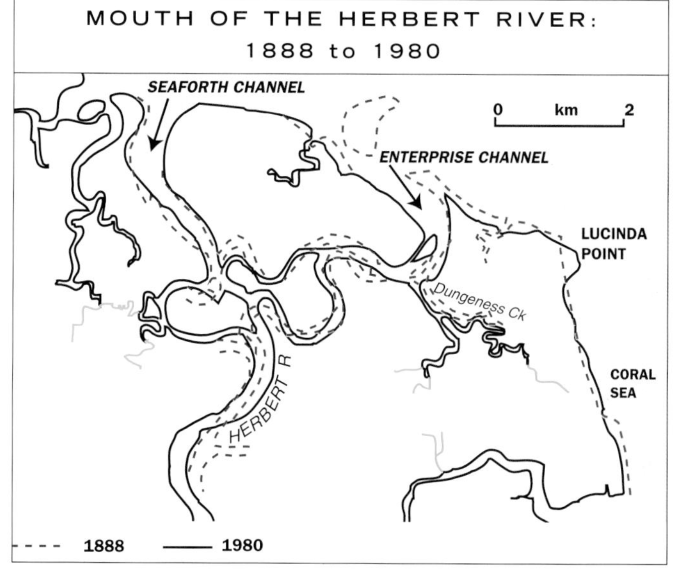

After Drummond & Associates[153]

Past history of the Herbert River

Over the last 6000 years the Herbert River has occupied many courses from Cattle Creek at the far southern end of the delta through the Trebonne, Palm and Gentle Annie Creeks to the present Seaforth Channel. The older courses end in foredunes, the more recent at the current beach front. On average, there has been a major course change every 1000 years and the mouth of the river has moved north more than 30 kilometres. The processes which caused the channel changes are still active today. Future course changes will cause damage to infrastructure and loss of farming land.

The development of new distributary channels is complex. Gradual changes occur while one channel develops and another is infilled until a threshold point is reached. Then a flood or other catastrophic event tips the balance—the flood is the trigger for the change, not the cause. The Enterprise Channel has been infilled with sediment in contemporary times and the Seaforth Channel has opened; the same process operated to cause the abandonment of Gentle Annie Creek.

Deltas are unstable places by their very nature, and problems can never be solved completely, only held in check. Where they are intensively farmed for sugar production, some of the flooding and unwanted changes can be managed by revegetating non-agricultural areas; modifying canefield layout to avoid preferential flow paths; harvesting and planting or ratooning early to ensure that there is a mature cane crop before the flood season; designing levees so that there is no concentration of outflow at washaways.

Drainage is essential for successful sugarcane growing as flooding and waterlogging adversely affect yields. Local Drainage Boards provide and maintain drainage for small groups of canegrowers and attempts have been made to coordinate their activities on a catchment scale, essential for satisfactory management of the whole area.

The Herbert River estuary

The Herbert River estuarine system is predominantly under marine influence, although some parts are under a sustained freshwater supply. Overall, it has an intermediate freshwater input, high catchment retention and low flow reliability, and in these characteristics it resembles the Proserpine, Barron and Lockhart Rivers estuary systems. These hydrologic characteristics are reflected in the mangrove species-groups which occur in the estuarine reaches—mangrove communities which are characteristic of marine-influence-dominated estuaries.

The Herbert River and its estuary are important from both a commercial and recreational fisheries perspective. Significant catch species include prawns, crabs, barramundi and other estuarine fish. The Herbert River estuary (Hinchinbrook Channel) is one of the

Sugarcane production in the Lower Herbert is based on a system of green cane harvesting and minimum till, to reduce soil loss. Trash is left on the ground for the same reason. CSIRO TROPICAL AGRICULTURE AND HERBERT RIVER CATCHMENT COORDINATING COMMITTEE KINDLY SUPPLIED ALL THE HERBERT RIVER PHOTOGRAPHS.

Mangrove communities in the Hinchinbrook Channel comprise species which are adapted to the strongly marine-influence-dominated estuarine conditions.

largest and most productive estuaries on the east coast of Queensland and is particularly important for larval and juvenile stages of fish and crustaceans.[153] Its extensive estuarine communities are of major conservation and scientific interest.

Prawns require freshwater flushes (produced by floods or spates in the rivers) to move them out of estuaries to continue their life-cycle in the marine environment. Flushing them out of the estuaries also enables the commercial crop to be harvested. Differing salinity levels in the estuary are critical at various stages in the life-cycle of barramundi: high salinity levels are needed during the November–December period to initiate spawning and facilitate the fertilisation of eggs; juvenile barramundi, which have strict requirements regarding salinity gradients, migrate up the freshwater reaches of the wetlands.

Hinchinbrook Channel has extensive seagrass beds which are of great importance as feeding areas for dugong. Because of its high conservation values and important commercial and recreational values, the Hinchinbrook Channel, including the Herbert River mouth, has been included in various Marine Parks and National Parks. All the mangroves are protected within these parks, while all parts of the estuary below low water are included in a Commonwealth Marine Park; and parts of the mangrove areas are included in the Wet Tropics World Heritage Area. Much of the estuary, including the Herbert River mouth, is a Fish Habitat Reserve.

Gentle Annie Creek and its estuary are connected to the lower reaches of the Herbert River. Because of their high productivity and importance to the maintenance of the local commercial and recreational fisheries, particularly the barramundi fishery, the estuary has been gazetted as a Wetland Reserve. The Palm Creek and Cattle Creek–Eleanor Creek wetlands have also been gazetted as Wetland Reserves. The freshwater wetlands and billabongs which are mostly associated with old courses of the lower Herbert River are a relatively rare resource. Most have suffered draining and filling, mainly for agriculture, and those that remain are important habitat areas for waterbirds and amphibians, as well as playing an important role as nursery areas for fish like barramundi.

The Nypa Palms National Park abuts the Herbert River. It contains stands of the mangrove palm *Nypa fruticans*, a species which grows in the upper reaches of estuaries where the water is brackish.

THE TULLY–MILLSTREAM HYDRO-ELECTRIC SCHEME

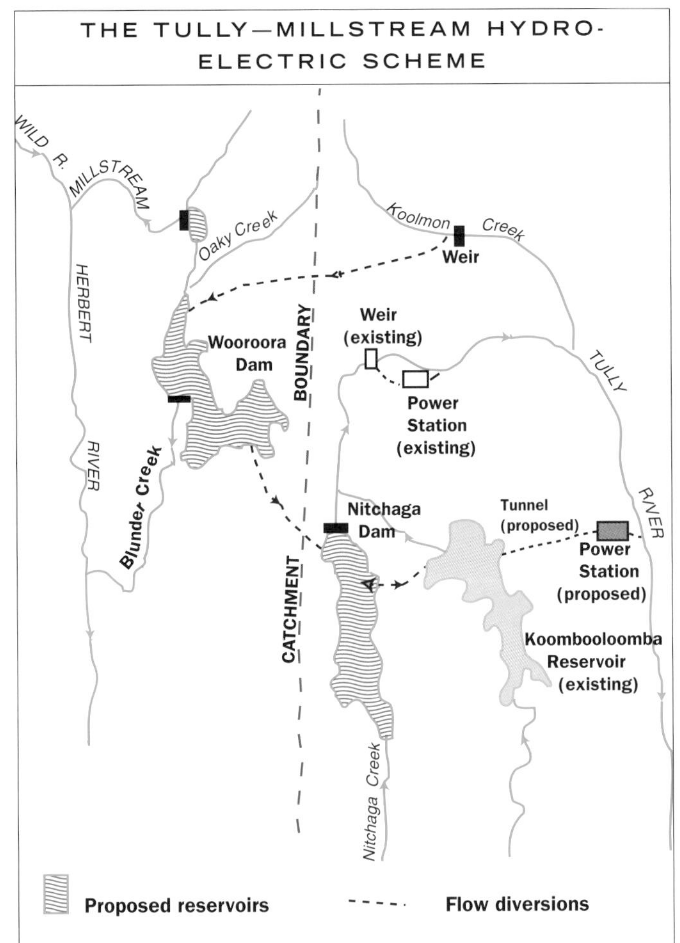

After Drummond & Associates[153]

The Millstream is a headwater tributary of the Herbert River. The Tully–Millstream Scheme was planned to use water from the adjacent Herbert and Tully catchments to generate hydro-electricity. Water was to be captured and stored on the tablelands at an elevation of about 800 metres, passed through tunnels and discharged through generators to the Tully River at an elevation of about 45 metres. A new 600-megawatt power station was to be constructed downstream from the existing 72-megawatt station near Tully Falls. Damming of Nitchaga Creek and Blunder Creek, and weirs on the Millstream and Koolmoon Creek would create two large reservoirs—the Nitchaga Dam in the Tully catchment and the Wooroora Dam in the Herbert, and a small dam on the Millstream.

The scheme was expected to have a negligible geomorphic effect on the Herbert River and not to cause any change in channel stability. About 4.6 per cent of the Herbert catchment would be involved, feeding 85 per cent of its water into the Tully catchment. The scheme was also not expected to impact on water quality at Ingham, particularly as it was proposed that no water would be taken during very dry years. Effects on the Millstream and on Blunder Creek will be considerable, however, due to dramatic reduction in flow downstream of the dams. Oaky Creek and the headwaters of Blunder Creek will become inlet channels for the Wooroora Dam.

LAKE BUCHANAN—AN ATYPICAL SALT LAKE

Lake Buchanan is a very large playa lake, 23 kilometres long and 7 kilometres wide (surface area 117 square kilometres), which lies at an altitude of nearly 300 metres above sea level in a depression among the hills which constitute the Great Divide in central Queensland. Its catchment area of nearly 3000 square kilometres is along the north–south trending ridges to its east and west, which rise to about 450 metres. The region has an annual rainfall of 60 millimetres, most of which falls during summer.[154]

A series of ridges lies to the west of the lake, parallel to its western shore, a product of the mainly east to west wind pattern. They represent beach ridges during high lake stands. Some of the depressions between the ridges retain water at times, forming shallow lakes. Lake Constant is the largest of these.

Most commonly Lake Buchanan is mainly dry, with a pool only 4 kilometres by 2 kilometres near the south-eastern shore. The slight upward slope of the lake floor to the north means that the north-western sector is the last to fill. Brief lake-full stages occur every few years, while occasionally the whole bed is dry. When that is the case, the usual pool area in the south-east has a thin halite crust. When there is water, the salt-tolerant water-plant *Ruppia* grows in dense beds and abundant crustaceans, including ostracods, comprise the zooplankton. The chemistry of the lake's brines differs from that of other 'typical' Australian playas whose waters approximate sea water in the proportions of their elements. In Lake Buchanan's brines the sulphate content is low (there is a total absence of gypsum) and the calcium-magnesium balance is different, with very high levels of strontium and fluoride. The water chemistry reflects the rock types in the lake's catchment, different from the marine sediments which dominate in the inland basins where the other playas occur, and where the inward drainage of regions has led to concentration of aerosol salt as well.

The lake's situation high in the landscape, straddling the modern Divide, was maintained throughout the Tertiary because the major palaeodivide in the region then was the Nebine Arc, with which the Great Divide coincides. (Other sectors of today's continental divide have migrated to some extent through geological time, altering drainage patterns.) The constant elevated situation probably explains the absence of the chemical evolution which led to gypsum formation in playas of the continental interior, and it explains the slow sedimentation rate which have been revealed by studies of the lake deposits.

These studies have shown that in the last 730 000 years there have been four major wet phases. In the first two, which were separated by a short dry interval, the lake was deep—considerably more than 2 metres. In the last two wet phases, also separated by a short dry phase, water levels were much lower. Prior to 730 000 years, an alternation of undated wet and dry periods is recorded in the sediments.

Lake Buchanan, an unusual salt lake, astride the Great Divide in central Queensland. A delta has formed in this northern end of the lake where a stream discharges, and clearly defined beach ridges are visible on the shoreline. JIM BOWLER

A CRATER LAKE IN VICTORIA: A HOLOCENE RAIN GAUGE

Volcanic activity in Victoria during geologically very recent times has left volcanic domes in the landscape and some crater lakes, some of which are entirely rain-fed, making them natural rain-gauges.

Lake Keilambete is a maar lake located four kilometres north of Terang, 190 kilometres west of Melbourne. The present water body occupies a circular area, with a diameter of 1.8 kilometres, and has a mean depth of about 11 metres.[116] The flat lake bottom is lined with impermeable muds. The salinity of the lake water is twice that of sea water. Wave-cut notches up the crater sides show previous lake levels and a study of the lake floor sediments reveals the history of the crater lake.

- **440–400 cm** The basal zone comprises ancient sediments which started to accumulate before 15 000 years ago. A soil layer, the Basal Soil, dated at 14 000 years ago, developed on these sediments during a dry period. The soil layer is leached and is overlain by an organically rich layer representing a swampy environment which persisted until 10 000 years ago, when the system became a shallow lake in which salt accumulated (due to evaporation in a closed basin).
- **400–375 cm** Basal Saline Sands, age 10 000 to 8000 years, containing calcareous ooids (which form in shallow, carbonate-saturated water), which represent a shallow lake environment.
- **375–200 cm** Lower Keilambete Mud, age 8000 to 4500 years. This zone of fine-grained dark muds, weakly calcareous, represents the main deep water facies. The lake rose substantially at 8000 years ago and continued to deposit muds until the first evidence of shoaling appears at 5500 years ago, when there was a relative fall in levels. When at its highest levels, around 6000 or 7000 years ago, the lake may have overflowed. This wet period was the wettest in the last 10 000 years.
- **200–0 cm** Upper Keilambete Mud, age 4500 years ago, up to today. Fine-grained dark calcareous muds with paler carbonate-rich bands and occasional sandy horizons comprise this zone. The sandy layer, at 3100 years ago, represents shallow conditions. The lake rose briefly after that episode, to fall again and maintain relatively low, fluctuating levels until 2000 years ago. In the last 2000 years there have been only minor fluctuations, and the largest reductions in water level have occurred in the last 100 years, the period of European occupation.

'FLIP-FLOP' CLIMATE CHANGES: COULD GREENHOUSE WARMING BRING BACK THE ICE AGE?

Global climate 'flip-flops' occur every few thousand years, from cooler to warmer mode and *vice versa*, according to William H. Calvin, a climatologist in the USA.[155] The contemporary global warming could be the trigger for a mode change—with ice age temperatures returning within a decade. Such abrupt cooling would be far more catastrophic than the creeping climatic change which is predicted with the Greenhouse effect. An abrupt cooling episode, a climate flip-flop, would cool the tropics by 9°F (5°C), and Europe, with 650 million people, could no longer grow its own food if its current Gulf Stream climate-warmer collapsed.

Ice cores from Greenland provide evidence of sudden climatic change within interglacials, showing that icy conditions have previously come on very abruptly, within a decade or two rather than over hundreds or thousands of years. (The first symptom of impending change is drought, followed by colder winters.) The Little Ice Age of historical times (from the early Renaissance to the end of the nineteenth century); the Younger Dryas event of 12 700 to 11 400 years ago; and an aborted switch at 8200 years ago which lasted only 100 years, are all examples of sudden onset ice periods. The last 8000 years have in fact been a period of unprecedented stability.

The 100 000-year cycle of glacial stages in the Pleistocene ice age during the last 2.6 million years is tied to changes in the Earth's orbit around the sun, so another glacial maximum is a long way off (officially and under natural conditions). But the interglacial in which we are living can suffer climatic swings which would have dramatic effects. The warming 15 000 years ago, after the last glacial maximum, started abruptly while most of the ice was still present, and the South Pole, in particular, has not yet emerged from the ice age.

Europe's climate today is anomalous, Northern Europe being 5° to 10°C (9° to18°F) warmer in winter than parts of Canada, the USA or Asia in similar latitudes. The populous parts of the USA and Canada are between 30° and 45°North. The agriculturally productive regions of Europe, supporting more than twice the population of the USA and Canada, are 15° further north. The anomaly exists because the cold, dry winds that blow eastward across the North Atlantic from Canada are warmed before they reach Europe by the waters of the warm North Atlantic Current, created by warm waters flowing northwards from the tropics merging with the east-bound Gulf Stream. This warmed water then travels up the Norwegian coast, with a westward branch warming Greenland's tip at 60°N. This is what keeps northern Europe warmer than comparative latitudes elsewhere. Were it to fail again, as it did partially in the Little Ice Age, the rest of the world would be chilled—and the associated results would be many—for example, tropical swamps would produce less methane, the Gobi Desert would whip up more dust into the air, Australia would be drier and more windswept. The cumulative effect of all the inter-related changes would result in a large-scale response, with global climate switching into a new mode of operation.

The sudden cooling scenario goes something like this: The North Atlantic Current, with a flow equalling the volume of 100 Amazon rivers, changes route—this has global implications because it is part of a long 'salt conveyor current' that extends through the southern oceans into the Pacific. (Salt concentrations in individual currents are determined by temperature.) Abrupt flips are orchestrated by the intermittent problem of ice melt in the North Atlantic Ocean, which causes a major rearrangement of atmospheric circulation. North–south currents redistribute equatorial heat into the temperate zones, supplementing the heat transfer by winds. When warm currents penetrate further than usual into northern seas, into the vicinity of Iceland and Norway, they melt significant quantities of the sea ice which normally reflects a lot of sunlight back into space—so the Earth becomes warmer. That causes further ice melt around the world, and sets in train another series of changes involving deep-sea currents.

The return loop of the North Atlantic Current runs deep below the ocean surface. Huge amounts of surface sea water sink at known down-welling sites every winter, with the water heading south when it reaches the bottom. When that annual flushing event fails for several years in a row, the 'conveyor belt' stops moving and heat stops flowing so far north—and the cold state of the flip-flop starts. (Surface waters are flushed regularly, even in lakes. Twice a year they sink, carrying their load of atmospheric gases downward, because water density changes with temperature. A lake surface cooling in autumn will sink into the less-dense-because-warmer waters below, mixing things up. The patterns of flushing in sea water are more complicated because sea water's salt content, not only its temperature, determines whether it floats or sinks. Evaporation concentrates salt, saltier water sinks. Thus salt circulation determines currents, and too much fresh water added to the surface of northern seas as a result of melting caused by Greenhouse warming would affect the warm north-flowing currents which keep Europe warm.) Oceans are not well mixed—they are a conglomeration of rivers and blobs of water (static accumulations confined between other layers) with different temperatures and different salt content at different levels.

Some dramatic examples of underwater rivers and blobs exist: Outflow from the Mediterranean from the bottom of the Straits of Gibraltar is 10 per cent saltier than the water of the Atlantic and sinks to the depths of the Atlantic, causing an Amazon-sized waterfall about 250 metres below the Straits. Another underwater ridge line stretches from Greenland to Iceland and on to the Faeroe Islands and Scotland. It pours the hypersaline water of the Nordic Seas (Greenland and Norwegian Seas) south into lower levels of the North Atlantic Ocean. This waterfall is more like 300 Amazons. Cold dry winds blowing eastwards from Canada evaporate surface waters of the North Atlantic Current, concentrating salt. Heavy surface waters sink in late winter. All three discrete water masses, pushed down by annual repetitions of these late winter events, flow south, along the bottom of the Atlantic.

The same thing happens in the Labrador Sea between Canada and the southern tip of Greenland. Salt sinking on this grand scale in the Nordic Seas fuels the Nordic Seas Heat Pump—producing a heat bonus of about 30 per cent beyond the heat provided by direct sunlight. Nothing like this happens in the Pacific—but it still feels the effects because the sink in the Nordic Seas is part of the worldwide salt conveyor belt. (The Atlantic is saltier than the Pacific, because the Pacific is larger, containing more water from rivers which dilutes its salt concentrations. The Atlantic would be even saltier if it did not link with the Pacific in long, loopy currents which carry the excess salt southwards from the bottom of the Atlantic, around the tip of Africa, through the Indian Ocean and up around the Pacific.)

There used to be a shortcut for currents from the Atlantic to the Pacific. Closing of the route 3 million years ago when South America and North America became joined by the Isthmus of Panama, forcing currents to make detours, may be what caused the Pleistocene ice age. The major change in ocean circulation, along with a climate that had been cooling for millions of years, led not only to ice accumulation most of the time, but also to climate instability with flips every few thousand years or so. There are a few indicators of the flushing failure which may herald a cold flip:

- **1 Diminished wind chill**, as happens in the Labrador Sea during the

North Atlantic Oscillation. This El Niño-like shift in the atmospheric circulation pattern over the North Atlantic, from the Azores to Greenland, often lasts a decade. At the same time as weaker winds give less salt-sinking in the Labrador Sea, Europe gets colder winters. An episode of the North Atlantic Oscillation started in 1996, and colder winters have followed in the Northern Hemisphere.

- **2 More floating ice**—which protects the sea surface from evaporation. Retained heat then melts the ice, and the cycle recurs about every five years.
- **3 *More rain in the northern seas***—this could dilute salt flushing (a situation predicted with global warming). Ice carried out of the Arctic Ocean could do the same; as could the melting of Greenland's glaciers. (Fjords of Greenland's east coast have been known to become dammed by glacier collapse. The breaking of such dams to release large slugs of fresh water might provide the trigger for the stopping of warm currents. Blobs of freshwater like the recently detected 'Great Salinity Anomaly' in the North Atlantic (containing 500 cubic miles of unsalted water) can travel about in the oceans, disrupting currents.

Evidence for Greenhouse warming, which so surprisingly could bring a little ice age when a critical threshold is crossed, is abundant and increasing. (See pages 28 and 29 in *Listen ... Our Land is Crying*.) The collapse of ice shelves around the Antarctic Peninsula goes on apace. In 1998, another 200 square kilometres of ice broke away, and according to the British Antarctic Survey, two-thirds of the 12 000-square kilometre Larsen Ice Shelf now threatens to disintegrate, marking the 'beginning of the end for ice shelves'. Dr David Vaughan (British Antarctic Survey) mapped the 0°C temperature gradient over the peninsula each January as a limit for viability of the ice shelves. The line has moved much further south over the last 50 years. The peninsula has warmed by 2.5°C, which is several times the global average, and there has been a dramatic acceleration in ice shelf breakages.

Closer to home, increased Greenhouse gases and global warming threaten Australia's greatest natural treasure. The Great Barrier Reef faces direct damage from the underwater equivalent of acid rain, according to Dr Terry Done (UN Inter-governmental Panel on Climate Change and acting head of CRC for the Great Barrier Reef). Doubling of the pre-Industrial Revolution concentration of Greenhouse gases by 2070 is expected to reduce calcification rates by 20 per cent because of acidification of the water, reducing coral growth and strength. Other threats also exist under Greenhouse—increased cyclone damage, flooding which brings sediments and nutrients from the land, and heat stress. Already, the hottest recorded January to March (1998) has caused serious and widespread coral bleaching. Heron Island reported 10 to 20 per cent recently dead coral, and in 1998 up to 88 per cent of inshore reefs suffered some bleaching; what the permanent recovery rate will be from such episodes is unknown.

The 1980s and 1990s were the hottest decades recorded on the Reef and contained coral bleaching episodes in 1980, 1982, 1987, 1992, 1994 and 1998. All the major bleaching events also had a connection with El Niño, and under Greenhouse, El Niño events are expected to be more frequent and more serious.

WATER—SOME ASSORTED FACTS: global perspectives and figures

The **global water budget** is finite. It has not changed through time. Only humans, of all the species alive on Earth, have had the ability to interfere fundamentally with global hydrology. Sea water comprises 96.5 per cent of the Earth's water. Only 1 per cent of the remaining 3.5 per cent freshwater budget is accessible for use, and its distribution is uneven and not proportionate to population centres and requirements. For instance, 20 per cent of the fresh water is concentrated in the Amazon Basin where population is very sparse.

Two thousand years ago, global population was 3 per cent of what it is today. Human over-population is slowly killing the Earth and putting the survival of the human species itself at risk. Water demand increases by 64 billion cubic metres a year. By 2025 it is predicted that 35 per cent of the world's people will suffer from significant water shortage.

Already some of the figures are startling:

- In 1998, 31 countries were recorded as facing water stress, and this will rise to 48 countries in 30 years' time.
- Middle East countries are 30–40 per cent dependant on groundwater. Saudi Arabia mines fossil water for 75 per cent of its needs.
- China, with a population of 1.2 billion, has half the water it needs. The water table is falling by 1 to 2 metres per year as groundwater is extracted—the rate of extraction exceeding acquisition by 25 per cent. China regards the use of water in industry as more important than in agriculture or for domestic purposes. Because of inefficient, often third-world manufacturing conditions, China uses 9 times more water to process a unit of steel than the USA does.
- In northern China the situation is already critical. Water shortage is so chronic that water pressure in some cities is such that none rises above the ground floor in buildings; supplies are so short that people use bath houses once a week; and there is talk of diverting water from major rivers into those where supply is critical—environmental catastrophe waiting to happen.
- In Africa, women are estimated to spend 40 billion person-hours a year fetching water.
- India gets 90 per cent of its rainfall in three months and, with the size of its population, needs continuity of supply, so much water stress exists.
- Some rivers in Europe are too polluted to be used in industry.
- In the USA, few lakes are suitable for swimming. In the St Lawrence River, pollution is such that for many years visiting whales have become 'toxic waste dumps' and could never have been eaten.
- Globally, water-borne diseases kill 12 million people a year. Child mortality decreases by 50 per cent with supply of clean water.
- The biggest cause of water pollution globally is agriculture, but industrial and domestic sources are becoming more significant.
- In developing countries 90 per cent of domestic sewage and 75 per cent of industrial waste is still released directly into surface waters.
- Twenty per cent of freshwater fish species of the world are endangered.
- Diversion of the Nile caused extinction of 30 of the 47 native fish species, destroying the fishing industry. Lake Chad was reduced to only 2000 square kilometres.
- Damming the Mekong River with attendant loss of forests resulted in a two-thirds decrease in the fish available to feed the locals.

SECTION SIX

URBANISATION AND WATER

Australia's population is one of the most urbanised in the world, and has been so since the early days of European settlement. More than 60 per cent of the population is concentrated in a few large cities along the coastal strip, and overall, 86 per cent of Australians live in urban areas. However, within each city the wide spread of suburbia results in one of the lowest urban population densities in the world. The national aspiration is still largely to own a home on a 'quarter-acre' block or larger, and the consequent urban sprawl with its attendant problems in supply of basic services is almost out of control in the larger centres.

The effects of this urbanisation on the hydrologic cycle are large. The demand for water is very high, and water has often to be brought over long distances from catchments remote from the developed areas. This may require damming, diversion or inter-basin transfer of freshwater supplies. About 10 per cent of water used in Australia is for domestic purposes, 7 per cent for industrial and commercial purposes, and 1 per cent for other urban uses.

At the present level of use, and maintaining the present standard of living, the freshwater supplies in the most populated areas (and in the major foodbowl regions) are already totally committed. Much of the water used is returned at a much poorer quality compared with the original supply. Return flows from urban and industrial uses range from zero to about 50 per cent of the water supplied, with values for inland basins in general being higher than for coastal basins where, to our national shame, most of the wastewater is discharged into the ocean.[2] (The return of wastewater to rivers in inland situations all too

One of Brisbane's rivers has become a drain—graffiti, shopping trolleys and all—'rivers into drains' exemplified. PETER OLIVER

Lake Wendoree, a man-made lake in the city of Ballarat—transforming an urban landscape by damming a stream. M.E.W.

THE EFFECTS OF URBAN RUN-OFF ON URBAN BUSHLAND STUDIES IN MIDDLE HARBOUR, SYDNEY

Urbanisation typically increases the area of impervious surfaces to between 30 per cent and 50 per cent. In the case of Middle Harbour suburbs, where a considerable amount of natural bushland remains, the figure is likely to be at the lower end. The volume of stormwater run-off is greatly increased compared with that which prevailed prior to urbanisation. The effects of this hydrological change, and how best to manage the problems it creates, have been a long-term interest of Dr Harley Wright, particularly during his association with the Pollution Control Commission (now the Environment Protection Authority).[156, 157, 158, 159, 160]

The Middle Harbour valley has about 25 per cent of its catchment as natural bushland along its floor and lower slopes. Small subcatchments of ridgetop housing above the bush, ranging from a few to 20 hectares in size, each discharge stormwater from single pipes onto slopes of natural bushland. Discharge occurs at low points like the heads of small gullies where, before urban development, there was no discrete drainage channel and surface flows occurred infrequently, following severe storms. The large increase in stormwater now discharged into bushland has created distinct drainage lines down slopes which generally have thin soils on a sandstone base. On steeper slopes, the hydraulic capacity of the majority of the ephemeral drainage lines has seen a 5- to 10-fold increase in line with the increase in water volume. In a small two-hectare catchment at St Ives it was shown that urban stormwater discharged into down-slope bushland on about 100 days in the year, while in a nearby undeveloped bushland catchment of comparable size and relief, surface flow occurred only on about five days a year.

Drainage lines have eroded as a consequence of increased stormwater flow, and material has been carried downstream and deposited at the estuarine mouths of major local creeks. Studies indicate that the period of active erosion is vigorous in the early years following new developments. After some decades with little new development, say around 30 years, the hydrology stabilises with a higher capacity and low erosion rates. However, in Middle Harbour, and in Sydney generally, a policy of urban consolidation has caused a resurgence of major redevelopment with expanded areas of impervious, water-shedding surfaces. Accordingly, a fresh regime of erosion and sedimentation occurs, with even higher run-off volumes causing the hydraulic capacity to increase yet again.

Not only the quantity of water which flows into bushland, but also the quality, is altered by urbanisation of catchments. Great increases occur in the amount of phosphorus and nitrogen as well as in suspended solids and unpleasant additions like coliform bacteria. It is the high nutrient levels, however, which have the most immediate effect on the bushland into which the stormwater discharges. Phosphorus is the nutrient which poses the greatest threat to bushland soils because it is sparingly soluble and, once introduced through urban run-off, is a permanent pollutant. Australian soils are notoriously low in phosphorus, and the species-rich ecosystems which have evolved with them are adapted to these low levels; many ecosystems are, in fact, adversely affected by high phosphorus levels. In sandstone areas, phosphorus concentrations in urban run-off have been shown to be between 40 and 100 times greater than in waters in natural catchments, and soil concentrations are 10 to 20 times higher in areas affected by run-off. Sources of the phosphorus typically include dog and cat droppings, garden fertilisers and compost, and blue metal worn off roads. In non-sewered areas, septic tank effluent is an added source. Unfortunately, in sewered areas, overflow regularly occurs during periods of heavy rain because the system cannot cope with the large increase in volume from illegal roof connections and ingress of groundwater through leaky domestic pipes. This is a major, and usually overlooked, source of high nutrient levels in stormwater.

Urban stormwater flowing into otherwise undisturbed bushland is the major factor encouraging and sustaining weeds in these areas in the long term. Soil disturbance and changes to natural drainage reduce the competitiveness of native plants and facilitate their replacement by weeds.

The degradation of our urban bushland remnants is a matter of great concern and a situation has been reached where no more of this precious resource should be alienated for urban expansion, and what remains must be managed for its preservation. (New housing developments within the greater metropolitan region should be restricted to areas already cleared and no longer used for the purpose for which they were originally cleared, and decentralisation should be the aim of future major development.)

Measures to minimise the damage from stormwater run-off into natural bushland include:

- **Sediment control** Appropriate sediment control measures should be implemented before the stormwater enters bushland. A large proportion of pollutants such as phosphorus is normally adsorbed to suspended solids and is contained in organic matter in urban run-off. Silt, nutrient and weed content can be reduced prior to entry into bushland by flow-retardation basins, wet retention basins and wetlands—all of which require a lot of land and would be unsuitable on most of the Middle Harbour terrain. Mechanical traps are more compact and can remove gravel, sands and vegetative material, but are not very efficient in removing nutrients. Regular maintenance of traps is necessary. (A number of companies are manufacturing hi-tech traps which filter as well, at a cost.)
- **Minimising** the number of stormwater drainage outlets into bushland areas and ensuring that they are directed only to existing well-defined drainage paths.

Other strategies seek to restore the former natural frequency of direct flows. This can be achieved through the use of contour or catchment protection drains and down-slope carrier drains or low-flow pipes. Impervious watercourses (drains or pipes) in dry gullies can carry polluting stormwater through to permanent watercourses, minimising the uptake of phosphorus on the way. Lane Cove Council has installed a number of lined channels and low-flow pipes through bushland reserves to mitigate the accumulating damage from stormwater outlets which previously discharged onto bushland slopes. This has greatly facilitated the managed regeneration of native vegetation in the reserves.

frequently results in pollution and associated problems.)

The hydrological consequences of urban sprawl are many and varied. Extensive impervious areas result in high run-off in storms; gutters and drains carry rapidly moving water with high erosive power; stormwater drains carry pollutants from roadways, garbage from careless citizens; dog droppings and garden fertilisers nutrify water and promote algal blooms; flooding of natural waterways is usually enhanced, and sedimentation increased; roads and railways can impede natural drainage in flood-prone areas; rubbish dumps can leach toxins into waterways; acid sulphate soils can be disturbed, resulting in acid leaching which may continue for hundreds of years—the list is endless.

It is even possible that large cities can have a climate-changing effect, increasing rainfall locally relative to the rural surroundings. Local heat sources and the increase in particulate matter in the atmosphere, providing nuclei for the condensation of water vapour, are possible catalysts. (Figures are available for European cities in this context, showing up to 16 per cent increases, but not yet for Australia.)

The **'turning rivers into drains'** syndrome applies to all our within-city rivers. Smaller streams become piped or concreted drains and the major rivers are tamed to fit the urban scene. Wetlands disappear or are badly degraded, and the chances of a natural, healthy river with intact aquatic ecosystems surviving are small. To combat the problems and threats to river health in urban catchments, a great deal is now being done by concerned citizens. A brief account of catchment management problems and some solutions, for rivers which flow through suburbia into Sydney Harbour exemplifies some of the activities which can improve local waterways and water quality, and improve prospects for maintaining biodiversity.

An aerial view of Chatswood with Blue Gum Creek central. The fragile linear arteries of remnant bush along the creeks which drain into the Lane Cove River are the last refuges for urban wildlife. WILLOUGHBY COUNCIL

Gross pollutants in a stormwater trap below the Chatswood CBD. ALFRED BERNHARD

MIDDLE HARBOUR, SYDNEY: *some initiatives to decrease pollution*

Sydneysiders living close to the 'most beautiful harbour in the world' have become increasingly aware of the effects of urbanisation on the local waterways. Middle Harbour has a catchment containing, at present count, 35 suburbs, and receives the drainage from 16 fair-sized creeks and an uncounted number of ephemeral creeklets and gullies. It is in many respects typical of areas around all branches of Sydney Harbour and the other major waterways around which the metropolitan development has occurred. The problems it has in terms of terrain and of managing stormwater run-off, reducing pollution and protecting what is left of the natural environment, are shared by all the other catchments. The last decade has seen a great increase in awareness at both governmental and individual citizen levels of the need for action, and much progress has been made.

Total Catchment Management (TCM) has been progressively implemented in New South Wales since the introduction of the Catchment Management Act of 1989. The TCM program has been coordinated through the Department of Land and Water Conservation and has developed into the primary community–government partnership for achieving sustainable management of natural resources in New South Wales.

On 28 April 1998, the Environmental Protection Authority (EPA) directed all New South Wales municipal councils to prepare Stormwater Management Plans on a catchment by catchment basis. The production of the plans has been assisted by grants from the Stormwater Trust. The recognition that such plans should be based on a total catchment management approach, and that therefore their development involves the participation of councils, citizens and a host of other organisations and service providers, has been a great step forward. In the case of Middle Harbour the list of participants with the councils (Ku-ring-gai, Manly, Mosman, North Sydney, Warringah and Willoughby) includes the Roads and Traffic Authority, Sydney Water, Sydney coastal councils, National Parks and Wildlife Service, Rail Services, Department of Land and Water Conservation, the Waterways Authority and the Middle Harbour Catchment Management Committee. (Hopefully not a bureaucratic nightmare ...)

The Middle Harbour Catchment Stormwater Management Plan, prepared by consultants for the councils involved, appeared in draft form in May 1999, assessing the situation, setting goals and priorities. Meanwhile, councils have been carrying out remedial work in their sectors of the catchment. The activities of the Willoughby Council, described below, are an example of the sort of works being carried out. A major Sydney Water project—the Northside Storage Tunnel—is also outlined as an example of the sort of infrastructure project which becomes necessary when a city outgrows the capacity of its sewerage system.

WILLOUGHBY COUNCIL'S STORMWATER MANAGEMENT PROJECTS

A program of watercourse rehabilitation has been undertaken by the Council's Bushland team to address the impacts of urban run-off on bushland, and on creeklines in particular. Key aspects of the work include sympathetic treatment of creeklines and discharge points to reduce water velocity and stabilise the areas, using rock armouring and geotextiles. Retention, establishment and management of riparian vegetation, using locally indigenous species, is an important part of the work.

In addition to this sort of work on a large number of minor stormwater outlet points and drainage swales, several larger projects have been made possible with financial assistance from the EPA's Stormwater Trust Grants. At Scotts Creek, Chatswood, a typically degraded creekline has been restored to a more natural condition, as seen in the adjacent photographs.

The restoration works involved:

- Clearing of degraded slopes and creek bed.
- Removal of a wide range of rubbish.
- Stabilisation of eroding creek banks.
- Construction of a series of drop structures to create a multiple-pool system.
- Revegetation of the previously degraded slopes and waterway, assisted by Bushcare volunteers.
- Establishment of a trap so that gross pollutants will be captured and can be removed before the creek drops to bushland and Middle Harbour.

Water quality will benefit by the settling of suspended material, the absorption of nutrients by aquatic plants, and aeration. Habitats for a variety of macro-invertebrates will be enhanced. These sorts of measures, implemented in all catchments, will make an enormous difference to the health of rivers and the harbour, and inland water bodies.

Willoughby Council has provided public awareness and community education initiatives which have included interpretive walks for local schools, drain stencilling activities, community planting days, displays in commercial areas, and sediment and erosion control programs for building sites. Local businesses have been successfully involved in sponsorship of site restoration works—as in the case of the Flat Rock Landcare Group (page 276 in *Listen … Our Land is Crying*). Monitoring of water quality by members of the local community has been an important part of the restoration process.

Scotts Creek, west Chatswood, before and after restoration. ALFRED BERNHARD

Armoured stilling basin in drainage line, to trap sediment. ALFRED BERNHARD

THE NORTHSIDE STORAGE TUNNEL PROJECT

From the brochures issued by Sydney Water:

> The Storage Tunnel is one of the 'early Action' measures identified by the EPA and Sydney Water as crucial to improving the water quality in Sydney Harbour. The Northside Storage Tunnel is being built to collect wet weather sewage outflows that currently go into Sydney Harbour without treatment. It consists of 19.8 kilometres of tunnels running from Lane Cove and Scotts Creek to North Head. The Storage Tunnel will capture wet weather sewer overflows from four sites: Quakers Hat Bay, Tunks Park, Lane Cove and Scotts Creek. These four overflow points represent the four largest current wet weather overflows that enter into Sydney Harbour.
>
> The Storage Tunnel will be able to store a significant volume of overflow (about 500 million litres) and will convey it to the North Head Sewage Treatment Plant for treatment. It will be built using four tunnelling machines, digging from North Head to Clontarf, and from Tunks Park to Clontarf, Scotts Creek and Lane Cove.

A great deal of opposition to the scheme was voiced, particularly from residents of Manly where excavations and activities near the Quarantine Station are a source of great concern for the local marine and terrestrial environment. But the situation regarding the entry of untreated sewage into the harbour, now that it is admitted to and publicised, has proved to be so serious that this immediate, if possibly (and hopefully?) short-term, measure is considered by environmental science experts (personal communication) to be warranted.

During wet weather, rainfall enters the sewers through leaks and illegal connections. (One might ask why there was no policing of the illegal connection of roof-water to sewers until comparatively recently.) Overflow points were provided in the sewerage system to avoid overload; two of these overflow points are located in the Scotts Creek system, where an average of 20 overflows occurs each year. One overflow is situated where a submain crosses the creek in an above-ground aqueduct. A hinged lid allows the system to 'blow' when excess of capacity is reached, whereupon wet weather sewage overflow sprays over the aqueduct and drops into Scotts Creek, decorating the vegetation with paper and, worse, waste. The second overflow on another submain discharges nearby, straight into the creek. The two overflows together dump 1600 million litres of untreated sewage into the creek each year—900 Olympic-sized swimming pools of the stuff—going directly into Middle Harbour.

It is said that the tunnel will reduce the overflow events at Scotts Creek by 80 to 90 per cent—from 20 events a year to two. The existing hinged lid will be replaced by a pipe which delivers the effluent during excess events straight into the creek instead of spraying it all over the environment—some consolation, admittedly. The air which has to vent when the tunnel fills with sewage is a problem which Sydney Water assures us will be handled by activated carbon filters. No suburb is too happy about having sewer vents in its backyard. It is claimed that an added benefit of the tunnel will be its availability to store sewage from the North Head Treatment Plant when the plant is shut down for maintenance or suffers mechanical failures.

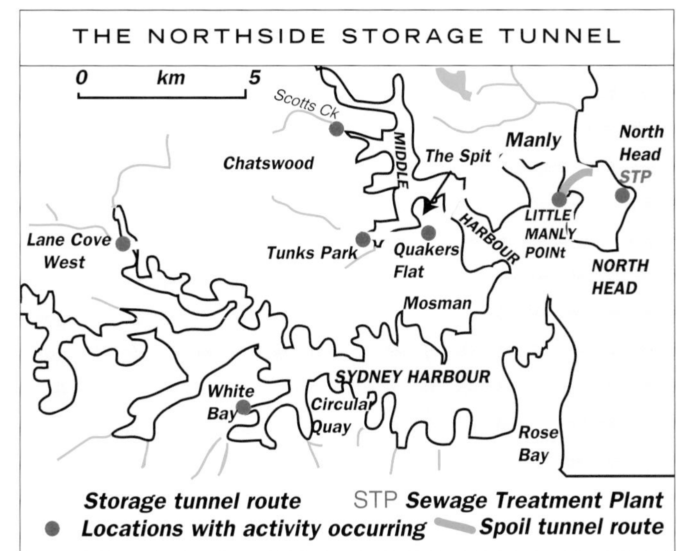

How clearly we see from all this that 'people = pollution'. We cannot go on increasing population density in any city without renewing urban infrastructure.

THE YARRA RIVER

The Yarra River plays an important role in the lives of the 2.5 million residents of Melbourne, being a feature of the city as it runs towards its estuary and its exit into Port Phillip Bay. The well-manicured fringes of the river, its parks and gardens as it runs through the CBD, are a far cry from the noisesome drain which ran through 'Smellbourne' in the early days of settlement when this river, like others running through towns, was simply an open sewer. Much of the beauty of the modern city is greatly enhanced by the river's immaculate urbanised image.

The Yarra River catchment provides most of Melbourne's water supply and supports a variety of agricultural activities which contribute about $140 million annually to Victoria's revenue. Above Warrandyte, where a gorge separates the upper river from the urban lowland stretch, the Yarra is a Heritage River. Since 1998, a comprehensive ecological study of the river has been initiated in order to maintain and improve the health and function of this important waterway and its tributaries. No such biological study has been done before and it will be some time before this one is complete.[161]

The Yarra rises in the vicinity of Mt Gregory on the Great Divide

RIVERS INTO DRAINS: THE FATE OF SMALLER RIVERS AND CREEKS THAT RUN THROUGH URBAN AREAS.

Bendigo Creek, beautifully paved with basalt blocks by convict labour—a river tamed in an urban environment in the early days of settlement when the town suffered serious flooding. Major floods have occurred as recently as the 1940s. M.E.W.

Bendigo Creek in a park in the city centre. M.E.W.

and flows generally westwards. Its catchment has an area of about 4000 square kilometres, and the total river length, including tributaries, is about 2000 kilometres. Relief varies from 1250 metres near Mt Arnold to sea level in Port Phillip Bay. Rainfall varies across the catchment, from 1600 millimetres in the mountains to 600 millimetres in the western sector. The headwater streams flow through narrow valleys into the Upper Yarra Reservoir. Below the dam, the Yarra eventually emerges into a wider valley below Westburn. Midway between the Little Yarra River and Wóori Yallock Creek confluences it swings north until it is joined by the Watts River from the east. It then resumes its westward direction and flows through the Healesville, Yering and Warrandyte Gorges (separated by confined floodplains) and into a wide valley between Templestowe and Fairfield Gorge. At Dights Falls, a natural cascade at the lower end of the gorge, a weir has been constructed. Once the Yarra has flowed over the Falls it becomes tidal.

As about 34 per cent of the Yarra's flow is diverted to supply Melbourne, the flow at Warrandyte is about half what would occur under natural conditions. It appears that the flow regime has not been changed significantly despite the diversion of so much water; this is explained by the fact that similar proportions of flow are diverted from tributaries throughout the year and all the diverted water is completely removed from the system. How halving the flow, even without altering the flow regime, affects the river ecology is yet to be assessed. Regular de-snagging, removing the large woody debris—from the channel—has certainly impacted on the in-stream biota.

Based on land use and catchment form, it is possible to identify six distinct segments of the main stem of the Yarra:

- **Uplands** From the headwaters to Westburn, mountains and narrow valleys, devoted to water supply and forestry. The main impact of European settlement has been flow diversion. De-snagging has been carried out since 1930 and has probably had an effect on the biota. In the lower reaches of this section riparian vegetation has been cleared for agriculture and only remnant patches of damp to wet sclerophyll forest remain. Stream Watch assessment of the waterways in the uplands catchment has rated conditions generally as good, with a score of 9 out of 10 to 10 out of 10 for water quality.
- **Rural unconfined** From Westburn to the Woori Yallock Creek confluence the floodplain remains relatively unmodified, without levees or drainage works, so that the river can leave the channel and flood out during high flows. The channel has broad bends and large amplitude meanders as it crosses the floodplain which varies from several hundred metres to several kilometres wide. Large areas have been cleared for agriculture and urbanisation; gold mining was an important activity in the past and the sediment it delivered to some tributaries is still entering the Yarra in some reaches; and upstream flow diversion has impacted on the river. The Stream Watch assessment of the main stem of the Yarra in this segment is rated as moderate to poor. Water quality is rated as 8 out of 10, but turbidity is a problem during low flows, with the sediment being mainly contributed by the Little Yarra River and Woori Yallock Creek.
- **Rural confined** From the Woori Yallock Creek confluence to the upper end of Warrandyte Gorge (including the Henley floodplain), the floodplain has been modified by levees, drainage and channelisation in order to confine floods to the channel as far as possible. Changes since European settlement have been widespread—clearing and the development of agricultural industries and urbanisation. Gold mining has taken place in the Steeles Creek catchment. Extensive flow diversion occurs in this

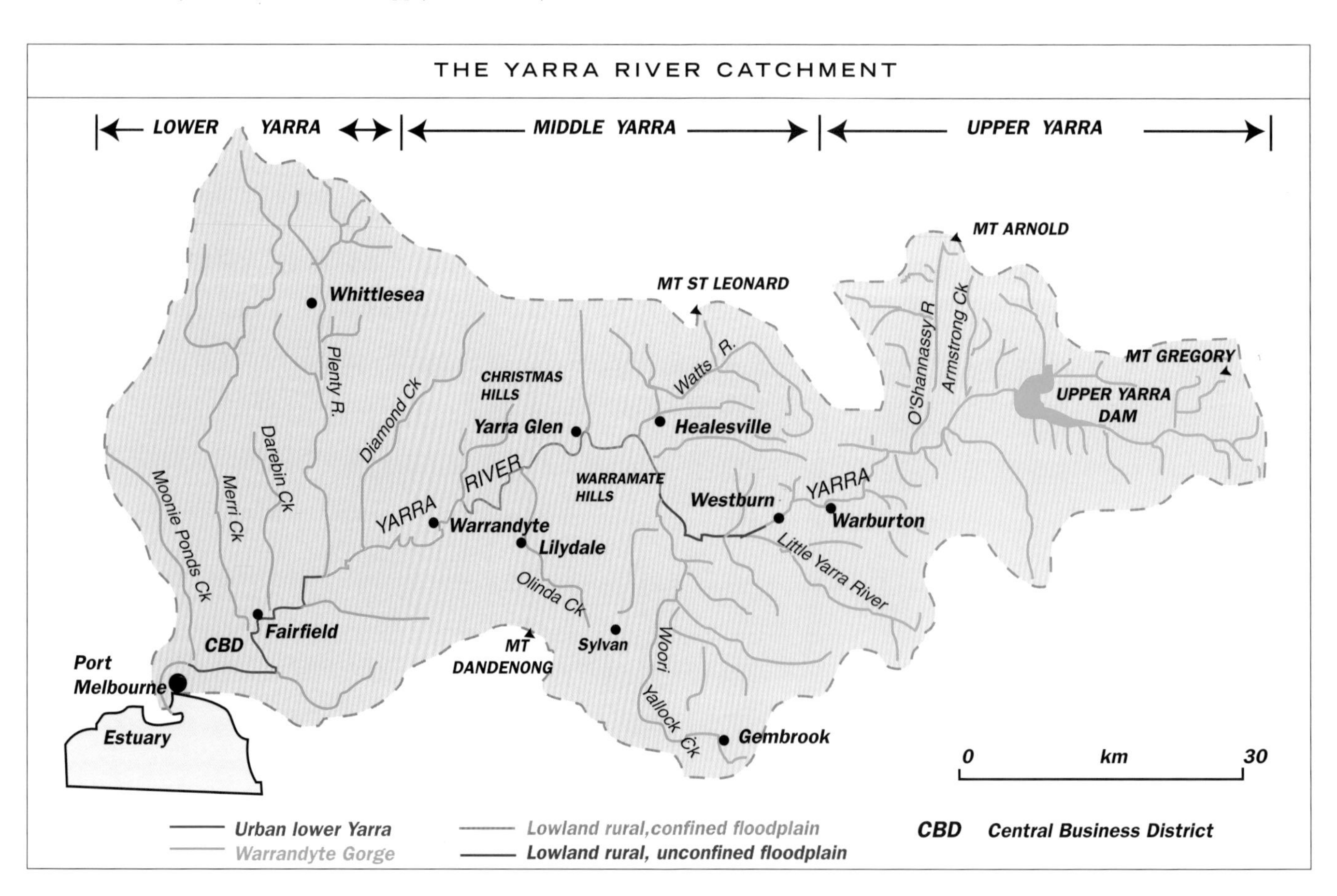

part of the catchment: Badger Creek flows are diverted into the Upper Yarra Aqueduct, the amount of the diversion varying from 64 per cent to 80 per cent (February to March and August 1994); flows from Watts River are stored in the Maroondah Reservoir and a proportion of the water is used in the Melbourne water supply system (amount diverted: 59 per cent September, 85 per cent May to June); water pumped from the Yarra at Yering Gorge is diverted to Sugarloaf Reservoir, an off-channel store, for use in the Melbourne water supply system. On average, about 50 000 megalitres a year is pumped from the Yarra at Yering Gorge. Stream Watch has rated the waterway condition of the main stem of the Yarra in the rural confined sector as poor. Water quality varies in the different stretches of this part of the river, rated between 6 out of 10 and 8 out of 10. Run-off from agriculture; pastoral areas; effluent from septic tanks; areas of bank erosion; and sediment from mining all contribute to water pollution.

- **Gorge** Warrandyte Gorge down to Fitzsimons Lane, a steep segment which lacks broad floodplains. Changes upstream have affected this section, particularly the reduction in flow. Urbanisation has had a local effect. While land adjacent to the river is mainly reserves used for recreation, the tributary catchments have been increasingly urbanised. Water Watch rates the gorge sector of the Yarra as in good condition, but water quality is only 7 out of 10, largely because of run-off from housing developments, poorly constructed roads and drains and raised nutrient levels with attendant raised *E. coli* bacterial counts as a result of sewage pollution.
- **Urban lowland** From Fitzsimons Lane to Dights Falls where urban land use dominates. Early records from 1897 indicate that deep pools existed in Fairfield Gorge. None were found 100 years later above the gorge and it is probable that sediment has filled many in the gorge. The modification of this segment of the river has been great, taming it to suit the city. Water Watch rates the river condition as moderate and water quality as only 5 out of 10.
- **Estuary** Below Dights Falls the stream bed is bedrock with some deep pools. It grades into sand further down and finally crosses a delta of sand and silt built out into Hobsons Bay. The estuarine sector of the river is controlled by the local geology, with the river following the boundary between Silurian sedimentary rocks and Quaternary volcanic rocks. The estuary itself is a salt wedge estuary, with the tip of the wedge moving up and down with tide and flow conditions. Prior to European settlement, the estuary did not extend nearly as far upstream. A bedrock bar formed the Queens Street Falls, restricting the penetration of salt water upstream. Channel widening, construction of embankments, removal of some bedrock controls, channel straightening and dredging have all contributed to the enlargement of the natural estuary, allowing the salt wedge to move upstream as far as Dights Falls. The present-day estuary seems to be silting up, and it is anticipated that infilling will continue until its size is again what it was pre-settlement. High concentrations of surfactants—pathogens like *E. coli, Salmonella* and *Pseudomonas*—are found in the estuary, making it unsuitable for recreational pursuits. Sediment samples contain DDT, polycarbonates and other non-biodegradable pollutants.

Fundamental changes have been made to the hydrology of the Yarra River in the last 150 years as can be seen from this summary—its flow has effectively been halved, and its water quality has declined steadily downstream, from pristine in the headwaters to grossly polluted at its exit to the sea. In spite of the halved flow, the flow regime and the seasonal highs and lows have not changed significantly; in addition, the geomorphic response (the change in physical form) of the river would seem to have been minimal. Many of the other rivers described in this book have suffered great physical change when their hydrologic balance has been disturbed.

The Yarra serves to emphasise how different rivers can and do have individual responses to the same set of man-made changes in their catchments. It has proved to be resilient, to have maintained a balance which is indicative of channel stability. The low inter-annual flood variability in this catchment compared with that in most other eastern Australian river catchments may have a lot to do with the river's resilience. It will be interesting to know how its ecosystems have fared when the current investigations by the Cooperative Research Centre for Freshwater Ecology and other bodies interested in Melbourne's water supply have completed their investigation of the river's biota.

The present reasonably satisfactory state of the river (when compared with many others at least) is surprising when the impact of mining in the catchment is considered. In the past, many sections of the river and its tributaries have been sluiced, dredged and diverted by miners, and the nearby areas have been cleared of timber during these operations. Changes in planform locally, increased sediment loads, heavy metal contamination of waters, and invasion of weeds resulted from the mining activities, and only the resilience of the river made the detrimental effects largely transient.

THE NATIONAL LAND AND WATER RESOURCES AUDIT
A program of the Natural Heritage Trust

The Natural Heritage Trust Act of 1997 states: 'Australia's natural environment is central to Australia's non-material well being and to Australia's present and future economic prosperity. Accordingly, present and future generations of Australians will benefit from the ecologically sustainable management of the natural environment.'

Reports on Ecologically Sustainable Development and the major State of the Environment Report of 1996 emphasised the need to achieve sustainability in environmental, economic and social terms. The same aims underlie the Natural Heritage Trust Initiative which has instigated the National Land and Water Resources Audit, designed to improve decision making in land and water resource management. The Audit will do this 'by compiling nationally compatible data sets of parameters which provide a measure of the status of the nation's land and water resources, and which over time will identify changes to that status, either positive or negative'.[162]

The strategic plan for the audit covers the four-year life of the project, 1998 to 2001. It has been developed following widespread consultation with Commonwealth, State and Territory agencies and with community groups, setting down priority areas and approaches. It will provide a framework for long-term monitoring and assessment, and hopefully for better management.

Such a national approach to problems is necessary and is to be applauded. It will be seen, eventually, whether the audit leads to actions and whether the funding provided for rehabilitation of the soil and water resources is used to solve the problems or largely to support bureaucracies and expensively produced reports and brochures. If the audit is genuine, it cannot fail to reveal stark facts about the degradation of resources which up to now have been watered down with the excuses such as 'the degree of degradation cannot be measured' or 'there is insufficient published evidence to determine the extent . . .'—the sort of half-hearted presentation of unpleasant facts nicely sanitised, as they were in previous official publications designed not to rock the boat. As I see it, when the boat is rapidly sinking and there is still a little time in which to fix the leaks, the truth has to be paramount and action has to be immediate and decisive. A new millennium is a good time to be able to say 'this is how it is and this is what we are doing about it'—looking forward, not back.

THE SALINITY AUDIT OF THE MURRAY–DARLING BASIN

The Salinity Audit of the Murray–Darling Basin, prepared by the 'Commonwealth, New South Wales, Victorian, South Australian, Queensland and Australian Capital Territory Governments working with the Community' became available after this book had gone to the publisher. It tells an horrific story, but one already well-known and well-documented among scientists and other interested parties, if not among the wider public—who may still be unaware of its implications. It makes horrendous predictions for salinity levels in the rivers and soils of the Basin over the next 50 to 100 years. It is essential reading for everyone interested in the welfare of the nation. This brief summary of some of its findings is included here because salinity issues in the Murray–Darling Basin were largely dealt with in *Listen . . . Our Land is Crying* and have not been repeated in this volume.

The audit emphasises that the Basin is geologically and climatically prone to concentrating salt in the landscape, a statement of the obvious when the flatness of the terrain, the saucer-like basin structure with only one outlet to the sea, and the high evaporation rates coupled with uncertain rainfall, are considered. It admits that inappropriate land uses, replacing natural systems, have resulted in massive hydrological imbalances that will take 'up to several hundred years to stabilise'. While natural vegetation used nearly all the rainfall, recharge of groundwaters results from failure of land-use and agricultural systems to do so, and rising groundwaters mobilise the salt which was previously maintained at depth in the water table below the root zone. 'Much of the salt mobilised does not get exported through the rivers to the sea' but stays in the landscape or is diverted into irrigation areas and floodplain wetlands.

> "Currently, of the 5.1 million tonnes mobilised [annually], 3 million tonnes is retained. By 2050, the salt mobilised will reach 8.3 million tonnes. However, 3.3 million tonnes will be exported to the rivers. This will rise to 3.8 million tonnes per year by 2100, resulting in the build-up of salt in the landscape of about 5 million tonnes per year."

Dryland salinity is an ever-increasing scourge and threat.

While the Audit quotes a 1987 estimate of the extent of rising water tables and salinisation of soils as 96 000 hectares of salt-affected irrigated land, and 560 000 hectares having water tables within 2 metres of the surface (that is, within the danger zone), other figures quoted leave one with the impression that no one wants to give an up-to-date figure. To quote directly from the Audit:

> "A 1993 study identified only 20 000 hectares as grossly affected by secondary salinisation, but it was probable that 200 000 hectares were actually affected at that time. The report indicated the likely future hazard for the Basin was of the order of 1 million hectares.
>
> Another study in July 1995, by the National Dryland Salinity Program, estimated 300 000 hectares of land within the Basin as salt affected in the form of dryland salinity. [The Audit] does not estimate the location and area of salt affected land... [it] establishes a trend line, river valley by river valley, for salt mobilisation in the landscape and its expression at the land surface and in the rivers. This is the predicted rise in salinity if there are no new management interventions to prevent it."

There have been so many investigations and so many schemes, and so much is already known, that to the cynical this is just more of the same—we do not want bureaucracies spending big money and producing glossy publications when that money is better spent on doing something about the problem on the ground. The only new angle presented by the Audit is 'the extent to which salinity levels are rising in tributaries of the Murray–Darling system . . . Future salt exports will shift from irrigation-induced sources to dryland catchment sources.'

Predictions of the increases of salinity of our rivers are at least now official, to give the Audit its due. The critical 800 EC units threshold (the salt content of the water) for suitability as drinking water will be exceeded in all the rivers in the near future, and some rivers have already attained that degree of salinity:

- The water of the lower Murray (measured at Morgan) will exceed the 800

EC threshold in the next 50–100 years. By 2020 the probability of exceeding 800 EC will be about 50 per cent. About 40 per cent of the salt will come from the Mallee dryland zone and 25 per cent from the tributaries upstream. (The cost of 1 EC unit rise is estimated to lie within the range of $93 000 to $142 000 per year; the total current economic impact of the river's salinitiy is estimated at $46 million per year.)

- The Macquarie, Namoi and Bogan rivers will exceed the 800 EC threshold within 20 years and exceed the 1500 EC threshold for irrigation crop and environmental damage within 100 years.
- The Lachlan and Castlereagh will exceed the 800 EC threshold within 50 years.
- The Condamine–Balonne, Warrego and Border rivers will exceed 800 EC before 2020.
- The Avoca and Loddon already exceed the 800 EC threshold on average.
- Major wetlands of the Basin—the Macquarie Marshes, Great Cumbung Swamp, Avoca Marshes and Chowilla Floodplain—are likely to suffer severe impacts.
- Salt mobilisation across the major river valleys is on a very large scale. The annual movement of salt in the landscape will double in the next 100 years.
- There is a future hazard for many other rivers and those people dependent on them as a source of water. Average river salinities will rise significantly, exceeding the desirable thresholds for domestic and irrigation water supplies in many tributaries and exceeding critical levels in some reaches.

Obtain and read the Audit—and weep salt tears for the salt-induced desertification of our land!

IRRIGATION—A BLESSING OR A SCOURGE?

Government-sponsored irrigation schemes were introduced in Victoria, New South Wales and South Australia at the end of the nineteenth century.[2] Their rapid expansion went ahead even though none of the schemes was capable of repaying the capital invested. They were deemed to be successful if farmers could pay the operating and maintenance costs (and even these payments were decreased in lean years). In many cases their introduction was preceded by considerable agitation for government action, and they were seen as politically expedient. (So, what's new?) There was even then a firmly held belief that irrigation is an essential prerequisite for national development, while alternative forms of government investment, like research into dryland farming, remained unexplored. The more cynical among us know only too well that large and grandiose schemes with major dam building will always triumph over reason and lesser schemes because of their hallucinogenic effect on voters and the next election.

Irrigation schemes in Australia are always expensive, and their ultimate cost in terms of the environmental degradation they cause is not considered in the equation. Large water storages are required because of the small yield and high variability of our rivers. Believe it or not, irrigation as currently practised does not overcome aridity. It is used more as an insurance against climatic variability. All of Australia's irrigated lands would still be productive under rain-fed farming, although irrigation obviously increases the productivity. Profitability is not necessarily improved, however, with the soaring costs of technology and other inputs, as many farmers will testify. Increasingly the massive agri-business enterprises are taking over, because only the very large-scale operations are financially successful—and with them the scale of water extraction from rivers which cannot sustain such use and remain healthy (or, in many cases, alive), becomes environmentally unsustainable.

About 90 per cent of the continent's irrigated land is in the Murray–Darling Basin, over the adjoining States of Victoria, New South Wales and South Australia (including the Murrumbidgee Valley, Riverina–Murray Valley, and areas along the Lachlan, Macquarie, Namoi, Gwydir and Barwon Rivers). Queensland also has extensive areas in the Bundaberg, Burdekin delta, Emerald, St George, Mareeba–Dimbulah and Dawson River districts. In Western Australia, irrigation is practised in the south-western corner and in the controversial Ord River Scheme in the north-east. The only major irrigation scheme in Tasmania is the Cressy–Longford Scheme 50 kilometres south of Launceston.

Irrigation accounts for more than 70 per cent of Australia's water use. About two-thirds of irrigation water comes from government schemes, and the rest

Irrigation at Warren, New South Wales. REG MORRISON

from private dams and groundwater sources; 50 per cent of the irrigated area is used for pasture, and a substantial proportion of the rest for fodder crops; 75 per cent of the irrigation water is applied in the Murray–Darling Basin.

Irrigation is an inefficient use of water. A loss of at least 50 per cent overall to evaporation (and locally this may reach 80–90 per cent in open canals and channels in arid regions in summer), the considerable seepage from channels, and the run-off of surplus water where surface flooding is practised, all contribute to enormous waste. In addition, the portion of the water which is not taken up by plants and goes down to the water table has serious consequences, adding to the rate of rising of saline water tables, particularly in the Murray Basin. Detailed accounts of this irrigation-related salinity problem are given in *Listen… Our Land is Crying*, pages 145–54. In the Murray irrigation areas, pipes are largely replacing open canals, with a great saving of water; and drains of various kinds are used to lower water tables locally, but on a regional scale water tables are still rising, and will continue to do so no matter what is done on the surface—because the hydrology of the closed Murray Basin has been upset and the man-made effects have to work their way through the system. Irrigation's contribution to the adverse hydrological changes is considerable.

Land damage from excessive irrigation may not become evident for several years. Leaching of top layers of soils occurs (and the soluble salts increase the concentrations in groundwater which is rising as a consequence of the irrigation). Soil structure can be damaged or destroyed in the long run, and the soil organisms are affected. The off-site problems of fertilisers and pesticides in irrigation water are well known.

SECTION SEVEN

THE MURRAY–DARLING BASIN

The Murray–Darling Basin comprises 26 major catchments and covers more than one-seventh of the Australian continent. It is Australia's largest river system but on a world scale it does not compare with major rivers like the Amazon, the annual flow from the whole catchment into the ocean being less than the Amazon's daily discharge. It is a perennial system because its main catchments are in relatively well watered regions. It is essentially inward-draining and only a very small proportion of its sediment load ever reaches the sea.

Tributaries of the Darling arise on the Great Divide and flow generally westwards into ever more arid terrain with little

MAJOR TRIBUTARIES OF THE MURRAY–DARLING RIVER SYSTEM

River boats, catering for tourists, still ply the waters of the Murray. This scene at Echuca has changed little since the early days, though much has changed in the river itself and in the surrounding landscapes. M.E.W.

contribution from run-off on their journey towards the Murray. They contribute only 12 per cent of the Murray's flow. Tributaries of the Murray in New South Wales also flow essentially westwards from the Great Divide, and those in Victoria, which drain the Great Divide where it curves westwards, flow essentially northwards. The Murrumbidgee system contributes only 13 per cent of Murray flow, thus 75 per cent is provided by tributaries of the Murray which lie upstream of the Murrumbidgee confluence. The Darling depends on summer rainfall, the Murray on winter rainfall, and the extreme climatic variability which is orchestrated by ENSO means that the whole Murray–Darling river system is highly variable.

The Murray–Darling Basin is Australia's major food bowl. At least 85 per cent of the Murray–Darling water is fully developed, with more than 70 per cent used for irrigation, and today less than 20 per cent of the natural flow ever reaches the sea.

Murray Mouth is frequently in danger of closing at times of slack water, and massive changes to the estuary have resulted from alterations to natural flow as a result of river regulation. In the 1930s, barrages were constructed at the Murray's mouth to control water flow and to reduce the size of the estuary, whose tidal influence extended as far as Murray Bridge, 60 kilometres away. Prior to the building of barrages, the mouth was 1.6 kilometres wide. Now, the river is 30 metres wide where it runs to the Southern Ocean. The rest has been filled in by land regrowth, a set of large sandy dunes splitting salt and fresh water and creating Mundoo Island.

REGULATION OF THE MURRAY–DARLING RIVER SYSTEM

River regulation structures were originally constructed along the River Murray to drought-proof the region and to improve the river for navigation and water supply purposes. The surprisingly early start to irrigation, which began in Victoria in the 1870s, was a response to the low and extremely variable rainfall in the Murray Basin which limited its productive capacity. Without a reliable water distribution system it could not have been developed as the food bowl of the nation.

The Murray–Darling Basin Commission (MDBC) is responsible for regulating flow in the River Murray, its anabranches and the Darling River downstream of Menindee Lakes. Water is shared between New South Wales, Victoria and South Australia, according to provisions of the Murray–Darling Basin Agreement.

Regulation of the Murray by building dams and weirs has resulted in large areas being permanently inundated. Lake Mulwala at the Yarrawonga Weir drowned forests of river red gums, which still stand like tree ghosts in the water. REG MORRISON

The junction of the Murray and the Darling at Wentworth. M.E.W.

In the late 1800s and early 1900s there was much wrangling over the control of Murray waters.[163] New South Wales claimed prior rights as the Murray had always been in its domain—South Australia and Victoria having by then been hived off from New South Wales. Victoria and South Australia sought a reasonable share, including navigation rights.

The first commercial use of the Murray–Darling river system was for navigation. From 1853, steamships operated and played an important role in settlement, carrying supplies inland and returning with produce. During average years the rivers were navigable for eight or nine months, for over 5000 kilometres, and for up to 6800 kilometres during floods, but they ran dry during severe droughts. Irrigation from the Murray, which started in Victoria in the 1870s, was seen as a threat to navigation, particularly by South Australia which, then as now, got what was left in the river when the other two States had taken what they wanted. When railways began to compete with river trade in the 1880s, navigation ceased to be a major priority for use of the river water.

A succession of dry years from 1895, which climaxed in the record drought of 1902, showed that some drought protection was necessary if the Murray valley was to be developed. It took until 1915 for agreement between the three States and the Commonwealth Government to be reached—and the River Murray Waters Agreement was enacted. Flow at Albury is shared equally between New South Wales and Victoria, and below Albury the two States retain control of the tributaries in their sectors; South Australia receives a guaranteed 'entitlement' from New South Wales and Victoria. The agreement also provided for the construction of a storage on the upper Murray (the Hume Reservoir); a storage at Lake Victoria; locks and weirs on the Murray extending up to Echuca; and locks and weirs on the Murrumbidgee or Darling. (New South Wales chose the Murrumbidgee.)

The Hume Reservoir and Lake Victoria were to store water during high flows and release it later to supplement low flows; the smaller regulatory structures were to make the Murray permanently navigable and provide steady pools for irrigation diversions. The structures were sited to enable irrigation by gravity where possible. The growing importance of irrigation and the decline of river trade resulted in amendments to the agreement in 1924 to give preference to building

The Darling at Menindee used to flow into the Menindee lake system during floods, but the lake water used to drain back into the river again. The lakes were not a reliable water supply so much engineering and modification has resulted in large bodies of permanent water. M.E.W.

structures for irrigation instead of navigation. Construction of barrages at Murray Mouth was a special concession to South Australian authorities, concerned that water extraction for irrigation would worsen saltwater intrusion from the ocean at high tide into Lakes Alexandrina and Albert, which were used for irrigation.

The Menindee Lakes on the Darling River system were incorporated into the storage program by the construction of banks, weirs and channels during the 1950s and 1960s. The lakes could not be relied upon as a water source in their natural state. They filled during floods but drained back into the Darling when the river fell. This water storage is now leased to the MDBC by New South Wales to supplement flow into the Murray.

The Snowy Mountains Scheme, which commenced in 1949 and was completed in 1974, impounded the waters of the Snowy and Eucumbene Rivers and diverted them inland via tunnels for the production of electricity and for irrigation and water supply along the Murray and Murrumbidgee Rivers. It increases inflow into the Hume Reservoir by about 580 GL per year.

The most recent stage of regulation of the Murray system was the construction of the Dartmouth Dam between 1973 and 1979. A dam at Chowilla in South Australia was proposed, and work had started, but it was decided that high evaporation rates would have increased salinity problems downstream, and a very large area round Lake Victoria and the Darling–Murray junction would have been submerged, so the project was abandoned and the Dartmouth went ahead instead. This dam was constructed to be a back-up in dry periods because the great increase in water use for irrigation had resulted in shortages. Releases from the Dartmouth are delayed until the Hume Reservoir is nearly empty.

WHERE DO YOU DRAW THE LINE WHEN A RIVER FORMS THE BORDER BETWEEN STATES?

The River Murray, about 20 kilometres downstream of Swan Hill, becomes two streams, separated by Beveridge Island, and rejoining after about 4 kilometres (measured by the median east to west length of the island). In terms of divided stream distances, the northern stream wanders for 11.2 kilometres, the southern for 5.7 kilometres, before they rejoin below the island. In the 1980s a farmer who leased the island from the Crown entered into a legal battle with the Victorian Government about payment of irrigation fees.[164]

Under the Water Act 1958 (Victoria), it is an offence to take water from the River Murray for irrigation without a licence or permit from the State Rivers and Water Supply Commission. The farmer was charged with illegally taking water for irrigation from the southern stream. His defence was that his land was in fact part of New South Wales, not part of Victoria, although he leased the land from Victoria, and the case went all the way to the High Court. New South Wales intervened to argue that Beveridge Island forms no part of New South Wales territory.

At the time when the Colony of Victoria was established in the mid-1800s, the northern stream, which follows the original watercourse of the river, was the main stream in terms of volume and width. From 1853, when the paddle steamer *Lady Augusta* navigated the southern stream in an 'experimental trip', the shorter length of the stream caused it to be used for navigation. Siltation since has decreased the flow in both streams but with more pronounced effect on the northern stream, so more water now passes down the southern stream. The island in between the streams has always been subject to periodic flooding and levees built over time have not completely solved the problem, so no permanent buildings exist on it and the lease is for grazing and other land use.

There had been problems about which State owned islands in the Murray before, and about which of the two streams into which the main river splits to accommodate an island was to be regarded as the river, and hence as the demarcation line between States. Pental Island had been the subject of a legal battle in the early days of settlement when rent was paid first to Victoria, then to New South Wales, then to both while legal argument was heard. In the Act of 1850 'the whole watercourse of the said River Murray, from its Source therein described to the Eastern Boundary of the Colony of South Australia, is and shall be within the Territory of New South Wales'.[164]

Back in 1873, the question arose whether Beveridge Island was part of Victoria or of New South Wales and was resolved by determining which of the streams, the north or the south, was actually the Murray River. A joint survey by officers of the Land and Survey Departments of Victoria and New South Wales determined that the northern channel was considerably deeper and discharged more water and it therefore was the river, and Beveridge Island belonged to Victoria.

The legal arguments involved in the 1982 High Court case were long and learned. It was held by the court that 'the Watercourse of the said River Murray' referred to in the 1855 Act was not the constantly changing stream of water but the contour feature within which the River Murray flowed and that, where the course of the River Murray constitutes the boundary between New South Wales and Victoria, the boundary line runs along the top of the southern bank with all territory to the north being in New South Wales. (There had even been a case in which a murder had been committed on the southern bank of the river in Victoria and it had to be decided which State should deal with the crime.) By determining the watercourse, and not the river as the intended meaning of the law in setting the border between States, the dynamics of ever-changing river morphology were recognised.

The Beveridge Island apellant lost his appeal to the High Court. Although within the watercourse, which would make an island technically part of New South Wales, legal argument resolved that when ownership by one State had been resolved by law in the distant past, the principle of prescription prevailed and Victoria owned the land.

Wind waves on a beach at Menindee Lakes. M.E.W.

Wind-tossed water, pelicans and drowned trees. The wide shallow lakes become choppy on windy days. M.E.W.

In addition to dams and weirs under Commission control, State authorities have built storages on tributary rivers. These include Glenlyon and Beardmore Dams in Queensland; Burrendong, Wyangala, Carcoar, Burrinjuck, Blowering, Windamere, Keepit, Chaffey, Pindari and Copeton Dams in New South Wales; and Eildon and Eppalock Dams in Victoria. Of all the Murray tributaries, only the Kiewa and Ovens Rivers in north-eastern Victoria have not been substantially regulated.

The Murray today is simply a well-regulated drain. It was studied in detail in *Listen ... Our Land is Crying* from the point of view of land use in the Basin, particularly in relation to salinity and rising water tables; the state of its wetlands; and its use for irrigation. In this volume other aspects of the Murray River are described to complete the picture.

The Darling is a river system starved and dying with its waters over-allocated for irrigation of cotton.

Salt—the menace which used to lurk beneath the soil, hidden at the water table—is now a nightmare, creating salinised deserts and unusable water. REG MORRISON

THE HEARTLANDS PROJECT
A joint project of the Murray—Darling Basin Commission and the CSIRO

Professor John Lovering, who retired from his post as Chairman of the Murray–Darling Basin Commission in 1999, addressed the National Press Club on 29 June. A press release was distributed at the presentation outlining *Heartlands*, and is used here to summarise the project, which is in Third Draft form and had yet to be finalised and funded in late 1999.

To quote directly from the press release (the emphases in bold type are mine):

Heartlands is a visionary project aimed at providing innovative solutions to Australia's fundamental and intractable land management problems. Sustained rural productivity is critical to Australia's economic well being, yet it is clear that without radical changes to land management, Australian landscapes can no longer cope. The scope of the problems confronting us is all too evident. We are seeing the **escalating salinisation of large areas of the landscape. Salt flow into rivers is rising, reducing water quality in the rivers that form a crucial part of our domestic, industrial and agricultural water supply. Woodlands now cover only about one per cent of their original distribution, and biodiversity is threatened**. The social, economic and environmental costs of these degradation processes are being counted in the Murray–Darling Basin, and elsewhere in rural Australia.

... The hallmark of this project is **an integrated approach, with inputs from science, economics and rural sociology.** The aim is to develop, test and deliver integrated solutions that will be practically possible and economically acceptable. To achieve this, *Heartlands* will not only tackle the range of biophysical processes underpinning potential adaptive management techniques; it will also address the economic costs and benefits of these landscape changes at the farm, local and catchment level. Community involvement is essential in order to assess the impacts on rural social structure of changed land and water management, as well as the practical capacity of rural communities to manage change.

Several MDB catchments will be selected, representing many of the typical land-use problems, together with a number of desirable features: an enthusiastic community, reasonable data history, and the likelihood of relevant land-use change. While dryland salinity is a significant focus, the project is expected to produce results applicable in broader natural resource management context.

The approach in *Heartlands* involves three related activities:

- *Developing* a scientific framework for predicting the hydrological, ecological and socioeconomic consequences of purposeful alterations to vegetation cover, especially through revegetation and reforestation.

- *Applying* this framework, in consultation with communities and land managers, in the design and implementation of both opportunistic and strategically planned land-use changes.
- *Measuring* and assessing the consequences of these changes in terms of environmental and economic impacts over the following years.

The expected outcomes from the project are:

- A significant impetus for change based on sound science, economic analysis, community involvement and an understanding of rural community needs
- Defined, tested and demonstrated strategies
- Quantification of trade-offs such as stream flow versus improved hydrology, water quality versus water quantity or farm forestry versus grazing
- Estimates of the extent and nature of change necessary for some defined natural resource changes (for instance, in dryland salinity or water quality)
- A high level of community understanding
- Systems of long-term monitoring

Extensive consultation with catchment communities and agencies has already resolved principal issues. Local communities have readily grasped the importance of the project and are eager to be involved. Other stakeholders, such as the relevant scientific organisations and government agencies, recognise that this is an opportunity to test and deliver large-scale innovative solutions. By building on individual projects already undertaken, such as the National Dryland Salinity Program and the National Land and Water Audit, and by securing the financial support needed for integrative research, the large-scale solutions to Australia's land-use problems can be delivered.

Information on the Heartlands Project can be obtained from the Project Coordinator, Andrew Bain. Phone 02 6246 5553, Fax 02 6246 5560 or E-mail Andrew.bain@cbr.clw.csiro.au

MOORNA: A PROPERTY IN THE SOUTH-WESTERN CORNER OF NEW SOUTH WALES AFFECTED BY ARTIFICIAL WATER STORAGE IN LAKE VICTORIA

LOCATION MAP:
LAKE VICTORIA AND MOORNA

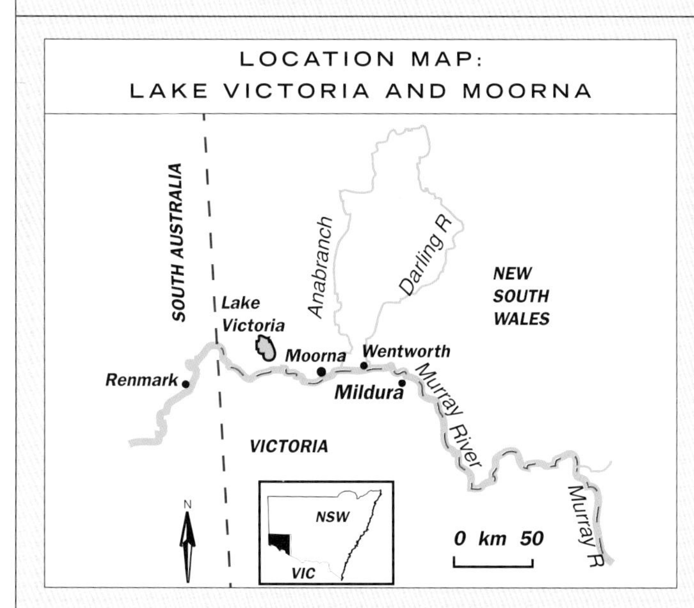

Moorna was established in the 1840s as head station of a group of sheep runs on the Murray River and the Anabranch of the Darling. In 1862 the holdings comprised 500 000 acres and easy access by river transport made it the shearing centre of the district, concentrations of 250 000 sheep resulting in the shearing season. Records show that rainfall since 1882 had averaged only 230 mm (9 inches) a year.

When the Moorna property was bought by the present owners, John and Annabel Walsh, in 1984, it comprised 27 108 hectares, of which 15 000 were freehold. The property at that time was showing ominous signs of salinisation. Like other Western Division lands it had suffered heavy grazing pressures in the early days of settlement and land degradation and reduction in carrying capacity had followed. Rising of water tables had occurred on a regional scale. (Full accounts of water table rise, salinisation of land and water in the Murray Basin are given in *Listen... Our Land is Crying*.)

In the case of Moorna, which lies between the major storage Lake Victoria (created in the early days of Murray River regulation) and the Anabranch, where dams and weirs had been built in the 1880s, the rising water table and the salt threat it imposed was greatly exacerbated by Lake Victoria's artificial alterations to the local hydrology. When piezometers were installed all over the property to monitor the depth of groundwater in order to initiate measures to combat salinity, it was found that the depth of the water table fluctuated according to the depth of water in the storage lake. The groundwater is twice as salty as sea water and was at an average depth of only 60 centimetres below the surface on Moorna. The extra fresh water in Lake Victoria and in diversion works into adjacent Frenchmans Creek was forcing salt groundwater into the landscape and also into the Murray River—a phenomenon well known to occur when wetlands connected to the river had substantial changes made to their hydrology, as in the Loveday and other wetlands which were used as saltwater disposal basins (see *Listen... Our Land is Crying*, pages 133–40).

The extra water diverted into Frenchmans Creek was affecting the two natural intermittent freshwater lakes on the property, Snake Island and Latina Flat, waterlogging and killing stands of red gum and changing the lakes from intermittent to permanently wet.

Moorna's chief income is wool, and low wool prices and the need for sufficient income to start salinity-control measures demanded a review of management practices. (Here, as everywhere else in the marginal lands, the only way farmers have been able to remain on their land and be financially viable has been by complete reassessment of their situation and practices and a realistic adaptation to working with the land and within the parameters set by it.) The percentage of lambs lost after weaning was unacceptably high at 25 per cent. It was found that the problem was their inability to cope with the high salt levels in the pasture while they were very young. So arrangements were made to agist them for six months in regions where salinity was not so high—costly but dramatically successful.

Because it had become necessary to augment the income from wool, it was decided to control the entry of water into the two small lakes on the property, draining them for several months each year and thus providing 405 hectares of lake bed for cropping. These measures are profitable and the wetlands are recovering now that they were not permanently waterlogged. River red gums are regenerating. Care is taken in the cultivation of the lake bed to minimise problems. Safflower, barley, oats, wheat and canola are organically grown, and chick peas are used in rotation to replace nitrogen.

In 1994, the full storage height of Lake Victoria was lowered by 3.8 metres to carry out repairs on regulation structures. The lower level exposed a number of Aboriginal burial sites which had been revealed by the erosion of the lake bed and shores. In consequence, management practices of Lake Victoria have been altered to accommodate these culturally significant sites. The lower water level this required has affected Moorna by decreasing the availability of water for lake bed cropping and wetland health. A total dilemma exists—full to overflowing lake means more salinity because groundwater level rises; low lake levels mean no irrigation water and no lake bed cropping and no money to finance rehabilitation.

The WEST 2000 strategy was launched in 1997 by Commonwealth Department of Primary Industry and Energy and State Departments of Agriculture and Water Conservation, with considerable community input, to aid farmers in the Western Division. It funds projects in 'economic development, structural adjustment, natural resource management and social issues in an integrated way, drawing on relevant government programmes such as the National Landcare Program and Rural Adjustment Scheme'. It has a dual focus—rural assistance and forward planning.

West 2000 aims to be the financial catalyst to help Western Division communities to deal with the problems confronting them:

"... drought, high feral and native animal pest populations, encroachment of woody weeds, and low commodity prices for wool, that have significantly increased the debt burden of many landholders. Landholders in the Western Division have limited opportunity to earn off-property income, and many properties in the region are too small to be profitable, thereby reducing opportunities for adjustment. The situation is compounded by few opportunities for diversification, the need to carefully manage total grazing pressure, areas of unstable soils and low, unpredictable rainfall."

Moorna has become very much involved in the scheme, its planning and local application. One of its major problems today is the build-up of kangaroo numbers, which negates the care taken to monitor grazing pressures and avoid over-grazing.

It takes a special brand of courage and determination to remain on the land and try to understand and work within the bounds set by the environmental problems that are all man-made and all based on hydrological changes resulting from European land-use and water-use practices. Much of the southern and western parts of the Murray Basin cannot avoid increasing salinisation over the next 50 to 100 years, even if all land-management on the surface is best practice from here on, because the effects already built into the hydrological systems will have to work their way through. The Walshs' fears that the whole of Moorna may be salt within 30 years may not be far from the truth.

THE DARLING RIVER SYSTEM

The Darling River system and cotton production under irrigation have become synonymous. The rapidly escalating cotton industry and its enormous water requirements present the greatest dilemma facing water and land management in Australia. The cotton industry is, on the one hand, a shining example of high-tech farming and, increasingly, of sophisticated agri-business; on the other, the nature of our rivers with their patterns of irregular flow and, on a world scale, very meagre capacity, is completely incompatible with the demands imposed by the industry.

The continuing saga of the extraction of massive amounts of water from inland rivers to satisfy the escalating demands of the irrigation industry is Australia's most serious, and ultimately potentially most disastrous, water-related issue. It is a battle between two essentially irreconcilable attitudes to land use. At its centre are socio-economic structures and attitudes which are now global, and which are antagonistic to the very nature of our unique continent. ('Globalisation', 'rationalist economics', the power of corporations to influence government, all for the glory of the almighty dollar, are destroying the environment and its life-supporting soil and water resources worldwide—robbing the Future for the sake of a highly profitable Present.) The lack of a vision and the will of government to enforce it so that today's activities do not impoverish future generations are the fundamental problems. The inherent greed of the individual, and in particular the agri-business corporations, has to be limited so that land and water resources, on which both the short- and long-term survival of the nation depends, are not degraded and destroyed.

On one side of the battle are the very vocal and influential big dollars and a nation's misguided building of a major sector of its economy on a totally unsustainable operation which makes huge money now and will ultimately leave desertified environments.

A chemical-loaded cotton field near Trangie. The irrigation water carries a toxic cocktail. REG MORRISON

Irrigation canals run into the distance, carrying precious water (of which a great deal evaporates) and chemicals through a landscape modified by technology to an astounding extent. And the river from which the water is drawn is out of sight, out of mind. REG MORRISON

A huge dam, plastic-lined and 10 metres deep, back of Bourke—insurance for the cotton farmer against the vagaries of a river which does not always run and a climate which is unpredictable. M.E.W.

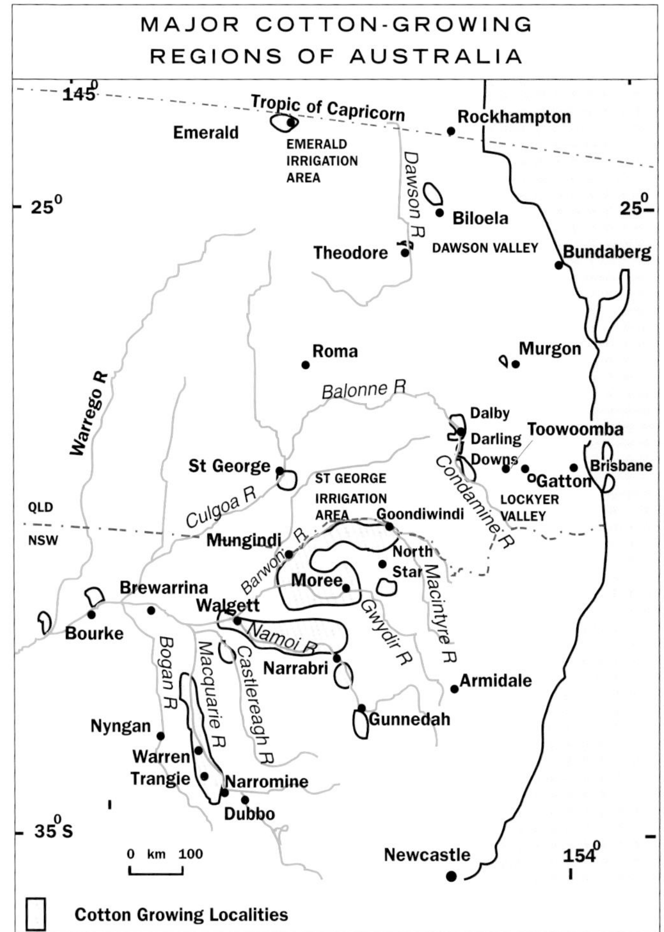

After Arthington[233]

River red gums, roots exposed when banks slump when draw-down of water being extracted is too rapid. M.E.W.

On the other are the comparatively voiceless and without-political-clout environmental issues which are so easily ignored in the headlong rush to 'increased productivity'.

The intractable nature of the problem is brought home to any traveller who visits towns like Bourke. There, the cotton industry has brought amazing change, immediately visible to anyone like me who visited the town even five years ago when it appeared a run-down, depressing place with a major problem with its Aboriginal population, unemployment and little hope of a future. Transformed, in a mid-1999 visit, it has come alive and cleaned up its act, and its neat and tidy houses and streets speak of civic pride and a town rejuvenated. The praises of cotton and the other irrigated industries that have wrought the change are sung loud by guides who join coach tours to show visitors the district.

One can only marvel at the technological wizardry that produces the laser-levelled fields that stretch to the horizon in all directions with regular rows of recently harvested cotton stubble; the enormous covered bales of cotton waiting to be collected and taken to the gins; the unbelievably large dams brim-full of water; the networks of irrigation channels; the ubiquitous vineyards with their efficient watering of each vine by buried plastic pipes, reducing water wastage; orchards similarly reticulated; even computerised water distribution over whole properties ... all triumphs of engineering, best-practice farming and land management and justified by prosperity, job production, the contribution of billions of dollars to the national economy and quasi-sustainable, at least for long enough to pay off the huge investments and make a huge profit for the operators.

But what of the river that supplies the water? The downstream impoverishment, where floodplains are dying, wetlands are disappearing, and the river itself is doomed, its ecosystems deprived of the sustenance that comes from floodouts and wetlands. The slumped banks, roots of giant river red gums exposed and clinging precariously; the turbid brown water which does not allow light to penetrate and promote the plant life whose photosynthetic activity fuels the food chains; the pesticide, herbicide and defoliant chemicals and the fertilisers which poison the water; and the destruction of natural rhythms of flood and low water which are intricately connected to life-cycles of animal and plant life in the river, floodplains and wetlands? And what of the livelihoods of dryland farmers and graziers downstream whose operations are impaired, and of towns and properties whose water supply is river-dependent?

The Bourke tourist guides who sing the praises of the irrigation industry justify the using of the water and the appalling wastage from evaporation in open dams and drains (which may at times represent 80 per cent of the water in irrigation channels) by saying it is better used for cotton and the rejuvenation of Bourke than just sent to evaporate in the Menindee Lakes. (What they don't say is that that option is also a man-made one and that the need for storage there is because of the impoverishment of the lower reaches of the Murray River by the over-regulation of that river.)

And what of the land itself? The soils which lose their structure and ultimately can take no more fertiliser, becoming acid and unproductive; the organic and microbiological component, meagre to start with in this driest of all continents, decimated and destroyed. And what of the pests which develop tolerance to all the chemicals and become rampant, or of the effect of those chemicals that enter the human food chains on human health? At the end of its cotton-growing usefulness, land capable of being returned to pasture would have to endure a long wait before grazing was safe, because endosulfan and other poisons remaining in the soil contaminate beef grown on contaminated pastures. Already there is evidence of the great distance to which over-sprays can contaminate land far from the nearest cotton property. Near Dubbo, one property more than 15 kilometres from the boundaries of the nearest cotton farm has suffered endosulfan contamination of its cattle and rejection of its beef for export as a result of spray drift onto its land, and as a result has now applied for a licence to grow and irrigate cotton—adopting the principle that if you can't beat them you join them.

Some specialist scientists, who know what is really happening in this regard and who admit that there is only a limited life for the irrigation enterprises because of the problems that emerge increasingly after a honeymoon period, admit that these concerns are not exaggerations of the situation or rabid greenie ravings—they are well-founded and factually backed. It is a matter of national shame that scientists in government-funded organisations are gagged and not allowed to publicise the unpalatable truths or rock the boat. Their funding or jobs depend on their compliance.

THE BARWON–DARLING RIVER SYSTEM

The Barwon–Darling is a large semi-arid river system draining the inland slopes of the Eastern Highlands.[165] The 650 000 square kilometre catchment is of low relief, with elevations ranging from 1000 metres in the headwaters to 50 metres in south-western New South Wales, and with 60 per cent of the area at less than 300 metres. The Barwon–Darling, the principal river, receives major contributions from the Border Rivers—10 per cent of its water comes from the Gwydir; 25 per cent from the Namoi and 35 per cent from the Macintyre (making 70 per cent of the water). The Condamine–Balonne–Culgoa–Bokhara complex provides about 20 per cent, with minor contributions from the Bogan and from the western tributaries, the Warrego and Paroo, completing the picture.[166]

Catchment and drainage evolution in the Barwon–Darling Basin is associated with the long-term tectonic and landscape evolution of south-eastern Australia. The headwater drainage features are modified from Gondwanan landscapes. During and prior to the Jurassic, drainage was directed north-west to the Eromanga–

THE BARWON—DARLING RIVER SYSTEM AND ITS DAMS

The Darling River at Brewarrina in October 1994, reduced to green-algaE-infested pools. REG MORRISON

The Barwon at Walgett—dry—not sending much water into the Darling. REG MORRISON

The debris-strewn bed of the Barwon River (which is the upper Darling) at Walgett. The Darling is entirely dependent on the water it receives from its headwater and tributary streams, as there is virtually no run-off in the arid lands through which it runs. REG MORRISON

Surat Basin from a continental divide on the New Zealand Subcontinent, 200 to 300 kilometres to the east and parallel to the present coastline.[167] Components of this ancient north-north-west flowing river system are evident in the Macquarie and Bogan Rivers. Rifting to form the Tasman Sea between 80 and 60 million years ago resulted in the formation of a new continental margin and the formation of the present drainage divide (the Great Divide). When the Murray Basin started to sink about 60 million years ago, the Canobolas Divide developed between the Murray and the Eromanga Basins. This divide effectively separated the Murrumbidgee–Lachlan and Macquarie catchments and the drainage systems of the Murray and the Barwon–Darling.

The upwarping of the eastern margin, and of the Canobolas Divide, changed the flow direction of the Barwon–Darling system from a north-east to south-west flowing drainage system. The drainage system is ancient, dating from early in the Tertiary (about 65 million years). It has been suggested that further upwarping of the Great Divide in the northern extremities of the catchment resulted in the development of the Paroo, Warrego, Culgoa and Maranoa systems, which would make these the youngest drainage networks in the Barwon–Darling catchment.[168] Tectonism in the region during the Cainozoic has modified drainage, particularly the north-east to south-west trending Darling and Cobar structural lineaments, which are fault lines which constrain the river. The Darling lineament constrains the Darling downstream of Bourke and the reach upstream of Walgett is constrained within the Cobar lineament.

Three broad river reaches can be identified along the Barwon–Darling system in New South Wales:[165]

- The river **between Mungindi and Walgett,** where the floodplain and channel are constrained within the Cobar lineament. Floodplain development has been restricted, particularly along the western edge of the valley. The river flows in a general south-west direction following the strike of the lineament. The geomorphology of the floodplain–riverine environment is also affected by the Macintyre–Gwydir fan complex which extends from the eastern highlands westwards. Wetlands in this sector are small and are associated with former courses, such as abandoned anabranches, avulsions and cut-offs, and are on the eastern margin of the valley. The in-channel environment is variable with pools and rocky reaches.

- From **Walgett to Bourke** the river is not constrained by a lineament but is influenced by the megascale alluvial fan morphology emanating from the Gwydir, Namoi and Macquarie Rivers. The river flows for 450 kilometres westwards, through alluvial plains, is highly sinuous and has many wetlands. Former courses show the active and unrestricted nature of the system in this reach. However, the planform has not changed significantly since the 1850s, so the palaeochannels relate largely to more pluvial times in the past. Several large tributaries join the river in this reach—the Bogan, Bokhara, Castlereagh, Culgoa,Macquarie, Namoi and Narran—resulting in greatly increased channelcapacity downstream. Near Bourke the channel reaches 60 to 80 metres in width and depths of up to 20 metres. Channel cross-sections are complex, with the presence of three or four bench-like surfaces.

- From **Bourke to Wentworth** the river is constrained by the Darling lineament. Below Wentworth, where the Darling enters the Murray Basin, it is still constrained within the lineament. The river flows in a general south-west direction. In general, the channel below Bourke is deeply incised with a canal-like form and depths of up to 25 metres and widths between 60 and 80 metres. The alluvial valley and floodplains are minimal because of the lineament and the influence of the east–west trending Canobolas Divide, which creates a large 'choke-like' feature for the drainage of floodplains from the upper Barwon–Darling catchment. Many smaller fault lines and structures also affect this reach, and many of the larger anabranching wetlands are structurally controlled (like Tallywalka Creek). Below Tilpa, the Darling River is actually flowing through sediments of the Murray Basin, and benches reappear in cross-sections.

The old lock and weir at Bourke: (above) water overflowing in May 1999; (below) a toxic algal bloom like thick pea soup in the same place, just below the weir in October 1994. M.E.W. AND REG MORRISON

Eleven major dams have been constructed on the headwater streams of the Barwon–Darling river system; along with the massive Menindee storages on the anabranch 200 kilometres above the junction with the Murray. In addition, 17 major weirs exist along the course of the river, impounding up to 40 per cent of its length. Off-stream storages which pump water from the river are uncounted. Bearing in mind the fact that very little, and decreasing, run-off occurs as the rivers flow westwards into increasingly arid country, the significance of the extent of regulation is always under-estimated. The Darling itself is an arid-land river, typically running through mallee and spinifex landscapes. The whole catchment is subject to climatic, and therefore flow, unpredictability orchestrated by ENSO, and the rhythms of life and breeding cycles of the instream and floodplain biota are intricately connected to this variability.

Most of the dams and weirs were constructed before the Environmental Planning Act of 1979, which requires formal environmental impact assessments. As a result, it is not at all surprising that the current level and manner of water regulation and use does not reflect the ability of the environment to sustain the changes to the stream flow regimes that have resulted. Up to the present, the missing concepts in such management as there has been for sustainable use are the understanding that floodplains and wetlands are inseparable parts of healthy river systems; that it is not enough to allocate a certain proportion of flow, that variability is essential for a healthy river; that maintaining constant levels, as occurs with weirs, is not in the best interests of the river; that with reduced flow overall the removal of pollutants is incomplete; and that people downstream of the dams and regulators and of the irrigation schemes, in particular, also depend on the river water and have rights.

For as long as we think of rivers merely as channels carrying useful water for our exploitation, and not as living entities which

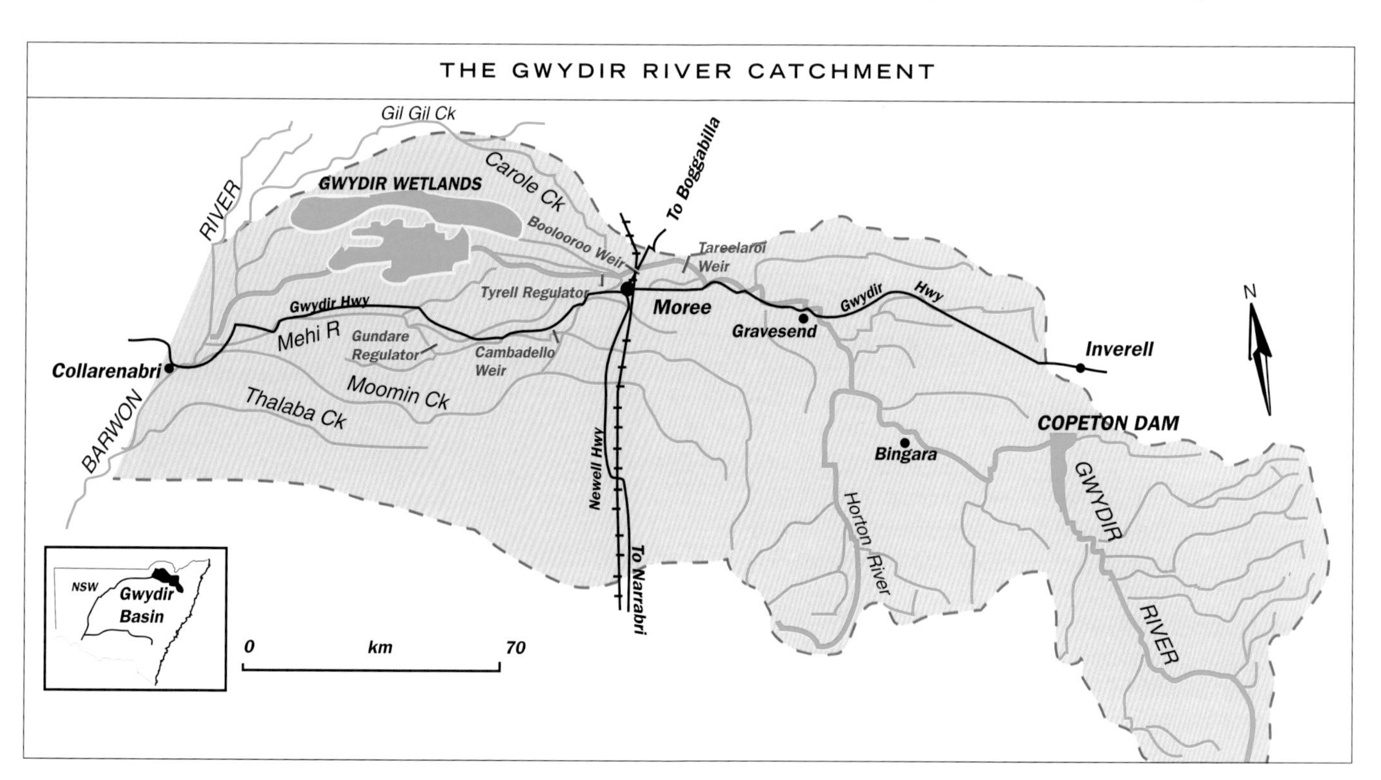

comprise finely balanced systems as a result of co-evolution with climate and landscape through great lengths of geological time, there is little hope that they will not become, simply, drains.

THE GWYDIR RIVER CATCHMENT AND WETLANDS

The Gwydir River rises in the Northern Tablelands of New South Wales in the vicinity of Armidale and Guyra. The Horton River, a major tributary, rises in the Nandewar Ranges and joins the Gwydir downstream of Bingara. The land in the vicinity of Moree and east of the Barwon River comprises a vast alluvial floodplain. The low gradient of the floodplains west of Moree has caused the Gwydir River to distribute into various effluent streams which include the Mehi River, Carole Creek, Moomin Creek and the Gingham and lower Gwydir Watercourses (which are known collectively as The Watercourse). Prior to river regulation in the 1970s these watercourses formed huge terminal wetlands which were a breeding habitat for waterfowl from all over Australia. Water only left the Gwydir Valley in very big flood events.

Downstream of Moree, an obstruction of logs, debris and silt, known as the Gwydir Raft, effectively blocked the original river channel from the early 1900s onwards. The Raft progressively accumulated—a product of land clearing and catchment alteration by European land use upstream—and its head is now situated some 30 kilometres east of the point of original formation. Its progression significantly altered the distribution of floodwaters between the Gingham and lower Gwydir Watercourses. During the earlier part of the twentieth century most floodwaters flowed to the south, into the lower Gwydir Watercourse, and water entered the Gingham Watercourse only during major floods. The situation reversed in the 1940s when water flowed more readily to the north, and the Gingham Watercourse now receives the majority of flood waters.

The Watercourse contains one of the most extensive wetland systems in north-western New South Wales, and its national and international significance as a waterbird habitat has been recognised since early this century.[236] Massive breeding events took place as recently as in the summers of 1995–96, 1996–97 and 1998–99. The Watercourse is also the focus for a multi-million dollar agricultural

IRRIGATION AND THE FATE OF THE GWYDIR WETLANDS

Information on problems faced by wetlands downstream of major water extractions for cotton irrigation is provided by farmers in the Gwydir wetlands. Here we have a unique response to problems—privately owned land being managed under Ramsar Convention agreements which give them World Heritage status and (hopefully) protection. The following accounts of two of these farmers' observations are presented with minimal editing so that they remain first-hand accounts.

Howard Blackburn of Crinolyn on the Gingham Watercourse supplied this account:

> I came to the Moree area in 1982 and bought the watercourse property Crinolyn which is 85 kilometres west of Moree joining Watercourse Road and the lower end of the Gingham Watercourse. It was very dry at the time but useful cattle feed could still be found in the form of dry water-couch and other dry grasses. Cattle still had a good time here. The irrigation industry was well established and expanding. Local watercourse graziers were told that 80 per cent of the natural flooding would be retained with the advent of irrigation and river regulation. They were gullible enough to believe it and no written documents were forthcoming.
>
> It rained in 1983 and we witnessed an amazing change as the entire area became inundated with floodwater, not deep, fast-flowing, destructive flooding but rather shallow, slow-moving and highly productive in terms of stock feed and the wildlife that emerges from this truly amazing area.
>
> As the irrigation industry expanded beyond the capacity of its allocation water from Copeton Dam, it was given access to so-called surplus water, or water that entered the river downstream of Copeton Dam. This was the really destructive element for the Gwydir wetlands as now the irrigation industry had access to all but stock and domestic allocation water and uncontrollable floodwaters. On-farm water storages began to appear and by 1990 natural low river flows into the wetland had all but ceased and all river flows that could be extracted were extracted by the irrigation industry. Water was pumped at every opportunity, even off-season, and put into on-farm storages and sometimes left to evaporate if not required later.
>
> Gwydir Valley on-farm storages can store in excess of 350 000 megalitres of water in 1999. Despite the Gwydir water resource being grossly over-allocated, there is still new development taking place today.
>
> A survey of watercourse grazing properties several years ago indicated stocking rate reductions of up to 70 per cent since river regulation. This kind of income reduction has caused many to change their enterprise to a more mixed-farming program, and cereal cropping has become popular. The problem with cereal cropping in an ephemeral wetland is that after the land has been cleared, the wetland wildlife habitat has gone, even if it does get wet again.
>
> This problem is currently being exacerbated by those who wish to dry out the Gingham Watercourse to control water hyacinth. As soon as it is dry enough, someone will want to clear and cultivate it. It should be allowed to dry out in the natural wetting and drying cycle, but I don't believe it should be dried out unnaturally. We have seen enough of that scenario in the past, and no doubt the future holds our fair share of natural droughts.
>
> In late 1995, the Gwydir Environmental Flows Committee agreed to a few alterations to river flow rules. The area was drought-declared at the time and the watercourse had also endured a man-made drought because

Where the Gwydir runs into the Ramsar wetlands. BRUCE SOUTHERON of river regulation for years. It was agreed that the first 100 000 megalitres of the next substantial flow should go to the wetlands on a once-only basis. Three other points were agreed to:

- The watercourse be given priority of flows up to 500 megalitres per day
- Tributary flows to be shared between irrigators and watercourse on a 50:50 basis when they exceed 1000 megalitres per day in excess of the in-stream requirements
- The Environmental Contingency Allocation (ECA) of 25 000 megalitres to be held in the Copeton Dam.

A few short weeks after these rules were agreed to, it started to rain and triggered the huge bird breeding events of 1996. It was a truly amazing event.

Since 1995, there has been ongoing lobbying and criticism from the irrigation industry about so-called wasted water in the wetlands. The battle for water continues and will get more intense as demands increase on a limited resource.

Four landholders (including myself) have offered portions of our land for Ramsar listing in the hope this may be of some use in the ongoing struggle for a fair share of water. One can't help wondering what future there is for wetlands when we see [the] many areas in this country that are being ruined in a very short period of time. One that springs to mind is the Narran Lake in north-western New South Wales which is Ramsar listed and fed mainly by water from Queensland's Balonne River system which is undergoing massive irrigation development. This can mean only less water for the lake—and State and Federal governments don't seem to be doing much to help the plight of the lake.

Bruce Southeron of Old Dromana recorded his account of the lower Gwydir channel and wetlands:

I came to the watercourse, west of Moree, in 1954 as a horse breaker... The lower Gwydir Channel was known then as the Big Leather Channel and this name has been chosen for our Ramsar block. We own and run cattle on the land where the river breaks into a floodplain. We have watched a hungry cotton industry kill much of this land...

In the fifties, the country was still getting over the war years and a lot of fencing was bad or non-existent, so a lot of miles were ridden to muster stock. The cattle were fairly wild and some exciting rides were had trying to beat the mob into the sag. And if they made it, with wild pigs and nesting waterbirds—mainly duck—jumping out from your horse's hooves, it tested the skill of both horse and rider. (Sag is marsh club rush or *Bolboschoenus fluviatilis*, and we are told the largest area of it in New South Wales, and probably in Australia, is here on the lower Gwydir. But it is still b— sag to us when the mob disappear into it.) Sag has suffered from the drying out of the wetland.

I estimate that there was 10 000 acres [1 hectare = 2.7 acres] of it, at least. Now only 2000 acres remain, but with the new water policy it is fighting back and I am confident we can preserve what is left of it. The wetland carried on reasonably well until the early 1980s even though the Copeton Dam held water from the mid-seventies. We had floods and droughts, which is the nature of this country.

There was approximately 100 000 acres that flooded—regularly close to the river, and once every two or three years further out. The whole area was grazing both sheep and cattle and a flood meant twelve months' feed, if we didn't over-graze it or grasshoppers or fire didn't get it. The grazing industry looked for a flood and the bountiful food it produced... but mosquitos were so bad they could kill a dog left on a chain overnight and snakes invaded dry areas and homesteads...

At the first sign of droughts State-wide, large mobs of sheep and cattle moved into the watercourse country...

In the late seventies, Copeton Dam started to hold water and, with the irrigation industry slow to commence, it soon filled. The spillway started to erode and the water level was lowered to enable repairs. At the same time a river improvement program was under way and work was being done on streams in the catchment. The water was sent into the wetlands instead of down the streams and for the next four years 9000 acres of the property I managed was permanently flooded, becoming a paradise for birds, but killing thousands of coolibah trees, some hundreds of years old. The grazing industry suffered with the land too wet for sheep and cattle, the latter suffering worms and copper deficiency. The country dried out in 1979 and then the real rot set in... We began to see the first effects of what Copeton Dam had in store for us when the irrigation industry started to take water out of the river system and that coincided with a run of very dry years. With drought becoming the norm, large areas close to the river produced very little feed and the lippia weed invaded.

From then on, the irrigation industry started to use water at an alarming rate, and so the wetlands receded. With the then Water Conservation and Irrigation Commission pushing the irrigation button and handing out licences to the extent that the capacity of the Copeton Dam was over-allocated by approximately 100 per cent, an impossible situation developed and the river could not supply the amount of water necessary to satisfy a hungry and developing industry. The powers that be, in their wisdom, then allowed 'on-farm' storage and I don't think any of us envisaged the size of their dams and the pumps that filled them, so that other than at times of a large flow in the river, very little water ever got past the pumps.

I am quite sure this is what put the death knell on the wetlands. At the same time we were entering a very dry period... Between droughts and poor stock prices and a falling wool market, people started ploughing more country for wheat... When debt is high and banks are pushing for their drop of blood, people become desperate.

As the wetlands dried out, lippia weed invaded, and it remains a major problem. When only 2000 out of 100 000 acres of wetland remained in 1995 the EPA looked at the situation in horror and the new agreement came into being, following a change in State Government at this point... It caused a roar throughout the irrigation industry... but I am confident that the agreement can work and we can preserve what is left of the wetlands. That is why we have become involved in Ramsar listing...

When I stated that cotton was killing our land, I was referring to the extraction of water from our river. The Gwydir is only one such river... We do not know what damage the sprays are doing. At the present time we are caught up in endosulfan trouble. Cotton does not adjoin our land on any boundary, but we had very high endosulfan readings in cows that were fat-tested, and in grass samples. The endosulfan spray has drifted at least 15 kilometres from the nearest cotton. And if it is drifting so far, what of other sprays being used [and not tested for]?

Up until now the cotton industry has bulldozed and bluffed its way through its problems with money its only interest, and at the expense of the whole nation... and while it claims a good neighbour policy, we are yet to see it... But the irrigators are beginning to talk to the cattle people—something that has never happened in the past—and a working agreement has developed which can only be good for the land and the district, so I have hopes for the future.

industry. A grazing industry has existed since first settlement in the 1830s. There has been extensive land development for dryland cropping in the last 25 years.

The completion of the Copeton Dam in the upper catchment in 1976 enabled the establishment of an irrigation industry on the lower Gwydir River floodplain in the vicinity of Moree. The subsequent construction of weirs and regulators allowed water that traditionally flowed into the Watercourse wetlands to be diverted into the Mehi River, Carole Creek and Moomin Creek for irrigation. Approximately 90 000 hectares of land have now been developed for irrigation in the Gwydir Valley. The Watercourse wetlands have contracted in area from around 200 000 hectares to a present area of about 10 000 hectares because of the diversion of water for irrigation.

Since river regulation, land use along the Gingham channel has been changing as landholders switch their operations from largely grazing enterprises to farming and grain production. This has been partly brought about by droughts, including a five-year period beginning in the late 1980s when the irrigation industry demanded all the available water. Thanks to the efforts of the present New South Wales Government's more equitable water-sharing arrangements, a reduced but very viable wetland has seen the return of large breeding flocks of colonially nesting waterbirds which feed over the entire area.

However, the dry years took their toll as some landholders cleared the floodplain of green and dead timber and vegetation including coolibah, belah, river cooba, lignum and natural pastures. Habitat areas were drastically reduced, in some cases right into the heart of the Watercourse. Most landholders have welcomed the return of flooding and a more equitable water regime. But those who have farmed the bed of the Watercourse in the dry years are now finding it difficult to accept that it is no longer an option.

Some of the more militant farmers are now constructing illegal banks and earthworks to divert water away from its natural flow path, and prevent it from entering their cropping areas. There have been other instances where floodwaters have been illegally harvested, denying adjacent landholders access to the benefit of natural flooding. The area is a designated floodplain and all above-ground structures are supposed to be licensed and to have gone through due process. The earthwork structures of all sorts are creating problems in areas where water does not usually flow and a host of new difficulties have occurred as a result. The Department of Land and Water Conservation has been slow and reluctant to act even though the area is a designated floodplain.

An insight into the whole situation, including the impact of irrigated cotton on the rivers and wetlands, is provided in the personal accounts of two farmers' experiences.

The Horton River

John and Pam Hall of Moreena, near Bingara, have supplied information on the Horton River from their time living on its banks since 1954, on which this account is based.

The farm buildings are on top of a high natural levee bank next to the Horton River which has cut a deep trench through the layers of river gravels and alluvium that tell a history of alternating high energy and quieter phases which probably relate to climate changes in the Late Pleistocene but which have not been dated. Deep downcutting occurred in the major flood of 1964, and another memorable flood occurred in 1984. Half-way down the walls of the trench is a terrace on which sediment is deposited as the modern river falls during floods. The junction of the Horton with the Gwydir is about 14 kilometres away and, during floods, water backs up in the Horton and some bank erosion results. Pam Hall explains that 'the Horton "surfs" like Hawaii when it is rising, but levels out at its peak; it falls like pulling out the bath plug until "half a bank", after which it falls slowly'. It only deposits sediment while it is falling slowly at 'half a bank' so during major floods, when water ran under farm buildings, only water and not mud had to be contended with.

On the whole, the river seems to be in reasonable health and tree planting (some funded by Greening Australia) has been undertaken to stabilise banks.

The water in the Horton was clear in 1954 and a healthy population of catfish, yabbies and water-scooters lived in it; waterbirds were abundant—duck, wood ducks, teal, chestnut teal, spoonbills, herons, cormorants, darters, dotterells, etc. Catfish numbers have since been seriously reduced and replaced by European carp but, fortunately, lately there have been increased numbers of Murray cod. Waterbirds are now few and far between, possibly because they have gone to the large water storages in the cotton-growing areas. The turbidity of the river has increased through time and increased cultivation of the land has resulted in siltation.

THE COOLIBAH WOODLANDS OF THE MURRAY–DARLING BASIN

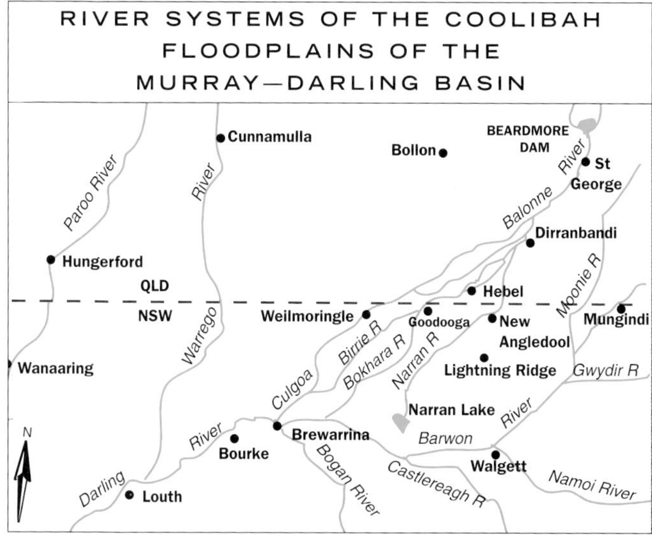

After Freudenberger[169]

'Coolibah—sometimes a wetland, usually a woodland.'[169]

Coolibah woodlands (*Eucalyptus microtheca*) are widespread on floodplains in the north of the Murray–Darling Basin. They demonstrate the important principle, and a hitherto largely neglected aspect, of river management—that floodplains are in fact an important and integral part of Australia's river systems. They have high conservation and production values.

The coolibah floodplains have been extensively grazed by livestock which consume primarily native forages, since the early days of settlement. Over time, a large proportion of the woodlands and associated native grasslands has been cleared for dryland and irrigated cropping and intensifying production uses increasingly threaten the survival of the natural ecosystems.

The native biodiversity, including the coolibah, relies on

intermittent flooding, which has been reduced in frequency, duration and extent due to flow regulation and abstractions. The areas of permanent and intermittent wetlands on many coolibah floodplains have been significantly reduced.

Water abstractions also threaten the economic viability of downstream landholders who rely on flooding for increased pasture and livestock production in this semi-arid environment. Clearing in the eastern portion of the Murray–Darling Basin has left small and fragmented patches of coolibah with a highly disturbed understorey.

The coolibah is an example of the products of droughts and flooding rains—it requires floods to regenerate, long dry periods to survive, and prolonged inundation kills it. There have been only four major recruitment events (times when new plants become established) since pastoral settlement—in the 1890s, mid-1950s, mid-1970s and in 1983—along the Balonne–Culgoa floodplains.

Three subspecies with distinct distributions are recognised: subspecies *coolabah* on blacksoil floodplains east of Wilcannia towards Narrabri and north; subspecies *excerata* on floodplains north of Moree and into southern Queensland; and subspecies *arida* along sandy or gravelly creeklines north and west of Menindee. Scattered populations of black box (*Eucalyptus largiflorens*) also occur on black soil floodplains and become more prevalent along the Darling floodplain south of Louth.

Coolibah woodlands and associated grassy floodplains cover 14 per cent (4.4 million hectares) of the Western Division of New South Wales and 770 000 hectares of coolibah floodplain occur in the Channel Country of Queensland; more than 1 million hectares through the mulga region of south-western Queensland; and 115 000 hectares through the brigalow region further east. In the northern reaches of the Paroo and Channel Country, coolibah is replaced by yapunyah (*Eucalyptus ochrophloia*) or grades into gidgee (*Acacia cambagei*) in the mulga region.

Coolibah woodlands are irregularly flooded by the northern river systems of the Murray–Darling Basin, including the Paroo, Warrego, Barwon–Darling, and the Balonne (which forks into the Culgoa and the Ballandool, Birrie and Bokhara, which flow intermittently into the Darling above Bourke. The Balonne also forks into the Narran which ends in a terminal lake system). The floodplains are remarkably flat. The lower Balonne has an average fall of 1 metre over 2 to 3 kilometres. Coolibah woodlands also extend east up the lower reaches of the Namoi, Gwydir and Moonie Rivers.

Great variability is seen in the flow of rivers and in the duration and extent of flooding. For instance, the Barwon–Darling river system had flow events sufficient to inundate 50 per cent of wetland areas of various reaches only twice at Walgett, three times at Bourke and six times at Wilcannia over the twenty-year period 1972–92.

Production value of the floodplains

The coolibah floodplains of the Murray–Darling Basin have been highly productive semi-arid rangeland for the past 150 years. Ninety-six per cent of the Balonne Shire in Queensland is grazed by 1.3 million sheep and 79 000 cattle. Average stocking rates are one sheep or dry sheep equivalent per hectare. (In dry mulga land to the west the stocking rate is only 0.1 dry sheep equivalent per hectare). About half of the floodplains in the lower Balonne are grassland. Flooding is essential for productivity—83 per cent of the area is flooded one in five years (1.38 million hectares). Such a flood is worth $143 000 per enterprise over three years; minor floods, covering only 20 per cent of a property for one to two weeks, add about $74 000 per property. The value of a flood stems from increased and sustained pasture production, allowing a 44 per cent increase in stocking rate for sheep and 137 per cent increase for cattle in the year after a flood.

Rabbits have not been a major problem in core areas of coolibah floodplains of the Barwon–Darling and Balonne–Culgoa systems due to their clay-rich soils, which are poor substrate for burrows. Feral goats are also not abundant on the black soil plains, but both species are a problem on the more arid floodplains of the Warrego and Paroo with their friable red soils.

Dryland and irrigated cropping are also dependent on and responsive to flooding. Floods of 1983, 1988, 1989 and 1990 generated an increase in cropped area about 70 per cent greater than that cropped in 1983 in Brewarrina and Walgett Shires—mostly opportunistic dryland and lake bed cropping.

Irrigated cropping has increased in the last twenty years: now 4 per cent of the lower Balonne floodplain is irrigated; 10 000 hectares of cotton are irrigated in St George Irrigation District with water from the Beardmore Dam; 11 000 hectares are irrigated with water taken from the Balonne and stored off-stream; more irrigated cotton is being grown on former coolibah floodplains near Walgett and Bourke; and extensive cropping has been developed on floodplains of the Warrego and Paroo.

The coolibah floodplains are on relatively fertile alluvial soils with a high clay content that cracks and self-mulches as it dries. They are resilient soils, retaining their productive potential under heavy grazing, though the species diversity of grasses like the Mitchell grasses has decreased, particularly in the eastern woodlands. Decline in the best grasses is also linked to flood frequency and soil processes. Kangaroo densities are high throughout the floodplains, representing a large proportion of grazing pressure in many areas, contributing to over-grazing and posing considerable management problems.

Conservation values

More wetlands of national importance occur on the Paroo and Darling than on any other river system. Currawinya Lakes (Paroo system) can support up to 250 000 birds, including 10 000 endangered freckled ducks. Narran Lakes support the largest breeding colony of straw-necked ibis. Coolibahs are important for tree hollows. Floodplains of the New South Wales reaches of the Birrie and Culgoa have 19 mammal species, 112 birds, 23 reptiles, 3 frogs (45 per cent of the mammals and 25 per cent of the birds depend on hollows). Carp, foxes and feral pigs are the most significant vertebrate pests of the coolibah floodplains.

The native flora is diverse, and introduced weed species are not widespread, except for lippia (*Phyla canescens*) which now affects 300 000 hectares of floodplain grazing country in the Murray–Darling Basin, particularly along the upper Condamine, lower Gwydir and lower Macquarie Rivers. Reduction in flood frequency and in extent and duration of flooding has favoured its rapid spread since its first appearance on the Condamine in 1944. Inundation for a month kills it, but major floods are far less frequent these days.

Regulation of flow and abstraction of water

Diversions from the border rivers of Queensland and New South Wales have increased between 1988 and 1994 by 187 per cent and

THE BRIGALOW

The Brigalow, dominated by *Acacia harpophylla*, is characteristically open forest with trees 10 to 25 metres tall, with dark, furrowed stems and a silver-grey canopy. Brigalow occurs in subcoastal and inland eastern Australia, where rainfall is between 500 and 750 millimetres.

DISTRIBUTION OF BRIGALOW-DOMINANT VEGETATION

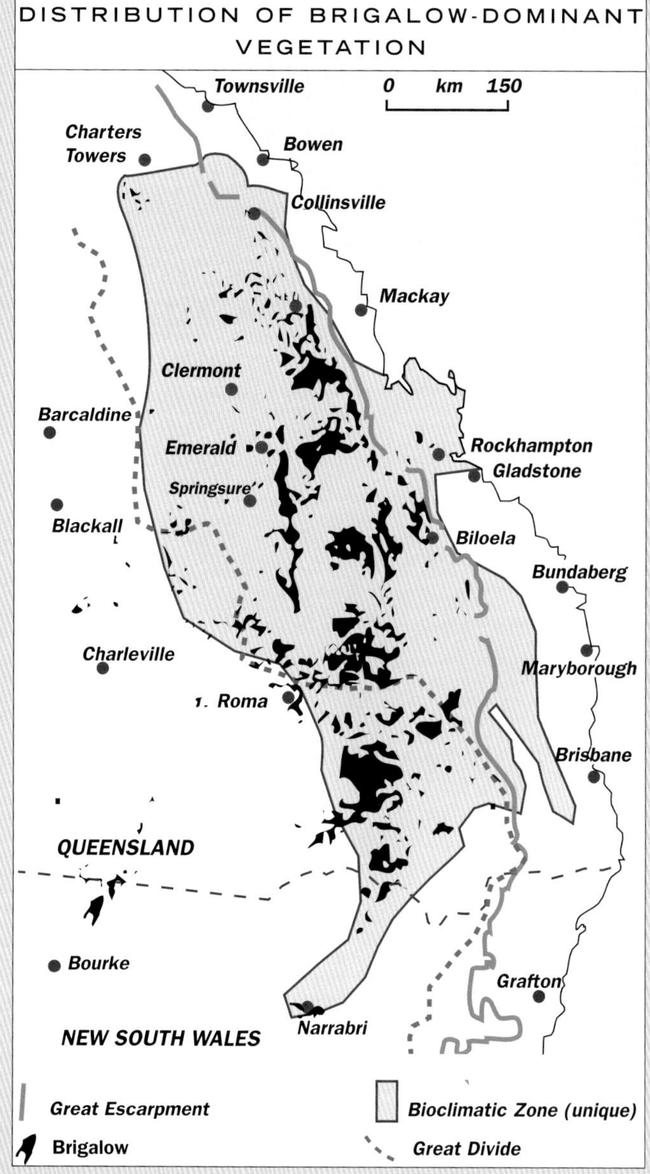

After Nix[135]

The brigalow belt extends from Collinsville in Queensland (20°30'South) to just south of Narrabri in New South Wales (30°30'South). The core area is 1200 kilometres long, 300 kilometres wide.[135] Brigalow-dominant vegetation occupied at least 5 million hectares originally and was co-dominant or a significant element in at least 10 million hectares within the core area.

Brigalow has an extensive lateral root system which stores starch and has the ability to sucker when damaged. Suckers have a low branching habit, and suckering can result in 25 000 stems per hectare which can stay at the sucker stage for 30 years before progressing to the whipstick brigalow stage by self-pruning of lower branches and self-thinning. A density of 5000 to 20 000 whipsticks per hectare is possible, with a height of 3 to 9 metres, and brigalow thickets are able to persist in this state for more than 50 years, making disturbed and regenerating brigalow a formidable enemy. Acacias are legumes, and brigalow has *Rhizobium* bacteria in its root nodules, able to fix nitrogen. The plants are extremely drought resistant. Seeds of brigalow do not have a hard seed coat, and germinate immediately after shedding. In the wet 1950s they were observed germinating in their pods before being shed by the trees. The seeds only remain viable for a year, and the seedlings produce deep taproots which are capable of absorbing water as saline as sea water and surviving. Only infrequent regeneration occurs by seed.

Brigalow grows on cracking clays, many of which have high sodium content at depth. Half of the brigalow-dominant region occurs on deep, gilgaied, cracking-clay soils. The gilgai are mainly of melonhole type, large depressions up to 3 metres in diameter, with elevated rims from 30 centimetres to 2 metres high.

The Brigalow occupies a unique bioclimatic zone: Summer rainfall is dominant, with 60–80 per cent of mean annual rainfall occurring from November to April. The summer peak is separated by autumn and spring troughs from the winter peak. The pattern is bimodal, with a discontinuous period of effective rainfall. In the long hot summers evaporation is high, with high intensity rainfall and high run-off. This is balanced by mild winters with lower evaporation and lower rainfall. Soil water deficits occur in autumn and spring; soil water recharge occurs in summer and winter, with a significant soil water surplus likely in summer. Flooding rains associated with cyclones saturate the landscape, recharge aquifers and maintain base-flow in the major river systems. Winter rainfall recharges soil moisture and reduces the severity of the driest season in spring.

Rainfall is highly variable, due to ENSO. Major streams in the brigalow domain have the most variable flows of any major region in Australia, including the ephemeral rivers of the arid zone.

By 1832, the grazing potential of southern brigalow lands had been assessed by Oxley (1818 and 1825); Cunningham (1827); and Mitchell (1831–32) as far north as the Darling Downs and westward to the Barwon. By 1840 the Leslie brothers had reached the Darling Downs and by 1844 virtually all lands in the southern one-third of the domain had been occupied. In 1844 to 1845, Leichhardt's expedition traversed the northern two-thirds of the brigalow lands and pastoral settlement followed, slowly at first but quickening as the 1850s gold rushes created markets for livestock. By 1861, just two years after Queensland had separated from New South Wales, all the lands in the brigalow belt had been occupied.

Prickly pear invaded the brigalow lands from 1860, affecting 4 million hectares. The drought of 1899–1903 aided its spread—plants were chopped up and distributed by rail as drought relief, and each piece was capable of striking and creating a new plant. By 1920, 25 million hectares were infested. It is said that 3000 tonnes of arsenic pentoxide were used in attempts to poison the plants. The *Cactoblastis* moth was introduced in 1914, unsuccessfully; and with success in 1924, and it had destroyed most of the pear by 1935. At one stage during the infestation, the government was willing to give away pear-infested brigalow lands.

Massive clearing of brigalow occurred after World War II because bulldozers and other machines were readily available. Chaining was widely used, followed by poisoning of regrowth. The weed killer 245T was popular from 1952, and 120 000 hectares had been cleared by 1964. The Brigalow and Other Lands Development Act of 1962 sponsored clearing and settlement—the land had to be cleared, burnt, sown to improved pasture and stocked within a prescribed period. By 1990 almost all of the northern brigalow lands were substantially cleared.

Brigalow being cleared. JIM GASTEEN

Pulled brigalow in the Pegunny district. JIM GASTEEN

A melonhole in brigalow country, a sedge-filled boggy depression with a pronounced rim. M.E.W

The brigalow had just been cleared, the ground ploughed and prepared. The rains came and the soil washed away, so that Jim Gasteen found himself literally knee-deep in the mud. Erosion from unprotected soil in regions with heavy tropical-style rainfall is enormous and unstoppable—and these brigalow soils lose their fertility in a couple of years of cropping JIM GASTEEN

The hydrology in cleared regions had been altered: dryland salinity was appearing and soil structure decline was widespread. Erosion, gullying and invasion by feral plants like the dreaded rubber vine, and prickly acacia (*A. nilotica*), followed.[170, 171]

38 per cent respectively (and are still increasing). The annual volume of water diverted from this region had reduced the average annual flow at Bourke by 36.5 per cent by 1994.[169] The average annual pre-development flow of the lower Balonne had been reduced by 17 per cent, and the flow into the Narran Lakes by 31 per cent. (Flows to Narran Lake have been decreased by 40 to 60 per cent, according to the Australian Society for Limnology.[50]) Abstractions have increased the numbers of small flows and reduced the number of large flows, causing loss of diversity of flow heights and durations. As a result of abstractions and clearing for cropping, wetlands have been lost: 50 per cent of the Macquarie Marshes and 90 per cent of the Gwydir wetlands, where only 1000 hectares is left out of an area of 20 000 hectares that used to be flooded annually.

Water quality

The river systems of the coolibah floodplains are typically turbid with relatively high nutrient levels. Flow regulation can influence the frequency, duration and intensity of toxic algal blooms. Pesticide pollution is a threat downstream of intensive cropping. The salinity of the lower Balonne and its bifurcations is generally lower than the Barwon–Darling, but there is a rising trend in salinity in all these rivers. A substantial decline in water plants has occurred in rivers of the coolibah floodplains, affecting the food chain for fish. Reduced flooding also affects the nutrient levels and hence the quality of fish habitat.

Clearing

Extensive clearing of coolibah floodplains has occurred in the last 150 years but, particularly in the east, a great deal of clearing is recent (mainly driven by cotton farming). East of the Macintyre and Barwon Rivers in New South Wales an average of 55 per cent of the remaining native vegetation was cleared during the 1970s and 1980s—much of it coolibah floodplain. On the northern floodplains of the Western Division of New South Wales about 200 000 hectares was cleared or thinned, but extensive tracts of coolibah woodland still remain in the Western Division.

(The area sown to cotton in New South Wales rose by 43 per cent between 1988 and 1991. Not only the effect of extra abstractions but also the impact of clearing has to be considered when assessing the long-term implications. Biodiversity loss, soil structure and fertility changes and surface and subsurface hydrology changes have to be included.)

MELONHOLE COUNTRY

Much of the eastern Darling Downs and the districts to the north and south of Chinchilla are referred to as 'melonhole country'.[171] Soils are vertisols with high clay content. Gilgais result from the uneven swelling of the clay layers—clay content is usually higher than 40 per cent and associated with high levels of calcium carbonate. Two main forms of gilgai are recognised—round and linear. Round are most common, comprising near-circular mounds of cracking clays up to 15 metres in diameter and 1.5 m high, with a depression 20 to 50 centimetres deep in the centre. The cracking clays under the mounds usually have higher pH, more soluble salts, higher carbonate and less organic matter.

As the soils form, leaching of soluble elements occurs from mounds to depressions. Soils in the bowl-shaped depressions are up to 1 metre deep in the centre, thinning to the margins, with higher organic content—hence they are darker. Soils on the mounds have a few centimetres of dark, granular clay overlying grey or yellow to red subsoil which extends as a tongue below the surface, giving the bowl effect in cross-section.

Swelling and shrinking cycles are related to the amount of water lost or taken up at different seasons. Tops of mounds dry first while depressions are still ponded, then depressions dry and cracks develop, to depths of 1 metre, 5 to 10 centimetres wide at the top.

Shallow gilgai country usually carries open eucalypt forest with grass and herb understorey. Larger, deeper gilgais are associated with dense forests of brigalow, brigalow and belah, or gidgee. In these forests the trees are confined mostly to the mounds and where ponding is common, the depressions may support only sedges and nardoo.

Clearing, levelling and cultivation of gilgai country tends to accentuate the differences in soil conditions between mounds and depressions. Crops do better on the depression areas. Zinc deficiency can occur on the mound soils because of higher calcium and pH levels. Differences even out after 5 to 7 years of cropping.

Melonholes offer immense surface water storage so run-off seldom occurs, and, where it does, it is slow-moving and causes little erosion.[170] Water generally flows from one melonhole to the next in a definite natural drainage network. Road cross-drainage structures have often been linked into this network. Melonhole areas are difficult to crop. Paddocks dry out unevenly and cultivation and planting machinery have problems; some farmers have levelled these areas, which again presents problems. Levelling results in greatly increased run-off and natural drainages are altered. Flooding and erosion can result on adjoining properties; road drainages can be overloaded or found to be in the wrong places. A drop in productivity can result from the mixing of subsoil into soil profile. (See photographs of melonhole country in the Wimmera, page 237.)

Lowering of roads, strip cropping, removal of fencing and other practices may become necessary to combat the effects of levelling. A better alternative in gilgai country may be less-intensive land use. Forage crops or improved pastures may be a better alternative to grain crops.

Several different kinds of gilgai occur. In order of increasing dimension. These are: crabhole gilgai, normal gilgai, linear gilgai, melonhole gilgai, and tank gilgai. Half of the brigalow core area is melonhole country.

There is still no consensus on the exact method of formation of gilgais. Not all vertisols or cracking-clays develop gilgais while some texture-contrast soils with sandy to loamy surface horizons can have well developed crabholes, so it is more than just soil type which predisposes an area to their formation.

Floodplains of the Condamine River and south to the New South Wales border

Widespread flooding occurs regularly on the floodplains of the Condamine River and on the brigalow lands west of the Darling Downs on floodplains of the Moonie and Weir, which connect with those of the Macintyre River along the Queensland–New South Wales border. Before the region was cleared for agriculture, the floods spread out over the landscape, revitalising the native vegetation, moving slowly as a creeping tide which caused little erosion. How the clearing and agriculture have altered the floodplains, and the serious problems of soil loss which have arisen as a result, were the subject of a study in *Listen … Our Land is Crying* (pages 98–100).

The problems of managing these valuable agricultural areas are difficult to resolve. The climatic conditions, with rainfall coming in severe storms, and the erosion which is inevitable if the rain falls on bare ground, require modification of farming practices to reduce risk. A great deal is being done using strip cropping, rotations which minimise times when bare soil might be exposed to heavy rainfall, and measures to minimise diversion and concentration of floodwaters. The aim of all the engineering practices is to restore the flat landscapes over which slow floods can creep as they used to do over natural landscapes before flow was interrupted by roadways, fences, irrigation structures and standing crops. The areas where melonholes abound present special problems—and these areas are widespread in the brigalow country.[170]

FLOODPLAINS OF THE DARLING DOWNS AND BRIGALOW LANDS TO THE WEST

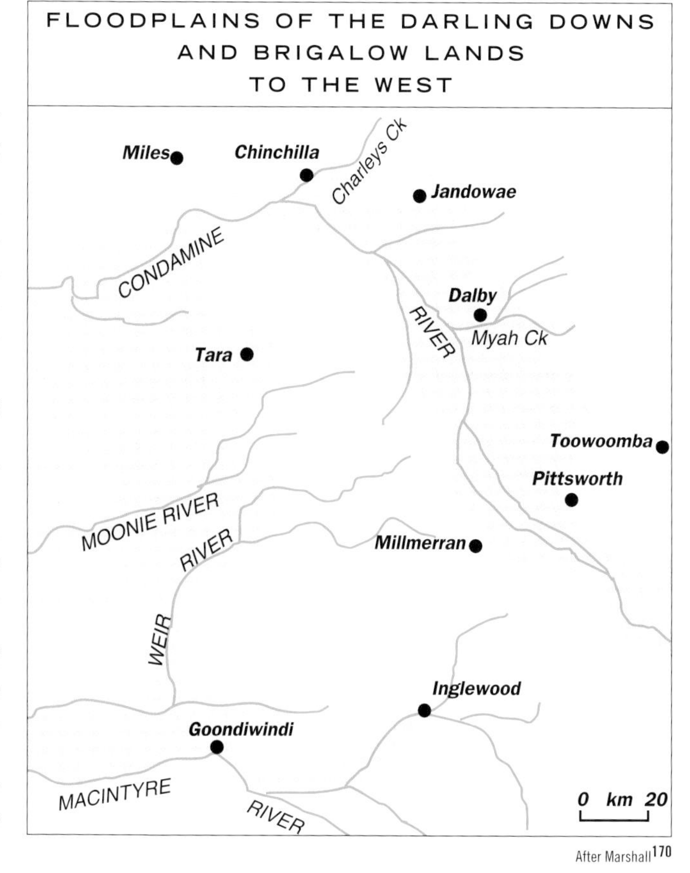

After Marshall[170]

BOGGOMOSSES

Boggomoss is the name given to the mound-springs of the Taroom District in Queensland, which are fed from shallow aquifers in Precipice Sandstone in the Great Artesian Basin where the aquifers intersect the land-surface.[172] They are prevalent east of Taroom, along the Dawson River, near the Glebe Weir and south-east to Cockatoo Creek. They vary in size from small muddy swamps to raised mounds up to 150 metres across, growing sedges, ferns, mosses and trees, including eucalypts and melaleucas. Their name derives from the resemblance of their accumulations of organic matter to the raised peat bogs of northern Europe.

Bores tapping the Precipice Sandstone aquifer in the boggomoss region are shallow—from 20 to 120 metres. Surveys by the Queensland Museum and the Department of Environment and Heritage discovered new and rare species in the boggomosses, including two endemic land snails. The boggomoss springs are probably central to the dispersal and survival of local wet-adapted biota,[57] and several plant species in their vicinity are classified as vulnerable or endangered. There is concern that many of the boggomosses will be inundated if the proposed Nathan Dam is constructed on the Dawson River. It would require the construction of high, strategically placed levees to protect them. Recent investigations suggest that a reduction in the maximum height of impoundment from 185 metres to 177 metres would significantly reduce the impact of inundation on threatened plant species, and would save some of the springs.

The Dawson Valley Development Association was formed in 1991 to lobby for the dam—proposing new coal mines in the Theodore area, expanded irrigation and a cotton gin and peanut processing plant, increased tourism and possibly a coal-fired power station and coal mine at Wandoan. It is hard for protection of the natural biodiversity and its environment to compete with economy-driven schemes when it comes to lobbying governments, even when impounding water for irrigation of cotton on other rivers has been shown to have devastating downstream effects.

Price Creek Springs, 1997: a boggomoss mound-spring. LINDSAY RIGBY

The Dawson River

Settlement of land along the Dawson River in Queensland started 150 years ago, following exploration by Leichhardt. The Jiman Aboriginal people were virtually annihilated and dispersed from the area when the Europeans took up the land. Clearing of the catchment (brigalow) has been extensive, particularly on the watershed, resulting in massive siltation of the river. A weir was constructed in the 1970s below one of the largest and deepest waterholes in the upper Dawson. As a result, the tree-lined billabongs and reed-clad banks of the river above the weir have been replaced by tonnes of silt—and dead eucalypts.

RIVERREACH: IMPLEMENTING SUSTAINABLE RIVERINE MANAGEMENT IN THE QUEENSLAND MURRAY-DARLING BASIN

RiverReach is a community based riverine management program jointly funded by State and Commonwealth Governments through the Natural Heritage Trust. Its function is for rivers what Landcare is for the land.

The State of the Rivers Report for the upper Condamine River revealed that much of this river system was experiencing moderate degrees of degradation through erosion, sedimentation and loss of riparian vegetation. Other rivers of the Queensland sector of the Murray–Darling Basin were known to be in similar condition. Opportunities for riverine management have been identified through the strategic plans of Catchment Management Associations.

RiverReach:

> 'will implement sustainable riverine management through an extension program which promotes best practice in management and rehabilitation of riverine systems. It will do this in a planning framework through the support of community based on-ground works designed to maintain or improve the condition of riverine systems. In the longer term, RiverReach will also increase the community skills base for riverine management, as well as demonstrating the practical aspects of on-ground works. Extension based planning and community awareness and education relating to riverine management are important elements of RiverReach.'

The proposed Nathan Dam on the Dawson will be an 880 000 ML dam, flooding 72 kilometres of riverine environment and 15 000 hectares of prime alluvial land in the Taroom Shire. Many farmers will be displaced from the fertile alluvial plains.

THE PAROO AND WARREGO RIVERS

The Paroo and the Warrego have their headwaters in south-western Queensland and drain enormous catchments of 73 600 and 69 100 square kilometres respectively.[174] The Paroo is 600 kilometres long, the Warrego 900 kilometres. Like other arid-land rivers, they have highly variable flow—from droughts when only a few lakes and waterholes in their catchments contain water, to times of massive floods when together, as in the 1990 flood, they created 686 000 hectares of wetland. The two river systems are connected by Cuttaburra Creek and both contribute to the filling of the Paroo overflow lakes. The importance of the wetlands and lakes for maintaining biodiversity in the Murray–Darling Basin cannot be over-estimated. The wetlands are the most extensive remaining in the Basin and are among the most important waterbird breeding areas in Australia. The Paroo River, the last free-flowing river in the Murray–Darling system, is in almost pristine condition, and should be preserved in that state for all time. Floods are relatively frequent

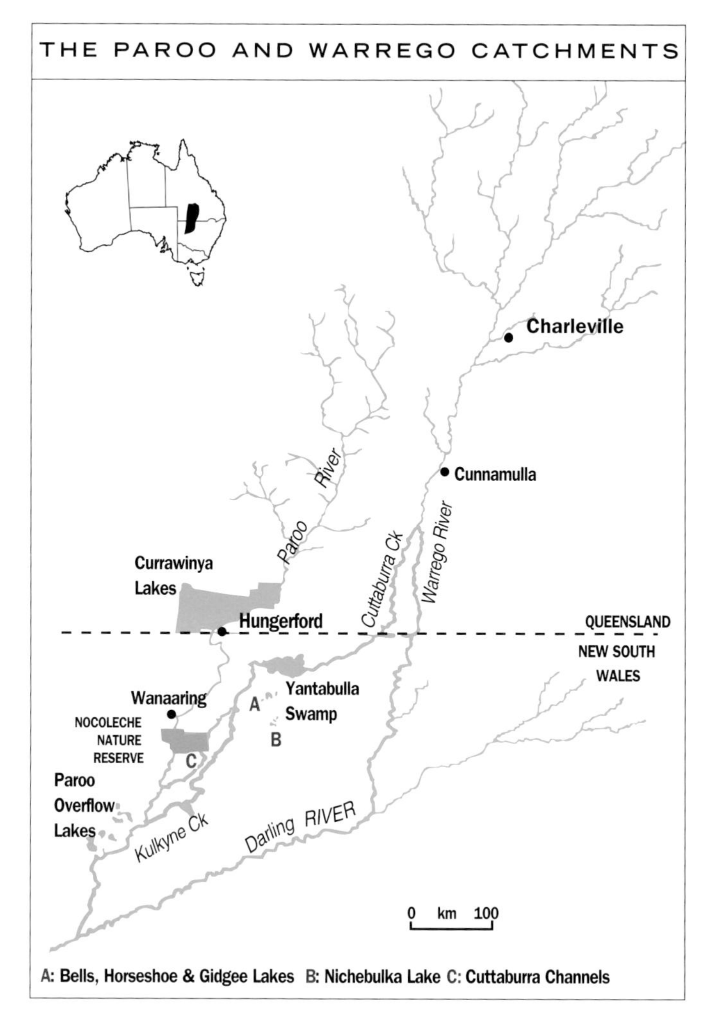

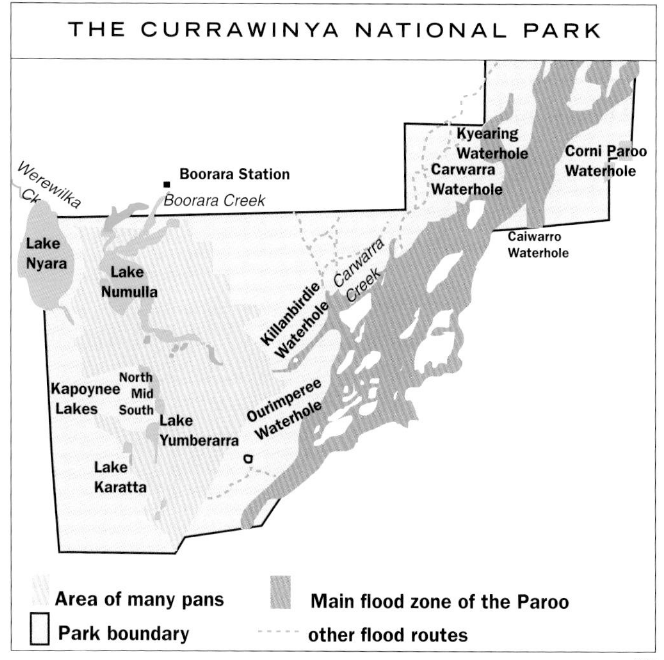

After Timms[173]

on the Paroo, compared with other inland river systems, a few occurring each year, with moderate floods every two or three years and major floods about every five years.[174]

Both the Paroo and the Warrego connect only to the Darling and contribute to its flow at times of high flow. The first major wetlands on the Paroo are the Currawinya Lakes. The river fills Lake Numalla, and during particularly high flows water reaches Lake Wyara, which is a salt lake. The Paroo then flows south, flooding areas of canegrass and lignum before reaching an extensive area of freshwater lakes and lignum called the Paroo Overflow. The largest of the lakes here is Peery Lake. During floods, water flows through a series of the lakes and ultimately reaches the Darling.

South of Cunnamulla, during times of large floods, the Warrego's effluent stream—Cuttaburra Creek—flows south and fills Yantabulla Swamp, a large lignum and sedge wetland, and proceeds onward through a number of channels and Kulkyne Creek to join the Paroo.[173]

Apart from wetlands reliant on river flows, there are hundreds, or perhaps thousands, of small wetlands which fill from local rainfall. There are also small catchments which are independent of the main rivers and terminate in large wetlands. In New South Wales alone it has been estimated that 87 000 hectares of potential breeding habitat for birds are present on the Paroo and Warrego river systems. The most important wetlands for waterbird breeding in the Paroo system are the Currawinya Lakes just north of Hungerford. In 1992 and 1993, an estimated 100 000 or more birds were counted on Lakes Numalla and Wyara. Lake Wyara consistently had about ten times more waterbirds than Numalla, a fact related to available food. Salt Lake Wyara has more invertebrates and macrophytes. A substantial proportion of Australia's freckled duck population is found on these lakes.

Although the Paroo is not regulated, grazing in the catchment has resulted in some changes to the river system, with increased sedimentation of some lakes and wetlands. Changes have been made to floodplains with levees and roads obstructing flow. Feral animals, pigs, foxes and cats in particular, cause problems for native animals and for the breeding waterbirds. Carp are an increasing problem here as in other rivers and lakes. The noogoora burr is a weed species which is increasingly invasive.

Some water resource development has occurred on the Warrego River. About 2500 megalitres is extracted at the Cunnamulla Weir for about seven irrigation farms, and more licence applications are in the pipeline. It can only be hoped that applications for irrigation licences for the Paroo River, and additional ones for the Warrego will be refused. Australia should have the wisdom to protect for all time the last free-flowing rivers, knowing what happens to wetlands and biodiversity when river flow is diminished and altered. Guaranteeing 'environmental flows' only protracts the dying of the wetlands and the rivers downstream of major water extraction schemes.

Claypans on the floodplains of the Paroo River near Locoleche.
RICHARD KINGSFORD

THE MACQUARIE RIVER AND THE MACQUARIE MARSHES

The Macquarie River is a much-regulated river. Prior to the good rains of 1998, it was reduced to a few pools of green water in a dry channel between Burrendong Dam and the Macquarie Marshes.

The main catchment of the Macquarie River is west of the Great Divide, about 200 kilometres to the south-east of the Macquarie Marshes. It supplies about 88 per cent of the water for the catchment. Water flows in a north-westerly direction, confined to a narrow, steep-sided valley in the upper catchment, to reach Burrendong Dam. Built in 1967, the dam captures most of this water flow and then allows release for irrigation. The Windamere Dam, built in 1984 on a headwater stream, further reduces flow in the Macquarie system. More than half of the water is used to irrigate cotton—the industry in the Macquarie Valley expanded by 400 per cent in the 1980s. The river has many other management structures—weirs, bypass channels, block banks and levees all direct and control the flow of water. Water use for irrigation upstream of the Marshes peaked in 1993–94 at 543 000 ML per year.

The Macquarie Marshes comprise about 220 000 hectares when flooded. Its upstream margin is defined as beginning below the last regulator on the river at Marebone Weir, 40 kilometres downstream from Warren. The town of Carinda is in the far north-eastern corner of the Marshes. Two distinct areas are recognised—the Southern

Dead eucalypts: changes to hydrology, which result in some areas being permanently wet and some to drying out completely are causing heavy losses. REG MORRISON

*Dead belah trees (*Casuarina cristata*) in the Macquarie Marshes, which have died because of changes to the hydrology in the region.* REG MORRISON

The Macquarie Marshes from the air. CAROLANNE WOLFGANG

Birds in their millions frequent the Marshes, which are very important breeding grounds in good years. CAROLANNE WOLFGANG

Marshes, of approximately 25 000 hectares, and the Northern Marshes, which begins 10 kilometres north of the junction of the Bulgeraga Creek and the Macquarie River. Only some18 000 hectares in all are reserved in the Macquarie Marshes Nature Reserve, which is a Ramsar listed site. The reserve comprises three segments, the largest and smallest in the northern sector (160 square kilometres and less than 5 square kilometres) and the middle-sized one of about 50 square kilometres in the southern sector.

The wetlands consist of open water lagoons; mixed marsh and water couch communities; reedbeds; red gum, coolibah and blackbox woodlands. They represent the most southerly occurrence of coolibah (*Eucalyptus microtheca*) and the most northerly occurrence of extensive reedbeds (*Phragmites australis*). Upstream catchment activities and water diversions as well as local land use have had a major impact on the health of the Marshes. Damming of the river and reduction in the frequency and magnitude of flooding have caused vast areas to dry out and die. Areas adjacent to major permanently flowing channels at the same time suffer waterlogging and salinisation, as occurs when weirs and dams result in continuous flow. Current estimates are that at least half the Marshes has been lost and more is threatened—and the demand for more water licenses for cotton goes on.

CENTRAL AND NORTH WEST REGIONS WATER QUALITY PROGRAM

The Central and North West Regions Water Quality Program is a multi-disciplinary investigation of surface waters and, to a lesser extent, of groundwater, helping to address concerns over the potential impacts of irrigated agriculture, particularly cotton, on water quality. In the 1995–96 report, pesticide monitoring results over a five year period were published for surface waters in the Macquarie, Namoi, Gwydir and Border Rivers Basins.[142]

The program revealed the seasonal appearance of endosulfan and its break-down products in surface waters, closely tied to its use as the principal pesticide in both Stages 1 and 2 of the Insect Resistance Management Strategy (IRMS). The parent isomers first appear each year in late November to early December and there is a regular appearance of moderate levels from December to March each year.

Drought had impacted on the four valleys by the 1995–96 season, particularly in the Macquarie River Basin, where only 40 per cent of the previous season's crop was planted due to water restrictions, but early season rains in the Gwydir and Macintyre river valleys had resulted in a significant increase in plantings there. In the Namoi valley, endosulfan that was not related to river flow was detected in surface waters. Aerial contamination was the suspected cause.

While aerial pathways have been shown to be important, water pathways during periods of intense rainfall were also important. Such events in the Namoi, Gwydir and Macintyre valleys in the 1995–96 season led to contamination. (The Macquarie experienced much less rain in January.) In the Namoi, a large fish kill was attributed to the release of pesticide-contaminated stormwaters from nearby cotton farms—the Liverpool Plains was shown to be a major source of storm run-off containing high concentrations and loads of pesticides and herbicides. (See *Listen … Our Land is Crying*, pages 192–201, for a detailed account of the Liverpool Plains, the Namoi catchment and geology and land use.)

In the Liverpool Plains, a rain event between 3 and 7 January saw 1.1 kilograms of endosulfan, 27 kilograms of atrazine and 5.3 kilograms of metolachlor exported out of the catchment into the Namoi River. In a second rainfall event between 20 and 28 January, a further 9.4 kilograms of endosulfan, 130 kilograms of atrazine and 1.5 kilograms of metolachlor entered the Namoi. A storm event in the Mooki River catchment between 4 and 10 January resulted in 1.3 kilograms of endosulfan, 170 kilograms of atrazine and 47 kilograms of metolachlor entering the Namoi. Drinking water guidelines for endosulfan concentrations were exceeded by 32 per cent of river samples collected from within irrigated areas between November and March and there was a marked increase in the percentage of samples from irrigated areas which exceeded the guidelines.

The general increase in endosulfan levels reflected the increased cotton crop in the more northerly valleys coupled with adverse weather conditions and the higher number of sprays because of higher pest pressures. (There was no change in endosulfan levels in the Macquarie that year in spite of the 60 per cent reduction in plantings, which highlights the problems of dealing with aerial pathway contamination.)

Moderate to high concentrations of atrazine were a common feature at sites both upstream and within irrigated agriculture for the Gwydir, Namoi and Border Rivers. (The Macquarie showed no atrazine in the 1995–96 season.) About one-fifth of the water samples from within irrigated areas exceeded the recommended drinking water guidelines between November and March. The Liverpool Plains was again shown to be a major contributor to atrazine in the Namoi valley when January storm flows in Coxs Creek and the Mooki River resulted in peak loads of 120 and 100 kilograms per day respectively being exported.

Because pesticides are transported by aerial spraying and by water run-off, an integrated approach to farm management and pesticide application is required to minimise pesticide impacts.

Almost-empty drums abandoned around bores were found to be a source of pesticide/herbicide leakage into the soil and into the groundwater on the Liverpool Plains. A program called 'Drum Muster', organised by the Gunnedah Shire Council, raised awareness of the issue and organised collection of empty drums, largely solving the problem. WENDY TIMMS

THE LIVERPOOL PLAINS WATER QUALITY PROJECT:
1996–97 Groundwater Quality Report

The Liverpool Plains, in the upper Namoi catchment, has traditionally been a producer of high quality wheat for domestic and export markets. It also comprises the largest dryland summer cropping area in New South Wales and the clay soils are considered to be some of the most productive agricultural soils in Australia. The area sown to dryland and irrigated cotton has expanded significantly in the last ten years, with most of the cotton grown in the Gunnedah and Boggabri districts.

The winter of 1996 saw 180 000 hectares of wheat and barley sown on about 15 per cent of the area. About 250 000 hectares of summer crops were planted in the 1996–97 season, comprising 213 136 hectares of summer cereals; 19 160 hectares of irrigated cotton; 8970 hectares of dryland cotton and 5480 hectares of other irrigated crops.

The main point sources of pollution on the Liverpool Plains are sewage treatment plants at Gunnedah and Quirindi, urban stormwater, industries, mining and intensive agriculture (cattle feedlots, piggeries and poultry). However, the contribution of these sources to declining water quality is minor compared with that of agriculture.[175, 176, 177, 178, 179, 180]

Agricultural systems are becoming more reliant on chemical use. Insecticides to control pests and herbicides to control weeds, particularly as minimum till practices (increasingly being used to control erosion) require chemical controls. In 1996–97, the chemical used in the greatest quantity on the plains was the herbicide glyphosate, with approximately 558 tonnes applied to control weeds and as a desiccant on sorghum. Other herbicides used were 470 tonnes of atrazine; 134 tonnes of metolachlor; 70 tonnes of 2,4-D; 34 tonnes of MCPA; 19 tonnes of trifluralin; 19 tonnes of prometryn; and 19 tonnes of fluometuron.

The most commonly used insecticides in the period were parathion methyl (56 tonnes); chlorpyrifos (53 tonnes); endosulfan (51 tonnes); thiodicarb (37 tonnes); and profenofos (30 tonnes). The bulk of these chemicals was used in the production of cotton. During the 1996–97 season the cotton industry used pesticides at approximately 15.5 kilograms per hectare. However, the greatest overall quantity of chemicals used was in the production of summer crops like sorghum, which required massive application of active ingredients during the year.

Terrifying figures—what a cocktail is going into the soil (and it does not include the fertilisers)!

Two main river systems drain the Liverpool Plains area—the Mooki River (including the Goran Basin) on the east, and the Coxs Creek to the west. The Mooki joins the Namoi above Gunnedah and joins Coxs Creek at Boggabri. The major streams and rivers on the plains are ephemeral, mostly flowing only during or immediately after rainfall.

The Liverpool Plains Water Quality Project aims to evaluate the impacts of dryland agriculture on the quality of both ground and surface water. Quality of groundwaters has been assessed since 1992, when traces of the herbicide atrazine were detected in bores, sparking a broad-scale project to assess many aspects of water quality, rather than just chemicals, and to relate results to land-use practices. Public water supplies on the plains are almost entirely sourced from groundwater, hence the immediate concerns.

Substantial evidence has been presented that agriculture has had a significant impact on the quality of groundwaters on the Liverpool Plains.[178] Declining quality has mainly been caused by herbicide contamination leaching through the soil profile beneath crops and from abandoned chemical drums dumped near boreheads. Shallow saline water tables and increasing salt loads in productive aquifers also result from agricultural practices, which have increased deep drainage through the soil profile and possibly induced vertical flow of saline groundwater by over-extraction. Increased nitrate levels in shallow groundwater can also be attributed to agricultural practices.

Quarterly groundwater sampling during the first year of the project found atrazine in 40 per cent of samples, with 11 per cent over the recommended healthy drinking water guidelines. Isolated cases were found where low contamination levels existed with diuron, fluometuron, metolachlor, simazine and trifluralin. Nitrate concentrations above background level were detected in 35 per cent of samples, with 7 per cent above drinking water guidelines for infants.

Considerable variation in concentrations of contaminants were found, spatially and temporally. Early observations indicate that atrazine contamination is widespread, occurring in close proximity to crops where it is used. Rapid leaching occurs through the cracking-clays which comprise most of the rich soils of the plains. Low ratios between atrazine and its metabolites indicate that little breakdown of the chemical occurs in the soil profile.

Saline groundwaters impact on water quality in two ways: salinity increase in productive aquifers may in time render them unsuitable for irrigation; and highly saline shallow water tables cause soil salinisation and declining crop yields. There is close correlation between groundwater quantity and quality. Long-term draw-downs may cause contamination of productive aquifers with highly saline groundwaters.

One of the strange new clay mound-springs which appeared on cleared and cultivated land in the Lake Goran Basin. WENDY TIMMS

Strange new-type mound-springs on the Liverpool Plains

The sudden appearance of two clay mounds oozing water in the middle of a vast paddock cleared for cotton growing has been of great interest to hydrogeologists. The area is within the Lake Goran catchment in the centre of the plains. The Quaternary Lake Goran Basin is filled with unconsolidated alluvium, derived from Tertiary and Jurassic basalts. The stratigraphy of the alluvium is not fully understood, but at this site, recent clays overlie the Garrawilla Volcanics of the Jurassic Oxley Basin and a weathered layer forms a confined aquifer at the top of the basalt.[238] The mounds appear to have originated at a high point in the underlying basalt where the clay confining layer became thin enough for the groundwater to rise to the surface under artesian pressure. The mounds have formed through saturation of the swelling clays, starting to be visible in 1990.

Clearing of the catchment, removing deep-rooted perennial

THE NITROGEN CYCLE
Why organic farming cannot solve the world population's feeding problems

Professor Vaclav Smil (Geography Department, University of Manitoba, Canada) recently wrote a lucid explanation of this subject and stated: 'Feeding humankind now depends so much on nitrogen-based fertiliser that the distribution of nitrogen on the Earth has been changed in dramatic, and sometimes dangerous ways.'[181]

Nitrogen is a minor constituent of living matter, but it is essential for DNA, RNA and proteins. Unlike carbon, hydrogen and oxygen which are readily available from food and water, nitrogen remains locked in the atmosphere. Lightning can cleave paired atmospheric nitrogen and make it available to plants on a small scale, but nitrogen fixation is mainly by bacteria (*Rhizobium* spp.) in roots of legumes, and cyanobacteria (free or symbiotic). The bacterial action makes nitrogen available to plants, and plants are the basis of 99.9 per cent of food chains.

Cropping removes fixed nitrogen. Traditional farming used crop residues and animal and human wastes or green-manure legume crops to provide the nitrogen required for plant growth. (*Azolla*, a water fern with symbiotic cyanobacteria, is used in some parts of the world, including the lowlands of Java, the Nile delta, north-western Europe, Japan and China; and cultivation in ponds is starting in Australia for 'organic' farming.) The traditional method, using wastes and green manure, can supply 200 kilograms of nitrogen per hectare of arable land—which will produce 200 to 250 kilograms of plant protein. Therefore a hectare of farmland with good soil, adequate moisture and a mild climate that allows continuous cultivation should be able to support up to 15 people. In practice this does not happen: in China five or six people were supported per arable hectare in the early part of the twentieth century; Japan had a slightly higher figure because of the additional fish protein which forms part of the average diet; and only five people per hectare were supported in north-western Europe in the nineteenth century.

The practical limit of five people supported per hectare, using traditional methods of replenishing soil nitrogen, exists because of the following limitations: weather and water stresses and pests; the need for land for fibre production; and the amount of productive land taken for green manures in order to keep up productivity because of the closed nitrogen cycle. (Rotating legume crops with staple cereals is practised to improve this situation, but legumes have lower yields and limitations persist.)

Global population almost quadrupled during the twentieth century. This could not have been achieved without the synthesis of ammonia and production of nitrogenous fertilisers. Only limited natural sources of nitrogenous fertilisers exist, such as the guano deposits on certain Pacific islands. The Haber-Bosch process, developed in 1913 in Germany in order to produce explosives for World War I, enabled the production of nitrogenous fertilisers. Less than 5 million tonnes of fertiliser was used annually worldwide until the Green Revolution started in the late 1940s. Rapid expansion followed and 10 million tonnes were used annually in the 1950s. Production efficiencies were developed in the 1960s and an 8-fold expansion in use had taken place by the 1980s.

In the 1960s, affluent countries used 90 per cent of the production; by the 1980s the proportion had fallen to 70 per cent; by 1988 the developing world was using 50 per cent of the global production of nitrogenous fertilisers.

Now 175 million tonnes of nitrogen flow into the world's croplands annually—incorporated in cultivated plants. Synthetic fertilisers provide 40 per cent of all nitrogen taken up by these crops, directly as plants, indirectly as animal foods. This represents 75 per cent of all nitrogen in consumed proteins (the rest comes from fish, meat, dairy). About one-third of the protein in humanity's diet comes from synthetic nitrogenous fertiliser. Land-scarce countries with high population density depend on synthetic fertiliser for their very survival. Affluent countries use synthetic fertilisers to grow feed for livestock because they prefer high-protein animal foods. (Despite the fact that 3 or 4 units of feed protein are needed to produce 1 unit of meat protein, still more of the world is going for more meat, while the amount of available arable land is decreasing because of land degradation.)

The harmful consequences of artificial fertiliser production are many. Nitrogenous fertiliser leaching causes high nitrate levels in surface and groundwaters which results in eutrophication (increased nutrient levels) of rivers, lakes and estuaries, causing algal blooms and other problems and imbalances. Acidification of soils as a result of high fertiliser use lowers productivity, and releases toxic elements. Nitrous oxide enters the atmosphere, where it survives for 100 years and absorbs 200 times more outgoing radiation than carbon dioxide does. It is a powerful Greenhouse agent and a dangerous component of the photochemical smog in cities.

To summarise the situation: **Crop rotations, legume cultivation, soil conservation and recycling organic wastes are all desirable techniques. But in land-short, over-populated nations they are not enough. There is insufficient recyclable nitrogen to produce food for 6 billion people. Two billion of those alive today owe their survival to synthetic nitrogen. The 2 billion which will be added in the next two generations will require continued expansion of nitrogenous fertilisers—*'in just one lifetime, humanity has developed a profound chemical dependence'*.**[181]

vegetation, altered the hydrology of the area and the extra recharge that has resulted has led to the underground water coming to the surface. A period of increased rainfall during the 1980s, and a wet year in 1990 when Lake Goran, in the centre of the Basin, flooded extensively, explains the appearance of the mound-springs.

Archaeological investigations of the site suggest that mound-springs were present in the area during the Pleistocene. A scatter pattern of stone tools and flakes around the site indicates that Aboriginal activity was concentrated here during the wetter interglacial which preceded the last glacial stage of the Pleistocene ice age, presumably because of the local availability of water.

THE MURRAY BASIN

The Murray Basin formed as a result of tectonic activity associated with Australia's rifting from Antarctica and subsequent movement north.[182] Upwarping of the Proterozoic and Palaeozoic rocks at the Basin's margins produced a saucer-shaped structure flanked by low mountain ranges. It contains thin horizontal sedimentary strata that have been accumulating since the Basin began to sink 60 million years ago.[183] It is a huge structure, covering 320 000 square kilometres. Terrestrial environments have occurred continuously in the eastern half of the Basin throughout its history, but the western half has been subjected to repeated marine incursions, the last of which occurred in the Pliocene, ending only 3 to 2 million years ago.

The major modern divisions of the Basin into the Mallee and the Riverine Plain result from the different geological histories of the eastern and western sectors, the former being influenced by alternating marine sediments, the latter by freshwater palaeoenvironments only. The Pleistocene ice age amplified the distinctions between the two regions because the Mallee, being to the west of the Plain, received the full brunt of the salt- and sand-laden winds of glacial stages (particularly the last glacial which had its climax only between 22 000 and 18 000 years ago) which turned much of the Basin into a salt desert.

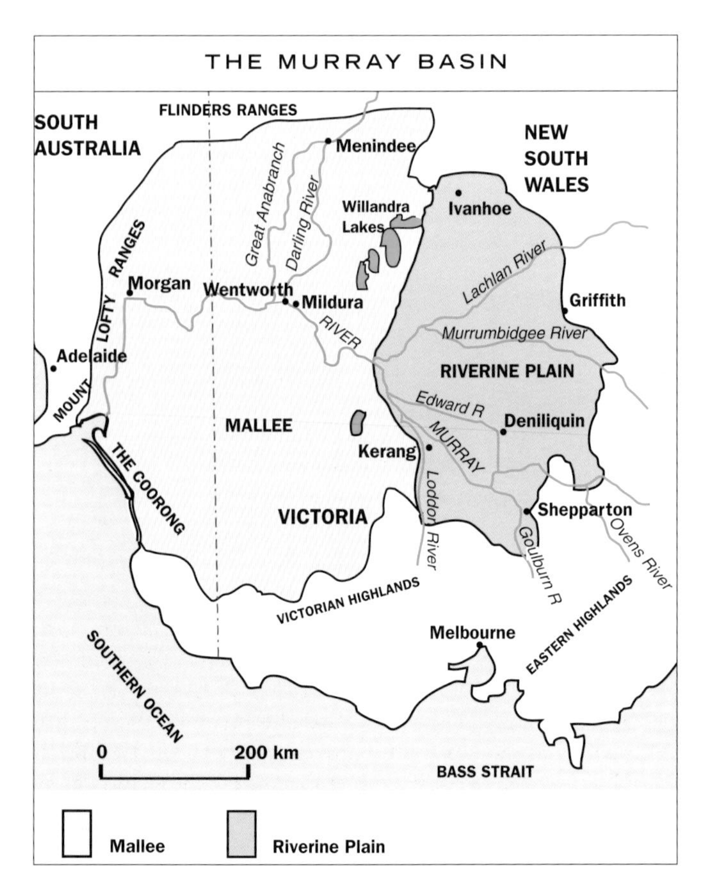

THE MURRAY RIVER

The Murray River is ancient and has existed as the major drainage in the Basin for at least the 60 million years during which it was sinking, and could possibly be equated to the southern part of the river systems which drained eastern Australia during the 'Congo-like' drainage stage between 100 and 90 million years ago (see pages 107–108). The Murray has seen vast changes. At times its lower reaches were curtailed by marine invasions into the western half of the Basin, the last occurring only between 5 and 4 million years ago. The huge freshwater Lake Bungunnia was created between about 2 million and 500 000 years ago, when the Murray River's exit to the sea was blocked by the uprising of the Pinaroo Block.

Salt and waterlogged pasture in the Murray Valley, Victoria. REG MORRISON

Lake Bungunnia covered the Mallee region, occupying an area of 66 000 square kilometres. It extended from near Lake Menindee in the north to near Boundary Bend on the Murray in the south, with lobes extending southwards along depressions between the north–south ridges. The lake was fed by rivers (including the ancestral Murray) which apparently discharged at least double the amount of water that modern rivers in the Basin discharge today. The start of the draining of the lake is dated by a geomagnetic reversal at 700 000 years ago. Following the development of a permanent outlet, water levels dropped rapidly and Lake Bungunnia fragmented into a number of smaller basins, of which Lake Tyrrell is the largest surviving remnant.

Following the disappearance of Lake Bungunnia, significant changes in sedimentation patterns became apparent in the Murray Basin, culminating between 500 000 and 400 000 years ago with the establishment of the aeolian landscapes of the Mallee with the dunes and salt lakes which characterise the region today. Recurring glacial stages of the Pleistocene ice age saw the dune systems reworked and salt and gypsum blown across the landscapes, with the last glacial maximum turning the Basin into a salt desert. Recovery from this stage has occurred in the last 14 000 years—the present interglacial in which we are living.

The water table which underlies the Basin's surficial sediments contains the salt of ages, concentrated and greatly increased when dry salt lake sediments were mobilised and blown across landscapes at those glacial stages. Groundwater discharge zones, where the water table intersects the surface, are indicated by salt lakes, occurring individually or in groups, and, on a more regional scale, by the amazing 'boinkas' or salt and gypsum plains like the Raak Plain in the northern Wimmera. Salt lakes, groundwater discharge zones, and salt groundwater entering rivers and periodically contaminating adjacent areas in climatically controlled events have been natural phenomena in the Basin at least in the last few million years. Like all natural systems, a dynamic balance had been established in ecosystems in which co-evolution of the biota and the environment had occurred over great lengths of time. The fragility of this balance, because of the low landscape relief and the correspondingly low

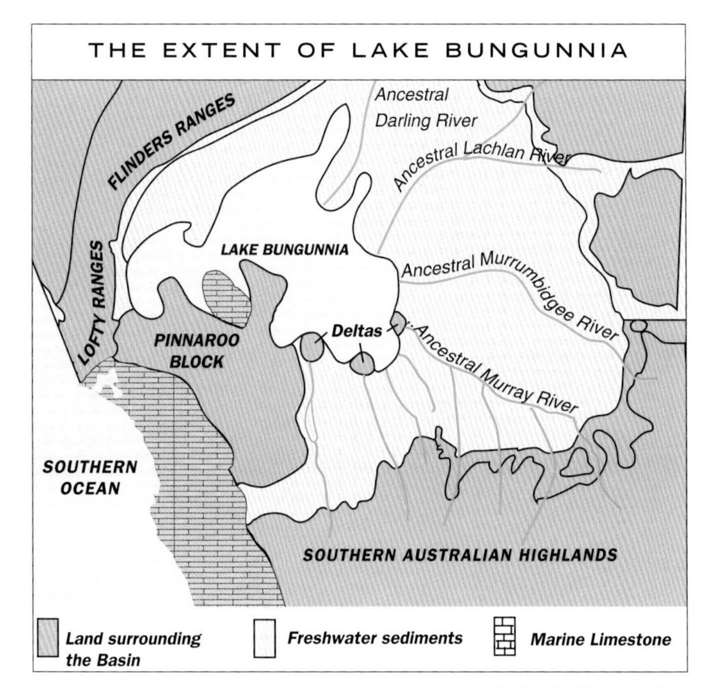

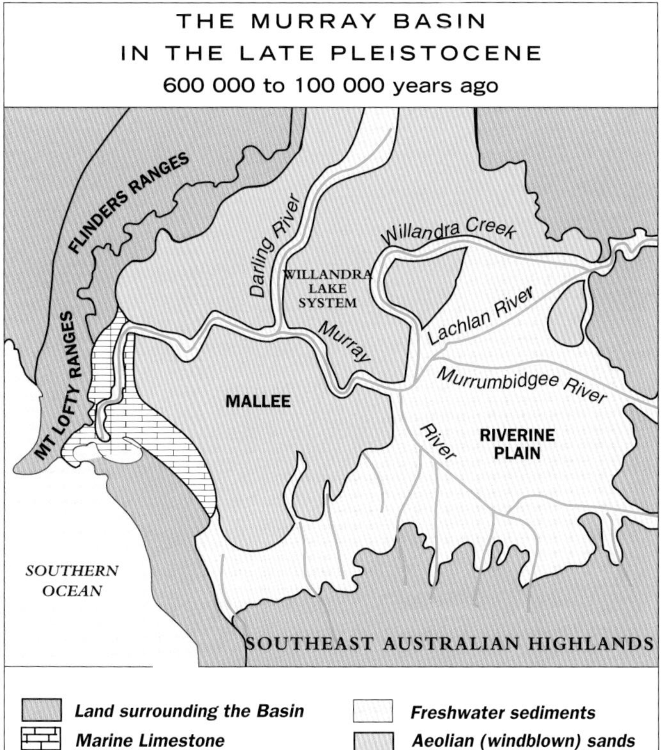

hydraulic gradients, which respond quickly to changes in water budgets, has been the modern Murray Basin's Achilles' heel. Now, recharge of groundwaters and pressurisation of underlying aquifers has resulted from European land use, much exacerbated by the addition of irrigation water, and boundaries of groundwater discharge zones have moved. The water table has risen close to or into the root zone in soils, and salinisation has resulted on a vast and ever-increasing scale. The saga of rising water tables and salinity is told in detail in *Listen … Our Land is Crying.*

A surprisingly recent geological event altered the course of the river in the Echuca region, when ancestral streams were disrupted by the Cadell Fault. The ancestral Murray and Goulburn Rivers flowed across the Cadell Block along the path of the well-preserved Green Gully channel. Following faulting and uplift of the land to the west (by 10 to 12 metres on the eastern edge of the tilt block) the ancestral Murray was diverted north through the present site of Deniliquin. The Goulburn was dammed and produced a large lake—Lake Kanyapella—with associated beach and lunette sand dunes along the north-eastern shoreline. Today, these are the Barmah sandhills. The lake drained eventually and the Goulburn River maintained a course south of the Cadell Block through the present site of Echuca. Only in comparatively recent times did the Murray divert to the south and establish its present course through the Barmah sandhills and across the floor of the Kanyapella depression. Movement on the Cadell axis started about 30 000 years ago, and the Murray's northward diversion occurred about 24 000 years ago. The modern channel of the Murray, its diversion to south of the Cadell Block, may be no more than 10 000 years old. (The Murray's original northward path at Deniliquin is today's Edward River. See page 153, *Listen … Our Land is Crying.*)

The Murray River traverses five distinct regions in its 2500-kilometre journey from the Eastern Highlands to the Southern Ocean:[163]

- **The headwaters, to Corowa, 450 kilometres downstream** From its source in three small springs on the side of Forest Hill, 40 kilometres south of Mt Kosciusko, the Murray flows west for a short distance across Cowombat Flat and then turns north until it reaches the Tooma River junction, where it begins its long journey westward. The Swampy Plain River joins it above the Tooma junction, and it is debatable whether the true Murray River starts from the Swampy Plain junction—the upper section is also known as the Indi River. Near Khancoban, the Swampy Plain River receives water diverted by the Snowy Scheme via Murray 1 and 2 Power Stations. West of the Tooma River, the Mitta Mitta, now diverted into the Hume Reservoir, used to join the Murray. The Kiewa River joins the Murray between the Hume Dam and Wodonga. The catchment above the dam contributes nearly 40 per cent of the Murray's water.

- **The Riverine Plains, from Corowa, 800 river kilometres downstream to the Wakool River junction,** just west of Swan Hill. The River Murray emerges onto the plains at Corowa, downstream of Albury. The plains consist of riverine sediments derived from the erosion of the Eastern Highlands and deposited by ancient rivers over many millions of years. The Murray meanders across the plains in a shallow channel. The gently undulating country becomes flatter westwards. A number of Victorian tributaries join the river in this section—the Ovens, Goulburn, Campaspe and Loddon; the only significant New South Wales tributary is Billabong Creek, which flows into the Murray via the Edward and Wakool Rivers. At a point east of Mathoura, about half-way between Albury and Swan Hill, the Edward River breaks away from the Murray and flows north to Deniliquin and then westwards to rejoin the Murray 500 kilometres downstream near Tooleybuc, forming a huge anabranch. Between the Edward and the Murray is a network of distributaries and anabranches, the largest of which is the Wakool River. The Wakool occupies part of the channel through which the Murray flowed before the Cadell uplift occurred. At the Wakool junction it is obvious that the Murray joins the larger capacity Wakool River and not vice versa. During floods, more than half of the water is diverted via the Edward River and the other anabranches. (When Captain Francis Cadell was giving evidence on whether more navigation on the Murray–Darling river system would be enabled by removal of timber and the clearing of channels, he said of the Wakool that '... having been augmented by so many streams falling into it, [it] becomes much larger than the main Murray; it is fully three times larger than the main river where it joins it again; the Murray seems a perfect ditch in comparison'.[240] On the southern side of the Murray are smaller anabranches, including Gunbower Creek.

- **The Mallee Trench, from the Wakool junction, 850 kilometres down to Overland Corner** A wide plain of marine origin is crossed by the river in a single, well-defined channel which cuts deeper into the plain as it moves downstream. This section is of Pliocene age, cut after the sea retreated for the last time. Downstream of the Wakool junction, the Murray is in fact back in its original main channel, cut before the Cadell uplift redirected it into its southern, modern channel. The Murrumbidgee and the Darling join the Murray in this section. The Darling brings water from southern Queensland and contributes about 12 per cent to the Murray's flow in an average year; the Murrumbidgee contributes about 11 per cent.
- **The Mallee Gorge, 280 kilometres from Overland Corner to Mannum** Here the river has cut down through hard limestone during periods of low sea level, forming steep cliffs along the river channel. The gorge is of Quaternary age. The river bed intersects the regional water table and salty groundwater enters the river through aquifers exposed in the cliff face. During times of low sea level, the river flowed out across the exposed continental shelf to beyond Kangaroo Island, and turbidity currents excavated deep submarine canyons.
- **The Lakes and the Coorong** The terminal lakes, Lakes Alexandrina and Albert, together with the Coorong, once formed a huge estuarine system. Barrages now separate the lakes from the Coorong and retain fresh water in the lakes. The Murray finally enters the sea near Goolwa, where it flows into Encounter Bay.

How has the planform of the Murray changed in the last 100 years?

A survey of the Murray between Albury and the South Australian border was carried out in the 1860s by the New South Wales Government. Later surveys, made in the 1920s and the 1970s, enable an understanding of just how much—or how little—real change has occurred. With the benefit of aerial photography and satellite imagery, the complex patterns of meanders, effluent channels and anabranches of modern and older palaeodrainages is clearly visible. It comes as a surprise to find how little change there has been when the three surveys are compared. The Murray's meanders have not changed shape to any extent since the 1860s and only a few show minor localised movements.

A study of world literature on rivers reveals that on the evidence presented the Murray is one of the most stable rivers described.[163A] Estimating the age of river red gums on its banks establishes, independently, that it has been stable for at least 200 years and probably for far longer. The river owes its stability to its low gradient and low energy.

Now that the complex patterns of anabranches and distributaries are mapped and the intricacies of palaeochannels and other features are known, it is clear why early explorers and surveyors, travelling the country on foot to try to construct the first maps, had such a difficult task. How would you decide which was a river's main channel when a maze of alternatives existed, when so often there was little or no flow in many of them to help decide, and when in times of flood the whole landscape was under water? The extreme variability of all the inland rivers, major ones like the Murray (pre-regulation), included, contributed to the difficulties.

The Campaspe River, where it joins the Murray at Echuca. M.E.W.

The Edward River at Deniliquin, a major anabranch of the Murray and part of its original course before the Cadell Fault caused changes. M.E.W.

The accounts of the early mapmakers, which prompted the question 'What if there were no river systems?' which was one of the catalysts for the writing of this book, are seen in their correct context when the information available today is taken into account. In the first place, the preconception of what a river system should be, based on the European experience which informed the explorers, was far from the reality of river systems in Australia. In this ancient land, because of its flatness, its inwardly draining characteristics, its increasing aridity with increased distance from the coast, and its extreme climatic variability, what comprises a river system is different.

Floodplains are an inseparable part of our river systems; anabranching and formation of distributaries on floodplains are a product of flat terrain and the accumulation of sediment on those floodplains; and the variablity of flow, which is the greatest anywhere in the world, results in river dynamics very different from those based on the concept of a river which flows from high ground to the sea, receiving contributions from tributary streams in their catchment on their way. The ephemeral nature

CHANGES TO THE MURRAY SHOWN BY SURVEYS OVER THE LAST 100 YEARS

DIMENSION	CHANGE
Planform	
Accretionary (gradual)	Insignificant amount of meander development; river length unchanged
Avulsive (sudden)	22 cut-offs between 1860s and 1920s shorten the river by 65 kilometres (5 per cent); the river becomes steeper
Cross-section	
Width	Eroded trees suggest widening
Depth	Comparing cross-sections suggests decreased depth
Long profile	
Amount of material	General erosion of bed before 1920s; deposition after 1920s Up to 3 metres deposition in deepest parts of channel, but great variation
Distribution of material	No trend in pool riffle amplitude; possible increase in spacing between pools

After Rutherfurd[163A]

of all our inland rivers, including the Murray before it was regulated, is an intrinsic characteristic. Flow stops in droughts, rivers dry to a series of pools.

The question which should have been asked was not 'What if there were no river systems?' but 'What constitutes a river system in Australia?' When one understands that, one can see why it appeared to pioneers that some rivers 'died in the plains' or ended in swamps, going nowhere in particular.

Before the major changes made to the Murray with the construction of the headwater dams, the diversion of water from the Snowy Scheme or even the modern array of locks and weirs, the bulk of the man-made changes to the river were related to navigation. The original channels of the Murray, the Murrumbidgee and Darling were full of large woody debris (snags) and their banks were so heavily wooded that the funnels of the early steamers constantly collided with overhanging branches. The mammoth task of de-snagging the rivers was largely the work of one man and his team. Francis Cadell, whose name graces the Cadell Fault and tiltblock which changed the course of the Murray, did much to establish a navigable channel within the Murray River.

IVANHOE BLOCK AND GROUNDWATERS OF THE RIVERINE PLAIN OF NEW SOUTH WALES

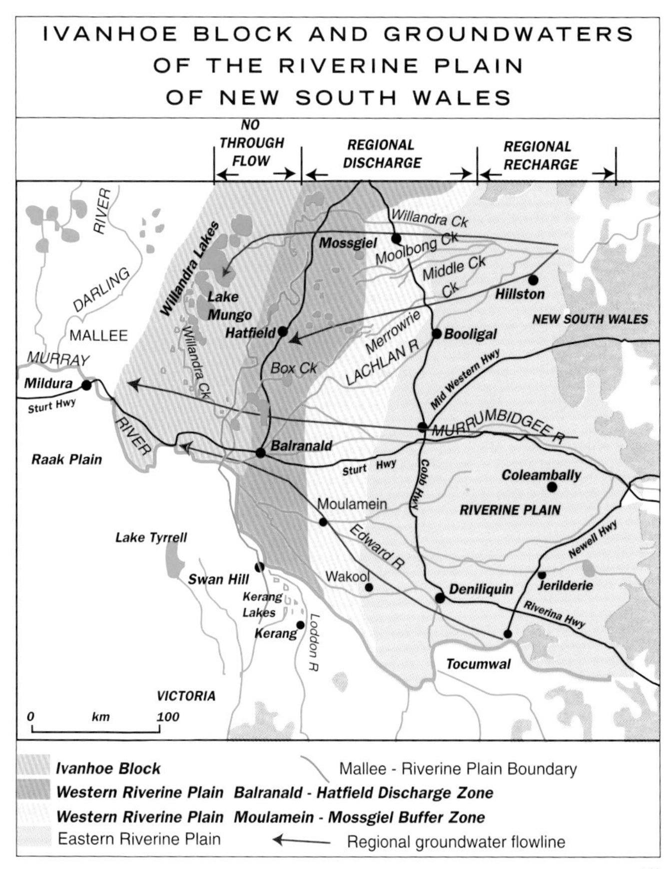

After Kellett[184]

THE RIVERINE PLAIN OF SOUTH-EASTERN AUSTRALIA

John Oxley was the first white man to look out across the Riverine Plain of New South Wales, and he was not impressed. He wrote to Governor Macquarie in August 1817: 'The country south of the parallel of 34 and west of the meridian of 146 30 was uninhabitable and useless for all the purposes of civilised man.' The featureless, flat landscapes stretching to the horizon were without interest to him. (The first settlers moved into the area in 1825, and by 1840 pioneer pastoralists had occupied all grazing land by free selection. Today, the eastern Riverine Plain is one of the most productive agricultural regions in Australia.)

Had Oxley proceeded westwards for another 200 kilometres, he would have seen the open, undulating grasslands of the eastern Riverine Plain merge into the vast saltbush plains of the western Riverine Plain, with a more varied landscape pockmarked by an extensive network of fossil gypsum playas and associated lunettes, increasing in abundance westwards. The western Riverine Plain is about 100 kilometres wide, and gives way abruptly to the elevated east–west longitudinal sand dunes of the Ivanhoe Block. Not only is this boundary a sharp physiographic one, as shown in the Landsat mosaic, but it defines a distinct vegetation break, from the saltbush of the western Riverine Plain to the mallee of the Ivanhoe Block.

The western ridge of the Ivanhoe Block forms the regional divide between the Darling and Lachlan–Murrumbidgee groundwater systems.[184] The parallel basement ridges of the block deflect flow

A BRIEF HISTORY OF NAVIGATION OF THE MURRAY AND ITS TRIBUTARIES

One of the most remarkable river journeys ever documented, and certainly the most remarkable Australian example, is that of Charles Sturt and his party, who made the first boat journey from the Murrumbidgee to the Murray and downstream to the river's termination in Lake Alexandrina—after which they rowed all the way back against the current!

Sturt had previously traced the course of the Macquarie River and had discovered the Darling. The purpose of the journey, which started from Sydney on 3 November 1829, was to determine the course of the Murrumbidgee. In the party were George Macleay, son of the Colonial Secretary of New South Wales; Sturt's personal regimental servant; two soldiers, a carpenter and two convicts, and several dogs. There was a small supporting land party. A whaleboat 7.6 metres long with a 1.5 metre beam was carried in sections and assembled at the point on the Murrumbidgee, near the present town of Maude, where Sturt took to the waters. The river was low and 'we were obliged to haul the boat up between numberless tree trunks, an operation that exhausted the men much more than rowing ...'

On 14 January 1830, having travelled for a week down the Murrumbidgee, they entered a 'broad and noble river' which Sturt named the Murray. On the way down the river:

> 'I thought it advisable to lay its course down as precisely as circumstances would permit: for this purpose I had a large compass always before me, and a sheet of foolscap paper. As soon as we passed an angle of the river, I took the bearings of the reach before us, and as we proceeded down it, marked off the description of country and any remarkable feature. The consequence was that I had laid down every bend of the Murray river, from the Murrumbidgee downwards. Its creeks, its tributaries, its flats, its valleys and its cliffs, and, as far as I possibly could do, the nature of the distant interior. The chart was, of course, erroneous in many particulars...'

Each night Sturt tied his observations in to the miles of latitude travelled according to star sights because 'the sextant would not embrace the sun in its almost vertical position at noon'.

The voyage was dangerous and difficult, and hostile Aboriginals were often encountered. Arriving in Lake Alexandrina, the plan was to row out to sea and round to the Gulf of St Vincent, where a ship was to pick them up, but this proved to be impossible. Since food was running low, they had to start back without delay, the way they had come, but this time against the current. (The dream of establishing a port at the mouth of the Murray persisted during the early days of river trade when steamships plied the rivers, but the dangerous conditions of the bars at the river mouth, and the south-easterly gales which prevailed, caused their abandonment.)

Of the return journey, Sturt wrote:

> 'We had now only to make the best of our journey, rising at dawn and pulling to seven and often to nine o'clock. I allowed the men an hour from half past eleven to half past twelve to take their bread and water. This was our only fare, if I except an occasional wild duck; but these birds were extremely difficult to kill, and it cost us so much time that we seldom endeavoured to procure any. Our dogs had been of no great use, and were now too weak to have run after anything if they had seen either kangaroos or emus; and for the fish, the men loathed them and were either too indifferent or too much fatigued to set the night lines. Shoals frequently impeded us as we proceeded up the river, and we passed some rapids that called for our whole strength to stem. A light wind assisted us on two or three of these occasions, and I never failed hoisting the sail at every fitting opportunity. In some parts the river was extremely shallow, and the sandbanks of amazing size; and the annoyance of dragging the boat over these occasional bars was very great.'

The detailed chart made on the downward journey was a great help on the return voyage.

> 'It cheered the men to know where they were, and gave them conversation. To myself it was very satisfactory as it enabled me to prepare for our meetings with the larger tribes, and to steer clear of obstacles in the more difficult navigation of some parts of the stream... The journey to the sea and back again [to the point at which the Murray turns south at Morgan] had occupied us twenty days. From this point we turned our boat's head homewards... Our attention was now directed to the junction of the principal tributary, which we hoped to meet in twelve days, and anticipated a close to our labours on the Murray in eight days more from that stage to the Murrumbidgee...'

A couple of days into the next stage, the boat was holed by a snag, but fortunately did not sink because a piece of the log which had holed it was stuck in the hole. It was repaired by the carpenter, and Sturt's accounts deal with Aboriginal threats, battling up rapids, and dread at the thought that rains might cause increased flow and increased currents to row against.

Twenty-three days after leaving Lake Alexandrina, the party re-entered the Murrumbidgee (on 16 March). This was three days less than it had taken them to go downstream. Their troubles were not ended, however. The ground party was not where they hoped to find it, and they had to continue their way up the Murrumbidgee. By now their provisions were almost exhausted. In addition, rain in the mountains had caused a rise in the river and they were rowing, and making little progress, against a stiff current. Weakened by lack of food, and suffering scurvy, it is incredible that they survived at all.

> 'On the 8 and 9 April we had heavy rain, but there was no respite for us. Our provisions were nearly consumed, and would have been wholly exhausted if we had not been so fortunate as to kill several swans. On the 11th, we gained our camp opposite to Hamilton's Plains, after a day of severe exertion ... We were still between eighty and ninety miles from Pontebadgery, in a direct line, and nearly treble that distance by water. The task was greater than we could perform, and our provisions were insufficient. In this extremity I thought it best to save the men the mortification of yielding, by abandoning the boat; and on further consideration I determined on sending Hopkinson [a soldier] and Mulholland [a convict] whose devotion, intelligence, and indefatigable spirits I well knew, forward to the plain.'

The two returned six days later, bringing with them 'such supplies as they thought we might immediately want.' They were both in 'a state that beggars description'. Ten days later the party arrived at Pontebadgery (having burnt the boat) and there 'found Robert Harris [with the land party], with a plentiful supply of provisions. He had everything extremely regular, and had been anxiously expecting our return. He had been on the plain two months, and intended to have moved down the river immediately, had we not made our appearance when we did.'

It was not until the 1850s that riverboat trade on the Murray–Darling–Murrumbidgee system was developed. It was to play an important and colourful part in the lives of pioneering settlers for eighty years, and no name is more synonymous with the early riverboats than Francis Cadell, who was also largely responsible for preparing plans to channelise the Murray and de-snag the rivers and remove the obstacles and dangers to navigation.

REG MORRISON

RIVERBOATS ON THE MURRAY

Much has been written, and films and documentaries have been prepared, on the 'romantic' days of riverboat trade; stories beyond the scope of this book. Careful reading of the original accounts by Cadell and others, written at the time, provide insights into the status of the rivers. Before the main channels were altered by removal of reed beds and vegetation, cutting down trees which overhung the streams in narrow reaches and caused problems by entangling masts, and literally making a satisfactory channel through the flooded red gum forests where those occurred, the river system had more channel diversity. The meanders in some reaches were so sharply curving that only boats of short length could negotiate their loops, and cut-offs were engineered in many places where even those suitably designed boats had problems, to eliminate some of the acute wanderings. All the boats were shallow draft because of the fluctuations in depth of the rivers and the abundance of snags before they were controlled.

Before regulation of the rivers, travel along them was only possible in the years when floods were sufficient to enable penetration to any distance. In years of big floods, boats could leave the river channels and proceed literally across country, visiting outlying stations, collecting their wool and delivering their supplies to the door. Alterations of all sorts made to the channels to improve them for navigation led to flow changes and channel incision.

Captain Cadell was to advise the government on how to make the improvements, and was personally involved in the design and operation of the snag boats which cleared the channels.

The *Border Post and Albury Telegraph* of 1 December 1858 (and several following weekly editions) carries a full account of the questions and answers in the examination of the 'evidence given before the Committee appointed by the Legislative Assembly of New South Wales, to report upon the advantages likely to accrue from the navigation of the Murray River system, and upon the best means of clearing the channel of the river and its tributaries—Thursday, November 12th, 1857... Captain Francis Cadell called for and examined'.

Several days of intense questioning of Cadell about his experience on the river resulted in money being allocated for the de-snagging programs. In the course of the examination, Cadell had to recount what he had learned from his first voyage in the *Lady Augusta* (which developed into a race between his boat and the *Mary Ann*), from Goolwa to Swan Hill in 1854. He was asked about every aspect of the economics of riverboat trade and the future prospects for wheat and agricultural production in the Murray region (where some irrigation had already commenced, but he was not optimistic about the future of large-scale development of agriculture). The examination makes fascination reading.

The snag boat built to his design operated a steam-driven saw which cut off trees and snags at a uniform depth below the water. In some billabongs, created by cut-offs of acute meanders, which still exist, below the water tree trunks remain in situ at a uniform level, neatly sliced off by Cadell's underwater saw.

south in the deeper aquifers. The western Riverine Plain is the groundwater discharge zone for the eastern Murray Basin in New South Wales as a consequence of the impeding effect of the block. On the basis of salinity of groundwaters it is divided into two regions, an eastern buffer zone, less threatened by salinity, and a western strip, where aquifers are very saline. Based on this subdivision, the eastern Riverine Plain has a low risk of salinisation in its non-irrigated lands; the buffer zone has a medium risk; and the discharge zone and the lower Willandra Lakes have a high risk, if water tables continue to rise.

If the Ivanhoe Block did not exist, groundwater of the Murrumbidgee and Lachlan alluvial fans would discharge into the Darling River, permitting export from the Basin of salt accumulated along the regional flow lines.

EVOLUTION OF THE MODERN RIVERINE PLAIN

The Riverine Plain lies mainly in New South Wales, with a south-eastern extension in Victoria adjacent to the Murray River between the Goulburn and Loddon Rivers. The research on palaeochannels on the New South Wales Riverine Plain was a response to the need for accurate soil maps for the region's expanding agriculture. Initial investigations in the 1950s established that the ancient river patterns defined distinctive channel and levee soil associations that extended as a vast network over the surface of the plain.[185, 186] Subsequent research has clarified the picture. Palaeochannels in the Victorian sector were largely mapped in response to the need for groundwater and location of the deep leads which were so important in mining activities.

The Riverine Plain consists of alluvium spread by rivers that rise in the south-eastern highlands.[183] In New South Wales the rivers flow westwards from the Great Divide; in Victoria they flow northwards from the south-eastern highlands. The plain covers 77 000 square kilometres. Much of the alluvium was spread by an extensive distributory network of ancient sand-bedded streams very dissimilar to the meandering channels of the Murray, Murrumbidgee and Lachlan Rivers of today. There were also meandering channels similar in form to the present rivers, but which in many cases had obviously carried up to five times as much water as today's rivers. Modern dating techniques have enabled the establishment of the chronology

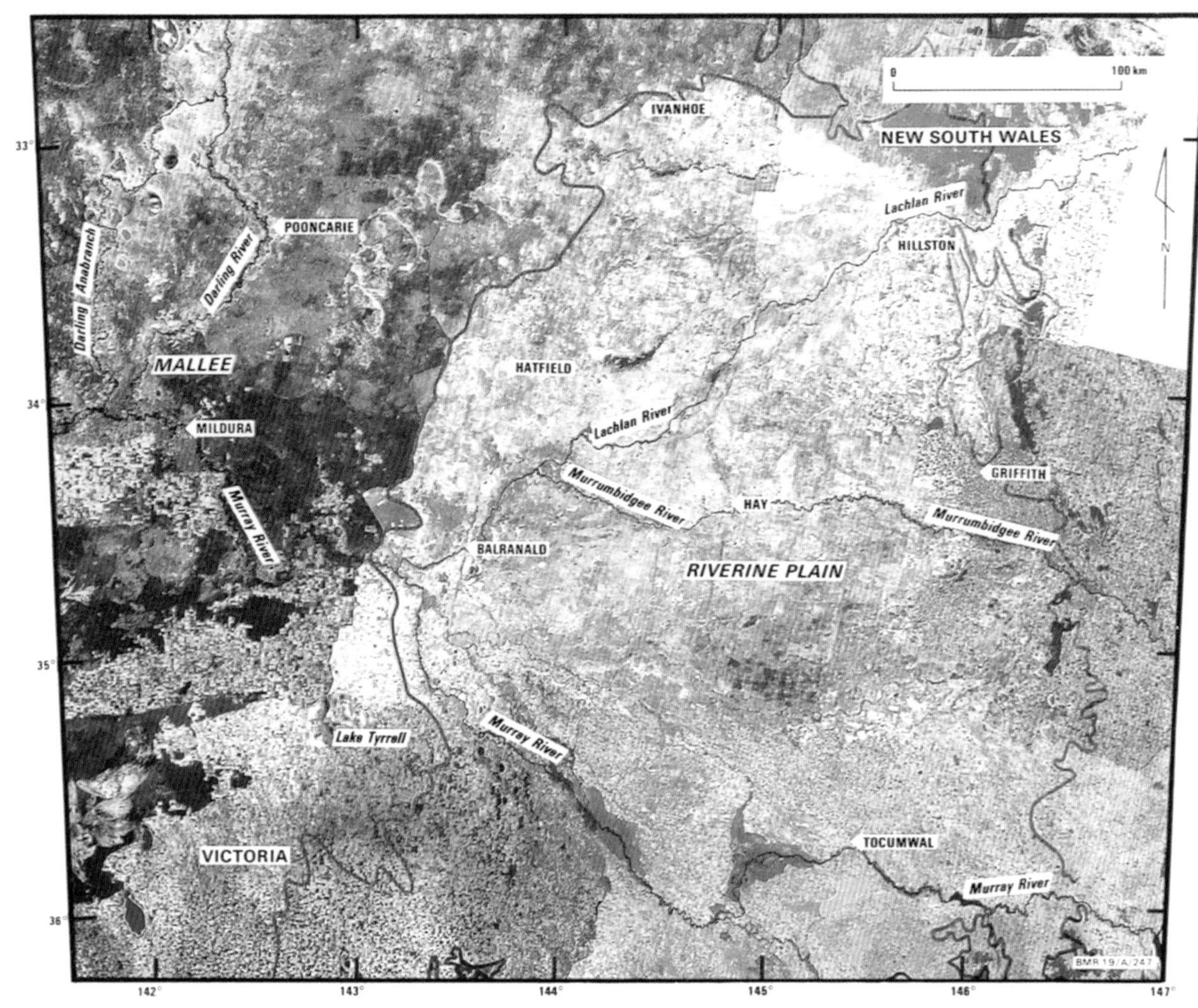

LANDSAT MOSAIC OF THE CENTRAL AND EASTERN MURRAY BASIN
Landsat satellite image acquired by ACRES, the Australian Centre for Remote Sensing, a business unit of AUSLIG, Australia's national mapping agency.

of 100 000 years of deposition of the sediments spread by the river systems, ancient and modern, as well as those which have accumulated in lakes, or were aeolian.

The region has been geologically stable, with little recent tectonic activity on the Riverine Plain except for one episode during which a narrow block of land was raised along a fault line between the present-day sites of Deniliquin and Echuca. Disruption of the ancestral Murray and local drainage resulted from movements along this north–south-trending Cadell Fault, mainly occurring from 30 000 years ago. In Thule Lagoon in Green Gully, however, dates of ancestral stream sediments show that the disruption to that branch of the palaeodrainage by the fault commenced about 60 000 years ago. The Murray was diverted around the top of the fault block via what is now the Edward River and only changed course southwards through the Barmah sandhills, making the major river bend which runs south to Echuca, within comparatively recent times, possibly as recently as 10 000 years ago.

THE WILLANDRA LAKES

The Willandra Lakes in western New South Wales are World Heritage sites. They have been much studied from both hydrological and dune-building perspectives.[91] Their lakeshore, beach and wind-blown sediments can be related to water level changes in the basins, and sediment cores from the lake floors provide information on vegetation, aquatic fauna and climate change.

The Willandra Creek, on which the lakes lie, is a distributory of the Lachlan River. The lakes filled about 50 000 years ago after a long period during which they had been dry. While they were full, water was passing through them southwards and down to the Murrumbidgee west of Balranald (the Mungo phase). A period of low water and deflation of lakebed sediments occurred at 36 000 years ago, but the lakes were full again by 32 000 years ago and remained reasonably high until about 25 000 years ago. Drying started then, except for a brief recovery at about 19 000 years ago, and the major last glacial-stage arid spell of 18 000 to 16 500 years ago saw major dune construction (the Zanci phase). Since that time the top lakes in the chain, Mulurulu and Garnpung, have been the only ones to receive water from the Lachlan, and then only when flood peaks overflowed from the main river channel.

The sequence of events recorded in the Willandra Lakes is repeated at Lake Frome in South Australia, and at Lake Tyrrell in Victoria, and in playas elsewhere—a major wet phase between 50 000 and 36 000 years ago; a transitional phase from 35 000 to 25 000 years ago; dry environments between 20 000 and 16 000 years ago; and a return to conditions much like the present since.

SCOTIA SANCTUARY: saving endangered wildlife

Scotia Sanctuary is the latest Earth Sanctuaries project devoted to the preservation of biodiversity and the saving of endangered species. It is situated in far western New South Wales on the South Australian border, about 80 kilometres to the west of the Silver City Highway at about the halfway point between Wentworth and Broken Hill. It comprises two old sheep stations with an area of about 65 000 hectares.

Like all Earth Sanctuary projects, its aim is to provide a feral-free environment in which native species can breed and flourish, and where endangered animals which have been produced in captive breeding programs can be released to resume a natural life. The special feral-proof fencing which is erected round the property, and the feral animal eradication programs inside the fenced areas, are the first stage in the establishment of a sanctuary. Already by 1998, brush-tailed bettongs, plains rats, hopping mice and bilbies had been brought back to Scotia. In August that year, six brindled nailtail wallabies were flown in from Queensland where the sole remaining population contains only 700 individuals. Four of these animals were females and already had joeys in their pouches. By January 1999, the joeys had all left the pouch and had been replaced by the next generation. By mid-1999 the Scotia population of brindled nailtail wallabies had grown to 22 individuals. Clearly, the breeding program was off to an exciting start.

The numbat, another mammal species which once lived in the region but which has been locally extinct since the late nineteenth century, was returned to Scotia Sanctuary in late 1999. Tweny-five numbats were translocated from Earth Sanctuaries' Yookamurra Sanctuary in the Murray Mallee in South Australia, where a successful breeding program was based on 15 animals obtained from Western Australia in 1993.

The Australian Geographic Society has provided funds for a bat survey at Scotia, which will provide valuable information on these little-studied animals.

Native biodiversity, particularly in the marginal Western Division lands, has been decimated by over-grazing and other human-induced changes to the environment since European settlement. In many cases, species numbers are so low and populations are so fragmented that unless something is done now, extinction is only a moment away. Scotia's location therefore makes it particularly valuable. Natural Heritage Trust funds have boosted the amount available from private contributions and Scotia, like the other Earth Sanctuaries properties, is already a success. There can be no more worthwhile cause than prevention of extinction and the restoring of natural ecosystems.

Anyone interested in obtaining information or making a contribution to Earth Sanctuaries should contact the Foundation, PO Box 1135, Stirling SA 5152.

A nailtail wallaby at home in Scotia Sanctuary.

The New South Wales Riverine Plain

A period of major river flow in the ancient drainages of the Riverine Plain, recorded in the deposits of stream sediments, declined between 100 000 and 85 000 years ago. (This corresponds with the pluvial period when rivers were spreading vast sand sheets in the Lake Eyre Basin between 120 000 and 95 000 years ago.) Renewed prior stream activity occurred in the north of the plain between 50 000 and 40 000 years ago, and this correlates closely with evidence from inland playa lakes, particularly the Willandra Lakes in western New South Wales and Lake Tyrrell in north-western Victoria, and with a limited phase of active deposition of river sediments in the Western and Gilbert Rivers in Queensland.

This phase was followed by a transition to more sinuous mixed-load channels with discharges greatly in excess of present rivers. Discharge of the ancestral Murrumbidgee near Darlington Point is estimated to have been about five times that of the modern river. Large meandering remnants of these channels are particularly well-preserved along the present margins of the Murrumbidgee floodplain between Narrandera and Carrathool and also along Yanco Creek.

Recent research on the Murrumbidgee River palaeochannels has refined the general palaeodrainage picture for the region between the Murrumbidgee and Yanco Creek. It has been possible to distinguish four systems of decreasing age and relate them to climatic regimes:[182]

PRIOR AND ANCESTRAL STREAMS BETWEEN THE MURRAY AND THE MURRUMBIDGEE RIVERS

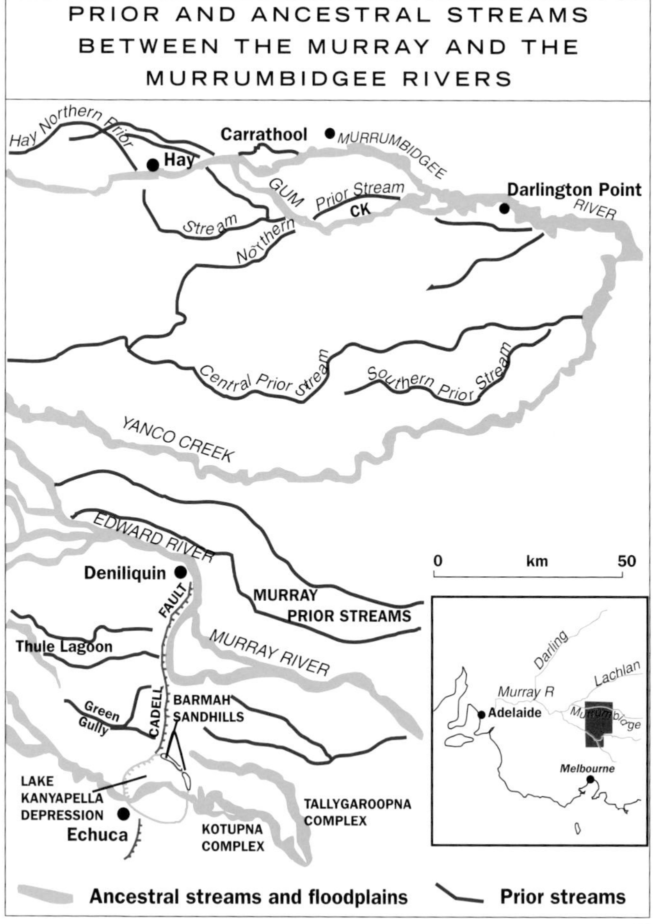

After Page *et al*[183]

- **Coleambally palaeochannel system** is the oldest, situated in the vicinity of the Coleambally Irrigation Area. The two arms can be traced back to a single palaeochannel that emerges from the western margin of the modern Yanco Creek floodplain north of Morundah. Both arms have classic prior stream form—low sinuosity, low levees. Ages for channel sediments are between 100 000 and 85 000 years.
- **The Kerarbury system** This complex system comprises the northern channels above and below the present river, with a south west trending, branching arm across the centre of the region which disappears under the modern deposits of the Edward River near Moulamein. This system is believed to have operated before 50 000 years ago and its age corresponds to the subpluvial phase of enhanced river activity in inland and southern Australia, when Lake Urana on the plain had a lake-full period.[3A]
- **The Gum Creek system** corresponds to the sinuous ancestral Murrumbidgee river. In the region upstream of the Yarrada Lagoon it exists as a meander belt incised 1 to 3 metres below the surface of the surrounding plain and controls the course of the present Murrumbidgee River. The wavelength of meanders is greater than in the modern river and reflects greater discharge than in the modern system. Dates for this system are between 35 000 and 20 000 years.
- **The Yanco palaeochannel system** branches south from the present Murrumbidgee floodplain about 15 kilometres west of Narrandera and trends south-west and west across the plain towards Moulamein, where it terminates abruptly at the edge of the modern Edward River floodplain. The northern part of the system is incised about 2 metres below the general level of the plain and is occupied by the present channels and floodplains of Yanco and Billabong Creeks. In its south-western section, downstream from Wanganella where the modern streams no longer use its channel, its palaeochannel width averages 225 metres, suggesting that it was a

A section of the sediments in the Kerarbury Pit showing cross-bedding near the top, below the deep soil zone, and a basal clay-rich zone. KEN PAGE

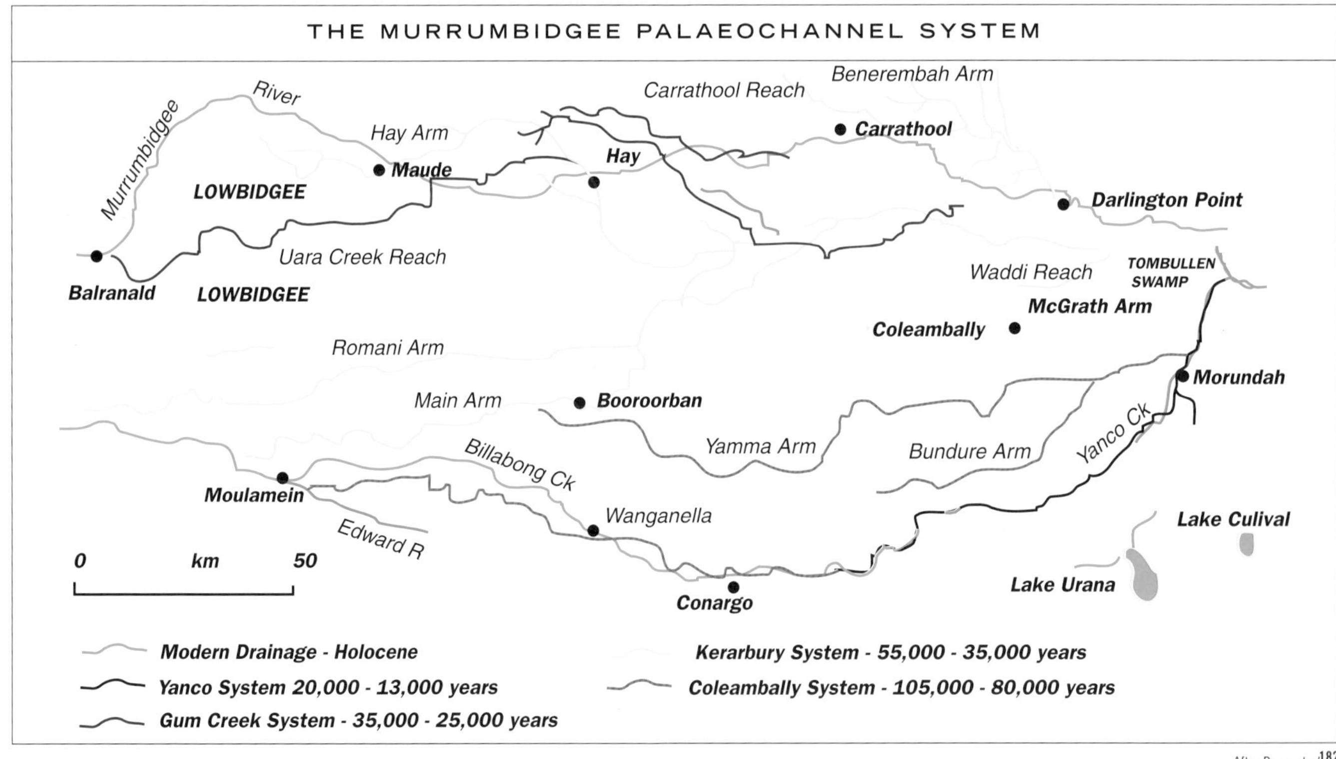

After Page *et al*[182]

The Kerarbury Pit—its sands and gravels are being commercially exploited.
KEN PAGE

powerful river at times between 20 000 and 13 000 years ago. Peak flows in this system are correlated with snow melt on catchments during the last glacial stage of the Pleistocene ice age.

The present flow regime in the Murrumbidgee was established about 12 000 years ago. Today, major salinity problems exist in the upper catchment. Wagga Wagga has urban salinity problems (as do Yass and a number of other towns).

THE MURRUMBIDGEE RIVER

The Murrumbidgee River is one of Australia's largest rivers. It rises in the Snowy Mountains and flows through undulating terrain to Wagga Wagga. Between Gundagai and Wagga Wagga it has formed a well-defined floodplain approximately 1 to 2 kilometres wide. Downstream of Wagga Wagga the gradient decreases and the floodplain width increases to between 5 and 20 kilometres. At Narrandera the Murrumbidgee enters the Riverine Plain and from there follows a highly sinuous course to its junction with the Murray River below Balranald. The Murrumbidgee drains one of Australia's largest catchments, 84 000 square kilometres in extent, which has three distinct sections:[182]

- The **upper catchment** is mountainous and hilly (20 500 square kilometres), and is separated from the middle by two large reservoirs, Burrinjuck and Blowering. These dams trap most of the sediment derived from the upper region and effectively isolate it from the rest of the river. In steeper headwater regions the channel grade is typically controlled by bedrock, separated by intermittent deposits of angular gravels, small cobbles and sand. Deposits of fine-grained material in this system can be found in the low-gradient channels, either in stable in-stream bars, where it is trapped by densely growing reeds, or on small transient bars in the channel bed.
- Downstream, the river enters the **mid-catchment section**, of 13 500 square kilometres, which is characterised by rolling terrain dissected by numerous gully networks. The major tributaries of

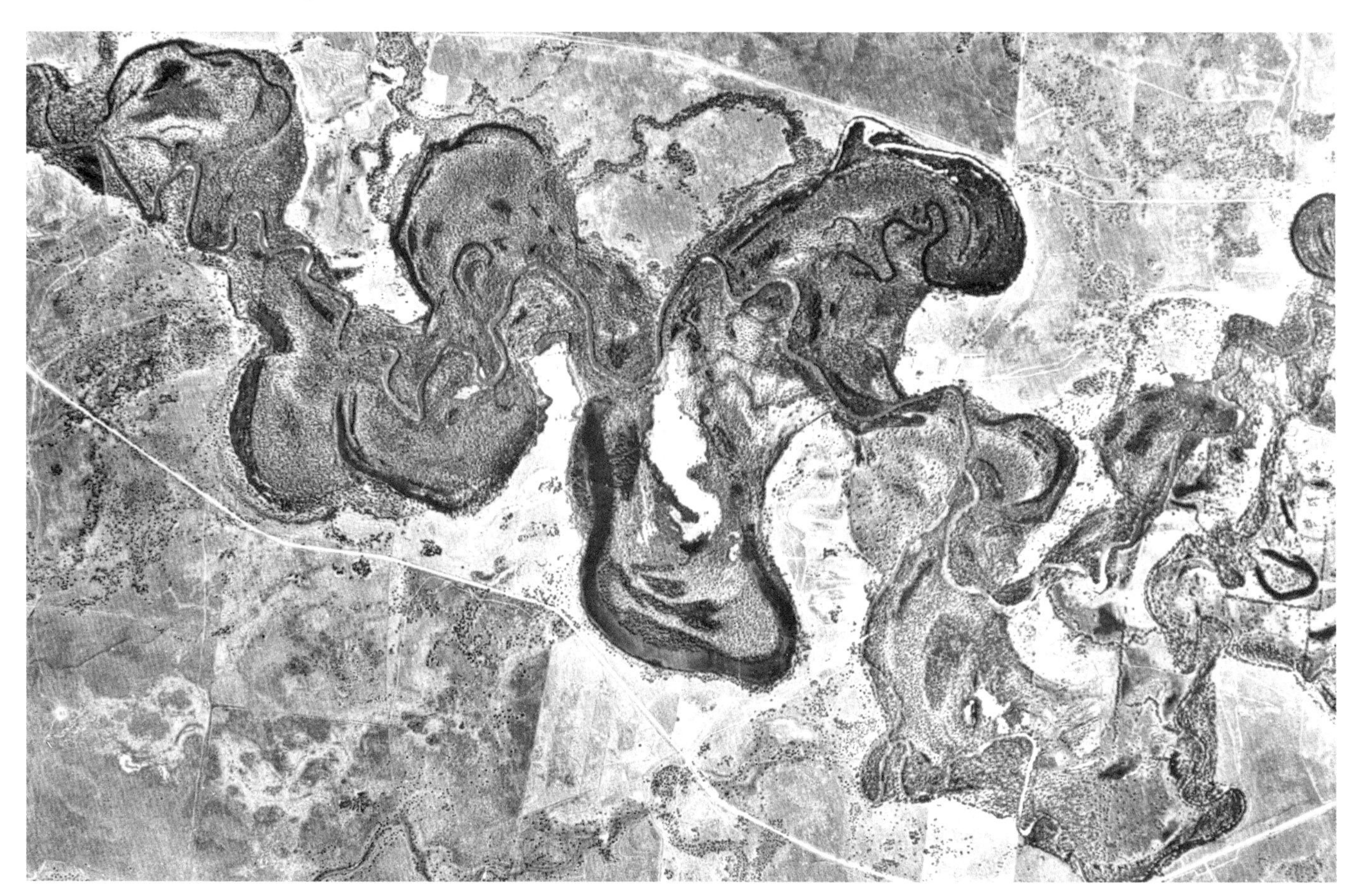

Aerial photograph of the Gum Creek palaeomeander scars and cut-offs along the modern Murrumbidgee floodplain west of Darlington Point.
PHOTOGRAPH KINDLY SUPPLIED BY DR KEN PAGE

the Murrumbidgee join the river in this region, each with a subcatchment of about 1000 square kilometres. The lower reaches of these tributaries, and the main channel of the river itself, are characterised by deposits of coarse-grain to fine-grain sand beds. Three major types of land use occur in the tributaries in the mid-Murrumbidgee—wheat and cereal growing (22 per cent), pasture (59 per cent), and forest (18 per cent). About 20 per cent of the catchment has been assessed as having a potential for sheet or rill erosion of up to 5 tonnes per hectare, per year. The incision of river and stream channels in this region has been shown to be a massive source of sediment which chokes lower reaches of the river.[187] (New and very sophisticated technology is now available which can determine the source of sediments precisely.) [188]

- Downstream of Wagga Wagga the river enters the **lower catchment region** (50 000 square kilometres), where it becomes highly sinuous as it crosses the Riverine Plain to its confluence with the Murray, some 1500 kilometres from its headwaters. The average annual flux of sediment at Wagga Wagga is estimated to be 580 000 tonnes. The only major tributary in the lower Basin is the Lachlan River, which joins the Murrumbidgee between Hay and Balranald. The Lachlan passes through an extensive area of low gradient marshes before it enters the Murrumbidgee. Flow is dissipated in this region and consequently the Lachlan only contributes flow to the Murrumbidgee during infrequent large flood events.

In recent years there has been widespread concern about the 'dirtiness' of the Lower Murrumbidgee and the occurrence of toxic algal blooms in weir pools. It has been widely believed that these algal blooms were due to nutrification from agricultural fertilisers and sewage treatment plants, feedlots, and so on, which increased phosphate levels. A report by the CSIRO Division of Water Resources[189] on research commissioned to find out what the real situation was, made some interesting findings:

- It was confirmed that most of the suspended sediment in the Lower Murrumbidgee is derived from the tributaries between the Burrinjuck and Blowering Dams and Wagga Wagga. Gully and channel erosion have mobilised the sediment, more than 90 per cent of which is derived from subsoil sources.
- A large proportion of the suspended sediment (about 60 per cent), is only transported a relatively short distance before it is deposited on the floodplain between Wagga Wagga and Narrandera. Less than 20 per cent of the suspended sediment which passes Wagga Wagga reaches the Murray River.
- The Tumut River, Burrinjuck Reservoir and Billabong Creek are not major sources of sediment.
- Phosphorus in the river is derived predominantly from diffuse natural soil sources. Most of the phosphorus (80–97 per cent) which passes Wagga Wagga is bound to the suspended sediments, and the study also showed that concentrations on suspended sediments have not varied significantly over the last 40 years. This is a particularly interesting finding and emphasises the need to control river bank and floodplain erosion and degradation in order to prevent toxic algal blooms and promote river health.

The Murrumbidgee River is a major supplier of irrigation water and flow is heavily regulated by both the two major dams and a series of weirs below Wagga Wagga. The flow regime before regulation was highly variable with a winter maximum. The combination of regulation and extraction for irrigation has resulted in a major change, with flow now being spring and summer dominated when water demand by irrigators is highest. Floods on the river are concentrated in winter and spring. The impact of regulation on flood events varies according to their magnitude. There has been a

THE UPPER MURRUMBIDGEE CATCHMENT

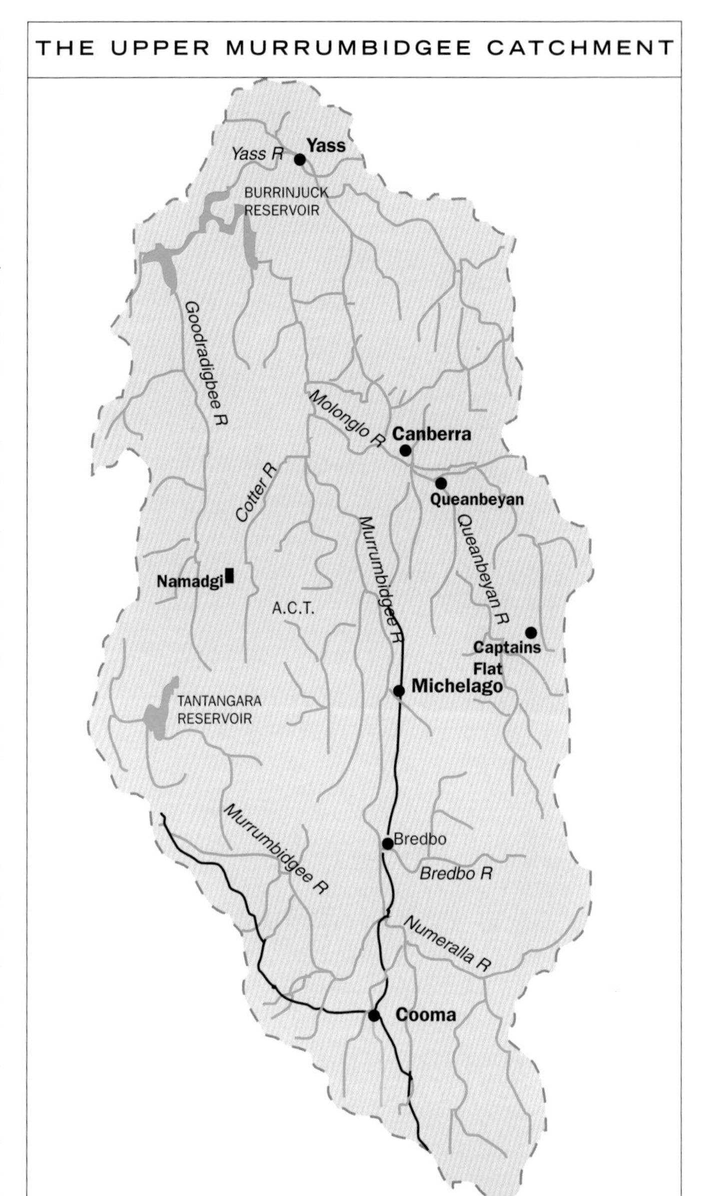

A 'chain of ponds' type of pond in the headwaters of the Numeralla River, in Dry Creek, preserved when the river changed course in the July 1991 floods.
BARRY STARR

CONFLUENCE OF THE NUMERALLA AND MURRUMBIDGEE

Top: The old Department of Lands Parish map showing the confluence of the Numeralla River and the Murrumbidgee shows a 'little river', the Numeralla, as described by Lhotsky in 1834. Middle: A 1994 aerial photograph shows that by then it had become as major as the Murrumbidgee. Lower: By 1990 the effect of reduced peak flows in the Murrumbidgee, a result of the Tantangara Dam, had reduced the overall width of the Murrumbidgee and its willow-clothed banks.

After Starr *et al* [190]

reduction in the occurrence of small and medium flood events, as flow from the upper catchments can often be controlled by the storage reservoirs. For larger floods, the flow response of the river is similar to pre-regulation regimes, especially in the mid and lower reaches where the weir gates are raised during these events.

The Bredbo and the Numeralla Rivers in the upper Murrumbidgee catchment which have undergone massive changes since European settlement

The small creeks in the headwaters of the Bredbo and Numeralla Rivers were 'chains of ponds' when the first explorers and settlers came to the region.[190, 191] The Dry River still has ponds in its lower reaches, preserved because in the July floods of 1991 the river changed its course and left them isolated. In the floor of the gully that caused the isolation, a new pond has developed.

Most of the headwater streams suffered the rapid changes described in other rivers which have their headwaters in the Southern Highlands. The introduction of grazing animals, followed by over-grazing, rabbits and droughts, led to serious erosion and stripping of upland valley fill. This in turn led to siltation of lower

A new pond developing in the floor of the gully which caused the re-routing of the river in the floods and saved the beautiful pond illustrated on opposite page. BARRY STARR

CHANGES IN THE COURSE OF THE MURRUMBIDGEE 1863—1944

CHANGES IN THE COURSE OF THE MURRUMBIDGEE 1863 TO 1944:
The line of willows in the 1944 photo shows the position of the original river channel. The river apparently 'jumped' into its new course during floods in the 1920s. BARRY STARR

reaches, destabilisation, and deep incision of rivers. The erosion in many cases preceded the clearing of riverine vegetation and land for agriculture and happened within a decade or two of the arrival of Europeans and their animals. The rate of erosion declined after this initial devastating stage, and a new equilibrium of sorts was established. Many of the original deep gullies in headwater valleys which appear on early maps and documents of the 1880s and 1890s are today visible as scars, mainly filled and stabilised by time.

PRODUCTIVITY OF THE MURRUMBIDGEE IN THE EARLY DAYS OF SETTLEMENT

Wagga Wagga on the Murrumbidgee has recently celebrated 150 years since its establishment, and it is interesting to read early accounts of the river and to find out how much human-induced change has occurred in that period. Who better to inform us than Dame Mary Gilmour, who is as much an Australian icon as the once-mighty river which she knew as a small child and whose changes she recorded in *Old Days: Old Ways—A Book of Recollections* (first published in 1934 and re-released in 1986)?[195] This source of information is recommended to anyone wanting an unbiased insight into environmental change, biodiversity, Aboriginal management of natural systems, and the attitudes of early settlers who were prepared to learn from those First Australians—and a great deal more besides. Mary Gilmour was a brilliant observer who had no hidden agenda, and the picture she paints of the highly productive river and floodplain ecosystems is so much at odds with the situation as we know it today that it is a revelation.

When Mary Gilmour first knew the Murrumbidgee it teemed with fish and freshwater lobsters, so numerous that they could be caught by hand or so easily by the simplest fishing methods that enough to feed a crowd were landed before the fire had died down to coals suitable for cooking them. Some of the lobsters were large enough to feed a family. Such abundance was the result of the 'farming' of the river and floodplains by the Aboriginal people, not just an unenhanced natural state. When, a few years later, the abundance of fish was no longer there, the lobsters were practically gone and only a few yabbies were caught, Mary Gilmour's father explained 'when the Blacks went the fish went', because the whites did not practice the 'farming' and the relatively few settlers had depleted the stocks by their changes to the river and their fishing. The freshwater lobsters, in particular, had required Aboriginal management—leaving the breeding adults, protecting the young, monitoring the abundance.

Fish traps were an essential part of the river farming, supporting the huge gatherings of Aboriginal people which took place on different rivers in turn, and providing food for resident or transient tribes. It was an accepted principle that no one tribe owned the water and its contents. By careful management a sufficiently large supply of river food could be guaranteed for these important cultural occasions—imagine feeding 8000 or more people from the river and floodplain for several days, and not depleting stocks so far that none remained for future use! The site for the gatherings was determined by the available food supply—and not only the rivers were farmed with this in mind. No hunting of kangaroo, possum or wildfowl was allowed in the area for several years before a gathering to ensure that there would be enough food for thousands. (Mary Gilmour describes the corroborees as like census taking, with competitive 'games' involving skills in hunting, fishing and physical prowess, and she comments that excuses were made by the settlers to use the opportunity of so many blacks collected in one place for the terrible massacres which 'solved' the Aboriginal ownership problems of the land.)

The fish traps were of several sorts. On the long reaches of the larger rivers, stone traps were made and outcrops utilised. These were keyed so that the size of the fish that could go through them could be regulated. This prevented the small fish, which could leave the trap, from being eaten by the larger ones that remained. Small traps or barriers were placed on tributary streams and their maintenance was a public duty. They were made of stones or of logs or smaller woven timber, to patterns known only to their Aboriginal manufacturers. When floodwaters penetrated up gullies, timber baulks were put across to impound fish when the water receded, and some of these fish were distributed to billabongs or stretches of rivers to increase populations in other useful places. The traps, of all sorts, were always placed in locations which were just right for survival of such structures, implying a knowledge of engineering and hydraulics. When the Europeans came and displaced the traps, the places where they had been were found to be the only suitable spots for dams, weirs or diversions.

On the Murrumbidgee the fish traps were blasted out or otherwise destroyed to allow the wool barges through and this early change to the river was devastating.

The preservation of sanctuaries by the Aboriginal people, protected places in rivers, some billabongs, parts of floodplains and valleys and forests, allowed animals to breed and live in safety. In the Wagga region there was once a large sanctuary for emus at

Aboriginal fish traps in the bed of the Darling River at Brewarrina, New South Wales. REG MORRISON

WAGGA WAGGA: A CITY ENCIRCLED BY A FLOOD-PRONE RIVER

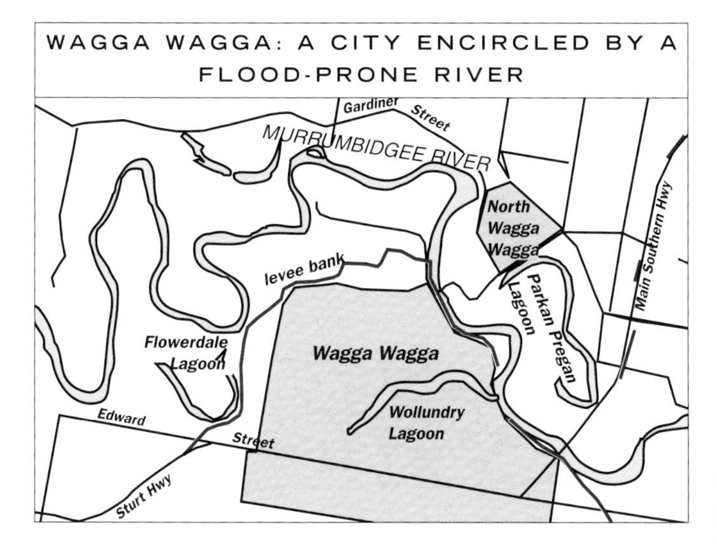

Eunonyhareenyha (which means 'breeding place of the emu'); the Pregan Pregan Lagoon was a sanctuary for pelicans, swans and cranes; there were duck sanctuaries in extensive and widespread swamps. In the early days the Europeans respected the sanctuaries to the extent that emu hunting with dogs was not practised in the nesting season and duck shooting was not annual in the swamps. But soon the restrictions vanished and the animal populations were decimated as white populations grew—and Aboriginal peoples too were decimated.

Mary Gilmour was an early champion of conservation, long before most Australians were aware of such a concept. She wrote, 'the sins one day to be repented [of] in this Australia of ours is the diversion of waters from the great fish and bird sources', and she goes on to list the wetlands of the Darling, Bogan, Barwon and Lachlan; the Deepwater and Ganmain Marshes on the Murrumbidgee towards Narrandera; Morangarell on the Bland; Tooyal on the Murrumbidgee and others near Deniliquin, all of which she knew personally. She would be turning in her grave if she knew what irrigation, largely for cotton, has done to wetlands like those of the Gwydir and Macquarie as well. She remembered the Castlereagh as a 'noble stream', a great and beautiful river, and mourned the degradation of all the major rivers which were so silted up by early in the twentieth century that river travel was well-nigh impossible.

How a third-generation farmer on the Murrumbidgee sees the way the river, its tributaries and lagoons have changed

In a privately published booklet of reminiscences, Max Leitch, an elderly farmer in the Wagga Wagga district, recalled what the river was like when he was growing up on his parents' property on the south side of the Murrumbidgee, and compared it with the situation as it was in 1985:[196]

- before Burrinjuck Dam was built, the river used to drop to very low levels in the summer months. Long streamers of 'green algae moss' were attached to every log and stick in the water, but the water itself was crystal clear, and the river bottom of clean hard clay or gravel was clearly visible at low water, even in the deep holes which occurred all along the river. These holes were up to 10 metres deep at low water. Before 1914, there was a series of dry years when all the pools and lagoons were nearly empty, but several floods filled them up later and fish and lobsters were again abundant from 1916 to the 1930s;
- the river changed its course frequently, its meanders swinging from one side of the channel to the other, in response to floods;
- after Burrinjuck Dam was built, the 'Conservation Department' was worried about algae at low water so they 'dumped huge quantities of Copper Bluestone into the dam', killing not only the organisms in the dam but also those in the river below, a disaster from which the ecosystems have never fully recovered. The use of more of the poison in springs and creeks to kill fluke larvae did additional damage. Introduction of European carp has completed the destruction and now lobsters are rare and stunted, native fish endangered, and the river waters are muddy, the deep pools gone.

The Bulgari Lagoon, an old cut-off meander of the Murrumbidgee, has an interesting history, according to Max Leitch. It covers about 50 hectares when full and is 10 metres deep in the centre. It has been dry twice since the first farmers came in 1840, once in 1914 and again in 1944–45. It is in the middle of the floodplain and fills from both ends when the river rises above 7 metres. It also has a local catchment. It holds water for ten years after being filled by floods. During its dry time in 1945 it was found that stumps of very large trees were buried in the deepest part, where they must have grown without permanent inundation for at least 40 years, and possibly far longer, to reach their final size. This implies a long dry period long before settlement, and it would be interesting [my comment] if a carbon date could be obtained for such tree stumps—they might be found to correlate to a glacial stage of the Pleistocene.

Water plants grew around the margin of the lagoon, in up to 3 metres of water, providing a haven for wildlife and a fish breeding area. Cattle used to wade in until just their heads were above water, to feed on waterweeds. The carrying capacity of the lagoon was calculated by the farmer to be 50 cows and calves, much higher than that of the surrounding land. However, carp have changed the productive lagoon into a useless mudhole not even suitable for stock water, destroying the plant life. When Max Leitch requested permission to net the carp, his application was refused because the law says that this stretch of the Murrumbidgee and its attendant waters is a line-fishing area only.

An interesting insight into how land use affected a tributary of the Murrumbidgee is given in his account of Houlaghan's Creek in 1914. The creek joins the Murrumbidgee about 10 kilometres west of today's city outskirts after running southwards through the Junee, Ganmain and Coolamon regions, which were major wheat growing areas at the time. At that time the Department of Agriculture advocated growing wheat on long fallow. The land was ploughed and left to lie as long as possible while weeds were controlled by constant working with scarifier and harrows. When the 1914 drought broke and a heavy storm dumped 100 millimetres of rain over the catchment, the rolling country with all its bare prepared land eroded and the river ran liquid mud. It rose 2 metres in a few hours and fell just as rapidly.

> 'The mud killed every living thing in the river. Fish lined the banks with their heads out of water in the morning, and were all dead and floating upside-down by lunchtime. Father and two employed men started pulling some out before breakfast. Some were to be cleaned to eat fresh, some were salted and smoked, and a waggonette load was taken around the farms to be given away.

THE LAKE GEORGE MINE AT CAPTAINS FLAT: episodes of river and floodplain pollution

The Lake George Mine, derelict since 1962, is located adjacent to the Molonglo River at Captains Flat, 70 kilometres upstream from Lake Burley Griffin which is in the heart of Canberra. The mine has a long history. The ore body was discovered in 1874 and an initial period of mining occurred between 1882 and 1899. The second and most productive mining operation began in 1939 and continued until 1962.[192, 193, 194] Four million tonnes of ore were mined by underground methods to produce lead, zinc, copper and gold concentrates. The 3 million tonnes of tailings produced were retained in dams on the banks of the Molonglo River. Mine and mill wastewater containing high concentrations of zinc was discharged into the river, and consequently water and sediment between Captains Flat and Canberra became heavily polluted with high levels of zinc. During floods in 1939 and 1942, tailings dams collapsed and catastrophic contamination of the river resulted. In the 1942 event, 30 000 cubic metres of tailings were discharged into the flooded Molonglo and were deposited on the river floodplain some 12 kilometres downstream, damaging large areas of pasture and resulting in a long-term problem with raised levels of zinc and other contaminants in the soil.

The large heaps of spoil at the Lake George Mine.

After mining ceased, the mine workings flooded and eventually overflowed from the northern ventilation shaft. Within the mine workings, the water is in contact with the exposed ore body and continually acquiring heavy metals and acid. Sealing of this shaft in 1966 resulted in a significant decrease in pollution of the river, but soon afterwards water found its way out through a collapsed adit and this spring is now a permanent feature of the mine site and the main outflow of mine drainage water. The red and yellow spoil heaps and the dramatic pit became a tourist attraction—everyone appalled by the photogenic degradation.

Seepage from the mine workings.

In 1973, a joint Commonwealth–New South Wales report proposed remedial action for the pollution of the Molonglo River because accelerated and ongoing erosion of mine waste dumps and continuing flow from the northern spring would continue to affect the river and, ultimately, Lake Burley Griffin. Not only zinc levels were the problem—acid and heavy metals would continue to be released from river bed sediments; dump instability would lead to further catastrophic pollution of river and floodplains; and the construction of the proposed Googong Dam would reduce the diluting effect of the Queanbeyan River.

Remedial measures were carried out in 1976 (costing $2.3 million), aimed at securing the dumps to prevent their collapse and sealing them to prevent erosion and leaching; and minimising the flow of mine water from the main spring. Re-formed dumps were designed with safer slopes and covered by a clay seal followed by a layer of crushed rock to provide drainage and prevent capillary rise of heavy metals; a revegetated layer of soil covered the dumps; and drainage was planned to prevent recharge of waters in the workings. The official word is that the remedial measures have been largely successful, and with ongoing maintenance over the two decades since, there has been little erosion and a decrease in river pollution. For some 30 kilometres downstream, however, zinc levels are greater than the recommended safe levels for consumption, and aquatic ecosystems are affected. A great reduction in aquatic invertebrates and in fish has been noted. A chronic input of zinc to the Molonglo continues from a number of point sources and, especially under low flow conditions, the water quality just downstream from Captains Flat remains poor.

The mine pit. LAND & WATER, QUEANBEYAN

The unofficial verdict is that little was achieved by the remedial action except a decrease in the number of tourists who used to visit the mine site and that the whole area 'leaks like a sieve and water still permeates the mine workings', some of it redirected from surrounding properties whose owners see the mine area as a convenient receptacle for unwanted run-off. Very serious problems obviously still exist.

There was one huge Murray cod that was too big for two men to pull out of the water onto a sand bank. They estimated it to be well over 300 pounds, and its mouth was 2 feet wide. Someone put its head on a stump and it remained there for many years. The lobsters and shrimps crawled out onto the bank to die in a solid band three to four feet wide and a foot deep ... there would have been a ton of dead or dying fish to every one or two hundred yards of river bank.'

Max Leitch adds that the long fallow system caused such degradation that farmers were ready to walk off their farms in the 1930s because skeleton weed had taken over—but they found that sheep thrived on skeleton weed. A return to more grazing and less cultivation revived the land, and the arrival of subterranean clovers made the farms viable again.

GROUNDWATER IN THE LOWER MURRUMBIDGEE VALLEY

The Lower Murrumbidgee Groundwater Management Area is situated in the eastern Murray Basin, mainly between the towns of Narrandera, Booligal, Balranald and Jerilderie. It has a total area of 32 000 square kilometres and encompasses an area of low salinity and high yielding aquifers in the lower Murrumbidgee catchment commonly referred to as the 'Murrumbidgee alluvial fan'. The lower Murrumbidgee was identified as a high risk groundwater system during the Statewide program of Aquifer Risk Assessment in April 1998, which identified over-allocation, local draw-down and invasion of aquifers by saline groundwater as current risks.[197, 198]

Because of the risk of over-allocation, a 12-month moratorium on issue of licences for groundwater use for irrigation was put in place in September 1997, and extended for 18 months in September 1998. The previous groundwater allocation guidelines had recommended allocation of up to 650 000 ML and allowed for gradual depletion of the resource. The moratorium was introduced ahead of the 650 000 ML ceiling and total allocations now stand at 494 000 ML. Usage in 1997–98 was 241 000 ML. The estimated annual recharge of the aquifers is only about 250 000 ML, clearly indicating that the groundwater system is over-allocated. Surficial aquifers occur in the Shepparton Formation sands and clays of the upper 50 to 70 metres; the 'deep aquifers' which are the important groundwater resource are in the 50 to 70 metres of Calivil Formation below (deposited between 15 and 5 million years ago).

THE NATIONAL WATER REFORM AGREEMENT

In 1994, the Council of Australian Governments (COAG) developed a National Water Reform Agreement, which provides a framework for water reform. The States gave a commitment to implement management policies for natural resources based on principles of ecologically sustainable development. The New South Wales Government announced the first stage of Water Reforms in 1995. In August 1997, the second stage of Water Reforms was announced, including the establishment of groundwater management committees. A statewide Aquifer Risk Assessment Program was undertaken in 1998.

The Murrumbidgee Groundwater Bore, situated between the Colleambally Irrigation Area and Narrandera, is pumped at a rate of 2500 ML/year between September and May (pump rate 350 L/sec.). The volume of excellent drinking water currently used for pasture irrigation would be enough to provide 2 litres a day for each member of Australia's population. WENDY TIMMS

The lower Murrumbidgee has a great amount of groundwater in storage, mostly saline, but with an estimated 280 million ML of low salinity. The groundwater flows from east to west down the valley, moving at 7 to 10 metres a year along the gentle gradients. The major recharge for the deep aquifers is the Murrumbidgee River. Age of the water increases westwards and away from the river. Ages of 3000 years and younger were determined for deep aquifer groundwater near the river and upstream of Gogeldrie Weir. Away from the river, ages increase to about 7000 years beneath Whitton and Coleambally, and downstream they increase to about 15 000 years old beneath Carrathool. Younger ages of less than 3000 years were determined for groundwater in the Shepparton Formation. Recharge appears to occur slowly and continually with the effects of floods and high rainfall events being spread over time as groundwater moves slowly downwards and away from the area of recharge.

Groundwater age, storage and recharge characteristics have major implications for sustainable groundwater use.[197] The large volume in storage can give a misleading impression of large groundwater availability. The considerable average age of water in deep aquifers may give the impression that current recharge is negligible, while in fact the age is related as much to the large amount in storage as to the rate of recharge. Also, the amount of recharge which is currently occurring may have as much to do with the basin being 'full' as to the physical ability of water to move from the main recharge source (the Murrumbidgee River) to the deep aquifers. It is possible that the most important change that may occur in recharge over time may actually be a response to groundwater extraction and its ability to facilitate additional recharge.

Groundwater pumping affects groundwater levels significantly over the short term. In the eastern half of the Management Area quite a large area, extending into the Coleambally Irrigation Area and fringes of the Murrumbidgee Irrigation Area, experiences seasonal fluctuations of 5 to 10 metres. However, little or no seasonal fluctuation occurs beyond Hay, along the northern boundary or in the south-east where little pumping occurs.

A great deal has yet to be learnt about groundwater recharge and use. A groundwater model for the lower Murrumbidgee is currently being updated to assist in estimating recharge amounts from various sources. The University of New South Wales is currently working on a Natural Heritage Trust-funded project to assess vertical aquifer connection at two locations.

A Groundwater Management Plan, when it emerges from all the research, discussion papers and reports, will aim to balance the needs of the environment with those of humans—sustainable use without degradation of the resource or the natural systems which depend upon it, and sustainable socio-economic structures which share and do not exploit the resource.

THE LACHLAN RIVER

The Lachlan River rises in the Southern Highland region between Yass and Gunning. It flows northwards at first before flowing west and south-west across the upper section of the Riverine Plain, with wide meanders, anabranches and a complex web of tributaries, wetlands and swamps. It terminates in the Great Cumbung Swamp north of Balranald, merging into the Lowbidgee wetlands and swamps where Lachlan and Murrumbidgee meet. Before the rivers were deeply incised, the Lachlan below the Cumbung Swamp met the Murrumbidgee only when floodout channels joined it to the Swamp in heavy rainfall events.

Valleys in the upper Lachlan catchment are old—Eocene basalts flowed down them, establishing their antiquity and persistence.[201] The upper Lachlan district is the hilly tableland area of the Great Divide, to the north-west of Goulburn. The district comprises the Shires of Crookwell and Gunning, and part of the Shire of Boorowa. The area is drained by the Lachlan, Abercrombie and Crookwell

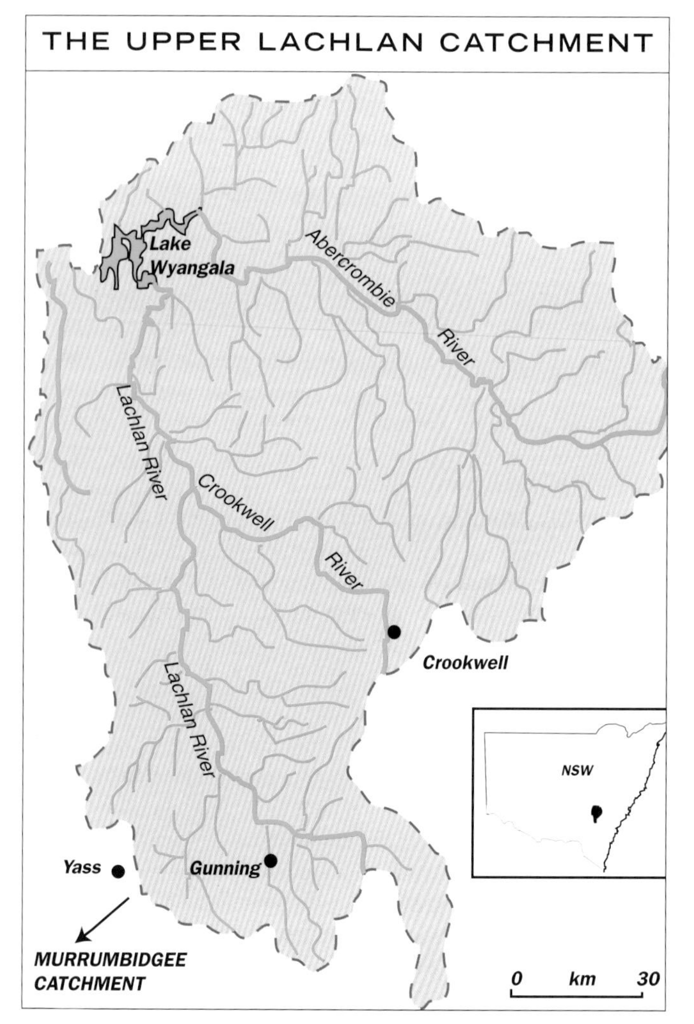

Serrated tussock has taken over on vast areas of the Southern Tablelands.
BOB SPROULE

Rivers, which all feed into the Wyangala Dam. Early and heavy clearing of the region; cropping, particularly in the volcanic soils of Crookwell Shire; and heavy grazing pressures by stock and rabbits, have led to erosion, changes to river and streams with incision and siltation of channels, and to the invasion of large areas by the dreaded serrated tussock (*Nassella trichotoma*).

This uncontrollable weed was probably first introduced into Australia from South America in about 1900. In 1935, it was identified in New South Wales from plants collected near the Yass River and was first known as Yass tussock. Its spread was dramatic. In 1938, it was proclaimed noxious in the shires of Abercrombie, Gunning and Goodradigbee; and Mulwaree, Oberon and Lyndhurst followed in 1940. It is now declared noxious Statewide. It grows on soils of all types, from the most fertile to the poorest rocky ground. It has an incredibly efficient distribution system: the seed head breaks off near its base and is distributed by wind, water, animals and humans. Wind can carry seed heads up to 10 kilometres, and close to dense infestations, masses are blown into fences, windbreaks and gullies. The seed is also transported by rivers, and plants have been found on the banks of the Lachlan 60 kilometres from the infestations in the headwaters. Animals carry the seed in their fur, humans spread it by machinery. Animals can also carry it in their digestive tracts. One study revealed that wethers taken from an infested property passed an average of 4600 seeds per animal in the ten days after removal—and most of the seeds were probably viable. (Tussock passes so slowly through the sheep gut and is so nutrient-poor that animals confined to mature tussock pasture can die with full rumens.)

The construction of major water storages at Wyangala, Lake Cargelligo and Lake Brewster have replaced the naturally high and variable flows of the Lachlan River with more even flow. The Murray–Darling Basin Commission has allowed development of some irrigation, and the river flow has been reduced overall. Many private, off-river storages have been built on properties to augment irrigation allocations, further reducing the volume available for the 'environmental flows' which are essential for river health. Major floods in 1972–73 brought European carp into the river system and they have successfully multiplied to reach plague numbers. Today, carp represent 85 per cent of the fish in the Murray–Darling Basin. In some reaches of rivers there is one carp per square metre of channel. They breed prolifically, with each female capable of producing 1 million eggs twice a year, and they survive in conditions that kill other species—they are the last to die in droughts. Their

OXLEY STATION ON THE LACHLAN RIVER: managing marginal land

Oxley Station, 88 kilometres north-west of Hay, lies on the southern side of the Lachlan, near its entrance into the Cumbung Swamp (and its junction with the Murrumbidgee). The station has 54 kilometres of river frontage. The Lachlan in this region is accepted as the eastern boundary of the Western Division of New South Wales, country which is increasingly marginal for both agriculture and grazing the further one moves west. The history of Oxley Station gives a good picture of what landholders have had to deal with since settlement, and the problems and changes which have occurred through time.[235]

The station was originally named Thelungerin, and later Thelangerin West, being re-named Oxley when purchased by Thomas Darchy in the 1850s. By 1865 it had a homestead, woolshed, cottage and stockyard; by 1872 the homestead was described as spacious and comfortable, set in an excellent garden with roses, grapes and an immense fig tree, and 'a dam with sluice gates regulated the reception and discharge of water from the lake to the Lachlan River'. When Darchy died in 1877, his Oxley run of 30 000 hectares (and his Western Division properties) was managed by trustees until sold to the owner of an adjacent property, Robert P. McFarland, grandfather of the present owner, with exchange of contracts in 1906 and the final bill of purchase in 1911. The two properties were run as one and renovation of the homestead and addition of buildings followed on Oxley. Considerable areas of the property were withdrawn in 1919 for post-war soldier settlement blocks, and 25 000 acres were sold in 1956 to help pay for probate on the death of Grandfather McFarland that year, reducing Oxley Station to its current size of 10 125 hectares.

In the 1920s, the McFarland properties ran 30 000 sheep. The 1940s drought decimated the stock, reducing it to 1200 breeding ewes. The 1950s brought some recovery over time, but to nowhere near the original carrying capacity. Even now, Oxley is still suffering from the effects of past droughts and overstocking. Scalds are still visible today where stock remained too long on drought-damaged pasture and wind erosion followed, but other huge areas of scald have been reclaimed and revegetated with native shrubs and grasses. In 1956 Robert P. McFarland died and his son Andrew worked the two properties for 27 years to pay off the estate death duties.

Andrew McFarland's son Bob and his wife Errolly took over the running of Oxley in 1968. Bob McFarland has seen many changes in the stretch of the Lachlan which runs through Oxley. As a child he knew it as a clear, deep channel shadowed by stands of river red gums. He and his wife have watched its degradation over time. It has become a shallow, turbid stream, impossible to navigate—with fallen timber from eroded banks blocking progress due, they believe, to the Booroola strain of introduced carp which arrived in the big wet years of 1973–74.

The 1970s were reasonably good years, but by mid-1980, with wool prices falling and production costs rising, the McFarlands were not making a living from the land. (This timing corresponds to the 'running out of the Green Revolution' scenario, when technology and land-use improvements ceased to make headway against the mounting degradation of environments, not only in Australia but elsewhere in semi-arid regions of the world.) A radical change of management was decided upon which involved pasture improvement, changing the quality and quantity of wool production, and generating off-farm income.

Feral pigs were a major problem on Oxley, as were damage to stock and property by illegal hunters. With the adjacent wetlands ideal for pig breeding and shelter, feral pigs were present in their thousands, invading the farmland, damaging pasture, digging up crops, killing young and weak stock, breaking down fences and harbouring disease. (Each sow has a potential breeding rate of five litters, each containing up to 10 piglets, in every two-year period.) The solution to this problem was not the poisoning/trapping/shooting programs usual in the early seventies, which were expensive, labour-intensive and seldom very successful. Instead, Bob McFarlane allowed recreational city shooters, who signed an agreement which protected the environment, to come onto his property legally. Around 4500 people have been through the property, and at least 10 000 pigs have been shot, which has helped the bank balance and counteracted the fall in wool prices—a novel on-farm, off-farm income! (A more recent project involved forming a company—the Four Seas (NSW) Pty Ltd—in 1991, processing European carp to make 'Charlie Carp' garden fertilisers and other products.)

Having reduced the feral and sheep grazing pressures on Oxley, the next step was to improve pastures and to start repairing the rangelands and wind scalds. (Irrigation was not an option—on the sodic soils, with their pH of 8 to 9, it would bring salt to the surface, and there are a lot of soils with 67 per cent sodium.) This they have done with planting of deKoch saltbush, with the help of agronomist Andrew Sippel, who established a nursery at Narromine and now produces more than 2 million saltbush plants a year.

The introduction of deKoch saltbush has improved the carrying capacity of Oxley. Having 'plantations' available has enabled resting of broader paddock areas, maintaining pasture cover and reducing wind erosion. Regeneration of scalds and of saline seeps has been possible with the saltbush. In addition, a major selective breeding program has changed the type of wool produced by Oxley's merino sheep to 'elite' type, produced by sheep with 'soft rolling skin' characteristics. Wool cuts have increased, with three times the usual cut from 5000 head, and the soft-prickle-free wool is in high demand. Now with a son, Andrew, taking an interest in the property, they are using Holistic Resource Management principles in the running of their enterprises, with increasingly controlled grazing practices, reduced paddock size, fencing according to vegetation types, and extending stock watering systems to distribute grazing pressure more evenly.

All this innovative and inspiring activity by the McFarlands is enough to give one hope that when others follow with adaptations to their land use suited to their situations, the goal of attaining regeneration of soils, vegetation and biodiversity might eventually be attainable. (Bob McFarland emphasises that they aim at more than 'sustainability', which would be maintaining the present situation—they aim for *improvement* on that state.)

---

feeding habits actually damage the river banks and create turbulence. They suck the mud and cause erosion around roots, even resulting in trees falling into the stream. Their activities almost certainly contribute to algal blooms by their nutrification of the waters. The turbidity they cause reduces light penetration in the water, and water plants suffer. Food chains, and the whole river ecology, change when the 'rabbits of the rivers' take over.

The region between the Lachlan and the Murrumbidgee comprises the extensive agricultural lands of the Central West of New South Wales, mainly given over to dryland farming. Rainfall is fairly uniform, not exceeding 500 millimetres a year, and evaporation is about four times that figure, so the potential for salinity problems is obvious. Large areas of the land surface comprise cracking-clay alluvial soil. Gilgais (crabholes, melonholes) are features of many areas, as they are on other floodplain cracking-clay soils elsewhere in the continent, such as the brigalow lands of the Condamine (see page 203). Poor infiltration into these soils results in overland flooding in rainfall events; roadside notices warning of flooded roadways in apparently flat countryside with no river in sight as one drives through the region seem inappropriate—but such regional surface flooding, often exacerbated by raised roadways which cause back-up of floodwater, and increases its erosional power, is very much a fact of life.

The heavy clay soils present problems for agriculture. Conventional cultivation practices compact the soil, causing loss of soil structure. Decreased infiltration results, and waterlogging in low areas, similar to the processes seen on similar soils on the Wimmera Plain (see page 237). A research farm at Morangarell, south-east of West Wyalong, is experimenting with raised-bed agriculture to improve drainage, and is showing good response; another research farm at Monteagle, about 20 kilometres north of Young, is experimenting with agroforestry, planting trees and saltbush. (Morangarell, on Bland Creek, is mentioned in memoirs of Dame Mary Gilmour as a place where she witnessed a gathering of about 300 Aboriginal people in 1897, attended by the last Gundagai chief.[195] It was a smaller than usual gathering because people from across the Lachlan were unable to attend due to the drought. Gatherings of up to 500 on Morangarell were documented by the Reverend George Grimm, who commented that these were just local gatherings as there would not have been sufficient fish in the Bland to accommodate larger numbers and the adjoining station owners would not have allowed unlimited hunting on their properties for fear that cattle might be speared, and reprisals would lead to another frightful massacre of Aboriginal people.)

Groundwater is generally saline, in confined, low yielding aquifers at various depths in the alluvial sediments (as, for example, in the Jemalong–Wyldes Plains irrigation district.) Significant quantities of good quality groundwater, however, are available from the deep lead palaeochannels of the ancestral Bland and Lachlan Rivers (and others).

The Wyalong Goldfield, a historic source of primary gold in quartz veins (1894–1920), lies to the west of the Bland Creek palaeovalley, which was probably first incised into bedrock early in the Tertiary.[199] Alluviation of the palaeo-Lachlan River system in the Late Tertiary buried the Bland Creek palaeovalley, and it is now considered likely

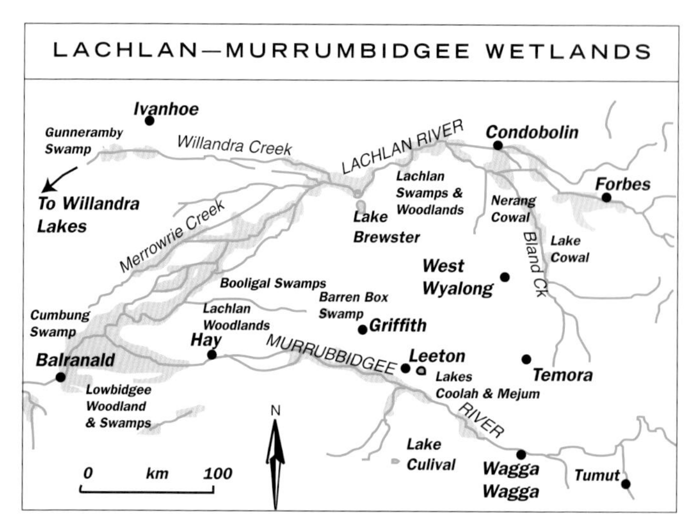

Willandra Creek, a major anabranch of the Lachlan, in line with the westward-flowing river before it sweeps south-west towards the Murrumbidgee, leads to the Willandra Lakes, of which Lake Mungo is one. The anabranch today only occasionally carries excess floodwaters.

that alluvial gold sourced by erosion of the vein deposits will be found in sediments in newly detected parts of the palaeochannel systems (as has been the case further south at Temora).

In the upper Lachlan catchment, as in the Yass River catchment, salinity and erosion are major problems. Landcare groups are active, doing what they can to rehabilitate stream banks, exclude livestock from riparian zones, and undertake widespread planting of trees, particularly in the elevated recharge areas of tributary catchments. Boorowa suffers urban salinity, with the salt water table so close to the surface—3 metres under the main street—that the foundations of houses and buildings are crumbling, particularly in lower areas. The town water supply is affected by salinity in the Boorowa River catchment, and roads crumble at ten times the rate they should, because of salinity. The adjacent shire of Harden–Murrumburrah spends half its annual road budget on salinity-related road repair.

The rising saline water tables result from land clearing and upset of the hydrological balance, an automatic response to increased recharge. Changes to the sorts of crops grown in recent years have improved the situation in some areas, but in many areas, particularly where the situation is amplified by irrigation, the rising water table syndrome is out of control.

A stretch of the Lachlan between Forbes and the Wallaroi Creek junction shows the complex meanderings and anabranches of the modern river.[200] South of the river, Lake Cowal, Nerang Cowal and Bogandillon Swamp lie along a palaeodrainage line which joins Wallaroi Creek near the junction.

Lake Cowal, with an area of 10 500 hectares, and Nerang Cowal, 4100 hectares, are recognised by the Department of Land and Water Conservation as wetlands in the Lachlan Basin. Lake Cowal is listed on the Register of the National Estate (which makes its impending degradation as described below all the more surprising).

Lake Cowal is an ephemeral lake, filled either from Bland Creek or from flood breakouts from the Lachlan. It empties initially by outflow to Nerang Cowal and Bogandillon Creek to Bogandillon Swamp and the anabranches of the Lachlan south of Condobolin. When the lake level falls below the Nerang Cowal sill, the lake dries by evaporation. Over the last 50 years, the lake has been substantially full in seven out of ten years. In the first half of this century it was dry for substantial periods. (Increased run-off from cleared land in the

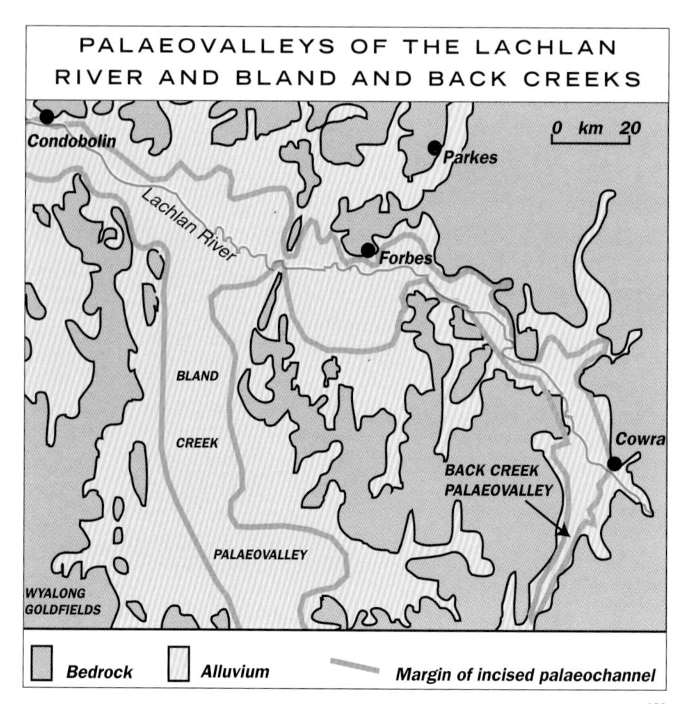

After Lawrie *et al*[199]

THE LACHLAN BETWEEN FORBES AND WALLAROI CREEK

Wallaroi Creek
Bogandillon Swamp
LACHLAN RIVER
Forbes
MANNA MT
JEMALONG - WYLDES PLAINS IRRIGATION DISTRICTS
Nerang Cowal
Burcher
Newell Hwy
WAMBOYNE MT
N
Lake Cowal
Marsden
Bland Ck
Mid Western Hwy
West Wyalong
NSW
Parkes
Map Area
Sydney
0 km 20

After Northern Mining[200]

last 50 years probably accounts for some of this difference, and the slight increase in rainfall in eastern Australia since the 1940s for the rest.) When the lake is full, it is a significant waterbird habitat and fishing spot. When dry, the lake bed is grazed, and parts of it have been cropped at times.

The lake has received a lot of attention and publicity in recent years because of the proposed Lake Cowal Gold Project. Considerable environmental damage, affecting areas far beyond the mining lease, could be done in this region where palaeorivers lie beneath the surface, ready to carry pollutants far from the point of entry. (A local resident tells how, not long ago, a dam was built on a property close to the lake in a spot chosen with care because the situation with subsurface palaeodrainages was known. The dam filled, everyone was delighted, and then suddenly it emptied completely—it had developed a vast plug-hole which took the water away into some underground conduit!)

The mine is to be sited on part of the lake and wetlands, which is seen by scientists who cannot be dismissed as 'greenies who protest about everything' as an environmental disaster just waiting to happen. No matter how thorough the environmental assessment has been or how stringent the regulations, there is much that is not known. So many factors and possibilities exist which are beyond the scope or expertise of current investigations that no one can be absolutely sure what will happen when a project with such levels of disturbance and so many potential hazards is embarked upon. The New South Wales Government gave permission for the project as an election 'carrot' dangled in front of the rural electorate just two weeks before the State election of 1999—the fact that three or more years previously, the project was absolutely unacceptable for all the

MURRINGO CREEK

Murringo Creek, a tributary stream in the headwater systems of the Lachlan River, runs north, parallel to the Boorowa River. The area was surveyed in 1850 and has been subjected to 140 years of grazing and clearing. In Murringo township (halfway between Boorowa and Young) the deep incision of the modern channel can be clearly seen on the Murringo Common. Active bank and streambed erosion have deeply incised the creek into its valley. Bank collapse is a major concern. As in Tarcutta Creek, alterations to the catchment since settlement have destabilised the drainage. Now a concerted effort is being made by the local Landcare group—the Murringo Common Rivercare Project—revegetating the area and restricting stock access. Native trees have been planted on the banks, and reeds have been re-introduced to the channel.

Landcare groups have been revegetating the catchment of the creek, which is deeply incised with collapsing banks. M.E.W.

OLD MAN SALTBUSH: a sink for carbon and a repairer of man-made deserts

Old man saltbush (*Atriplex nummularia*) has long been recognised as one of the best native fodder plants. Along with other palatable species which were once widespread in the semi-arid and marginal lands, it was rapidly eaten out in the early days of settlement when over-grazing by hordes of introduced animals was turning large tracts into unproductive wastelands. Early maps are dotted with 'Old Man Plains' and 'Saltbush Plains'—names which had disappeared on later maps and where now no *Atriplex* remains. As early as 1879 it was noted in an agricultural gazette that the plant was disappearing over wide areas, and its demise was regretted because it was known to afford green feed at times when other plants were dry, and thus was instrumental in maintaining the strength and uniformity of wool fibres.

Now the plant is coming back into favour again, planted to: rehabilitate worn-out land; as a satisfactory sink for carbon; to lower saline water tables because of its deep root systems; and to improve stocking rates and wool and meat quality. The plant is enjoying renewed popularity because of its many other valuable qualities, which include: [231]

- Ease of establishment over a wide range of soil types and moisture levels. Plantings provide a continuous high quality feed supply throughout the year, unaffected by seasonal shortfalls in rain, and provide flexibility of grazing management.
- The high protein content (22 per cent) results in healthy animals. Sheep produce finer micron wool while maintaining fleece weight. Protein supply is maintained during critical grazing periods.
- The prolific feed produced enables increased stocking rates—at least a five-fold increase over other native pasture. An added advantage is being able to predict accurately how much food will be available at any season.
- Growing the plants in hedge-like rows provides shelter for animals and increases productivity. Feeding from shrubs above the ground reduces problems with worms, and the chenopodium oil contained in the foliage of the old man saltbush is also a pest deterrent.
- As a low fat stock feed, saltbush produces excellent meat. The metabolised energy from old man saltbush is only slightly lower than from feed grains (oats and barley).
- Maintenance costs are low, with a one-off cost for establishment, and plants can live for 100 years.
- Saltbush plantings are efficient users of irrigation water, providing the same productivity levels as improved pasture, but using one-third ofthe water.
- The root system is three-tiered, with a penetration of up to 5 metres and spread up to 10 metres wide. The plant has access to deep-seated moisture, minerals and nutrients which are unavailable to other plants. The minerals brought up from depth are deposited on the surface by grazing and become more readily available. Surface roots collect moisture from light rainfall and are useful in binding the soil and preventing erosion.
- The ability of the plants to thrive in saline to extremely saline environments is an added advantage—they can work to lower saline water tables and improve productivity of irrigation-damaged lands.

Stocking rates on established old man saltbush plantations (more than three years old) were calculated by Dick Condon, formerly Agricultural Adviser for the Western Division of New South Wales, to be as follows:

Old man saltbush planted on a property, providing good pasture, restoring the soil, lowering the saline water table, providing income where the degraded land it replaces was not—and a sink for carbon. ANDREW SIPPEL

RAINFALL	DRY SHEEP EQUIVALENT UNITS (per hectare per year)
300 mm	6–7
400 mm	8–11
500 mm	13–16
600 mm	17–20

On a per annum grazing basis and allowing only for soil moisture from immediate rainfall:

A major old man saltbush nursery has been established in Narromine by Andrew Sippel, supplying literally millions of seedlings to farmers. From propagation to field establishment and grazing management, the business has grown over the last ten years and the results are evident all over the country. Vast improvements in soil structure and organic matter levels follow establishment of saltbush plantations. Oxley Station, where saltbush planting and holistic resource management are practised, is said to stand out like a fertile oasis in the largely degraded surrounding area.

A business such as this nursery and farm advisory service, providing remedies for degradation and hope for farmers, and promising a future for generations to come, gives hope that some of the damage done to our land can be reversed. It is a wonderful example of working with the land and within the parameters it has set, returning, if not to nature, at least closer to natural systems.

best environmental reasons, simply did not get a mention—all doubts had evaporated.

Hopefully, the time will come, though too late for the Lake Cowan region, when it is realised that there are some environments which are too fragile to abuse, where the long-term consequences completely outweigh the immediate financial gain.

THE CHANGING FACE OF THE SALTBUSH PLAINS OF THE WESTERN RIVERINA

Sheep are no longer providing a living for many graziers on the saltbush plains of the western Riverina. The escalation of changes to land use in order to survive financially is presenting the same dilemma which has been described for the irrigated cotton industry on the Darling.

From Urana in the east to Moulamein in the west, over the last few years a conservatively estimated 50 000 hectares of 'undeveloped' country has gone under the plough—some for irrigated crops, some for dryland cropping. **Underlying this 'development' is the widely accepted belief that the country involved is only shrubland (cottonbush, dillon bush and the like) or grassland and that therefore clearing presents no problem.** This quite erroneous belief is actually enshrined in the Western Riverina Grassland Plan (an amendment within the Native Vegetation Conservation Act 1996) which allows the clearing of grassland or shrublands with a shrub cover of less than 5 per cent without approval as long as there are no areas of 'high conservation value' present and as long as at least 15 per cent of these community types are left uncleared on any individual property.

While it is hard to blame the 'family' landowners who have to survive, once again the problem, and the ultimate threat to the environment, assumes different proportions when agri-business enterprises take over, amalgamating properties and doing things on a regional scale. Then it ceases to matter whether in the long run the land is desertified, provided the run is long enough for the business to recoup its investment and make a healthy profit.

One of the immediate casualties of loss of saltbush plains vegetation is loss of biodiversity, already very much affected by the almost universal degradation of the vegetation from the over-grazing pressures which had to be inflicted for the grazier even to survive financially to this point. Destruction of the plains wanderer's habitat will make this threatened species (listed under both the Threatened Species Conservation Act in New South Wales and the equivalent Commonwealth Act) an extinct species. While this bird is a visible and charismatic example which receives mention, what of the other

EMISSION TRADING, CARBON CREDITS AND THE SEQUESTRATION OF CARBON FOR THE REDUCTION OF GREENHOUSE GASES

At the first United Nations Conference of the Parties (UNCOP) in Rio de Janeiro in 1991, chaired by Maurice Strong, it was recognised that human activities were contributing to the build-up of Greenhouse gases and that the situation had to be addressed. In 1997, at the Kyoto Conference (UNCOP3), agreement was reached between nations and the **Kyoto Protocol** resulted. This document set out a plan for nations to reduce their Greenhouse emissions to a level set at 5.2 per cent lower than their 1990 levels. Australia claimed fossil fuel dependency, requested a level 128 per cent *above* its 1990 baseline emissions, and settled for 108 per cent. (And many of Australia's scientists and conservationists felt shamed by the stance taken by our government.)

Subsequent UNCOP meetings—UNCOP4 in Buenos Aires in 1998 and UNCOP5 in Bonn in 1999 (with a meeting arranged for The Hague in 2000) emphasise that the serious global warming situation is being recognised.

The Kyoto Protocol set out a range of parameters for developed and developing nations, two of which have particular significance for Australia—**emission trading** (offsetting emissions which are above the set limit but cannot so far be decreased, by financing measures which have a beneficial impact elsewhere) and the development of **sinks** (ways of sequestering carbon and locking it away in systems).

Considerable emission trading funds are available worldwide—large mining and other industrial corporations which are unable to comply with the reductions set by the Protocol are seeking involvement in forestry projects which will balance their CO_2 budget. Australia, very limited in the areas where massive reforestation would meet our aim of attaining the necessary offsets, has enormous opportunity if a modification of requirements in the Protocol from the planting of **trees** (reforestation) to **replacement of vegetation** can be achieved at the Bonn meeting. The role of old man saltbush, which is suitable for revegetating the semi-arid and arid zones of our land, and even the salinised areas the extent of which is always increasing, could be most significant. Vast areas of inland Australia were saltbush plains before they were altered by grazing and clearing, and the re-introduction of *Atriplex nummularia* into these areas would accomplish more than carbon trading requirements. (See page 230 for information on old man saltbush and the considerable expertise which has been developed for its propagation and establishment on degraded lands, and its commercial advantages.)

It has long been recognised that removal of the deep-rooted perennial vegetation (saltbush and native grassland or shrubland, *not* forest, over most of the used portion of the continent) has been responsible for the changes to local hydrology and rising saline water tables. What has not been recognised is just how much carbon is in fact tied up in saltbush, which has both very extensive near-surface and deeply penetrating root systems, and which restores organic carbon and nutrients (often drawn from deep in the soil profile below the reach of lesser plants) to soils. The restoration of vegetation over the vast areas of the continent which are now sparsely covered or scalded would benefit the Greenhouse gas reduction, while restoring the degraded land, and the cost of transformation could be met by the overseas funds available in carbon trading. The benefits do not stop there. The restored saltbush plains would restore viability to the sheep industry because the fine wools which are produced on a saltbush diet are sought after in world markets, and the superior quality and flavour of meat would revitalise the lamb trade.

A tremendous opportunity exists for Australia to turn a situation of disadvantage into one of huge advantage (and it is a limited window of opportunity in respect of available funds from other nations, which will be spent in other continents if Australia does not act fast). This sort of situation is very much in keeping with the philosophy of the Heartlands Project (see page 190), which seeks integrated projects specifically designed for Australia's particular and unique problems.

We need all the visions and rays of hope that can be found because so much of our environmental situation is grim—and the truly Australian thinking in a project like the one outlined above is an inspiring example.

species which are part of the already struggling web of life of the saltbush plains, which is the only web capable of maintaining natural sustainability? How sustainable is the use of groundwater in these enterprises? What will be the effect of the chemicals which get into it in the irrigated areas? How will the fragile, ancient soils cope with cropping and the rising water tables which will follow the removal of the extraordinarily deep-rooted saltbushes and even the perennial grasslands when crops with quite different patterns of water use replace them? High evaporation rates, the natural salinity of soils which increases westwards, uncertainty and variability of rainfall, all contribute to making the enterprises high risk in terms of the environment (and ultimately unsustainable).

In *Listen ... Our Land is Crying*, pages 155–57 carry an account of Steam Plains, a property south-west of the Colleambally Irrigation Area which had been grazed almost to extinction by the 1930s. This property is now being converted to irrigated cotton production. It is typical of many further west which are now changing over from unproductive sheep to agriculture, with or without irrigation. The fragility of the country under European-style land use was emphasised in the study of Steam Plains, and the need for caution in such ventures as are occurring on the saltbush plains now is clear.

COTTON ON THE LOWER LACHLAN

No new water licences have been issued on the Lachlan since 1979, due to over-allocations in the past. Many existing licences are sleepers, but with the sheep industry at an all-time low, more landowners are looking to alternative sources of income. The new irrigated cotton venture near Hillston on the lower Lachlan may be just the beginning of a swing to irrigated agriculture—and it would not take much expansion in this agriculturally marginal country on the edge of the Western Division to create a serious environmental situation.

Merowie, owned by the Twynam Pastoral Company, a major landowner in New South Wales, is an amalgamation of several properties—with licences for river and groundwater extraction. As with river water, no further licences are to be issued for groundwater, and those issued in the past have a 'mining the aquifer' component, as the extraction rate has exceeded the recharge rate (as in the lower Murrumbidgee). The Department of Land and Water Conservation is trying to find solutions to the situation. The groundwater at Hillston is generally of good quality and suitable for irrigation.

A pump station on the banks of the Lachlan extracts water from a pool in the nearly dry river to fill the dam on the other side of the bank. When the big pumps are in action the river has been known to flow backwards.
DAVID MARSH

According to Rob Collins, manager of Merowie and chairman of the Lachlan Cotton Growers Association, the total area of irrigated cotton on the Lachlan was 5000 hectares in 1998; 13 500 hectares in 1999; and rising to a possible 18 000 hectares in the future if all available licences are exploited. On Merowie, 300 hectares were grown in 1996; 800 in 1998; and 2000 in 1999. In 1998, Merowie used 80 per cent river water and the balance was bore and rain water. The aim is to use 50 per cent river water and 50 per cent other. The average yield of cotton prior to 1996 was too low and too variable to be economical, but since then new varieties and technologies have improved yields steadily. Twynam will keep on growing cotton back to back until falling yields dictate a rotation. Maize, wheat and canola are also grown. Soil organic matter is said to be increasing under cotton, as the stalks are smashed up and returned.

As with cotton everywhere, environmental problems relate to pesticide, herbicide and fertiliser application. Insecticide spraying regimes involve a minimum of 10 passes (endosulfan) for conventional cultivars and five passes minimum for genetically modified cotton. Neighbouring farms which graze cattle have a mandatory 'e' (endosulfan) rating which incurs a penalty in price at the abattoir. A responsible attitude to pesticide use prevails under the current management at Merowie, and the aim is to reduce chemical use generally. Experimental growing of irrigated lucerne in blocks close to the cotton is aimed at attracting predator insects which require year-round greenery and which may assist in the biological control of pests. (We see here, again, the dilemma which exists because of the conflict between best-practice, technologically up-to-date farming and the prosperity it brings locally, and the fundamentally incompatible demands on the river, groundwater and fragile environments.)

A number of regulations apply to cotton growing: it can be irrigated as long as it does not move closer than 40 metres to the river; tail-waters must be contained on the farm; and the layout must have the capacity to accept 25 millimetres of run-off without escape. However, there is no statutory obligation to monitor what is happening to the soil and to nutrients in the ground—and studies on the Liverpool Plains have shown that the build-up of substances and their contamination of groundwater is a serious area of concern. The requirement to plan layouts to cope with 25 millimetres of run-off does little in sudden big rainfall events, when run-off would certainly flow to the river. Contaminated groundwater would also be a source of pollution for the Lachlan if leakage of pesticides, herbicides and fertiliser residues was found to be occurring.

A factor which appears to have received no attention in the planning of the management of Merowie, or any other irrigation scheme in semi-arid landscapes, is the level of salinity of river water. The Lachlan is known to receive a great deal of salt from its upper catchment, where the Boorowa River, for instance, discharges more than 100 000 tonnes annually into it. The high evaporation rate in semi-arid areas ultimately causes salinisation of soils under irrigation—irrigation in places even with only seasonal aridity has a limited life, in consequence.

While one large enterprise under good management may get away with extracting the water it needs and containing its threats to environmental health locally, an expansion of such enterprises would be undesirable. If everyone jumps on the bandwagon (and who could sit in judgment when times are hard and farmers are trying to survive?), a situation will develop where over-exploitation is inevitable. It is enough that the Darling and its tributaries should be potentially emulating the Aral Sea disaster—surely lessons have been learned and big business will not be allowed to compromise the future in

THE RIVERINE PLAIN IN VICTORIA AND WESTWARDS TO THE MALLEE

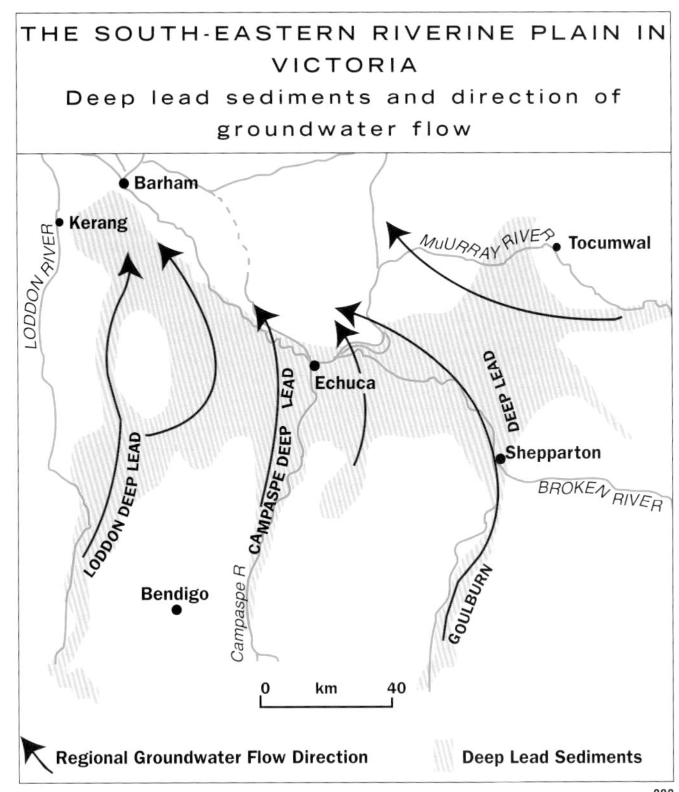

After Macumber[202]

THE LODDON, CAMPASPE AND AVOCA RIVERS

The Goulburn, Campaspe and Loddon Rivers, major tributaries of the Murray River, traverse the Riverine Plain of Victoria, flowing northwards from headwaters in the south-western extremity of the Great Divide.[202] The Great Divide forms a low watershed, less than 100 metres above sea level, separating Murray Basin drainage from the southward-flowing rivers of the Otway Basin. Westwards of the Loddon Plain lies the Mallee, with its largely aeolian landscapes. Streams which pass northwards across the Mallee do not reach the Murray River but flow instead into terminal lakes. The Avoca River normally terminates in Lake Bael Bael and the Avoca Marshes; the Avoca River distributaries, Lalbert Creek and Tyrrell Creek, flow into Lake Timboram and Lake Tyrrell respectively. The Wimmera River runs through Lake Hindmarsh to terminate usually in Lake Albacutya, but occasionally continues northwards through the Wyperfeld lake system towards Wirrengren Plain (the final lake on the Wimmera river system, last filled in 1923). The northern Mallee has major regional groundwater discharge areas, principally in the Sunset Country and the Tyrrell Basin.

Lake Hindmarsh, the terminal lake for the Wimmera River.
MALCOLM ANDERSON

Prior to the drought of 1967–68, little was known of the deeper flow systems which lie below the modern Victorian Riverine Plains. A great deal was known about 'deep leads' (palaeodrainages) in the upland valleys, because drilling had occurred on a large scale in the late nineteenth and early twentieth centuries to map them for mining. It was known that very large water reserves were contained in the auriferous leads. Eocene and Late Cainozoic basalts had flowed down some of the river valleys of the ancestral Avoca and Loddon. Former courses are marked by volcanic tongues, which buried the gold-bearing deep leads. The descendants of the original blocked rivers have been diverted laterally, often causing twin laterals to develop, one on either side of a lava 'tongue'.[17]

With the drought of 1968, detailed investigation of the downstream continuation of the Loddon Valley palaeodrainage commenced, and the subsequent drilling program traced the well-defined valley-fill system northwards, where it was designated the Calivil Formation. (It is overlain by the Shepparton Formation which forms a varyingly impervious aquitard or capping layer.) The Calivil Formation has since been shown to be the most significant regional aquifer in Victoria for transporting groundwater from highland catchments across the Riverine Plain towards regional groundwater discharge zones further down-basin, both on the plains and in the Mallee.

Regional groundwater discharge occurs in the lower regions of the Loddon Plain (where gypsum and salt harvesting takes place).

Commercial salt extraction near Kerang, Victoria. M.E.W.

The Kerang–Swan Hill region is the major groundwater discharge area of the system.

In general, the hingeline between groundwater recharge and groundwater discharge zones in a gently sloping system is approximately at the midpoint of the basin. Prior to 1973, the hingeline in the Loddon system was close to Calivil. Anomalously high rainfall was experienced in the catchments in 1973–74 and the plains were subjected to sheet flooding via the Loddon River and Serpentine and Bullock Creeks. Pressure in the Calivil Formation aquifer began to rise and bores in the southern sector became artesian; further heavy rain and flooding the next year again increased pressures, this time in the northern sector as well but declining the further north travelled, with bores at Kerang very close to, or actually just, flowing. Pressures peaked in 1975 and have since settled at plateau levels higher than the pre-1973 levels. The 1983 drought, one of the worst on record, had virtually no effect on pressure levels, and return of rains saw recharge and a new high plateau level. The hingeline between recharge and discharge had moved more than 20 kilometres south by 1975 and a slow southward advance has continued since.

Virtually the whole of the central and lower Loddon Plain lies within a major irrigation district, where salinity problems appeared soon after irrigation commenced in the late 1800s; today, much of the plain suffers from high water tables and salinity. The problems inherent in large-scale flood irrigation within a zone of regional groundwater discharge in a semi-arid region are waterlogging and soil salinity, leading to eventual destruction of the vegetation. Such problems rapidly developed on the Loddon Plain, after only 19 years. Today, many formerly grassed, dry or ephemeral lake beds are either ephemeral or permanent salt lakes. One-third of the district is severely salt-affected and virtually the whole of the Loddon Plain has a saline water table at less than 2 metres. Irrigation water has been a major contributor to the water budget.

It is considered quite likely that during the last glacial stage of the Pleistocene ice age the whole of the Loddon Plain (and the rest of the Riverine Plain to its east) would have been a groundwater discharge zone. Not until approximately 8000 years ago, when water tables declined, would the area have been revegetated by eucalypts. (Evidence of their reappearance is found in pollen records in stream sediments and shallow depressions on the plains dated at about 8000 years.)

Barr Creek, in the Kerang district, drains areas subject to groundwater seepage as a result of the pressurising of the system. It overlies a Murray Valley deep lead in the area where it is joined by the Loddon valley system. It is the single largest contributor of salt to the Murray River from Victoria, carrying a staggering (average) 170 000 tonnes per year. Eighty per cent of the salt in Barr Creek is of groundwater origin.

There are two distinct flow regimes in Barr Creek: during the irrigation season it carries large volumes of irrigation drainage waters; during winter, flows are lowest with a large base-flow component. Salinity levels inversely follow the flow rates.

Eastward of the Loddon Plain is the Campaspe Plain, which also overlies a deep lead system containing a Calivil Formation aquifer. The palaeodrainage system is partially interrupted by a bedrock high in the vicinity of Rochester and as a result it has a more complex hydrological system than the Loddon, which has a gently sloping palaodrainage without such a median barrier. Pressurisation of the aquifer increases towards Rochester and the water table rises to about 5 metres. Once past the obstruction, it falls again, to rise towards the groundwater discharge hingeline which lies between Torrumbarry and Gunbower. By this stage the Campaspe deep lead system has merged with an east–west trending Murray Valley trunk system, which is joined, in turn, by the Loddon Valley system further west. The rises in water table have been less in the Campaspe system than in the Loddon, and with increased pumping of aquifers since the 1983 drought, a quasi-stable situation had resulted by 1991.[202] A very wet year could change the situation, however.

The Goulburn Valley and the Murray Valley palaeodrainages have also shown water table and pressure rises over time. It was predicted in 1991 that the whole Goulburn Valley would become a groundwater discharge zone within 70 to 90 years, and that the Murray Valley province will have artesian pressures within 50 to 100 years (and both much earlier, if major recharge results from increased rainfall events).[202]

If the long-term trends in groundwater systems continue, it is inevitable that the Riverine Plain of Victoria will eventually become a single large discharge zone—its Tertiary/Quaternary aquifer will be full. This must lead to land and water salinisation. Irrigation areas will be faced by a situation like that on the Loddon Plain today, where a more or less stable saline water table oscillates always within capillary reach of the surface, rising and falling with the seasons. Some of the streams within the newly established discharge zone will take on a role similar to that of Barr Creek, which is essentially an effluent groundwater drain. On the basis of a 1991 assessment, regional groundwater discharge zones will first appear in the central Goulburn Valley and the Campaspe Valley, if pumping fails to suppress the pressure rises. These conditions will gradually extend throughout the irrigation districts. Conditions on the plains are not helped by the fact that dryland salinity and erosion are rampant in the headwaters of the rivers on the Great Divide and salt is being discharged into the river systems.

REMNANT VEGETATION IN NORTH-EASTERN VICTORIA

Formation of the eastern section of Victoria's Northern Plains began at about the start of the Tertiary Period, following the uplift of the eastern and central highlands, the consequent formation of the Murray Basin and the subsequent deposition of alluvial materials by the mighty ancestors of the present-day Murray, Ovens, Goulburn and Broken Rivers (and the ancestors of the Broken, Boosey and Nine Mile Creeks). The sedimentary sequences laid down over all those millions of years are covered by recent material deposited in the last 35 000 years, with a surficial capping of wind-blown material which blanketed the region during the last glacial maximum. Meandering just below the surface of the contemporary landscape are prior streams and their associated levees, creating a pattern of soils with different characteristics, as elsewhere on the Riverine Plains.

The whole region was rich in biodiversity, from the Barmah and other red gum forests along the rivers to the box–ironbark forests, the woodlands and the grassy plains. Lists of the species of plants and animals recorded from the start of settlement and through the early days make the region appear like a natural Eden.

Early squatters and their flocks and herds between 1837 and 1869 brought changes through grazing, but with the introduction of the 1869 Land Act, the entire Northern Plains region was taken up by selectors and almost totally cleared of its native vegetation for

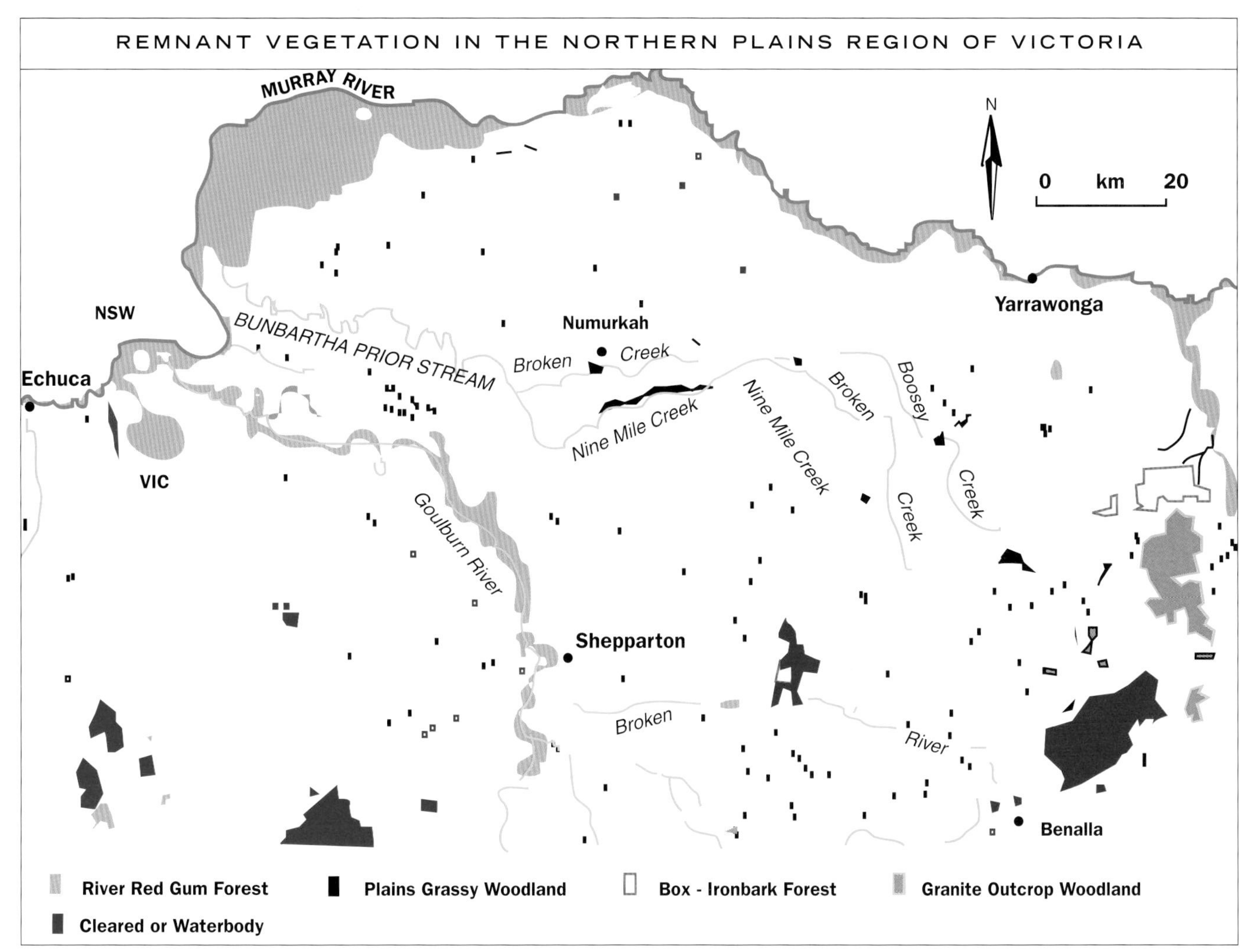

After Robinson & Mann[203]

pastures and crops, mostly within a decade. It was fortunate that there was legislation in place which protected river frontages and their vegetation along all permanent streams, otherwise there would not be the remnants of natural vegetation which exist today.

The Broken, Boosey and Nine Mile Creek systems in north-eastern Victoria retain more natural vegetation than is preserved on the rest of this most northerly portion of the plains adjacent to the Murray River. Remnants of almost-extinct grassy woodland and grey box forest; 27 species of threatened plants and three endemic species; nine threatened bird species; a threatened squirrel glider and a rare barking frog; 13 species of native fish, of which nine are listed as threatened, and many other significant plants and animals, all depend on the region. The Goulburn Valley Environmental Group applied for funds from the National Estates Program of the Australian Heritage Commission to undertake a survey along the entire length of the creeks—a total of about 450 kilometres. Very little was known about the natural environment in the eastern Northern Plains and much of the creeks system is bordered by public frontage, making changes in land management relatively straightforward if the survey proved that the region should be added to the National Estate. (The very comprehensive report on the flora and fauna, the conservation values of the remnants, and the current threats to survival, emphasises that action has to be taken now and a unified policy developed if what is left of the region's natural ecosystems is to continue to exist. A verdict on protection and conservation management is awaited.)

In the 1870s, the grey box (*Eucalyptus microcarpa*) woodland along the Nine Mile and Broken Creeks between Drumanure and Numurkah was so dense that 'Mr Brenion ploughed a furrow for his children to follow ...', and other parents blazed trees along the way to mark the route from their selections to the local school so that their children wouldn't get lost. Further west along the Broken Creek, selectors in the Barwo and Yalca districts likewise blazed trails through the dense white cypress pine (*Callitris glaucophylla*) forests to avoid becoming lost, and local farmers were paid £1 to guide visitors through the dense groves. One hundred and thirty years later, only 1 per cent of the tree cover of those pine and box forests still survives in the eastern part of the Northern Plains, and the woodlands along the Broken, Boosey and Nine Mile Creeks represent the single largest remnant of this ecosystem in northern Victoria. Now, nearly all vegetation types which survive are considered endangered.

The three creeks and their associated tributaries form part of the Broken River catchment which extends from near Benalla and the Killawarras in the south-east to Barmah and the Murray in the north-west. Altogether, the creeks drain a subcatchment of about 3700 square kilometres, with boundaries near the Murray between Cobram and Yarrawonga in the north; the Warby Ranges in the east; along the northern edge of the Broken River between Goomalibee and Pine Lodge in the south; and along the western and southern

POST-SETTLEMENT ALLUVIUM AND GOLD MINING IN VICTORIA

An example of the amount of off-site disturbance caused by gold mining is seen in the Bendigo district of Victoria. All the early activity was directed towards puddling (washing) gold out of alluvium, and puddling machines were introduced to speed up the process.[204] Gold was discovered in 1851 and by 1856 there were 1500 puddling machines in the area. In 1858 a Royal Commission was established to deal with the problems of sludge run-off. The problem was serious. On one station 70 kilometres from the diggings, an area of many square kilometres was covered by sludge. In places near the diggings the sludge can be several metres deep.

Later gold mining worked bedrock which was ground to powder in batteries. Battery sand is a distinctive sediment of uniform very fine grain size and low fertility. It can be found covering large areas of floodplains. It is often more than a metre thick and may extend many kilometres downstream of old mining towns.

Relatively intact buloke woodland near Gerang Gerang. LISA MORCOM

edges of Pine Lodge Creek and the Broken Creek in the west.

Each of the three creeks occupies a prior stream course of the Broken river system, and they all lie in the Broken River palaeovalley, which was formed about 100 million years ago by a river system much larger than today's, following the uplift of the central and eastern highlands. Over time, the palaeovalley filled with alluvial material and the river's course was constantly changing. The present-day channel between Benalla and Kialla was established in relatively recent times, when the three creeks became established in its old paths. The Broken Creek is the most direct descendant of the Broken River. It merges with the Boosey Creek near Katamatite and with the Nine Mile Creek near Dunbulbalane and then near Mundoona, finally flowing into the Murray River at the Moira Lakes in the Barmah Forest. (Nine Mile Creek was originally an anabranch of Broken Creek.) Downstream of its final junction with Nine Mile Creek, the present-day course of Broken Creek follows one of the prior streamcourses of the Goulburn River (the Bunbartha Prior Stream, which was flowing until about 21 000 years ago). The meanders and sandy levees along this stretch of the creek make it clearly distinguishable from the remainder of the creeks system. Other wanderings of the palaeo-Goulburn lie in the valley between that section of the creek and the Goulburn River.

THE WIMMERA PLAINS OF VICTORIA

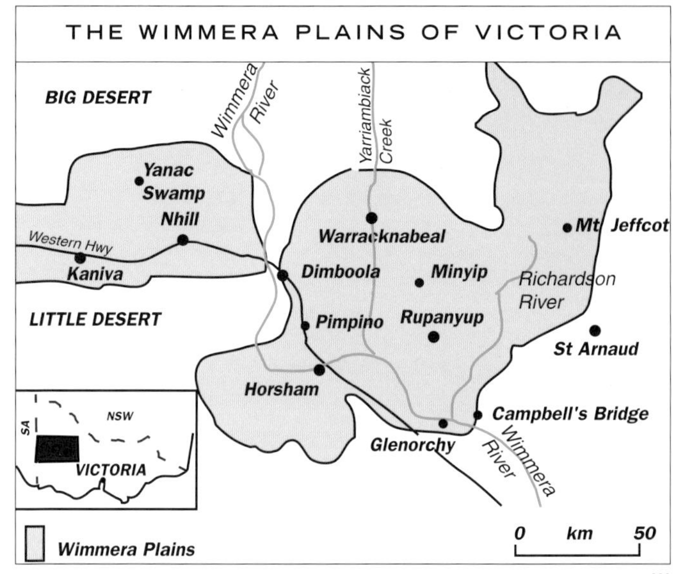

After Macumber[202]

THE WIMMERA AND NORTH-WESTERN VICTORIA

The Wimmera Plains occupy 12 500 square kilometres of western Victoria.[205] About 95 per cent of the plains is freehold and has been cultivated, and the 150 years of European settlement has had a profound effect on the native vegetation. Remnants are restricted to small blocks on private and public land, including road reserves. Historical records show that the fertile soils of the Wimmera Plains formerly supported grassy woodlands on rises and flats, and grasslands on shallow depressions and clay-pans—two ecosystem types among the most threatened in Victoria. Buloke (*Allocasuarina luehmannii*) grassy woodland is regarded as an endangered plant community in Australia.

Before European settlement, Aboriginal burning might perhaps have contributed to keeping the open grassy nature of the vegetation. Squatters arrived from 1841 and by 1851 all but the poorest land was taken and sheep grazing was the dominant land use. In the 1860s, with increased fencing and sinking of many wells, increased stocking rates became general. Rabbits began to reduce the carrying capacity of the land in 1876. There was little cropping until the 1890s when use of superphosphate started. A great increase in its use has occurred since the end of World War II, and land use in the region now is mainly cropping.

Landscapes in the Wimmera comprise low linear ridges separated by plains and depressions. Rainfall varies between 350 and 500 millimetres, while the mean annual potential evaporation at Horsham is 1450 millimetres. Shallow Tertiary marine sediments underlie much of the area, left by several penetrations of the sea into the sinking Murray Basin. As the seas regressed, north-north-west to south-south-east trending sandy beach ridges were left as strandlines; they are noticeable features of the landscape. Aeolian dunes, which are aligned west to east, are evidence of the dry and windy glacial stages of the Pleistocene ice age.

Mitre Lake and Mitre Peak, seen from Mt Arapiles on the Wimmera Plain, showing the almost total clearing of the landscape. M.E.W.

Melonholes in farmland on the Wimmera Plain. Farmers who have used conventional preparation methods, repeatedly crossing their land with ploughs and cultivators, end up with compacted soil, waterlogging and flooding. Those on adjacent properties who have used minimum-till methods have been able to sow their crops. The contrasts between fields managed in the two different ways is clear in these aerial photographs, which are stills from a video made by WCFA to promote better land management. CALVIN MUELER

Soils of the Wimmera Plains are generally grey, self-mulching cracking-clays in the western sector, with red duplex soils more common in central parts. The two types occur, however, as a complex mosaic across the whole region. The grey clays develop cracks, and gilgais are common. One elderly farmer remembers 'hearing the water running in the crabholes' on his grandfather's farm when 'some crabholes were that big, you could've hidden a truck in them'. The clay soils are neutral to alkaline at the surface, and strongly alkaline at depth. They set hard when dry, but are friable when moist and sticky when wet. They have a high water storage capacity—but the availability of the water to plants depends on a number of factors, including soil structure. The less common brown self-mulching clays are similar but gilgais are less common. The red duplex soils have a texture contrast between surface and subsoil at 10–30 centimetres. Surfaces set hard after wetting and drying. Soils are sodic, alkaline, with low phosphorus and nitrogen levels and low to moderate fertility.

Soils that were originally fertile have shown a productivity decline over time. Under cropping, levelling has largely eliminated gilgais in some areas but run-off and surface flooding are a widespread problem because the compacted soils do not let water soak in. Conservation farming has become almost universal on the plains. WCFA, the Wimmera Conservation Farming Association, is active. Minimum-till and no-till farming are now widespread, combating the soil structure loss which results from the use of heavy machinery on soils which compact into a solid, non-absorbing mass. Some dramatic contrasts have been seen after rains, where adjacent properties practise minimum-till and conventional land preparation. Fields in the first show uniform crop growth and only localised waterlogging, while in those using conventional methods, surface flooding occurs and may last for months, and irregular crop growth is seen (if the farmer can get the machinery in to the paddock to plant!).

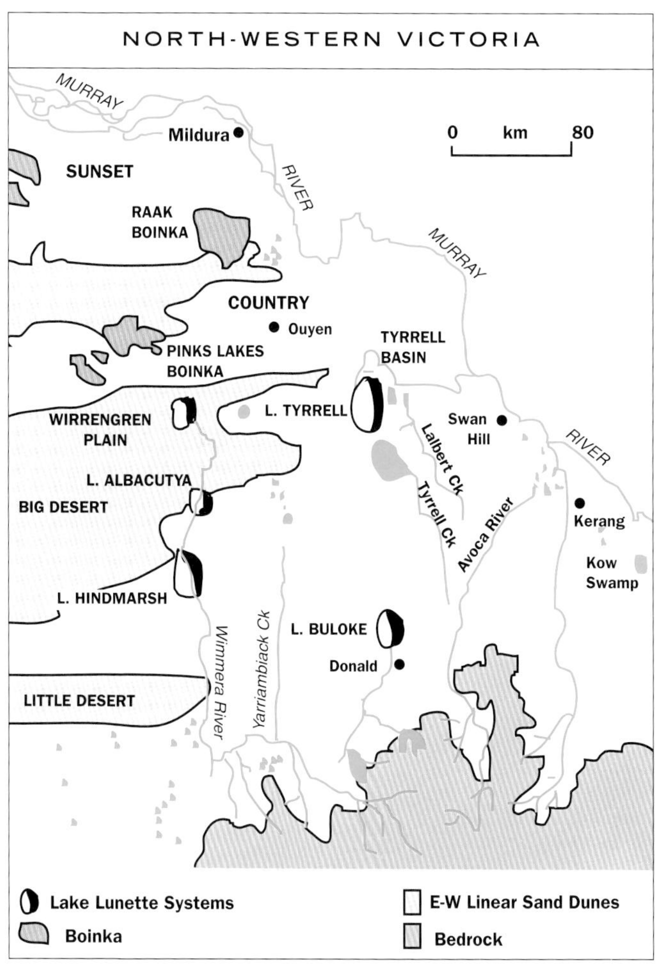

After Macumber[202]

THE WIMMERA RIVER AND YARRIAMBIACK CREEK

Lake Hindmarsh, the terminal lake for the Wimmera River under the present climatic regime, is a perennial lake, always having some water in it. When the Wirrengren Plain was under water at 7400 years ago, all lakes along the Wimmera River were full. The line between groundwater recharge and groundwater discharge occurs at the northern end of Lake Hindmarsh. At Wyperfeld the regional water table is about 10 metres below ground level. It rises to about 4 metres at Wirrengren Plain. It is non-saline and the *Callitris* forest on the plain is sustained by this shallow groundwater. Rising water tables reach the surface further north, near Underbool, and beyond this point groundwater-fed lakes are saline (they are on the other side of the groundwater discharge boundary).

The Wirrengren Plain is the largest of a number of now dry former lake beds at the terminus of the Wimmera River. They were freshwater lakes during the high lake period in the Holocene when Lake Mungo and Lake Frome, and indeed Lake Eyre, were all full, between about 50 000 and 30 000 years ago. At the north-eastern edge of the plain an 11-metre high strandline contains shells of creatures which no longer live in the Wimmera River. (*Plotiopsis*, a warm water species, comprising 95 per cent of the fauna, has been dated at 7460 ± 120 years.) The lake dried up 7000 years ago.

The Wimmera River in the Little Desert. M.E.W.

The Wimmera River near the edge of the Mallee, where it had not run far enough to reach its terminal lake, Lake Hindmarsh, for three years, 1997 to 1999. M.E.W.

LAKE TYRRELL

Lakes of north-western Victoria have provided interesting information on previous climatic regimes and changes to regional hydrology through time. Lake Tyrrell, Victoria's largest salt lake, has been investigated and its lake bed and dune sediments dated.[202]

The present-day lake reaches a maximum depth of only 0.7 metres during even the wettest years, but a high level beach ridge 13.5 metres above the lake bed shows that in previous times it was a major freshwater lake. *Coxiella* shells in the sediments indicate that 40 000 to 50 000 years ago the water had only one-tenth of the salinity which it now has. At that time Lakes Tyrrell, Timboram and Wahpool were amalgamated into a megalake—Lake Chillingollah. A string of small groundwater discharge depressions passing northwards from Waitchie to beyond the Towan Plains indicates the extent of the megalake.

THE FORMER EXTENT OF THE TYRRELL BASIN LAKES AT 40 000 YEARS AGO

LAKE
CHILLINGOLLAH
Lake Wahpool
Chillingollah
Lake Timboram
LAKE TYRRELL
Lalbert Ck
Tyrrell Ck
Sea Lake
0 km 10

After Macumber[202]

The change from high lake levels to an environment conducive to lunette formation represents far the greatest hydrological change in north-western Victoria over the last 50 000 years. In order to produce a lunette at Lake Tyrell, after drying of the lake, conditions would need to have been much more arid than they are today. Under the present regime, summer water tables fluctuate from being nearly at the surface to about 30 centimetres below the surface—far too high for lunette formation to proceed. The drier period in which lunettes were made corresponds to the last glacial maximum.

Evidence from the lake systems of northern and southern Victoria indicates that in the Early Holocene the dry conditions which had existed in full glacial times had passed. In south-western Victoria lakes began to rise, peaking at between 7000 and 5500 years ago. At Kow Swamp, high lake levels existed from about 13 000 to 8000 years ago; and at Lake Wirrengren the levels were high from before 7280 years ago. (Similarly higher lake levels than present-day are recorded for lakes in many parts of the world at this period.)

Lake Tyrrell, Victoria's largest saltwaste lake. PHIL MACUMBER

Salt patterns in the crust, Lake Tyrrell. M.E.W.

Kow Swamp: drowned eucalypt forest due to 'drainage management' for irrigation and flood control. REG MORRISON

KOW SWAMP

At the end of the Pleistocene and the beginning of the Holocene, Kow Swamp in northern Victoria experienced a lake-full stage, covering 25 square kilometres to a depth of 3 to 4 metres. Sand blown from the lake-full beach formed a low dune on the south-eastern shoreline—the Kow Sand. The lake shoreline was inhabited by Aboriginal people and burials were made in the dune, where the bones were preserved by the calcium carbonate which had been precipitated from groundwaters under high water table conditions. The shoreline was occupied from about 13 000 to about 8500 years ago and the lake dried at about 8000 years ago. The drying represents a natural fall in water tables at that time—a situation which contrasts with the steady rise and increase in salinity which have followed European activities on the plains.

PLATYPUS

Project Platypus in the Wimmera catchment is now *Rio Tinto Project Platypus.*[206] The Australian Platypus Conservancy is involved, also Grampians Water, Wimmera Mallee Water, the Natural Heritage Trust and the Mazda and William Brockland Foundations.

The Wimmera catchment is essentially isolated from other waterways, so its platypus population represents a unique and irreplaceable genetic resource. The landscape is ideal for an investigation of effects of land clearance, erosion and salinisation on the platypus populations.

Deteriorating water quality in the Wimmera catchment is due to vegetation clearing and unsustainable agricultural practices which have caused siltation and salinisation of the river. River ecosystems are threatened.

In the course of recent research 51 platypuses have been captured, fitted with transponders, and released. Ninety per cent of them remain in the upper reaches of the Wimmera where disturbance is least. General distribution has obviously declined dramatically.

THE GLENELG RIVER

The Glenelg River catchment lies in granite country in Victoria's elevated south-western region. Human impact on the river system through clearing, channel incision, river regulation and de-snagging has destabilised the river, and the abundant sand sourced from the granite terrain has resulted in the Glenelg and its tributaries being filled with more than 6 million cubic metres of sand over the last century. Most of the tributaries are incised streams which are filled with sand and are continuing to re-incise as the sand moves through. The supply of sand has decreased in the last 30 to 40 years and the sand slugs are moving downstream.

The result of siltation of rivers is loss of geomorphic complexity—channels become over-widened and there is little variation in cross-section; flat, featureless beds of sand lack pools, riffles and backwater zones; the substrate comprises uniform-size sediment; in-stream vegetation is minimal. Uniform habitat conditions replace diverse niches.[207]

The Glenelg provides an opportunity for researchers to determine whether complexity returns after a sand slug has passed through a reach of a river, and whether the reach will recover completely and return to its pre-sand-filled form, which had been expected to be the scenario. It appears that recovery is more complex than was anticipated. Bryans Creek and Deep Creek are both incised tributaries of the Glenelg River. In both, the level of sand is decreasing as the slug moves downstream. Both were chains of ponds before they were affected by human activities. It has been shown that they will not return to that state, and that they undergo a complex series of changes.

As the slug moves through, there is an initial increase of complexity as bars and pools are formed by the remaining sand, which is also colonised by reeds. However, the channel begins to incise again, destroying the developing complexity. Where sand has invaded a stream that was not incised, a quite different pattern of complexity has developed. Mathers Creek, for instance, has filled with sand, becoming a series of long pools and wetlands colonised by macrophytes. These new wetland areas appear to be permanent and stable.

It thus appears that there is no general pattern of behaviour—the recovery process will vary between streams and even between reaches, depending on the individual conditions in each stream.

LAVA FLOWS OF MT ROUSE, WESTERN VICTORIA:
how a volcanic eruption altered local drainage

The Western Plains of Victoria contain many extinct volcanoes, and are largely covered by a veneer of basalt.[208] There are about 200 points of eruption on the plains, ranging in age from several million years old to about 15 000 years for Tower Hill. The area is a distinct volcanic province characterised by numerous but small volcanoes, each of which was active only once and for a short time. This is in contrast to central volcanic type eruptions, which may continue for millions of years.

The general sequence of eruptions on the plains was first a highly explosive eruption to produce a maar; then effusion of very liquid lava which flowed down valleys and spread over plains if sufficiently voluminous; and finally the build-up of a scoria cone. Some points of eruption do not show all these phases, and occasionally other landforms, like lava cones, may be built. Mt Clay and Mt Vandyk are very old and deeply weathered; Mt Eccles is only about 20 000 years old and has hardly been weathered at all; in between are many volcanoes and flows with intermediate weathering and erosion. Mt Rouse is one of these. A new and revised date for Mt Rouse is 1.8 million years.

Mt Rouse is located just south of Penshurst and about 10 kilometres east of Hamilton. It is a composite volcano mainly built of scoria but with a few interbedded flows. To the south of the scoria cone lies a well-defined basalt-rimmed crater. Lava from Mt Rouse occurs in two large patches connected by narrow lava-filled channels. The lava flow extends 60 kilometres south to end on the coast just west of Port Fairy. Stony rises characterise the landscape and are interpreted to mean that a lava-tube mechanism was in operation when the lava was flowing, and indeed was necessary to produce such long lava flows. Only a small amount of lava flowed north, but one patch extended about 4 kilometres north-west and filled a valley head. Penshurst Creek flows along the northern edge of this lava lobe, away from the volcano and into Buckley Swamp, which was created when the younger Mt Napier volcano blocked the drainage in that area. Before that, the creek was probably a tributary of a river that flowed under the site of Mt Napier to the Harman Valley.

Along the Eumerella River, which marks the north-west edge of the lava flows, the lava simply banked up against older topography, having no convenient valley to flow down. Due south-east of the eruption point, a lava flow followed the valley of Spring Creek, and further south two other strips of lava followed Whitehead Creek and Whitehead South Creek. The main mass of lava flowed south through two main valleys, those of Back Creek and Moyne River. The Back Creek flow is the simpler, becoming more and more confined until it is only 200 metres wide. It remains less than 1 kilometre wide over a distance of more than 5 kilometres and then spreads out, overtopping low divides, and follows several different creeks. Islands of bedrock, called steptoes, protrude through the flow. The Moyne River flow is more complex, with two flows, the Wandilla and the Glenlevitt,

THE MT ROUSE LAVA FLOW AND ASSOCIATED DRAINAGE

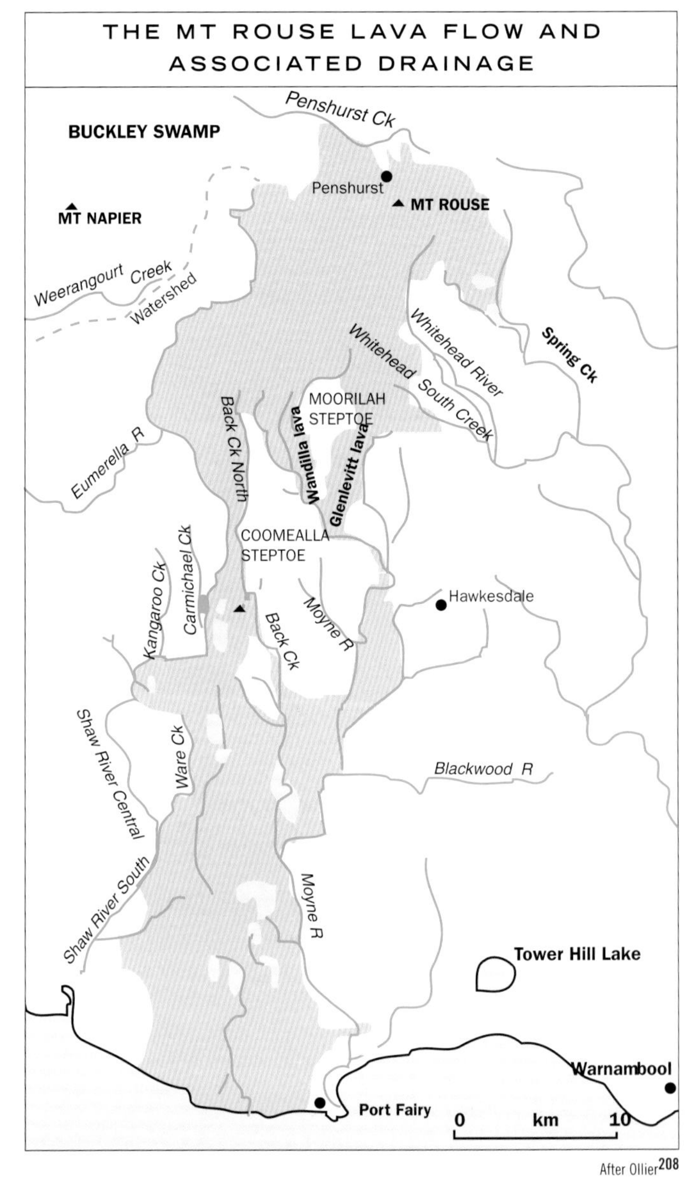

After Ollier[208]

THE MT NAPIER AND MT ROUSE LAVA FLOWS

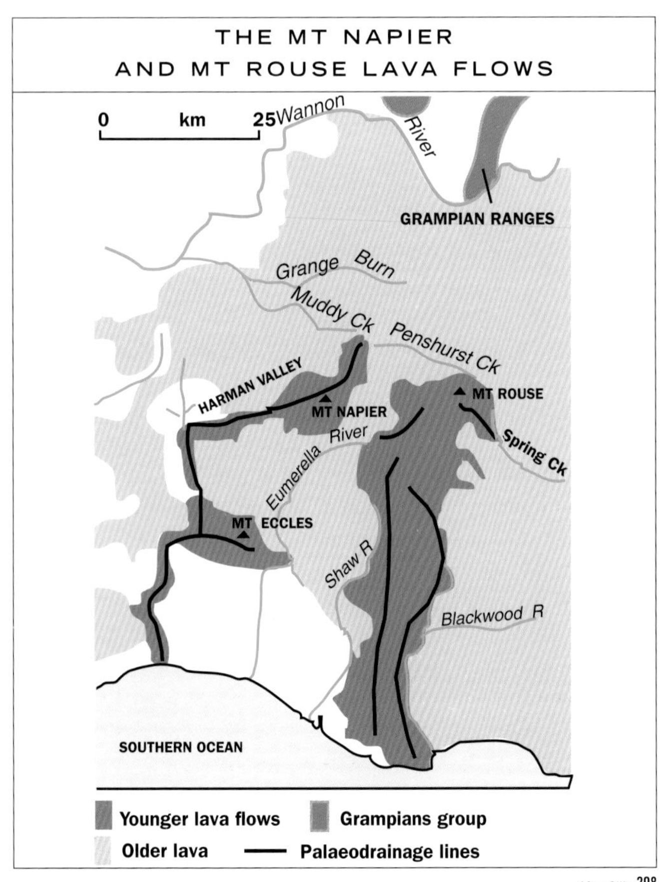

After Ollier[208]

Samphires on a salt marsh near Port Fairy, Victoria. M.E.W.

surrounding the Moorilah Steptoe and then uniting as the Moyne River lava flow. Over a distance of 6 kilometres this flow is less than 1 kilometre wide, but it widens out and joins the Back Creek flow. The Coomealla Steptoe, 19 kilometres long, separates the two flows. The united lava flow continues south, getting wider until at the coast, 60 kilometres from its source, it is 14 kilometres wide.

The Rouse flow is a textbook example of a lava flow which still shows all the features which tend to be seen randomly and individually in flows that have suffered further erosion:

- A **lateral stream**—the Eumeralla, which follows the edge of the flow for 15 kilometres.
- **Twin laterals** at all scales where flow divides around obstructions.
- **Successive diversion**, shown by the sequence starting at Shaw River North, which after being a lateral stream, is diverted into the Carmichael, which in turn is diverted into Kangaroo Creek, and finally becomes the Shaw River Central.
- A few rivers manage to **cross lava flows** from one side to the other—the Moyne River does it several times.
- The **inliers (steptoes)** are commonly bounded by streams, such as Back Creek and Moyne River bounding the Coomealla Steptoe, and the twin streams bounding the Moorilah Steptoe. Even quite small steptoes are bounded by streams on at least one side.

The flows from Mt Rouse were poured out over an originally flat topography with narrow, little-incised streams, and a generally southward slope. A simple dendritic pattern is assumed for palaeodrainage from a divide in the vicinity of Mt Rouse. Even this old drainage was on a landscape already modified by earlier lava flows.

THE MALLEE

The Mallee comprises the semi-arid plains of south-eastern Australia characterised by extensive sand ridges which support mallee—*Eucalyptus* species with a distinctive growth habit of multiple woody stems instead of a single trunk, and large underground lignotubers.[209] The growth habit of mallee is indicative of the nature of the region—the underground 'trunks' enables them to survive drought and fire; the leathery leaves, angled to the sun, reduce transpiration. Mallee lands are tough places where co-evolution between plants and environment has produced adaptations so that, in the natural state, saline water tables are kept at a safe depth.

Within the map area, the Loddon, Wakool, Murrumbidgee and Darling Rivers all join the main Murray trunk stream; and anabranches of the Darling and the Lachlan with their chains of lake-lunette basins add to the diversity of the region. Sands and desert dune fields, produced during the glacial stages of the Pleistocene ice age, cover most of the area, overlying a predominantly marine sequence of Tertiary sediments, deposited during times when the sea invaded the sinking Murray Basin. Evidence from drilling suggests that the marine transgression penetrated as far as the Willandra Lakes. (See *Listen … Our Land is Crying* for a full account of the geological history.)

The Mallee region demonstrates three types of drainage characteristic of arid and semi-arid environments:

- Firstly, throughout the greater part of the area, there is no contribution from surface run-off to overland flow. This *areic* aspect is reflected in the absence of channels from east and west of the Darling and from the entire area of northern Victoria north of Lake Tyrrell.
- Secondly, the streams in the southern region—the Wimmera River, Yarriambiack, Tyrrell and Lalbert Creeks—which rise in the better-watered hills to the south, are *endoreic* systems in that they lose their waters as they flow north, and terminate in a string of lake basins between the mallee sandhills.
- Thirdly, only the major rivers carrying waters from the wetter areas of the south-eastern highlands and from south-western Queensland succeed in crossing the plain. Both the Murray and the Darling tend to lose water as they traverse the region to the west, a common feature of *exoreic* drainage in arid and semi-arid regions.

Two other aspects of drainage in the Mallee are worthy of note:

- The absence of lakes on the Darling between Menindee and the junction with the Murray, while the Darling anabranch has an abundance of such features.
- The dry channel of the Willandra Creek which wends its way through the field of linear dunes south of Outer Arumpo is evidence of times when surface water was much more readily available. This channel last carried overflow water from the lake chain 16–18 000 years ago. It is interesting to note that such a minor landscape feature has survived for so long, a fact which emphasises the longevity of landforms in arid environments.

Lake basins in the Mallee present some of the best examples of lunettes found anywhere in Australia. These crescent-shaped dunes on the eastern sides of basins record the complex oscillations of past hydrological sequences. During periods when water was deep and reasonably fresh, clean quartz sand accumulated on eastern beaches and was blown into the downwind dune. Later, during more saline conditions associated with drying, gypseous clay pellets were transported by saltation from exposed lake floors to provide the smooth surfaces most characteristic of their present form.

The size of lunettes is related to lake size. Small lakes of less than 1 kilometre diameter have lunettes a few metres high; large lakes like Chibnalwood on Willandra Creek have lunettes 30 metres above lake floor; Lake Tyrrell has 40-metre high lunettes. Lunettes throughout the Mallee are relict features relating to Late Pleistocene climatic events. The last episode of lunette building was between 19 000 and 15 000 years ago.

Lake basins are typically smooth and elliptical, often kidney-shaped in outline, with the long axis north–south or north-north-west

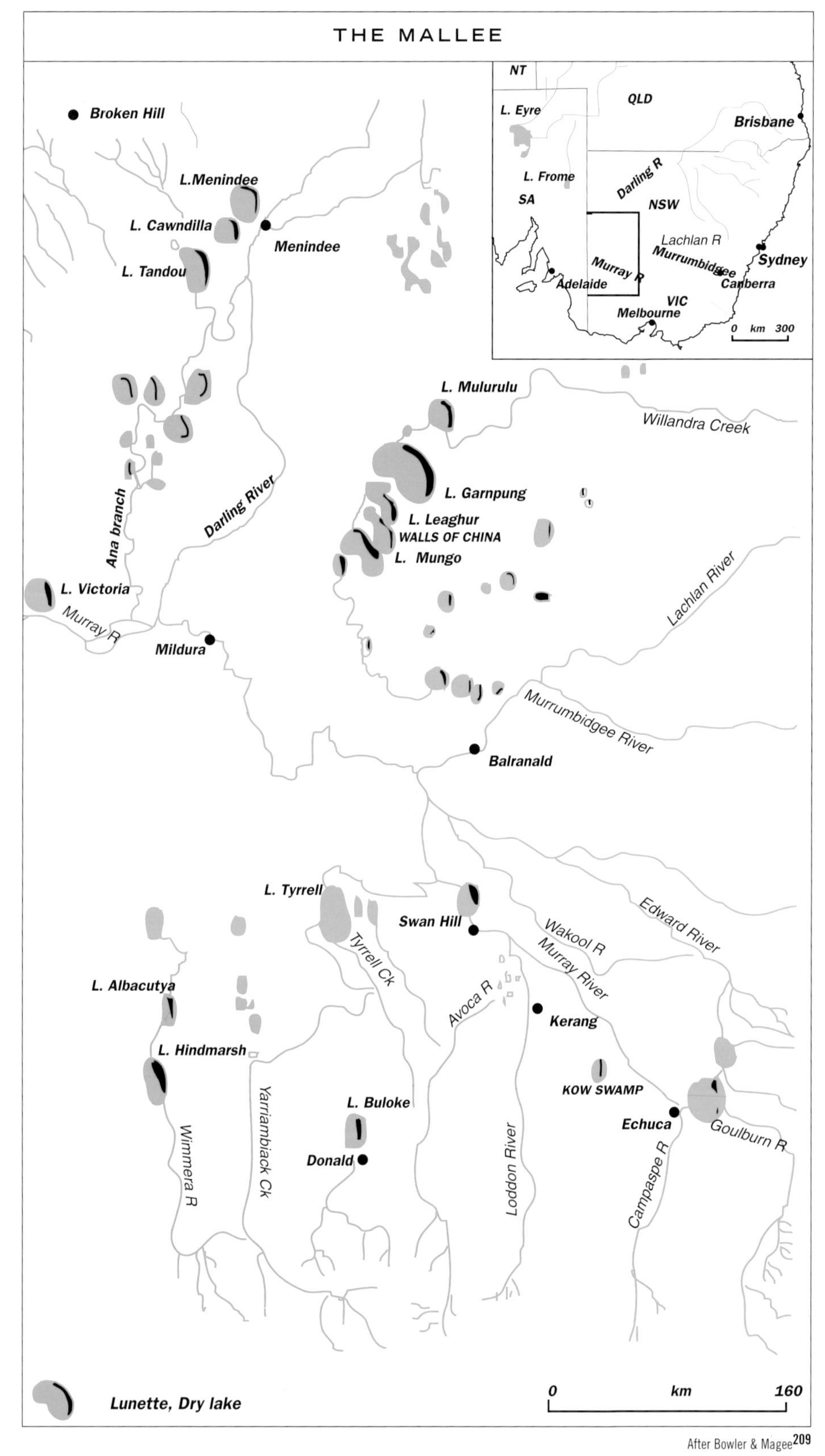

After Bowler & Magee[209]

terminal lakes of the Tyrrell Creek and Wimmera River. Other basins in areas where no surface water exists today are legacies of much wetter periods in the past. Most lake basins today are dry, some are filled artificially for use as storage basins—either for fresh water as at Menindee, Kangaroo Lake and Lake Charm near Kerang; or as evaporation disposal basins for saline groundwater as at Lake Tutchewop and Lake Tyrrell. Lake Tyrrell is Victoria's largest salt lake. It intersects the saline groundwaters in the region.

In the most westerly sector of Victoria, north of the Big Desert, and in New South Wales west of the Anabranch, irregularly shaped gypseous plains and lake floors occur. These are groundwater discharge areas, often displaying complex surface patterns due to wind sculpturing of their surface deposits. The Raak Plain is a dramatic example of one of these.

The Mallee extends into South Australia, along the Murray to Morgan and southwards in the south-western sector of the Murray Basin. Research into groundwater recharge in the South Australian sector has revealed interesting and disquieting facts.[210]

Water tables are deep down (30 to 60 metres) in most of the South Australian Mallee, but are shallower in the south-west where water tables are rising rapidly. Clearing of native vegetation has increased recharge in some parts of the Mallee by up to two orders of magnitude. Rising water tables have caused land salinisation problems in low-lying areas. South-east of Murray Bridge, salinisation is increasing and by 2005 another 20 000 hectares will be affected. All land less than 20 metres above sea level will be affected by 2030.

In the long term, rising groundwaters responding to clearing of natural vegetation will increase saline groundwater inflows to the Murray River. This inflow to the river is only from the south and east. Upstream of Morgan, to the north, there has been only limited clearing and the marginal rainfall makes further clearing unlikely; to the west, a geological fault (which caused the bend in the Murray and

to south-south-east. All large basins are associated with drainage lines which contributed the surface waters which were so important in shaping their outlines. The Murray has lunette lakes marginal to its channel at Robinvale, Hattah and Lake Victoria; lakes of the Anabranch owe their shapes to the past high-surface flow stage of the Darling; Willandra Creek lakes were shaped by high flow stages of the Lachlan. Lake Tyrrell and the Hindmarsh–Albacutya string are

makes it turn and run due south from that point) acts as a barrier and clearing is minimal because soils are stony and shallow. Computer models suggest that 50 years after the water table begins responding to the increase in recharge an increase of 70 EC units (a measure of total dissolved salts and salinity) will occur at Morgan.

Murray Group Limestone is in hydraulic connection with the River Murray, downstream of Overland Corner. It is recharged

The Raak boinka—a vast groundwater discharge region, pockmarked with salt lakes. PHIL MACUMBER

Lake Bellfield, one of the dams in the Grampians which store water to be piped to the Wimmera. M.E.W.

A small river on the beautiful south-western coast of Victoria, Campbell Creek runs through a reed swamp on its way to the sea at Port Campbell. M.E.W.

mainly in south-western Victoria, and groundwater moves under low gradients along flow paths of up to 350 kilometres before it drains into the river—an example of how effects can be the result of changes in catchments remote from the affected area. Salinities increase down-gradient from less than 1000 mg/L at the margin of the Basin to more than 20 000 mg/L next to the river. The limestone aquifer is overlain by a blanket of Loxton—Parilla Sands up to 50 metres thick.

Changes to regional hydrology which result from land clearing and other factors which cause groundwater rise may not have immediately detectable consequences serious enough to warn land-users that their practices are unsustainable. In the Mallee, major clearing began 50 to 80 years ago. Computer modelling using levels of clearing as at about 1990 indicates that as the pressure front from the clearing moves slowly and water tables under the Mallee are deep, deleterious effects will be evident in 10 to 30 years from now, depending on specifics of locations. Then the water table will rise steadily over the next 500 or 1000 years to reach a new stable position close to the surface.

A great deal of water is brought into the Wimmera–Mallee region from dams in the ranges to the south where the Wimmera River rises. Much of the Grampians' yearly average rainfall of 760 millimetres, to 1000 millimetres on higher peaks, is stored and diverted to the thirsty Wimmera–Mallee Plains in one of the world's largest gravitational channel systems. The construction of some 24 000 kilometres of open channels, carrying water up to 600 kilometres from its source and serving one-eighth of Victoria, was something of an engineering triumph. The losses to evaporation and seepage over time have been enormous. Recently, a program has been instituted to pipe much of the distribution–the 'steel tributaries' in some cases make a saving of 70–80 per cent.

Water has been diverted outside the Grampians since 1881, when 12 kilometres of open trough called the Flume channelled water to a tunnel hacked through the Mt William Range, then on to Stawell 30 kilometres away. The water has now been piped but still uses the original tunnel.

ARTIFICIAL WATERS AND BIODIVERSITY CONSERVATION IN SUNSET COUNTRY

Dams in the Murray–Sunset National Park in the Mallee are being infilled in a program aimed at restoring the natural arid-land conditions and dealing with feral herbivore and kangaroo problems.[80]

An extensive network of catchment dams was established during the 1920s and early 1930s to 'drought-proof' the area for pastoralism. As in the Great Artesian Basin and elsewhere, when artificial waters have been provided in arid and semi-arid environments, the advantage to introduced stock and their owners has been counterbalanced by disadvantages to the native biodiversity. Stock was able to graze throughout the year in areas that would only have supported it during times when rain had fallen recently, and feral grazers and native fauna were able to maintain and increase their populations and contribute to heavy grazing pressures. As a consequence, grazing-tender species disappeared from the vegetation, and with few areas sufficiently remote from water to be left ungrazed, many plant extinctions (local and complete) have occurred in the 70 per cent of our continent which is arid.

When the Murray–Sunset National Park was declared in 1991, stock grazing ceased but numbers of grey kangaroos and feral goats have continued to increase, reaching plague proportions. If the impact on the vegetation, and the grazing-sensitive species in particular, is to be counteracted, drastic measures are required. Restoring arid-land conditions without permanent surface water will restore natural curbs and balances. Research has shown that ephemerals are the component of arid-land vegetation hardest hit when they come within grazing range as a consequence of the availability of artificial waters. Also, some bird species like galahs and yellow-throated miners had benefited from readily available water and become so numerous that they out-competed many other species which were becoming rare in the region.

Closing of dams in the park started in 1997, with 32 filled in that year and 24 to follow. Where water is required for fire fighting and visitor use, closed concrete tanks have been installed.

SECTION EIGHT

TASMANIA

Tasmania, with about 3 per cent of Australia's population and less than 1 per cent of its land area, has 12 per cent of its freshwater resources. Average annual surface run-off is 53 million megalitres. (Run-off is what makes the rivers run and what fills the lakes and dams, so it is a useful measure.) Yet even within Tasmania the distribution of water resources is uneven, governed by regional geomorphology and the climatic patterns which are modified by the nature of the terrain.

Tasmania lies on the edge of the moisture-laden Roaring Forties westerly wind stream. This airflow is forced upwards when it meets the western, central and southern highlands and cools as it rises, releasing much of its moisture and bringing cloudy and wet conditions to the south and west of the State. Thus the west coast can receive six times the average annual run-off of the lower elevation

A frosty morning at Lake Dove. The Cradle Mountain–Lake St Clair National Park has many lakes, large and small, in hollows gouged by ice during the Pleistocene ice age. REG MORRISON

Lake Pedder, the icon which inspired the conservation movement world-wide in the 1970s—before the lake was drowned by the Impoundment. Photographed by Olegas Truchanas, who fought for its preservation and later lost his life in his continuing battle for protection of Tasmania's wild rivers.

Tasmanian midlands. Factors such as relatively high evaporation rates and ground permeability on the east coast also result in the south-east having only 10–15 per cent run-off from annual rainfall, compared with the 80–90 per cent run-off in the west. Rainfall patterns in the west are also less seasonal than in the east.

Tasmania's diverse landscapes reflect a complex geological history which has governed the development of drainage patterns. Its major rivers fall into two main categories—those which have largely been superimposed on the local geology (like the Gordon, King, Pieman and Mersey) and those which follow major geological lineaments (like the Derwent and Tamar).[211] Tasmanian streams are often young, high gradient and actively eroding, with many reaches that are steep, relatively fast-flowing, carrying cobbles and boulders, and in places structurally constrained right down to the estuary. This makes them different from mainland rivers—in fact, most rivers in Tasmania behave as 'normal rivers' should, even running to the sea!

Tasmania has thousands of freshwater lakes, including Australia's largest (Great Lake) and deepest (Lake St Clair). The many lakes which occur on the Central Plateau were created by glaciation. Tasmania had a small ice cap on the Central Plateau during the Pleistocene ice age and hollows gouged out by ice now contain lakes. Glaciers on the west coast mountains have also left a detectable record of advance and retreat in major river valleys during the major glaciations (the Linda Glaciation of the Early Pleistocene; the Henty of more than 125 000 years ago; and the Margaret Glaciation, which saw major temperature reductions at 30 000 years ago; the last glacial maximum about 18 000 years ago, and deglaciation as recently as between 11 000 and 9000 years ago.)[96]

THE HYDRO-ELECTRIC CORPORATION

The operations of the Hydro-Electric Corporation (the HEC or 'Hydro') have had a profound effect on a large percentage of Tasmania's lakes and rivers.[212]

HYDRO CATCHMENTS IN TASMANIA

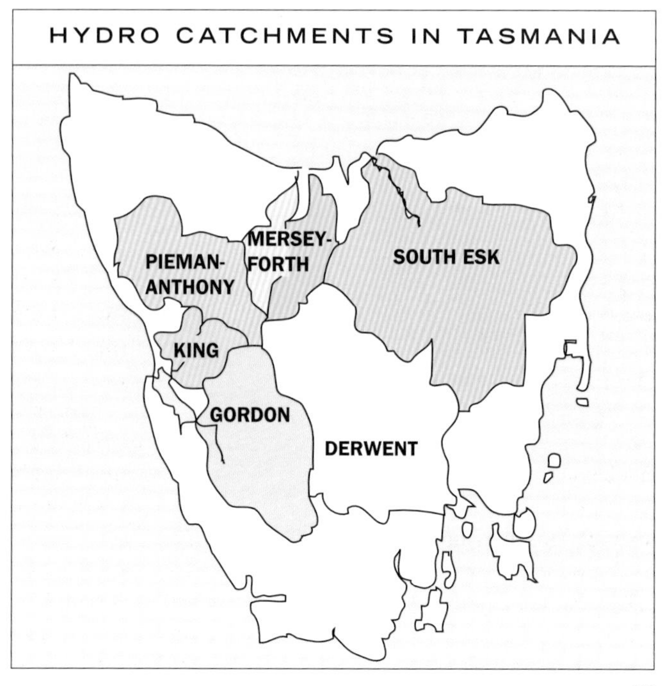

After Locher[212]

The electricity generating system was largely developed after World War II, complementing the massive dam-building program to drought-proof Australia and ensure water supplies. On average, one reservoir was built or expanded each year in Tasmania between the late 1940s and mid 1980s, to a point where Tasmanian storages for hydro-electric power generation can hold over 25 million megalitres, meeting 100 per cent of Tasmania's needs and with a large potential surplus. At present there are 27 power stations and 52 storages grouped into six major hydro-electric schemes. The last Hydro dam was built in 1993, ending the development of two-thirds of the potential hydro-electric resources in Tasmania. (The remainder are within World Heritage Areas or National Parks and one can only hope that this will protect them.) The effects on landscapes have been monumental:

- More than 1100 square kilometres of river valleys, wetlands and pre-existing lakes have been inundated for Hydro development;
- The catchment area feeding into Hydro storages takes up 21 500 square kilometres, or 36 per cent of the mainland Tasmania land area;
- One hundred and seven dams and weirs have been constructed, affecting 21 creeks, 25 rivers and 7 estuaries;
- At least 1200 kilometres of natural creeks and rivers are affected by hydro operations by way of diversion or addition of water and alteration of flow regimes;
- The generating system includes a further 212 kilometres of purpose-built water conduits including pipelines, tunnels and canals, which are often instrumental in moving water out of natural catchments into selected Hydro storages.

These developments have resulted in loss of natural river and wetlands, and the creation of new wetlands; decreases or increases in volume of discharge; changes and sometimes reversal in seasonal discharge patterns of streams; unusual pulses of released water; changes in flow variability and rates of rise and fall in water levels; and changes to the magnitude, frequency and form of flood events.

Among the inter-basin transfers resulting from Hydro schemes, the following affect well-known rivers:

- The headwaters of the Mersey River are diverted at the Parangana Dam into the Forth River on Tasmania's north coast.
- The upper reaches of the Huon River are diverted into the Gordon River, via Lakes Pedder and Gordon and the Gordon Power Station.
- The whole Great Lake catchment drained naturally via the Ouse and Shannon Rivers into the Derwent river system in the south of the State. It has been diverted northwards over the Great Western Tiers through the Poatina Power Station into the Macquarie South Esk river system and out into the Tamar Estuary on the north coast. The Great Lake developments have involved four successive stages of dam building or dam-wall raising.
- Significant reaches of the Derwent, Dee and Nive Rivers are frequently left dry due to diversions of water in the upper Derwent scheme.
- Additionally, there are a number of small diversions, like the three small tributary streams of the Franklin River which are redirected into Lake King William.

The Gordon Dam. HELEN GEE

Dam building did not deliver the promised economic boom in Tasmania, where electricity over-supply is estimated at 28 per cent, indicating that many Hydro schemes were unnecessary. Following cancellation of the Franklin Dam scheme in 1983, the Hydro was paid $330 million by the Federal Government as compensation. The funds were used to build two other dams—the King and Anthony schemes. Both of these have proved to be enormous failures, accounting for $1200 million of the total Hydro debt of $1700 million. Taxpayers in Tasmania (population 500 000) must now pay $140 million per year to service the debt without any return from energy sales.

With further large dam building in Tasmania out of the question, the HEC formed a subsidiary company, the Hydro-Electric Commission Engineering Corporation (HECEC), transferring much of its engineering expertise to South-East Asia. (The activities of this company in Laos are, fortunately, beyond the scope of this book.)

The effects on aquatic environments of all the changes due to HEC activities have obviously been considerable, but it is only when high-profile issues arise that public awareness of problems is raised. During the late 1980s and early 1990s, issues such as algal blooms in the Lagoon of Islands and fish kills in the Pieman River caused concern, and the Hydro set up a Consultancy Agreement with the Inland Fisheries Commission to address aquatic environmental issues. The main role of the consultancy has been to monitor water quality in Hydro lakes, to respond to other aquatic issues as they arise and to provide a link with the trout fishing community, which is a major user of Hydro waterways. Over recent times, the Hydro has undergone an institutional shift:

> 'from an era focussing on dam building (resource development), to a new era as an electricity generator with a strong corporate vision as energy and water managers (resource management). The Hydro is increasingly evaluating and seeking to understand its impacts on the aquatic environment. The challenge is now to find practical solutions which enable sustainable energy generation while improving lake and downstream river environments.'[212]

(What will happen if the **Basslink Project** goes ahead and Tasmania supplies power to Gippsland by cables under Bass Strait is open to speculation. It will certainly mean extra exploitation of the rivers which have been dammed to supply the grid and one can only hope that the project will prove to be economically unviable. Concern for the out-of-sight, out-of-mind rivers of Tasmania would not sway proponents of the project, if past history is any guide, and only money talks in these matters, as we have all found out.)

Ongoing environmental issues as a result of Hydro activities tend to reflect regional characteristics:

- The west coast is mountainous and forested and the Hydro schemes capture water which falls on the ranges, which are part of a highly metalliferous belt. Mining for copper, lead, zinc, silver and gold has the potential for water contamination. Many disused mines, as well as active operations, produce acid mine drainage and heavy metal contamination. The Pieman River is one example of a river with heavy metal contamination; the King River system has enormous acid drainage problems associated with Mt Lyell.
- On the Central Highlands, sparse alpine vegetation characterises the catchments and main concerns are water quality and matters affecting trout fishing. High turbity and nutrient levels have affected many Hydro and natural lakes including Shannon Lagoon, Penstock Lagoon, Lagoon of Islands and Woods Lake.
- The more populous Derwent, Mersey and South Esk catchments have a mix of rural and urban populations, and higher proportions of the land have been cleared for grazing and agriculture. All the issues of river and floodplain degradation following European land use are seen in these regions as well as the environmental changes brought by Hydro activities.

The consultancy agreement with the Inland Fisheries Commission has led to some positive environmental actions. Lake level agreements were made for Woods Lake and Lagoon of Islands to enhance water quality and protect threatened native galaxiid fish and optimise the trout fishery; a current-velocity barrier was established to prevent the spread of exotic red fin perch from Lake Gordon into the Pedder Impoundment; an outlet was installed in Penstock Lagoon to improve circulation and water quality; an elver ladder was installed at Trevallyn Dam in Launceston to enable upstream fish migration; and on the Liawenee Canal a barrier to fish migration has been constructed to prevent red fin perch from migrating into the World Heritage Area lakes.

The Mersey River and Hydro regulation

The Mersey River, in the central north of Tasmania, enters Bass Strait at Devonport. Its highland catchment was diverted into the Forth catchment by the HEC in 1973 for power generation.[213] At the time, no allowance was made for environmental flows in the Mersey downstream of the Parangana Dam. Instead, almost all of the water of the upper Mersey was, and continues to be, used to generate power at the four power stations on the Forth River, contributing to a scheme that meets about 16 per cent of Tasmania's energy needs. The mean and median flows left in the river were one-tenth of the natural flows, measured immediately below the dam.

Pressure for improved summer flow began to mount in the 1990s. By then significant agricultural development had occurred in the catchment, resulting in increasing water consumption, and a number of small farm dams had been constructed. By 1997, there were 137 licences to take water for irrigation or commercial use from the Mersey catchment. Recreational fisherman who wanted security for their trout fishing were among the most vocal in calling for a fair go for the river. Algal blooms and bacterial contamination—of tributaries more than of the main river—were causing alarm. The health of the tributaries and of the Mersey itself was declining, as evidenced also by a decrease in the macro-invertebrate population.

As is always the case, things have to get really bad and affect the health or convenience of a number of individuals before groups come together and exert sufficient pressure for something to be done about a problem. The Mersey River Working Group, formed in January 1996, has evolved into a catchment management team which works for the rehabilitation of the whole catchment as well as for environmental flows from the dam to restore the trunk river.

The Leven runs through mountainous, forested country in its upper reaches. DICK BURNS

THE LEVEN RIVER: *the success story of a river that flows free to the sea*

The Leven River is a short one by mainland standards, but like most Tasmanian rivers it can carry a lot of water, particularly in winter. It rises in the Black Bluff Range and travels some 80 or 90 kilometres to reach Bass Strait at Ulverstone. Its headwaters are in a timber-getting area where selective logging has occurred since settlement, and there are a few low-intensity farms (mostly cattle) along its route. Underneath the Dial Range, the Leven is a deep, peaceful, slow-moving river. It runs through steep-sided timbered valleys or open river-flat farmland at Gunns Plains. Appearances can be deceptive—after a night of rain it can rise 1 or 2 metres and the stretch between the Plains and Loongana becomes a truly wild river with waterfalls and gorges, deep waterholes for swimming in summer, and wild white water in winter. The HEC did have plans to build a dam in the Leven Canyon, which, fortunately, did not eventuate. The Leven runs clean and clear through natural vegetation despite the proximity of towns, and is a rare example of a small wild river which is readily accessible to the population centres

The Leven between Gunns Plains and Loongana. DICK BURNS

The Leven Splits, where the river cuts through a narrow slit in bedrock, upstream from the canyon. DICK BURNS

The Leven Canyon. DICK BURNS

of Burnie and Devonport on the north coast. It is unregulated, except for a small weir near sea level which allows water to be taken out to supply the nearby towns.

The protection of the Leven, keeping it in its wild state, has to a great extent been brought about by the activities of the local bushwalking club, the North-West Walking Club, with support from the Environment Centre and the Tasmanian Conservation Trust. In the 1970s the club cut and constructed a track that mostly follows the course of the Leven upstream. The Penguin–Cradle Trail is an 80-kilometre walk from Penguin, on the coast, to Cradle Mountain, crossing the ranges that feed water into the Leven (the Dial, Loongana and Black Bluff Ranges). The walk is listed as one of Tasmania's great bushwalks. The country through which it passes was not protected, so when in 1996 the Public Land Use Commission started looking at the status of all Tasmanian public land as part of the process leading to the Tasmanian Regional Forest Agreement, the Walking Club made submissions suggesting that all such land along the route of the trail be declared State Reserve.

While the outcome of the Regional Forest Agreement was not at all satisfactory (to put it mildly!) for conservation of Tasmanian forests, it was good for the Penguin–Cradle Trail and the Leven River. A political decision that mineral exploration should be allowed in almost all land reserved as part of the Regional Forest Agreement process meant that a State Reserve could not be proclaimed. But the Black Bluff Range, which provides most of the Leven's water, will become a Conservation Area and the public land in the forested valley through which the river flows from Leven Canyon to Gunns Plains has been recommended as a Regional Reserve.

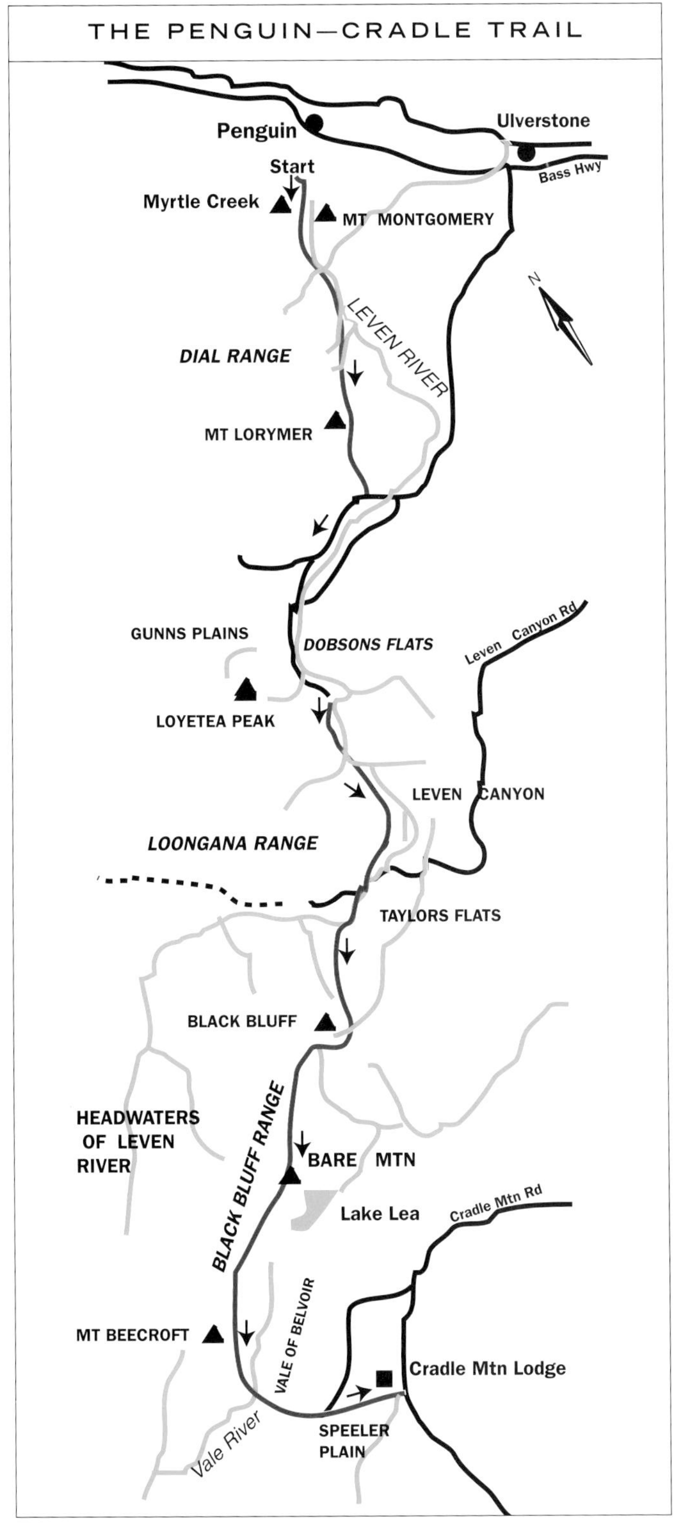

Map and information Dick Burns

THE PIEMAN RIVER

In the West Coast mountains, the Farrell Range and uplands to the east are deeply dissected by the Mackintosh and Murchiston Rivers. These rivers join west of Tullah to become the Pieman which drains westwards in a deeply entrenched, terraced valley that lies below 200 metres. The Pieman channel seems to be largely independent of structure, while its tributaries, the Stirling, Marionoak and Boco, follow the rock strike. The drainage of the upper Boco valley has been reversed, and together with the Bulgobac and Que Rivers, has been diverted through the Que Gorge into the Huskisson valley.[215] Cambrian Mt Read volcanics form a north–south trending belt from the Que River to Mt Read, fault-bounded to the west of Boco valley, and Lower Palaeozoic sediments form the West Coast Range to the east.

The surficial deposits of the upper Pieman and Boco valleys are the result of glaciations in the Pleistocene ice age. Evidence suggests that they were formed during at least four periods of glaciation that preceded the last glaciation in age. Boco drifts are likely to be Middle Pleistocene, while Bulgobac and Que drifts are likely to be Early Pleistocene or older.

Annual rainfall for the Tullah–Rosebery region is between 2000 and 2500 millimetres a year, sustaining rainforests.

Wide falls on the Pieman River. OLEGAS TRUCHANAS

LAKE PEDDER AND THE PEDDER 2000 CAMPAIGN

South-west Tasmania has been a focus for conservationists since the 1950s. The temperate wilderness of the South-West National Park, unique in the world and with an ancient biological lineage extending back into Gondwana 60 million years ago, deserved total protection. Gazetted in 1968, it contained the Lake Pedder National Park, which had been established in 1955. In spite of its National Park status, the wilderness was invaded and violated, its wild rivers tamed, its valleys flooded. The Gordon Stage 1 Hydro-electric Scheme created two vast dams in 1972, obliterating one-third of the South-West National Park.

It is interesting to reflect just how significant Lake Pedder was in focusing attention on environmental matters and bringing about a conservation movement whose influence is global, not just Tasmanian or Australian.[226, 227] Lake Pedder, because of its ethereal beauty and its setting in a region of true wilderness too beautiful and powerful to describe, was already an icon. Two thousand people were prepared to walk in to the remote wilderness to protest at the impending flooding in 1971. The world's first Green Party, the United Tasmania Group, was born out of the battle for Pedder, influencing environmental consciousness. Pedder remains an icon—a reminder of the value of bringing about the change in attitudes needed to have a strong new national ethic in which we recognise our mistakes, accept that our role has to be one of custodianship and not exploitation, and start to undo the wrongs inflicted on our continent.

Where the Serpentine River runs out of Lake Pedder—as it was before the Impoundment drowned it. OLEGAS TRUCHANAS

The original Lake Pedder was a small (9 square kilometres), shallow lake of outstanding beauty.[214] It owed its origin to glacial activity during the Pleistocene ice age, when glacial outwash from the flanking Frankland Range dammed the Serpentine River in its flat valley. A major feature of the lake was the beach along its eastern edge—a beach of pink quartz sand which rose steeply from the bed of the lake at 'the step' and then shelved gently to flanking dunes. It lay under water in winter, but as water levels fell during summer it was progressively exposed until it was wide enough (600 metres in width and about 2 kilometres long.) to be used as a landing strip by light aircraft. As the water retreated, a variety of small herbaceous plants sprouted from the moist sand, flowering progressively later down the widening beach.

The Serpentine River winds its way down its valley. OLEGAS TRUCHANAS

The prevailing westerly wind scouring through the valley of the lake caused the water to shape long flutings of pinkish sand along the edge of the beach with a series of prominent herringbone-pattern wedges known as submarine dunes or megaripples, in some ways resembling beach cusps. The bed of the lake was composed in the south of the same pink sand as the beach; to the north it comprised quartz pebbles, among which occurred ferromanganese concretions called Pedder Pennies. The beach was bounded to the east by a series of sand dunes, vegetated by shrubs and eucalypts. To the east of the dunes was a series of small lakelets, the largest of which, Lake Maria, gave its name to the Maria Complex.

The Serpentine River followed a sinuous course through the valley to Lake Maria, through which it flowed, leaving via a deep channel—Maria Creek—to fan out over Lake Pedder's eastern beach. At the western end of the lake, the Serpentine River continued its course to join the Gordon River.

The famous pink beach on Lake Pedder, the entrance from the Maria Complex Lakes in the background. OLEGAS TRUCHANAS

In the 1960s, plans (never presented for public debate) were announced to flood Lake Pedder as part of a hydro-electric power scheme. The Huon–Serpentine Impoundment of 240 square kilometres drowned the original lake when it filled in 1972–73. Although one of the Impoundment's dams was built on the Edgar Fault, safety issues associated with the risk of seismic activity were never adequately assessed. The enlarged Lake Pedder was intended to contribute approximately 11 per cent of the power-generating capacity of the then-proposed Gordon River power development. In the event, the storage volume of the Impoundment represents *only 3.2 per cent* of that available for the entire Gordon Scheme. The flooding was opposed by conservationists to no avail, even though the Federal Government was prepared to offer funds to Tasmania to consider an alternative system which would prevent the loss of the original lake. Power from the flooding of Lake Pedder has never been required.

At the time of the flooding, despite calls for professional assessments, practically nothing was known about the flora or fauna of the lake and its surrounds, except that a number of endemics had been described, particularly some species in the biota of the lake and of the pink sand beach.

The idea of restoring the lake to its original form has persisted. When the surrounding region was inscribed as a World Heritage Area, the Huon–Serpentine Impoundment was included, because the International Union for the Conservation of Nature expressed the hope that Lake Pedder would be restored. At its meeting in Buenos Aires in 1993, the IUCN passed unanimously a resolution calling for the restoration of the lake.

In 1994, the Lake Pedder Study Group commissioned a geophysical survey of the bed of the lake. A team of divers and botanists led by Professor Peter Tyler, a world authority on aquatic biology, confirmed that the pink beach and bordering dunes are preserved under the impounded waters. Startling footage made of the beach shows the tyre marks of the last planes to take off outlined in the few millimetres of sediment which has collected there on the pink sand. Since then, an international group, known as Pedder 2000, has formed with the aim of draining the reservoir and restoring the lake. As a result of 21 years' study by Dr Sam Lake (see *Reflections* 6, p.1), much more is now known about the local biota. It is sadly acknowledged that the endemic species which had been identified there may now be extinct, as the waters of the Impoundment now contain widespread non-endemic aquatic species. But the very large number of species of plants and animals which Lake Pedder shared with other lakes and with the region generally can be expected to eventually recolonise the area which has been denuded by the flooding.

The Lake Pedder Restoration Committee publishes the newsletter *Reflections*, keeping supporters informed about progress with the campaign, and supplying information on Australian and overseas related issues. Worldwide, moves are being made to decommission dams and restore natural river flows. It is recognised that too many dams have been constructed and that too much environmental damage results from them. A new international group called the Living Rivers Coalition, is successfully involved with the decommissioning of inappropriate dams built in the recent past.

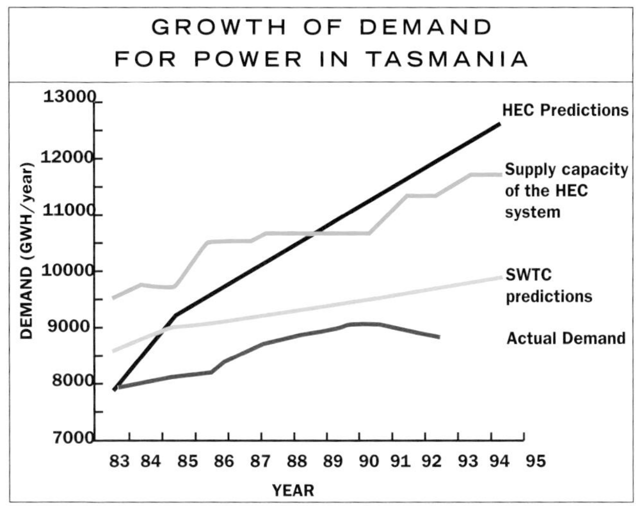

Predicted and actual

WHAT WILL HAPPEN TO THE GORDON POWER SCHEME IF PEDDER IS UNPLUGGED?

The Pedder Impoundment is not a power scheme. It has no generators. It is simply an extra water storage for the Gordon Power Scheme. If Pedder is unplugged, there will be no loss of maximum generator capacity; the peak power output of the Gordon Power Station will remain the same. Tasmania's entire power system has a 1300 MW capacity, and presently has a power surplus of 130 MW. This figure will rise to more than 200 MW when Comalco's smelter closes down in 2001. The Savage River and Mount Lyell Mines will also have closed down by then.

A canoe on the Pieman River. OLEGAS TRUCHANAS

HEAVY METAL CONTAMINATION OF WATERS

Mining and industrial processes frequently lead to contamination of waterways and the sea with heavy metals and acid drainage.

The North-west Coast of Tasmania has a history of such contamination.[216] In 1973, when the Department of Environment was formed, a survey was undertaken of the area from Table Cape to Port Sorrel and some 19 kilometres to seaward. Heavy metal levels significantly above those to be expected from natural geological sources were detected, and three industries were required to carry out a further survey in 1976 to monitor the dispersion of their waste products and the effects on the marine environment. Trioxide Australia Pty Ltd, North West Acid Pty Ltd and Associated Paper Mills Pty Ltd sampled sea-floor sediments, waters, plankton, shellfish and fin fish and confirmed the 1973 finding that ocean-dumped calcine waste and pipeline discharges had contributed to contamination of the marine food chain. Mercury levels had risen in the Wesley Vale area, but were stable in the Wynyard–Penguin area. (Calcine is a waste product from the manufacture of sulphuric acid; mercury originated at chlor-alkali plants at Burnie and Wesley Vale operated by the paper industry; lead, cadmium and other metals in sea-floor sediments had spread from the dump site 10 kilometres offshore throughout an area 10 kilometres wide.)

Direct comparison with the 1973 results was not possible because few sampling points coincided and currents and natural deposition processes had to be considered in analysis of the distribution patterns. Copper distribution was found to relate strongly to the calcine dump site; mercury was concentrated near Wesley Vale and Burnie; highest concentrations of copper, lead, iron and mercury occurred in bottom waters in the Wynyard–Penguin region, but surface waters did not contain higher concentrations. Zooplankton, shellfish and fin fish all had elevated levels of heavy metals, confirming that the food chain was contaminated. The levels were, however, within the range still considered safe for human consumption.

THE LAKE JOHNSTON NATURE RESERVE

The Lake Johnston Nature Reserve
The dead trees are the result of a fire in 1960 which swept up the slopes to the summit of Mt Read from the low country and wiped out a proportion of the Huon pines near Lake Johnston. None of the Huon pines is showing any sign of regeneration 40 years after the fire, and many other endemics show little sign of regeneration either, which indicates how sensitive alpine vegetation is to fire. (There is a similar situation at Cradle Mountain where dead trees, also killed by fire long ago, stand bare in the heathland vegetation.) PETER SIMS

Mt Read on the West Coast Range is a wet and windswept peak. It forms part of the catchment of the Henty River. A small area of 138 hectares was recently declared a Nature Reserve, the highest category of land classification that can be given in Tasmania, to protect the natural values of this region. Lake Johnston lies within the reserve. Mt Read is of special interest and importance (not only in Tasmania but in the world) because botanically it is unique. The list of its special botanical features is impressive:

- Huon pines growing on Mt Read are at the highest recorded altitude
- A stand of Huon pines near the lake may consist of, or be derived from, one or a few single individuals which may have been present on the site for more than 10 500 years
- A unique disjunct stand of implicate high-altitude Huon pine with very high conservation and scientific values
- A significant population of the endemic *Diselma archeri*, the Cheshunt pine, (family Cupressaceae) in its rare arboreal form, including the largest and possibly oldest specimen recorded
- A stronghold of the rare and restricted *Orites milliganii*, a member of the Proteaceae
- Ninety of the State's endemic plant species
- The highest number of endemic conifer species (at least seven), equal to those in Mt Field National Park
- The largest single patch of *Nothofagus gunnii* (deciduous Antarctic beech) in the State
- *Leptospermum nitidum* (a tea tree) up to 220 years old
- Fourteen rainforest community types, ten of which are unknown in other State Reserves
- In addition, tree studies on Mt Read Huon pines show the climatic record extending back 5000 years—the longest such record in the Southern Hemisphere.

The declaration of the reserve is long overdue and is to be applauded, but in the draft proposals for management it is proposed to include tourism. Up to 700 tourists a year would be taken into this fragile environment, to the top of the mountain and out onto a boardwalk to view the Huon pines. Surely there are many other places in the general vicinity which have as much to offer from a casual tourist's point of view. This new reserve must offer complete protection for what is known to be a unique and fragile place of inestimable value, which must not be put at risk in any way. It is to be hoped that the Minister responsible for setting the guidelines for National Parks will have the wisdom to rethink the proposal to open the area to tourism.

Environmental awareness and regulations governing management of wastes have both changed and hopefully improved since the 1970s, but a cavalier attitude still exists about dumping wastes into the sea. (When major cities like Sydney still persist in sending all their sewage, which also contains high levels of industrial chemicals and pollutants, into the ocean, the magnitude of the problem becomes apparent.)

The practice of dumping wastes into the open sea is not the harmless solution one would hope. When waste is going into closed estuaries or lakes, either directly or via rivers, the problems are obviously even more serious.

In highly metalliferous mining regions, like the ranges of Tasmania's west coast, some heavy metal contamination of river waters is natural—but few of the wild rivers along the west coast are untouched by additional pollution from mining, so widespread has been the development of small mines and diggings, now mostly abandoned. The very heavy rainfall along the ranges makes safe management of pits, shafts, dams and dumps very difficult. The impact of the big operation at the Savage River iron ore mine is not

The mouth of the Inglis River at Table Cape, showing the heavily silted estuary. M.E.W.

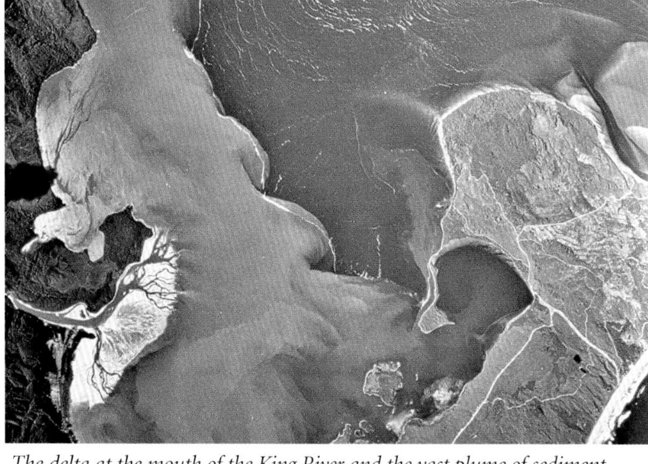

The delta at the mouth of the King River and the vast plume of sediment entering Macquarie Harbour are seen in this aerial photograph. A. DE BLAS

The smelter at Port Latta, where the pipeline from the Savage River Mine brings material for processing. Devastating pollution of the surrounding area results. M.E.W.

The King River, damaged by siltation and acid mine drainage. A DE BLAS

Devastation near the mouth of the King River. A. DE BLAS

limited to its local, remote area in the mountains. An amazingly long pipeline carries ore material to Port Latta on the coast, where it is processed, pelletised and exported to Japan. The acid rain created by the corrosive dust which emanates from the smelter is killing the local area; the waste products, leachate from dumps and associated problems and impacts are a nightmare.

Macquarie Harbour and the community of Strahan, the settlement within it, are at risk from increased contamination by heavy metals and acid drainage from mining activities in the area because of the closed nature of the harbour and its constricted flow patterns. Intensive mineral exploration began there in the 1850s, the first economic metal deposits being found in the King River catchment in 1879. When gold was found in the Queen River, 8 kilometres from Mt Lyell, in 1881, a minor gold rush followed.[217] The Iron Blow, a rich deposit of copper ore, was discovered on Mt Lyell in 1883 and over the next decade it became the largest copper producer in the British Empire and the largest in the Southern Hemisphere, worked by the Mount Lyell Company and the North Lyell Company. At least another 40 companies also worked the area. The Mount Lyell Mining and Railway Company Ltd was formed in 1893 and generated great wealth—by 1897, its budget was already as large as that of the Tasmanian Government and it was making hefty profits. In those early days, no concept of environmental protection existed. The operation of the smelters, involving cutting down trees for fuel and the creation of acid rain, which further denuded and poisoned the surrounding landscape, created the moonscape which characterises the region today.

The discharge of mine tailings into the Queen River at Queenstown began in 1922 with the introduction of flotation technology to pre-concentrate the ore. It was found that the tailings were able to mobilise smelter slag, so that was discharged into the river system as well. Although processes in the mine changed, tailings

ACID DRAINAGE FROM MINESITES

Acid drainage is a naturally occurring phenomenon where sulphide minerals in rocks, in the presence of oxygen and water, oxidise to produce sulphuric acid. It is a major environmental problem facing the mining industry today. Fractured or freshly ground mine waste is a source of rapidly appearing acid drainage which costs the US $20 billion, and Australia $60 million a year.

The acid can get into streams and rivers, affecting animal and plant life; it can also leach metals from rock, particularly broken-up or crushed rock resulting from mining processes, allowing them to contaminate the environment.

Acid drainage is worst in high rainfall areas. The Mt Lyell site and the Rum Jungle site (now abandoned) are typical examples, one situated in the high rainfall region of the Tasmanian west coast, the other in the monsoonal tropics of the Northern Territory. Mines in tropical regions with heavy rainfall events associated with monsoons tend automatically to have acid drainage problems.

Western Australia, with its arid to semi-arid climate and flat topography, tends to have more localised problems, which have been slower to develop. In the Collie Basin, the sulphur-bearing shales within the coal seams create problems in abandoned voids; the Beenup Mineral Sands project had potential problems with pyrite in some layers, largely avoided by awareness and appropriate handling, but has recently closed because of its mounting problems with acid sulphate soils; the iron ore mined at Mt Newman contains pyritic materials, but acid drainage problems there were dormant for a while because frequent cyclones have only returned to the region in the 1990s. (Cyclone Bobby caused widespread run-off, and now dumps are covered when cyclones are predicted, and stormwater is managed.) Gold in Western Australia is mainly greenstone-hosted, and carbonates in the rock provide a buffer, limiting the potential for acid drainage.[219]

The Captains Flat derelict mine adjacent to the Molonglo River just outside the ACT is a source of acid drainage and toxic metal contamination (see page 224).

The mining industry today is fully aware of the problems of acid drainage and has developed techniques for preventing or dealing with them.

Mine pollution in the King River near Queenstown from the operations of the Mt Lyell mine. REG MORRISON

A view to the denuded hills at Mt Lyell. REG MORRISON

The devastation caused by Mt Lyell involved destruction of vegetation on the surrounding hills. This picture of the town rubbish dump with the moonscape hills in the background says it all! REG MORRISON

continued to be discharged into Macquarie Harbour via the King River until the closing of operations in 1994. The Environmental Protection Act of 1973 should have applied, but the Mount Lyell Company was issued with a ministerial exemption with no specified limits and allowed to go on discharging mine tailings and mine water into the rivers.

The impact of acid drainage from the Mt Lyell operation was the subject of a study in *Listen … Our Land is Crying* (pages 248–50) and will not be repeated in detail here. However, some aspects of the pollution of the Queen and King Rivers, and of Macquarie Harbour, are relevant to the story of the difficulty of preventing environmental contamination by mining operations, in high rainfall areas in particular.[218]

The pollution from former mining operations at Mt Lyell is one of Australia's most visible mining disasters. Over 100 million cubic metres of tailings, slag and topsoil were deposited in the Queen and King Rivers and Macquarie Harbour as a result of mining operations. A 250-hectare tailings delta formed at the mouth of the King River. Because of the high west coast rainfall, seepage through the mining lease gives rise to metal-rich acid drainage on a scale possibly without equal in the developed world. The Queen and King Rivers have been virtually killed. Most of the acid and copper comes from the lease site; tailings in river beds and banks make a small contribution. Every day about 2 tonnes of copper are discharged from the lease site, which has

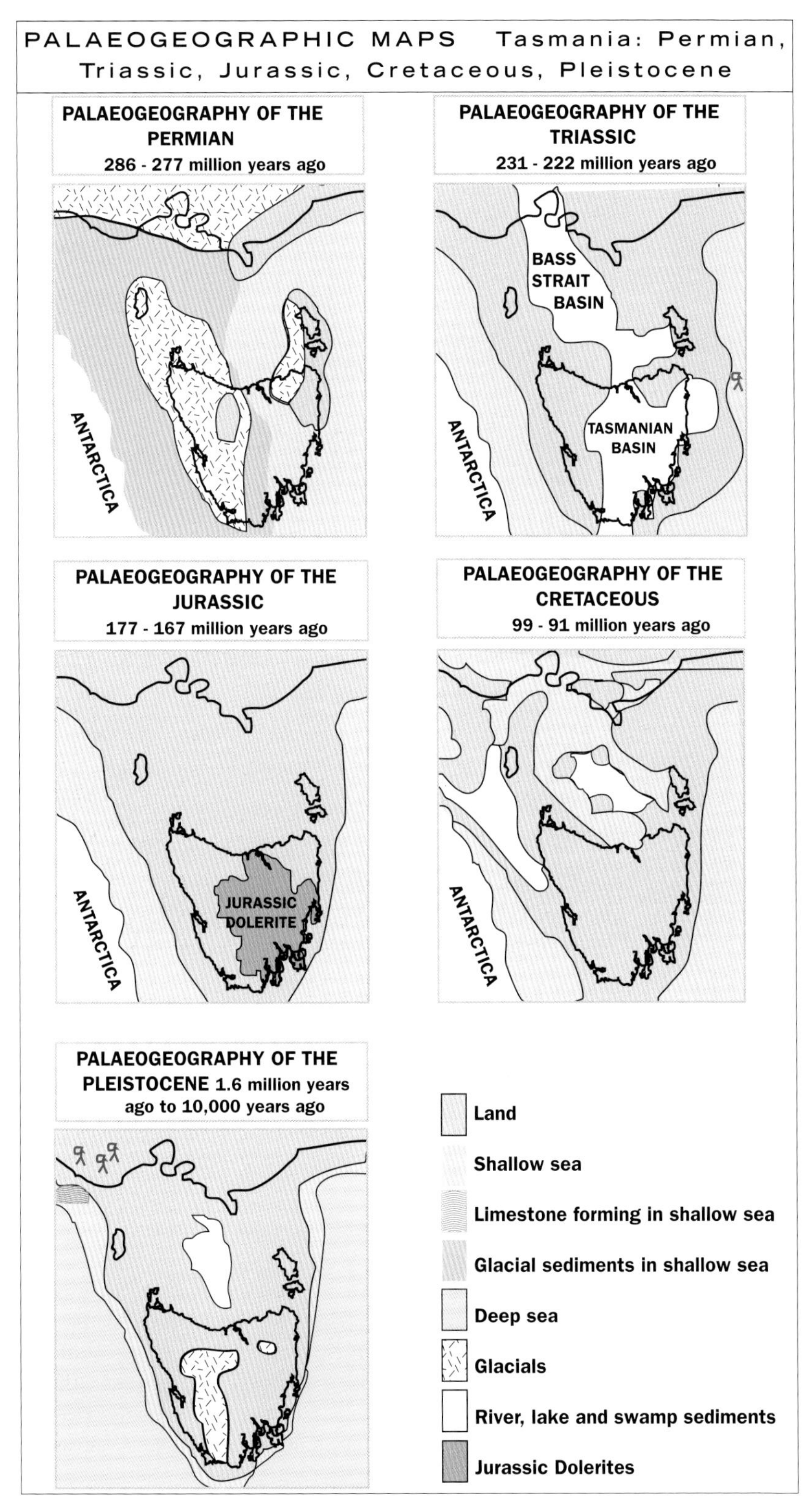

the potential to continue producing acid for thousands of years into the future. The same potential exists for the delta sediments if they are exposed to air. Covered by water and to some degree constantly redistributed by currents, their heavy metal content remains of concern because of its ability to enter food chains.

A study published in 1994 showed that heavy metals were present in significant amounts in the flesh of fish and shellfish.[217] Mercury in the tissues of some of the most commonly eaten fish was of concern because many residents of Strahan regularly eat fish several times a week, and mercury is a cumulative poison.

THE DERWENT RIVER

The Derwent, like the Tamar, occupies a broad valley reflecting the turbulent geological times when Australia and Antarctica were rifting in the final chapter of the break-up of Gondwana. The history before the commencement of Gondwanan rifting, which started in the Jurassic, hardly concerns the present story, but it is interesting to take a brief glimpse further back into the geological past and to realise that in the last major ice age (before the present Pleistocene ice age) Tasmania was near the South Pole and very little of what is now Tasmania was land. The time was around 290 to 280 million years ago, at the end of the Carboniferous and the start of the Permian Period. Shallow icy seas and glacial outwash from the huge ice sheets which covered most of the southern two-thirds of Australia and all of Antarctica left only small bits of land exposed, mainly in eastern Tasmania. The southern quarters of Africa and South America, as well as about half of India, were also under ice while all the Gondwanan lands were united in the supercontinent.

As the Permian Period warmed up, some volcanic activity started and increased through the Triassic.[220] Catastrophic volcanic activity was to follow in the Jurassic because Tasmania was in the hinge area between Antarctica and Australia as rifting commenced, and incipient rifts developed across the land (one major line now followed by the Derwent, one by the Tamar). Sheets of dolerite between 350 and 450 metres thick were intruded into the Permo-Triassic basin rocks about 180 to 175 million years ago, in up to five main phases. One of these sheets now caps Mt Wellington. Post-Jurassic uplift and erosion has removed most of the intruded volcanic rock; it is possible that as much as 1.5 kilometres has been removed from the roof of the great sheet exposed on Mt Wellington.[220A]

This was a violent time seismically, and the broad plains disappeared, leaving low fault-block ranges. Modern Tasmania was beginning to form. New rivers began to follow the valleys defined by the new ranges and fault systems. The Derwent, Coal and Jordan Rivers all follow paths defined at this time.

It is clear that a proto-Derwent is channelled into the rift axis from the Mid Jurassic, and there is evidence that from this time all drainage was rejuvenated periodically and incised into the valleys and structures which were first developed at the time of the dolerite emplacement. The uplift which elevated the Wellington Range was also triggered at this time. Both the river and the mountain have their origins in Jurassic rifting, but their present forms represent a complex and repeated application of rift forces. By about 100 million years ago, in the Cretaceous, a new thermal surge accelerated the rifting between Antarctica and Australia. The Bass Strait region was the active zone initially but later subsided while the rest of Tasmania, still attached to Antarctica and the New Zealand Subcontinent, rose and dipped

The Derwent Estuary from the air. DICK BURNS

Hop farming in the fertile alluvium in the Derwent Valley. M.E.W.

The Derwent, between New Norfolk and Bushy Park; trout fishing is a popular recreation. M.E.W.

TASMANIA'S RECENT EARTHQUAKE RECORD

WESTERN TASMANIA

Five major earthquake episodes, averaging about 5 on the Richter Scale, were documented between 1859 and 1924:

- At Circular Head in 1859
- In south-western Tasmania in 1880, when there was movement on the Lake Edgar Fault (on which one of Lake Pedder's dams is built!), the shock being felt in the Huon and Derwent valleys
- At Queenstown in 1908
- Along the west coast in 1911 and again in 1924, when loud rumblings and aftershocks were reported from Zeehan.

NORTH-EASTERN TASMANIA

North-eastern Tasmania, however, holds the record for shaking and quaking:

- Between 1883 and 1892 there were about 2000 events—a swarm of tremors and at least three major earthquakes
- In 1884, an earthquake in this region had a magnitude of 6.4 on the Richter Scale
- In 1885, there was an earthquake of 6.8 intensity
- In 1892 another earthquake with an intensity of 6.9, the strongest so far recorded in Australia, concluded the swarm series.

The three big quakes of 1884, 1885 and 1892 were felt from Hobart in the south to south-eastern New South Wales in the north. The epicentre appears to have been in the West Tasman Sea Zone, east of Flinders and Cape Barren Islands. (To this list must be added the 1854 quake at Glenorchy, not mentioned by Michael-Leiba.)

Minor earth tremors in Tasmania, as in mainland Australia, result from movement on faults due to the tensions generated by the continent's northward passage, pushing into the Timor region to the north.[222]

slightly to the south. There is also evidence that the uplift, and matching subsidence, is continuing (but at a much slower rate than was typical from Late Cretaceous to Mid Tertiary times). The stresses related to the opening of the Tasman Sea and the final separation of Australia from Antarctica impressed a repeated rift pattern on eastern Tasmania. The tensional direction was rotated from roughly north–south to north-west–south-east, rifts were widened and structures rejuvenated. Volcanic activity accompanied these changes and peaked during the Miocene, about 15 million years ago. Some vulcanism persisted until the Late Pleistocene.

Tasmania could easily have been torn right apart during all the rifting processes, situated as it was adjacent to two spreading systems. The crust was clearly thinned and, although under continuous extension, was uplifted. While uplift continued throughout the Tertiary, river systems were engraved. The drainage system of the Derwent cut narrow ravines into the rift axes and across lesser divides. Little of the original Tertiary valley fill survives. Late Tertiary and Pleistocene climatic changes transformed the shape of the river and its valley, and rising sea level after the last glacial stage drowned the valley, which had been deeply incised while sea level was low during the glacial maxima. Evidence that sea level was higher during the last interglacial is seen in valley-fill deposits 15 metres above present sea level. The height indicates that uplift of the area has also occurred because sea level rises have not been of that magnitude. A significant earthquake in the Glenorchy region in 1854 demonstrated that the region is still seismically active.

Recent research has shown that Tasmania is actually being uplifted, and this has been going on since the last interglacial (during the last 100 000 years).[221] Modern valley fills of the Derwent and

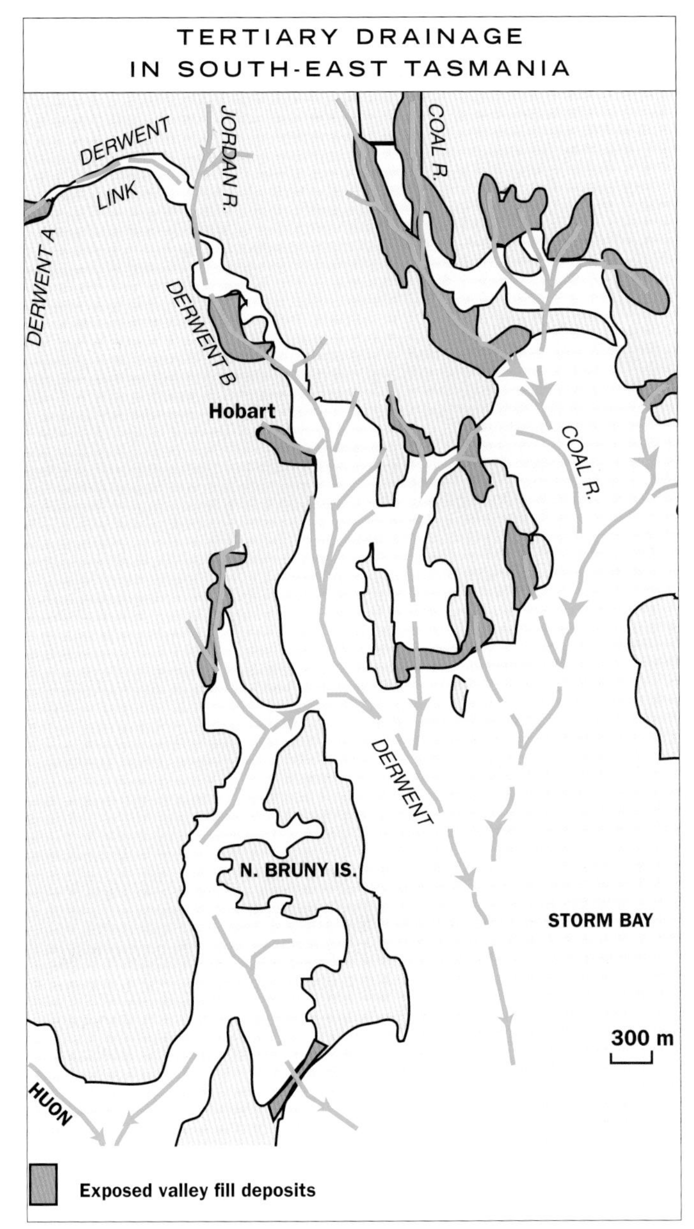

After Leaman[211]

Tamar systems are enormous and not fully compacted, which is a source of concern because of the possibility of earthquakes. When earthquakes shake unconsolidated sediment the effect can be thixotropic—the sediment acts like a liquid and structures on its surface can be engulfed. (Such a scenario has been seen in some earthquakes in Japan, where multi-storey buildings have disappeared into the ground.)

THE COAL RIVER

From its mouth near Richmond in south-eastern Tasmania the Coal River occupies a valley running essentially north–south, still within the ancient rift axis, to just beyond Rhyndaston in the central Midlands.[211] Then it curves to the east via a huge gorge which itself meanders—an unusual feature for a gorge, which implies that the river had cut deep meanders into a plain before that plain with its incised meanders started to be elevated, and that incision has simply continued during uplift. The incised meander systems prove that a long-lived river had existed *before* uplift and that the same river lives on. (This scenario is also the origin of the Grand Canyon in Arizona in the USA, except that uplift is much more recent and on a larger scale.)

The Coal was once a tributary of the Derwent, joining it before deep incision of the valley systems leading into the Derwent estuary concealed the connection. The Coal Plains were the granary of the colony in the difficult early years after European settlement. Low rainfall, soil depletion and salinity issues soon ended the intensive cropping and the area between Richmond and Campania has yet to recover. (Saline water tables characterise the midlands, and the problems of dryland salinity are the same as in parts of the mainland.) Irrigation schemes are starting to operate along the plains and it is too early to assess their success but, given experience elsewhere and the nature of the terrain and its history, it seems unlikely that any long-term irrigation industry is viable.

THE JORDAN RIVER

The Jordan is another tributary of the Derwent flowing in the rift system, though less control is exerted on the Jordan by the rift morphology, which is more complex in its case. The Derwent between Sandy Bay and Bridgewater occupies a major Jurassic rift axis. Further north this axis is occupied by the Jordan. Between New Norfolk and Ouse, the Derwent occupies another rift axis and wanders about within it. Although there is a major divide between the two axes today, this presented no problem to the migration of the Derwent in the rift system as the divide was rising while the river was down-cutting.

TASMANIA'S DISAPPEARING FORESTS

Logging old growth forests damages streams and rivers

The Federal Government has the power to protect areas of World Heritage value, and large tracts of Tasmanian forests have this protection. Despite this, the situation is far from satisfactory and much needs to be done, starting with an overall national change of attitudes to forest conservation, before it is too late and this wonderful heritage is destroyed and future generations are impoverished.

Tasmania exports an estimated 3 million cubic metres of woodchips a year, mostly to Japan. The ravaging of the last major

Liffey Falls. JIM FRAZIER

remnants of cool temperate Gondwanan forest on Earth goes on regardless, in places right up to the fringes of protected areas, putting their integrity at risk. The economics are suspect—the low price for *Nothofagus* woodchips, used for fax paper, barely covers the cost of the infrastructure roads and services involved in producing it for export. The ethics, or lack of them, involved in destroying a heritage which belongs to the world, let alone to future generations of Tasmanians and Australians, are shameful—yet government, for the sake of keeping a few foresters in work for a few years, has not the fortitude to stop the destruction. Because Tasmania is the most forested State with about 40 per cent of its area involved, and because the quite erroneous view prevails that forests regenerate after clear-felling, a shameful attitude to forest preservation exists at the highest levels of bureaucracies. Gondwanan forest, in its complex rainforest and wet sclerophyll expression, *does not* regenerate to reform its original mixture. The complex web of life which it comprises, the extraordinary antiquity and grandeur of its thousands-of-years-old giant components, and the untold interactions between its macroscopic and microscopic plant and animal species, make it a living entity—a surviving dinosaur deserving of complete protection. Surely 60 million or more years of existence in similar form entitles the ancient Gondwanan forest to reverence?

Even if only for the health and sustainability of Tasmania's rivers, the ravaging of the forests should cease. You cannot separate the wellbeing of rivers from the wellbeing of the forests. The statistics for the forests speak for themselves: under the Regional Forest Agreement, 46 per cent of Tasmania's old growth tall eucalypt forest is open to logging, of which 73 per cent (67 210 hectares) is in provisional coupes. The Forest Practice Code specifies that no logging should take place within 100 metres of major streams, but there are virtually no prosecutions for breaches, and many small watercourses are logged because they are not permanent streams. Silt makes its way from small disturbed streams into larger tributaries.

The Arthur River. M.E.W..

There are four major areas of the State containing unprotected wilderness of World Heritage value, containing the following rivers which are at risk of losing their wild and pristine character:

- **Southern Forests** Approximately 40 000 hectares of tall eucalypt forest is unprotected; the rivers threatened by logging are the Navarre, Counsel, Derwent, Florentine, Wedge, Styx, Weld, Huon, Picton and Esperance. Some of them are already carrying visible sediment loads.
- **Great Western Tiers** Approximately 200 000 hectares of rainforest, eucalypts and button-grass moorland are threatened, primarily by mining, affecting the Henty–Murchison river network.
- **Tarkine** About 110 000 hectares of tall eucalypt forest and rainforest are threatened by both mining and logging, affecting the Rapid, Arthur, Huskisson–Coldstream and Savage Rivers. An additional 220 000 hectares of rainforest threatened not by logging but by mining has implications for the Meredith, Donaldson, Sumac, Ramsay and Pieman Rivers.

Only 18 per cent of Tasmania's original tall eucalypt forest remains as old growth. Half of this is reserved from logging, half is threatened. Over 90 per cent of the original pre-European settlement tall eucalypt forests have been cleared, logged or burnt, or are threatened by logging today. As this figure, from the Regional Forest Agreement, includes the slivers of forests along creeks and roads as part of the total reserved, the real picture is even bleaker.

Many rivers in eastern Tasmania have been seriously impacted by logging in recent decades. In the north-east the Weld forests are currently being decimated as fast as possible because the logging industry sees the rising tide of opposition and wants to accomplish its mission before destruction of old growth forests is prohibited.

Lake Pedder morning. DICK BURNS

The wild Franklin River. HELEN GEE

Tuesday 14 March 2000 was declared International Rivers Day. The Living Rivers Festival in Tasmania, planned to coincide with International Rivers Day, had as its vision statement:

> ***'We celebrate the river as a powerful symbol of water and life and we aim to focus energy on the challenge of river restoration in the Century for Restoring the Earth.'***

Pedder 2000, the Lake Pedder Restoration Committee which organised the Festival, was a founding member of the **Living Rivers International Coalition** which arose out of an International Rivers Network workshop held in California in July 1998. (Helen Gee of Pedder 2000 has kindly supplied summary information for this account.) The aim of the Coalition is to restore rivers and the well-being of the communities which depend on them by working to change the operating patterns of dams, and to drain reservoirs and remove dams where appropriate, restoring the rivers and plains which they have impounded. Lake Pedder, submerged in the Impoundment and waiting to be released from its imprisonment, is still an icon. The proposals for, and commitment to, its restoration are an inspiration useful in focusing attention on other rivers already degraded or at risk of losing their pristine status. It is an example of the unnecessary damming of a wild river, and should warn against the exploitation of other Tasmanian streams.

The Franklin and the lower Gordon could have been added to the list of wild rivers tamed and violated if it had not been for the lessons learnt in the loss of Lake Pedder. What had happened to the Pieman River subsequent to the lost battle for Pedder added another dimension, emphasising the need to stand firm and fight the forces which were hell-bent on the development of more hydro power.

The **Pieman Scheme** was a major investment decision of 1971 involving an estimated $134 million dam that flooded huge quantities of valuable commercial timber, just as did the Gordon Dam in 1975. Neither of these projects was carefully and impartially examined and the Premier of the day urged the quick passage of the Bill for the Pieman Scheme, saying that there had been no protests. There had, however, been many written objections from citizens, the ACF and the Tasmanian Conservation Trust. Tasmania's famous photographer/ conservationist, Olegas Truchanas, had made several photographic journeys down the Pieman and its tributaries and had worked unstintingly on what was virtually a one-man active campaign. The Bill was passed and construction began, but soon slowed as there were no customers to buy the power it was expected to generate, and by 1982 construction was still incomplete and the estimated cost had risen to $700 million. A beautiful wild river had been dammed for no good reason, adding to the list of such sad projects in Tasmania. The Murchison and Mackintosh Rivers, both tributaries of the Pieman, have since been dammed, impacting on the western edge of the World Heritage Area.

The Franklin Gorge. HELEN GEE

The Franklin River Blockade. TONY MARSH.

The **Franklin River Campaign** goes down in history as a watershed for conservation, reaffirming the possibility of change and giving hope to a new generation of concerned citizens determined to protect the forests and rivers for future generations.

The **Franklin River** rises high in central Tasmania's Cheyne Range. Forty-five kilometres from its source it enters the first of a series of breathtaking gorges, separated by reaches curving south-west of the Frenchmans Cap massif. The features of the Great Ravine have now become legendary, named as they were by Dr (now Senator) Bob Brown after his historic descent of the river in 1976. The upper, middle and lower Franklin are distinctly different sections of a wild and wondrous river. In the late 1970s, the Hydro-Electric Commission had advanced plans to drown long reaches of the river behind a series of three large dams. The Blockade which followed is history—the strongest conservation controversy Australia has witnessed. Blockaders had come from all walks of life to participate in non-violent action: 1272 people were arrested, 450 of them remanded in Risdon jail; over 2600 Australians registered at Strahan during the Blockade when successive waves of rubber duckies assembled across the lower Gordon River in defiance of the dam builders. All State capitals saw demonstrations, the largest since the Vietnam War; international attention and support was generated; the issue became political and an election one which ushered in the Hawke Labor Government; 'Bob Brown of the Franklin River' became Australian of the Year. Stirring times and doings to protect a wild river!

Newland Cascade, Franklin River. HELEN GEE

The High Court of Australia delivered judgment on 1 July 1983—the Franklin and the lower Gordon Rivers were to run free. A standard had been set for non-violent actions around the country. By the end of the decade, the Tasmanian Wilderness World Heritage Area, incorporating the Wild Rivers National Park, was established, covering 22 per cent of the State.

Helen Gee of Pedder 200 kindly supplied information on forests and rivers.

THE GONDWANA FOREST SANCTUARY FOR AUSTRALASIA AND OCEANIA

A proposal to create, by inter-governmental treaty, an international sanctuary south of 40° South for the Gondwanan rainforests of Argentina, Chile, Tasmania and New Zealand, is taking shape. It is inspired by the Whale Sanctuary, south of Parallel 40, where these huge sea mammals find refuge from persecution—a concept which seemed impossible when it was proposed in the 1960s but which eventuated 30 years later.

The Gondwana Forest Sanctuary will primarily protect the southern beech forests (*Nothofagus* spp.) found in such environmental hotspots as Patagonia, Tierra del Fuego and the Tarkine in Tasmania. These ancient southern forests are now imperilled by the large-scale commercial logging, woodchipping, forest conversion and export activities of a host of multinational corporations.

The sanctuary would preserve all the primary forests and permit only sustainable uses of secondary forests. It would join the international whale sanctuary and be similar in its practical application to the UNESCO Biosphere Reserve System.

The Southern Hemisphere Gondwana Forests Sanctuary is part of a long-term strategy to protect southern forests. Organisations and businesses, and concerned individuals, are encouraged to endorse the proposal and become a signatory. Letters of support should be written to the heads of state of the countries involved: John Howard, Prime Minister, Parliament House, Canberra 2600; President Eduardo Frei, La Moneda, Santiago, Chile; Presidente Carlos Saul Menem, Casa de Gobierno, Balcalce 50, Capital Federal, Buenos Aires, Argentina; and Prime Minister Helen Clark, Parliament House, Wellington, New Zealand.

Unprotected forest in the Tarkine. M.E.W.

Information on the Australian branch of the Gondwana Forest Sanctuary can be obtained from the Steering Committee Secretariat, PO Box 33, St Helens, Tasmania 7216.

Run-Down

WHAT HAS THE MENTAL VOYAGE OF EXPLORATION WHICH HAS TAKEN US ALL OVER THIS INCREDIBLE CONTINENT TOLD US ABOUT SURFACE AND UNDERGROUND WATERS?

*The paperbark bushes (*Melaleuca lasiandra*) around this termite mound in the Tanami Desert are an indication of a palaeodrainage locally in which water is available below the desert sand. Surrounding areas cannot support more than spinifex.* REG MORRISON

- Australia is unique, the product of the co-evolution of environment and biota through enormous lengths of geological time. A great deal of the history of the land, of past climates and conditions which prevailed, can be interpreted from a study of modern rivers and drainage patterns and the palaeodrainages which preceded them.

- This ancient land has some of the oldest landscapes preserved anywhere on Earth and some of the most ancient drainage patterns. Many of our rivers occupy ancient valleys, incised by their ancestors multi-millions of years ago, or are superimposed on palaeoriver systems.

- The past history of the continent accounts for the nature of its rivers, its drainage patterns and its underground water resources. Its stability, lack of tectonic activity and its erosion through time into the flattest of all the continents explains the behaviour of its rivers. The flat landscapes and sluggish or inward-draining nature of all but the edges of the continent not only determine the form of rivers and the general hydrology, but explain the accumulation of salt and sediments and the problems encountered when the hydrological balance is upset.

- Rivers in Australia have the most variable flow patterns in the world. The climatic variability orchestrated by ENSO results in extremes. Rivers of the arid and semi-arid zones—85 per cent of the continent—are ephemeral. The Australian biota is adapted to this variability and in fact depends on it for balance and regeneration in ecosystems.

- What constitutes an Australian river system is different from the European model. Floodplains are as essentially part of the system as are the river channels and their banks. Alienation of floodplains impoverishes the river, and vice versa.

- Diversity of form, essential for biodiversity, characterises different reaches of rivers in their natural state. Where catchments are altered and sediments are released into streams, diversity is lost and the whole river system suffers. Regulation of rivers to drought-proof regions and suit our human needs has not taken into account the needs of the environment, and much of our water use is unsustainable.

- Arid land rivers get nearly all their water from distant upper catchments and damming or diverting water in the catchments has a profound effect downstream as there is very little local run-off. The Darling is a classic example, receiving no run-off in the arid terrain through which it travels, being totally dependent on its headwater streams and the major tributaries which join it themselves subject to similar limitations. Over-allocation of water for irrigation is a serious problem and so-called environmental flows' are inadequate and mostly do not take into account the real needs of the river ecosystems. Major rivers, including the Darling, are slowly dying.

- Changes resulting from our land-use practices and our use of water resources has turned rivers into drains. (A river is a living ecosystem in a state of dynamic equilibrium, whereas a drain is simply a conduit which carries water through a landscape.) Deep incision of stream beds, loss of chains of ponds or swamps in headwaters, loss of reedbeds, siltation and widening of channels, degradation of reaches downstream of dams and weirs, salination of river water—an endless list of changes results from human activities. Pollution with agricultural chemicals and from other anthropogenic sources is a serious problem. Toxic algal blooms result from increased nutrification.

- Groundwaters are a vital resource in this, the driest of all vegetated continents. They have been taken for granted, wasted on an unimaginable scale in the case of the Great Artesian Basin, and often over-exploited. Pressure loss over the whole GAB is indication of the serious situation.
- Over-allocation of licences for groundwater use results in draw-down of aquifers. There are many instances where the groundwater extracted is not being replaced. Groundwater is a finite resource. Fossil and sub-fossil waters, stored during times in the past when Australia was less dry than it is today, are being used at an alarming rate.
- Unless the past history of the continent is understood and taken into account, there is no hope of curing today's problems and achieving a situation which approaches sustainability ...

Above all, we have learned that, as with land use, our water use has to be limited by the parameters set by our ancient, venerable continent. The fragile hydrological balances, the ecosytems which depend on local hydrology to survive, and the climatic variability which adds another dimension to problems in the oldest, flattest and driest continent have to be accommodated in our thinking and our planning. Unsustainable water use leads to complete desertification in arid lands. We are well on the way to losing one-third of our arable land, meagre enough to start with, to water-misuse-related problems. If we are holding the continent in trust, borrowing it from our children and grandchildren, we have an enormous responsibility, and the solution of problems cannot wait any longer. We know what needs to be done. Let's hope that the Century for Restoring the Earth in this new millennium will see a transformation accomplished.

Floods in the Tanami Desert bring life. Seeds germinate almost instantaneously when the water comes. There is no time to waste and the brief interval when life-giving water is available is the time of regeneration and growth, punctuating long periods of inactivity. REG MORRISON

Glossary

aeolian blown by the wind
algae group of plants, single or many celled, non-vascular; aquatic, reproducing by spores
algal blooms periods when algae proliferate within a river or water body
alluvial deposits accumulation of sediment transported by water
alluvium unconsolidated sediments, silts, sands, gravels
anabranch part of a river that branches off the main stem and then rejoins or reaches a wetland
anastomosing channels channels branching to form a network
aquifer porous layer of rock that is saturated and can supply water to springs or wells
aquitard layer that retards but does not prevent the flow of water to or from an adjacent aquifer
archipelago chain of islands
arid zone region with annual rainfall usually less than 250 mm
artefact object made by humans
artesian water underground water that is stored under pressure
assemblage group of organisms and their numbers usually found in an ecosystem
atrazine herbicide used to control weeds
backwater area of still water connected to a river
bank erosion process by which banks of rivers lose soil
bank slumping occurs when sections of river bank (>2 metres) collapse
barrage construction across the mouth of a river to prevent entry of seawater
basalt fine-grained volcanic rock formed from a lava flow
base flow river flow sourced from dry weather groundwater inflows, not surface run-off
basement undifferentiated complex of rocks that underlies the rocks of interest in an area
basin area of continued subsidence of the earth's crust where sediment accumulates
bed-load sediment that moves by sliding or rolling along the bed of a stream or river
benthic animals or plants that live on or are attached to the riverbed
benthic ooze loose layer of fine organic matter found on the bottom of a wetland or waterhole
billabong waterhole which fills during floods, usually formed from an old river channel
biodiversity full range of organisms, including their genetic diversity
biomass measure of the total mass of a particular organism, whether alive or dead
biota all living things in an area—plants, vertebrates, invertebrates, fungi, bacteria
blue-green algae cyanobacteria
border rivers Macintyre, Dumaresq and Severn, on the Qld–NSW border
bore deep drill-hole tapping an aquifer
braided streams rivers which split into many meandering channels which separate and rejoin
Carboniferous Period 360 to 286 million years ago
channel bar formation process of forming sediment 'bars' in river channels
chemical weathering alteration, breakdown, leaching and decay of rocks by percolating water
chlorpyrifos organophosphorus-based insecticide
clay sediment particles 0.002—0.004 mm in diameter
claypan depression which holds water
COAG Council of Australian Governments
coarse particulate organic matter large fragments of decomposed organic matter (>1 mm in diameter)
colluvial deposit material transported to a site by gravity
community group of organisms of different species occurring in the same place or habitat
competition organisms or species competing for a common resource (food, nesting place, etc.)
Cretaceous Period between the Jurassic and the Tertiary: about 144 to 65 million years ago
CWD coarse woody debris
cyanobacteria blue-green algae; photosynthesisers which produce toxic blooms
decomposition biochemical processes that break down organic matter into simple compounds
deep lead palaeodrainage channel deposit
deflation wind erosion
delta fan-shaped mass consisting of the deposited load of a river where it enters the sea
depocentre area into which drainage brings sediments
detritus decomposed plant or animal material
Devonian Period 408 to 360 million years ago
diatom microscopic unicellular alga with shell made of silica
discharge amount of water that passes a particular point in the river
dissolved oxygen measure of the amount of oxygen in solution
distributory branch from the main river that does not rejoin
diversion water taken from a river or creek system
diversity number of species in a community or habitat
drainage water excess water that remains after irrigating a crop
draw-down unnatural lowering of water levels by extraction at a rate that exceeds inflow
dryland river river or stream in the semi-arid or arid zone where annual rainfall is <500 mm
dryland salinity soils become saline because of rising water tables or loss of surface soils
duricrust iron-or silica-cemented crust capping landscape features and slowing erosion
DWMP Draft Water Management Plan
dynamic equilibrium short-term changes either side of a stable long-term average
EC electrical conductivity units, measuring total dissolved salts and salinity
ecologically sustainable (usage which) ensures protection of biodiversity, ecological processes, soil and water
ecology scientific study of living organisms and their relationship to one another and the environment
ecosystem organisms in a community and their environment interacting as an ecological unit
effluent river or creek flowing out of another, or out of a lake
endemic organism found only in a particular area
endorheic rivers that do not reach the sea but run into wetlands or dry river beds

endosulfan insecticide used in agriculture
ENSO El Niño–Southern Oscillation climatic oscillation producing drought and floods
entitlement users of water are given a volumetric entitlement—a licence to extract water
environmental flow water for the needs of the provision environment
Eocene an epoch of the Tertiary Period: from 58 to 36 million years ago
ephemeral pools and rivers temporary wetland areas and rivers that only run some of the time
erosion the wearing away of the Earth's surface by running water, ice, wind and sea
escarpment steep slope forming a linear feature in a landscape
estuary broad mouth of a river where it enters the sea
eustatic related to world-wide changes in sea level
eutrophication increasing nutrient levels
evapo-transpiration total amount of water lost from soil, water bodies and vegetation
extraction removal of water from a river
fault fracture along which rocks have been displaced in a horizontal, vertical or oblique manner
feral introduced organism which has become naturalised
ferricrete iron-cemented sediments
fine particulate organic matter fragments of detritus <1 mm in diameter
floodouts areas which fill during flood periods
floodplain area beside a river that floods when the river overflows its banks
flood pulse description of a particular flood event in the flow record of a river
flow pulse short-term (days or hours) increase in flow passing down a river
fluvial sediment sediment that is washed along by the action of a river
food web network of food pathways within an ecological community
geomorphology study of the origin, character and development of land and rivers
gibbers small scattered stones forming a surface cover on deserts
gigalitre (GL) 1000 million litres
groundwater water below the surface of the soil
hardpan extremely hard to cemented soils, claypan areas
headwater streams streams in the upper catchment of a river
Holocene geological epoch within the Quaternary Period—the last 10 000 years
hot-spot volcanoes volcanoes which result from local weaknesses in the earth's crust
hydrological cycle the continuous interchange of water between land, sea, other water and air
hydrology study of the behaviour of water above or below the ground
hypersaline saltier than seawater
impoundment dam, weir or other water storage
inselberg steep-sided eminence rising abruptly from a plain
instream within the main channel of a river
Jurassic period between the Triassic and Cretaceous in the Mesozoic era: 208 to 144 million years ago
lacustrine sediment sediment laid down in a lake
laterite sediment impregnated with iron
leaching process by which materials are removed by percolating water
levee bank bank which stops a river overflowing or redirects its flow on a floodplain
lignotuber woody underground rootstock
lineament fault line
littoral relating to shoreline
local run-off flows derived from nearby sources
lunette crescent-shaped dune on the downwind side of a water body
maar lake lake in a volcanic crater
macroinvertebrate aquatic invertebrate, usually between 1–100 mm
macrophyte large aquatic plant, both vascular and non-vascular
malathion organophosphorus-based insecticide
mallee habit multi-stemmed growth of a eucalypt from a lignotuber
MDB Murray–Darling Basin
MDBC Murray–Darling Basin Commission
meanders loop-like wandering course of river in flat terrain
megalitre (ML) 1 million litres
Mesozoic Era Triassic, Jurassic and Cretaceous Periods
microclimate humidity, temperature, moisture prevailing in a small area
midden pile of shells and other litter marking an old Aboriginal campsite
Miocene epoch of the Tertiary Period: from 23 to 5 million years ago
morphological: relating to the form and structure of organisms
mound-spring natural spring where water in a pressurised aquifer comes to the surface
NPWS National Parks and Wildlife Service
nutrient material eaten or taken in for maintaining an organism or allowing it to grow
nutrient cycling path of nutrients through the ecosystem
off-river storage large dams which store water pumped from a nearby river
Oligocene: epoch of the Tertiary Period: from 35 to 15 million years ago
organic carbon carbon derived from living things
outflow water that flows from a wetland or part of a river
overbank flow river flow that leaves the main channel to reach the floodplain
overspray aerial spraying of pesticides or herbicides covering unintended areas
palaeo- prefix which denotes situation in geological time
Palaeocene first epoch of the Tertiary Period: from about 65 to 58 million years ago
palynology study of fossil spores and pollen
pan depression in the ground
peak flow highest flow for a specified time period
Permian Period geological period of the Palaeozoic Era: from 286 to 245 million years ago
pH measure of acidity or alkalinity: a pH of 7 indicates neutrality, 1 extreme acidity, 14 extreme alkalinity
photosynthesis process by which plants synthesise carbohydrate using solar energy
phytoplankton planktonic plant life
piezometer instrument that measures pressure
pisolites spherical, pea-like balls of ferricrete
planform bird's eye view of the river channel, and the extent of its meandering
plankton floating plants and animals, usually <1 mm in diameter
playa salt lake

Pleistocene epoch of the Quaternary: from 1.6 million to 10 000 years ago
Pleistocene ice age the last 2 million years (to Present) when glacials and interglacials alternated
Pliocene last epoch of the Tertiary Period: from 5.3 to 1.6 million years ago
pluvial period period with high rainfall, active rivers
point source source that can be pinpointed
prior stream course of an ancient river
Quaternary Period geological period comprising the Pleistocene and the Holocene
radiocarbon dating aging of part of plants or animals by measuring decay of their radioactive carbon content
Ramsar international convention recognising wetlands of international importance
recharge movement of surface water into the groundwater or an aquifer
refuge (pl. refugia) place where the biota can shelter from predators or from extreme environments
regolith mantle of weathered rock and soil overlying the bedrock on the earth's surface
regulation process by which stream flow is affected by dams, weirs, diversions etc.
riffle shallow section of a river with rapid, turbulent flow
riffle-pool sequence of fast shallow zone with larger sediments followed by slow deep zone
riparian associated with or living on banks of rivers or streams
run-off water that flows into rivers and streams across the land
salina saline spring or marsh; saline discharge area
salinisation accumulation of salts in the soil to a level that causes degradation
salinity amount of sodium chloride or dissolved salts in a unit of water
saltbush shrubs of the Chenopodiaceae family
savanna grassland with sparse trees and shrubs
scarp steep face on the side of a hill
scarp retreat process by which the edge of a plateau migrates inwards due to erosion
sclerophyll vegetation plants with hard leathery leaves, etc., modified to cope with drought and poor soils
sediment particulate matter transported by water and subsequently deposited
sedimentation deposition of particulate matter by water
seed bank seeds lying dormant in the soil
semi-arid area with 250–350 mm annual rainfall
silcrete sedimentary rocks impregnated with silica
silt coarse mud
snag branch or log that has fallen into a river
soak shallow depression holding water
southern beech *Nothofagus* species
storage dam or weir
suspension load fine insoluble particulate matter carried by water
swale shallow depression in undulating ground; low area between adjacent dunes
tailwater excess irrigation water draining away after crops are flooded in irrigation
tectonic relating to processes in the Earth's crust which create landscape features
terrane piece of continental crust attached to others whose origin is different
Tertiary Period from 65 million years ago to the Pleistocene, starting 1.6 million years ago
thermoluminescence dating dating by measuring light emission from heated sediments
topographic relating to geographical features
transpiration loss of water vapour by plants
Triassic Period geological period between Permian and Jurassic: from 245 to 208 million years ago
tributary creek or river that contributes its flow to another, usually more major, stream
turbid water which is opaque with suspended matter
vascular plant plant with specialised internal transport for nutrients—all the higher plants
water-mound aquifer sand body which has a saturated zone acting as a water reservoir
water table groundwater-saturated zone in the regolith
weir small dam across a river
weir pool area of water held back by a weir
wetland area that is either temporarily or permanently flooded
yabbie freshwater crayfish
zooplankton microscopic organisms that float (animal kingdom)

Bibliography

1. Vincin R. 1995: What if there was no river system? Abs. MD 1995 Workshop, AGSO, Wagga Wagga.

2. McMahon TA, Gan KC & Finlayson BL. 1998: Anthropogenic changes to the hydrologic cycle in Australia. *Bur. Rural Res. Proc.* 14, 36–66.

3. Cribb J. 1999: *Evidence of massive landscape change unearthed.* Press release. CSIRO, Land & Water

3A. Nanson G, Price DM & Short SA. 1992: Wetting and drying of Australia over the past 300 ka. *Geol.* 20, 791–794

4. Beard J. 1996: *Plant Life of Western Australia.* Kangaroo Press, Sydney

5. Van de Graaff *et al.* 1977: Relict early Cainozoic drainages in arid Western Australia. *Z. Geomorph. N.F.* 21 (4), 379–400

5A. Wyrwoll K-H. 1988: Time in the geomorphology of Western Australia. *Progress in Physical Geography* 12, 237–263

6. Waterhouse JD, Commander DP, Prangley C & Backhouse J. 1994: Newly recognised Eocene sediments in the Beaufort River palaeochannel. *Geol. Soc. W.A. Ann. Rev.* 1993–94, 82–85

7. Kern AM & Commander DP. 1993: Cainozoic stratigraphy in the Roe palaeodrainage of the Kalgoorlie region, Western Australia. *Geol. Surv.* WA Rep. 34, 85–95

8. Ollier CD, Chan RA, Craig MA, & Gibson DL. 1988: Aspects of landscape history and regolith in the Kalgoorlie region, Western Australia. *BMR J. Aust. Geol. & Geophys.*10, 309–321

9. Clarke JDA. 1994: Geomorphology of the Kambalda region, Western Australia. *Aust. J. Earth Sci.* 41, 229–239

10. Chapman A. 1996: *Report on a 1995 visit to Lake Boonderoo.* CALM Kalgoorlie

11. Chapman A & Lane JAK. 1997: Waterfowl usage of wetlands in the south-east arid interior of Western Australia 1992-93. *Emu* 97, 51–59

12. Minton C, Pearson G & Lane J. 1995: History in the mating: Banded stilts do it again. *Wingspan* June, 13–15

13. Minton C, Lane J & Pearson G. 1995: Update on banded stilt breeding event. *Wingspan* September, 9

14. Platt J. 1966: *Esperance Region Catchment Planning Strategy.* AgWA & National Landcare

14A. Johnson SL & Baddock LJ. 1998: *Hydrogeology of the Esperance–Mondrain Island 1:250 000 sheet.* Water & Rivers Comm. Report HM2

15. Clarke JDA. 1994: Evolution of the Lefroy and Cowan palaeodrainage channels, Western Australia. *Aust. J. Earth Sci.* 41, 55–68

16. Craig GF. 1998: *Oldfield Catchment.* Report for Oldfield Landcare Group, AgWA & NHT

17. Twidale CR. 1997: Persistent and ancient rivers—Some Australian examples. *Phys. Geogr.* 18(4), 291–317

17B. Twidale CR & Milnes AR. 1983: Aspects of the distribution and disintegration of siliceous duricrusts in arid Australia. *Geol. En Mijnbouw* 16, 373–382

18. Tapley I. 1990: Night thermal imagery reveals ancient drainage system of the Canning Basin. *Exploration and Water.* W.A. Govt. Res. Rev.

19. Pain CF & Ollier CD. 1995: Inversion of relief—a component of landscape evolution. *Geomorphology* 12, 151–165

20. Webb M. 1995: The strange death of trees in the Upper Fortescue Floodplain. A reconnaissance study and policy recommendation. Unpubl.

21. Twidale CR & Campbell EM. 1988: Ancient Australia. *Geo J.* 16(4) 339–354

22. Ollier CD, Gaunt GFM & Jurkowski I. 1998: The Kimberley Plateau, Western Australia. A Precambrian erosion surface. *Z. Geomorph NF* 32(2), 239–246

23. Wyrwoll R-H. 1979: Late Quaternary climates of Western Australia: evidence and mechanisms. *J. Roy. Soc. W.A.* 62(1-4), 129–142

24. Wende R & Nanson GC. 1998: Anabranching rivers: ridge-form alluvial channels in tropical northern Australia. *Geomorphology* 22, 205–224

25. Wende, R. 1999: Boulder bedforms in jointed-bedrock channels. In Miller AJ & Gupta A. (eds). *Varieties of Fluvial Form.* John Wiley

26. Fleming PM. 1994: The Gascoyne River basin, Western Australia. WMO Study, unpubl.

27. Twidale CR. 1998: Antiquity of landforms: an 'extremely unlikely' concept vindicated. *Aust. J. Earth Sci.* 45, 667–668

28. Bradby K. 1997: *Peel–Harvey. The decline and rescue of an ecosystem.* Greening the Catchment Taskforce, Mandurah, WA

29. Hatton T & Salama R. 1999: Is it feasible to restore salinity-affected rivers of the Western Australian Wheatbelt? *Proc. 2nd Aust. Stream Management Conf. Adelaide* 313–317

30. Schur B & Reeves H. 1999: The secrets of good agency-community partnerships: A review of stream repair programs on W.A.'s South Coast. *Proc. 2nd Aust. Stream Management Conf. Adelaide* 533–538

31. English P. 1998: *Cainozoic geology and hydrogeology of Uluru–Kata Tjuta National Park.* AGSO

32. Nanson GC, Chen XY & Price DM. 1995: Aeolian and fluvial evidence of changing climate and wind patterns during the past 100ka in the western Simpson Desert, Australia. *Palaeogeog., Palaeoclim., Palaeoecol.* 113, 87–102

33. Pickup G, Allan G & Baker VR. 1988: History, palaeochannels and palaeofloods of the Finke. River, Central Australia. Chapter 9 in *Fluvial Geomorphology of Australia.* Academic Press, Sydney, 177–199

34. Baker VR, Pickup G & Polach HA. 1983: Desert palaeofloods in central Australia. *Nature* 301 (5900), 502–504

35. Pickup G. 1989: Palaeoflood hydrology and estimation of the magnitude, frequency and areal extent of extreme floods—An Australian perspective. *Civil Engineering Trans.* 19–29

36. — 1991: Event frequency and landscape stability on the floodplain systems of arid central Australia. *Quat. Sci. Rev.* 10, 463–473

37. Jacobson G, Arakel AV & Chen Yijian. 1988: The central Australian groundwater discharge zone: Evolution of associated calcrete and gypcrete deposits. *Aust. J. Earth Sci.* 35, 549–565

37A. Jacobson G & Jankowski J. 1989: Groundwater-discharge processes at a central Australian playa. *J. Hydrol.* 105, 275–295

37B. Lau JE & Jacobson G. 1990: Aquifer characteristics and groundwater resources of the Amadeus Basin. In Korsch RJ &

Kennard JM (eds). Geological and geophysical studies in the Amadeus Basin, central Australia. *BMR Bull* 236, 363–379

37C. Jacobson G. 1988: Hydrology of Lake Amadeus, a groundwater discharge playa in central Australia. *BMR J. Aust. Geol. & Geophys.* 10, 301–308

37D. Jankowski J & Jacobson G. 1990: Hydrological processes in groundwater discharge playas, central Australia. *Hydrological Processes* 4, 59–70

37E. Jacobson G, Lau GC, McDonald PS & Jankowski J. 1989: Hydrogeology and groundwater resources of the Lake Amadeus and Ayers Rock region, Northern Territory. *BMR Geol. & Geophys. Bull.* 230

38. Patton PC, Pickup G & Price DM. 1993: Holocene palaeofloods of the Ross River, Central Australia. *Quat. Res.* 40, 201–212

39. DeDekker P, Correge T & Head J. 1991: Late Pleistocene record of cyclic aeolian activity from tropical Australia suggesting the Younger Dryas is not an unusual climatic event. *Geology* 19, 602–605

40. Calf GE, McDonald PS & Jacobson G. 1991: Recharge mechanism and groundwater age in the Ti-Tree Basin, Northern Territory. *Aust. J. Earth Sci.* 38, 299–308

41. Nanson GC & Knighton AD. 1996: Anabranching rivers: Their cause, character and classification. *Earth Surface Processes & Landforms* 21, 217–239

42. Nanson GC & Huang HQ. 1999: Anabranching rivers: Divided efficiency leading to fluvial diversity. Chapter 19 in Miller AJ & Gupta A (eds). *Varieties of Fluvial Form.* John Wiley

42A. Drexel JF & Preiss WV (eds). 1995: *The Geology of South Australia, Vol 2: The Phanerozoic.* Mines & Energy, South Australia

43. Magee JW, Bowler JM, Miller GH & Williams DLG. 1995: Stratigraphy, sedimentology, chronology and palaeohydrology of Quaternary lacustrine deposits at Madigan Gulf, Lake Eyre, South Australia. *Palaeogeog., Palaeoclimat., Palaeoecol.* 113, 3–42

44. Magee JW & Miller GH. 1998: Lake Eyre palaeohydrology from 60ka to the present: beach ridges and glacial maximum aridity. *Palaeogeog., Palaeoclim., Palaeoecol.* 144, 307–329

45. Ullman WJ & McLeod LC. 1986: The Late Quaternary salinity record of Lake Frome, South Australia: Evidence from Na^{+} in stratigraphically preserved gypsum. *Palaeogeog., Palaeoclim., Palaeoecol.* 54, 153–169

46. Gibling MR, Nanson GC & Maroulis JC. 1998: Anastomosing river sedimentation in the Channel Country of central Australia. Sedimentology 45, 595–619

46A. Nanson GC, Rust BR & Taylor G. 1986: Coexistent mud braids and anastomosing channels in an arid-zone river: Cooper Creek, central Australia. *Geology* 14, 175–178

46B. Rust BR & Nanson GC. 1986: Contemporary and palaeochannel patterns and the Late Quaternary stratigraphy of Cooper Creek, south-west Queensland, Australia. *Earth Surface Proc. & Landforms* 11, 581–590

47. McTainsh GH. 1989: Quaternary aeolian dust processes and sediments in the Australian region. *Quat. Sci. Rev.* 8, 235–253

48. Bunn SE & Davies PM. 1999: Aquatic food webs in turbid, arid-zone rivers: Preliminary data from Cooper Creek, western Queensland. In Kingsford RT (ed.) *A Free-Flowing River: The ecology of the Paroo River.* NSW NPWS

49. Taylor P. 1999: Draft Impact Assessment study for Currereva Agricultural Development Project. Unpubl.

50. 1999: *Critical review of the Draft Impact Assessment Study (DIAS) of a Development Application for intensive agriculture at Currareva near Windorah on Cooper Creek* Australian Society for Limnology.

51. Nanson GC, Young RW, Price DM & Rust BR. 1988: Stratigraphy, sedimentology and Late Quaternary chronology of the Channel Country of western Queensland. Chapter 9, in Warner RF (ed.) *Fluvial Geomorphology of Australia.* Academic Press, 151–175

52. Wasson RJ & Galloway RW. 1986: Sediment yield in the Barrier Range before and after European settlement. *Aust. Rangel. J.* 8 (2), 79–90

53. Miller GH, Magee JW & Jull AJT. 1997: Low latitude glacial cooling in the Southern Hemisphere from amino-acid racemization in emu eggshells. *Nature* 385, Letters, 241–244

53A. Cock BJ, Williams MAJ & Adamson DA. 1999: Pleistocene Lake Brachina: A preliminary stratigraphy and chronology of lacustrine sediments from the central Flinders Ranges, South Australia. *Aust. J. Earth Sci.* 46, 61–69

54. Woodroffe CD. 1993: Late Quaternary evolution of coastal and lowland riverine plains of South-east Asia and northern Australia: An overview. *Sed. Geol.* 83, 163–175

55. Torgersen T, Luly J, De Deckker P, Jones MR, Searle DE, Chivas AR & Ullman WJ. 1988: Late Quaternary environments of the Carpentaria Basin, Australia. *Palaeogeog., Palaeoclim., Palaeoecol.* 67, 245–261

56. Blake DH & Ollier CD. 1971: Alluvial plains of the Fly River. *Z. Geomorph. N.F.* 12, 1–17

57. Noble JC, Habermehl MA, James CD, Landsberg J, Langston AC & Morton SR. 1998: Biodiversity implications of water management in the Great Artesian Basin. *Rangel. J.* 20(2), 275–300

58. Nanson GC. Pers comm.

60. Nanson GC, Price DM, Short SA, Young RW & Jones BG. 1991: Comparative Uranium-Thorium and thermoluminescence dating of weathered Quaternary alluvium in the tropics of northern Australia. *Quat. Res.* 35, 347–366

61. Woodroffe CD & Chappell J. 1993: Holocene emergence and evolution of the McArthur River Delta, south-western Gulf of Carpentaria, Australia. *Sed. Geol.* 83, 303–317

62. Baker VR, Pickup G & Polach HA. 1985: Radiocarbon dating of flood events, Katherine Gorge, Northern Territory, Australia. *Geol.* 13, 344–347

63. Baker VR & Pickup G. 1987: Flood geomorphology of the Katherine Gorge, Northern Territory, Australia. *Geol. Soc. Am. Bull.* 98, 635–646

64. Woodroffe CD, Mulrennan ME & Chappell J. 1993: Estuarine infill and coastal progradation, southern van Diemen Gulf, northern Australia. *Sed. Geol.* 83. 257–275

65. Nanson, GC, East TJ & Roberts RG. 1993: Quaternary stratigraphy, geochronology and evolution of the Magela Creek catchment in the monsoon tropics of northern Australia. *Sed. Geol.* 83, 277–302

66. Morton SR, Brennan KG & Armstrong MD. 1990: Distribution and abundance of ducks in the Alligator Rivers Region, Northern Territory. *Aust. Wildl. Res.* 17, 573–590

67. Woodroffe CD, Chappell J, Thom BG & Wallensky E. 1989: Depositional model of a macrotidal estuary and floodplain, South Alligator River, Northern Australia. *Sedimentology* 36, 737–756

68. Cook G & Setterfield S. 1995: Ecosystem dynamics and the management of environmental weeds in wetlands. *Proc. Wet Dry Tropics Workshop, Jabiru*

69. Woodroffe CD, Thom BG & Chappell J. 1985: Development of widespread mangrove swamps in mid-Holocene times in northern Australia. *Nature* 317, 711–713

70. Woodroffe CD, Chappell J & Thom BG. 1988: Shell middens in the context of estuarine development, South Alligator River,

Northern Territory. *Archaeology in Oceania* 23, 95–103

71. Mulrennan NE & Woodroffe CD. 1998: Saltwater intrusion into the coastal plains of the lower Mary River, Northern Territory, Australia. *J. Envir. Management* 54, 169–188

72. Mulrennan ME & Woodroffe CD. 1998: Holocene development of the lower Mary River plains, Northern Territory, Australia. *Holocene* 8(5), 565–579

74. Woodroffe CD & Mulrennan ME. 1993: *Geomorphology of the lower Mary River plains, Northern Territory.* ANU & NT Conservation Comm.

75. Woodroffe CD, Chappell J, Thom BG & Wallensky E. 1986: Geomorphological dynamics and evolution of the South Alligator tidal river and plains, Northern Territory. *ANU Mangrove Monograph* 3

76. Habermehl MA. 1996: Groundwater movement and hydrochemistry of the Great Artesian Basin, Australia. *Geol. Soc. Aust. Ext. Abs.* 43, 228–236

77. Blick R. 1997: Managing rangelands better by managing artesian water. In Copeland C & Lewis D (eds). *Saving Our Natural Heritage?* Halstead Press, Sydney

78. Blick R. 1994: Some notes on conservation of rangelands resources. In Morton SR & Price PC (eds). *R&D for sustainable use and management of Australia's rangelands.* LWRRDC, Canberra

79. Landsberg J & Gillieson D. 1995: Looking beyond the piospheres to locate biodiversity reference areas in Australia's rangelands. *Proc. 5th Internat. Rangeland Conf.* 304–305

80. Bennett B. 1997: Water points: Where pastoralism and biodiversity meet. *Ecos* 92, Winter, 10–14

81. Veevers JJ. 1993: Mid-Cretaceous tectonic climax, Late Cretaceous recovery, and Cainozoic relaxation in the Australian region. *Geol. Soc. Aust. Spec. Publ.* 18, 1–14

81A. Ross Blick, ACF 1997: Environment vision for the Great Artesian Basin (draft 12.10.98)

82. James CD, Landsberg J & Morton SR. 1996: Maintaining biodiversity in Australian rangelands. *Conf. Pap. Third Bienn. Aust. Rangel. Conf. Port Augusta*

83. Walker KF, Sheldon F & Puckridge JT. 1995: An ecological perspective on large dryland rivers. *Regulated Rivers: Research and Management* 11, 85–104

84. Puckridge JT. 1998: Wetland management in arid Australia. The Lake Eyre Basin as an example. In Williams WD (ed.) *Wetlands in a dry land: understanding for management.* Environment Australia, Canberra. 85–96

85. Thoms M & Cullen P. 1998: The impact of irrigation withdrawals on inland river systems. *Rangel. J.* 20(2), 226–236

86. Mussared D. 1997: *Living on Floodplains.* CRC Freshwater Ecology, MDBC, Canberra

88. Ollier, CD. 1992: A hypothesis about antecedent and reversed drainage. *Geog. Fis. Dinam. Quat.* 14(2) 243–246

89. Ollier CD. 1995: Tectonics and landscape evolution in southeast Australia. *Geomorphology* 12, 37–44

90. Jacobson G, Jankowski J & Abell RS. 1991: Groundwater and surface water interaction at Lake George, New South Wales. *BNR J. Aust. Geol. & Geophys.* 12, 161–190

91. Bowler JM. 1986: Quaternary landform evolution. *The Natural Environment.* OUP Chapter 5, 117–147 in Jeans DN (ed.)

92. Ollier C. : Drainage patterns & their interpretation. Draft paper for new book.

93. De Deckker P, Chivas AR, Shelley JMG & Torgersen T. 1988: Ostracod shell chemistry: A new palaeoenvironmental indicator applied to the regressive/transgressive record from the Gulf of Carpentaria, Australia. *Palaeogeog., Palaeoclim., Palaeoecol.* 66, 231–241

94. Nanson GC & Doyle C. 1999: *Proc. 2nd Aust. Stream Management Conf. Adelaide* Landscape stability, Quaternary climate change and European degradation of coastal rivers in Southeastern Australia.

95. Lhotsky J. 1947: *A journey from Sydney to the Australian Alps undertaken in January, February and March, 1834.* Blubber Head Press, Hobart

96. Eyles RJ. 1977: Changes in drainage networks since 1820, Southern Tablelands, New South Wales. *Aust. Geogr.* 13, 377–386

97. Tooth S & Nanson GC. 1995: The geomorphology of Australia's fluvial systems: Retrospect, perspect and prospect. *Progress in Phys. Geog.* 19(1), 35–60

98. Warner RF. 1987: The impact of alternating flood- and drought-dominated regimes on channel morphology at Penrith, New South Wales, Australia. *IAHS Publ.* 168, 327–338

99. Kirkup H, Brierley G, Brooks A & Pitman A. 1998: Temporal variability of climate in south-eastern Australia: A reassessment of flood- and drought-dominated regimes. *Aust. Geogr.* 29(2), 241–255

99A. Young RW, Nanson GC & Bryant EA. 1986: Alluvial chronology for coastal NSW. Climatic control or random erosional events? *Search* 17 (10–12), 270–272

99B. Nanson GC & Erskine WD. 1988: Episodic changes of channels and floodplains on coastal rivers in New South Wales. In Warner RF (ed.) *Essays in Australian Fluvial Geomorphology.* Academic Press, Sydney

100. Tranter D. 1999: Wingecarribee Swamp: A case study. *Proc. Water Wet or Dry Conf. Sydney 1998*

101. Young RW & McDougall I. 1993: Long-term landscape evolution: Early Miocene and modern rivers in southern New South Wales, Australia. *J. Geol.* 101, 35–49

102. Page KG & Carden YR. 1998: Channel adjustment following the crossing of a threshold: Tarcutta Creek, southeastern Australia. *Aust. Geogr. Studies* 36(3), 289–311

103. Gippel CJ & Collier KJ. 1998: Degradation and rehabilitation of Waterways in Australia and New Zealand. In DeWaal LC *et al* (eds). *Rehabilitation of Rivers. Principles and Implementation.* John Wiley

104. Reinfelds I, Bishop P & Rutherfurd I. 1998: Relative impact of clearing of vegetation, de-snagging, artificial meander cutoffs and high magnitude floods on the morphology of the Lower Latrobe River, Gippsland, Victoria. *Proc. 8th Bienn. Conf. Aust. & N.Z. Geomorphology Group, Goolwa, SA*

105. Bain M & Tilleard J. 1999: A proposed plan for the rehabilitation of the Snowy River in Victoria. *Proc. 2nd Aust. Stream Management Conf. Adelaide* 27–32

106. Gippel C. 1999: Developing a focused vision for river rehabilitation: The Lower Snowy River, Victoria. *Proc. 2nd Aust. Stream Management Conf. Adelaide* 299–305

107. Smith T & Starr B. 1999: Willows—friend or foe? An historical Perspective. *Proc. 2nd Aust. Stream Management Conf. Adelaide* 573–577

108. Bobbi C. 1999: River management arising from willow removal. *Proc. 2nd Aust. Stream Management Conf. Adelaide* 69–73

109. Brooks A. 1999: Lessons for river managers from the fluvial Tardis. *Proc. 2nd Aust. Stream Management Conf. Adelaide* 121–128

110. Cohen T. 1999: Channel recovery mechanisms in a forested catchment, Jones Creek, East Gippsland: Lessons for river management in south-eastern Australia. *Proc. 2nd Aust. Stream*

Management Conf. Adelaide 181–186

111. Brooks A. 1999: Large woody debris and the geomorphology of a perennial river in southeast Australia. *Proc. 2nd Aust. Stream Management Conf. Adelaide* 129–137

112. Mitchell P. 1990: Environmental condition of Victorian streams. *DWR Vic. Report*

113. Thexton E. 1999: Rehabilitation of the lower Genoa River, far East Gippsland, Victoria, with assisted regeneration. *Proc. 2nd Aust. Stream Management Conf. Adelaide* 623–628

114. Jones RN, Bowler JM & McMahon TA. 1998: A high resolution Holocene record of P/E ratio from closed lakes, Western Victoria. *Palaeoclimates* 3(1-3), 51–82

115. Currey DT. 1964: The former extent of Lake Corangamite. *Proc. Roy. Soc. Vic.* 77(2), 377–387

116. Bowler JM. 1981: Australian salt lakes. *Hydrobiologica* 82, 431–444

117. Brierley GJ *et al.* 1999: *Post-European changes to fluvial geomorphology of Bega Catchment, Australia: Implications for river ecology.* In press.

118. Fryirs K & Brierley G. 1998: *River styles in Bega/Brogo Catchment: Recovery potential and target conditions for river rehabilitation.* NSW DLWC

119. Brooks AP & Brierley GJ. 1997: Geomorphic responses of lower Bega River to catchment disturbance 1851–1926. *Geomorph.* 18, 291–304

120. Fryirs K & Brierley G. 1998: *The use of river styles and their associated sediment storage in the development of a catchment-based river rehabilitation strategy for the Bega/Brogo catchment, South Coast, NSW.* NSW DLWC

121. Brierley G. 1998: River changes since European settlement of the Bega catchment, NSW. *Rivers for the Future* Spring, 16–29

122. Fryirs K & Brierley G. 1998: The character and age structure of valley fills in Upper Wolumla Creek catchment, South Coast, New South Wales, Australia. *Earth Surface Processes & Landforms* 23, 271–287

123. Brierley G & Fryirs K. 1998: A fluvial sediment budget for Upper Wolumla Creek, South Coast, New South Wales, Australia. *Aust. Geogr.* 29(1), 107–124

124. Brierley GJ & Murn CP. 1997: European impacts on downstream sediment transfer and bank erosion in Cobargo catchment, New South Wales, Australia. *Catena* 31, 119–136

125. Young RW & McDougall I. 1985: The age, extent and geomorphological significance of the Sassafras basalt, south-eastern New South Wales. *Aust. J. Earth. Sci.* 32, 323–331

126. Nott JF. 1992: Long-term drainage evolution in the Shoalhaven catchment, south-east highlands, Australia. *Earth Surface Processes & Landforms* 17, 161–174

127. Nott JF, Young RW & McDougall I. 1996: Wearing down, wearing back quantitative evidence from the Shoalhaven catchment, southeast Australia. *J. Geol.* 104, 224–232

128. Cho G, Georges A, Stoutjesdijk R & Longmore R. 1995: Jervis Bay: A place of cultural, scientific and educational value. *Kowari 5.* ANCA, Canberra

129. Young ARM. 1988: Quaternary sedimentation on the Woronora Plateau and its implications for climatic change. *Aust. Geogr.* 17, 1–5

129A. Young RW & Young ARM. 1988: 'Altogether Barren, peculiarly romantic': The sandstone lands round Sydney. *Aust. Geogr.* 19(1), 9–25

130. Nanson GC & Young RW. 1981: Downstream reduction of rural channel size with contrasting urban effects in small coastal streams of South-eastern Australia. *J. Hydrol.* 52, 239–255

131. Nanson GC & Hean D. 1985: The West Dapto flood of 1984: Rainfall characteristics and channel changes. *Aust. Geographer* 16(4), 249–258

132. Warner RF. 1991: Impacts of environmental degradation on rivers, with some examples from the Hawkesbury–Nepean System. *Aust. Geographer* 22(1), 1–13

133. Nanson GC & Young RW. 1988: Fluviatile evidence for a period of late-Quaternary pluvial climate in coastal south-eastern Australia. *Palaeogeog., Palaeoclim., Palaeoecol.* 66, 45–61

134. Nanson GC, Young RW & Stockton ED. 1987: Chronology and palaeoenvironment of the Cranebrook Terrace (near Sydney) containing artefacts more than 40,000 years old. *Archaeology in Oceania* 22, 72–78

135. Nix H. 1994: The Brigalow. Chapter 10 in Dovers S. (ed.) *Australian Environmental History: Essays and Cases,* OUP

136. Nanson GC. 1986: Episodes of vertical accretion and catastrophic stripping: A model of disequilibrium flood-plain development. *Geol. Soc. Amer. Bull.* 97, 1467–1475

137. Ollier CD. 1982: Geomorphology and tectonics of the Dorrigo Plateau, N.S.W. *J. Geol. Soc. Aust.* 29, 431–435

138. Haworth RJ, Gale SJ, Short SA & Heijnis H. 1999: Land use and lake sedimentation on the New England Tablelands of New South Wales,Australia. *Aust. Geographer* 36(1), 51–73

139. Haworth RJ. 1998: Preliminary report on 'An inventory of wetlands in The New England region and an assessment of their environmental Health, past history and present status'. Unpubl.

140. Gale SJ, Haworth RJ & Pisanu PC. 1995: The Pb_{210} chronology of Late Holocene deposition in an eastern Australian lake basin. *Quart. Sci. Rev.* 14, 395–408

141. Haworth RJ. 1994: European impact on lake sedimentation in upland Eastern Australia. Thesis, UNE. Unpubl.

142. Cooper B. 1996: *Central and north-west regions water quality program.* DLWC report TS96.048

143. Haworth RJ & Ollier CD. 1992: Continental rifting and drainage reversal: the Clarence River of Eastern Australia. *Earth Surface Processes & Landforms* 17, 387–397

144. Easton C. 1989: The trouble with the Tweed. *Fishing World* March.58–59

145. Loneragan NR & Bunn SE. 1999: River flows and estuarine ecosystems: Implications for coastal fisheries from a review and a case-study of the Logan River, south-east Queensland. *Aust. J. Ecology* 24, 431–440

146. Day D. 1989: Resources development or instream protection? *The Environmentalist* 9(1), 7–27

147. Fleming PM *et al,* (eds). 1991: *Burdekin Project Ecological Study.* CSIRO Inst. Earth Resources & DNDE Canberra

148. Fleming PM. 1985: The Burdekin Dam and irrigation project: Some environmental consequences with particular reference to drainage and flooding. *Proc. 5th Afro-Asian Conf. Townsville*

149. Lord DB & Van Kerkvoort 1982: *The effects of major harbour construction on longshore sediment movement—Coffs Harbour, NSW.* Report DPW Coastal Branch

150. Fleming PM & Loofs. 1991: *Flood generation and transmission in the Burdekin and Haughton Rivers, North Queensland.* CSIRO Water Resources Tech, Memorandum 91/15

151. Grasseni F, Jacobson G & Jakeman AJ. 1991: Major Australian aquifers: Potential climatic change impacts. *Water Internat.* 16, 38–44

152. Johnson AKL & Murray AE. 1997: *Herbert River Catchment Atlas.* CSIRO Tropical Agriculture, Townsville

153. Ian Drummond & Associates. 1993: *Stream management plan: Herbert River and District.* Herbert River Improvement Trust Report

154. Chivas AR, DeDeckker P, Nind M, Thiriet D & Watson G. 1986: The Pleistocene palaeoenvironmental record of Lake Buchanan: An atypical Australian playa. *Palaeogeog., Palaeoclimatol., Palaeoecol.* 54, 131–152

155. Calvin, William H. 1998: The Great Climate Flip-Flop. *Atlantic Monthly* 281(1), 47–64

156. Wright H. 1982: A phosphorus budget for Australia. Unpubl.

157. Wright H. 1988: The longterm threat to bushland from urban runoff—Minimising the damage. *Proc. 'Caring for Warringah's Bushland' Symp.*

158. Wright H. 1998: High phosphorus loads in urban runoff and soils—Implications for conserving natural bushland and wetlands. *Proc. 2nd World Congress of Building Officials*

159. Wright H. 1992: Notes for observing the promotion of weeds in urban bushland due to runoff. Unpubl.

160. Wright H. 1999: Managing urban runoff in bushland. Unpubl.

161. Davis J, Breen P & Hart BT. 1998: *The ecology of the Yarra River: A discussion paper.* Tech. Rep. CRC Freshwater Ecology

162. National Land and Water Resources Audit: *Draft Strategic Plan—1998 to 2001.* LWRRDC

162A. Murray–Darling Basin Ministerial Council. 1999: *The Salinity Audit of the Murray Darling Basin.* MDBC Canberra

163. Mackay N & Eastburn D. (eds). 1990: *The Murray.* MDBC Canberra

163A. Rutherfurd I. 1990: Ancient River, Young Nation. In *The Murray* MDBC Canberra

164. High Court of Australia document: 1982

165. Thoms MC. 1995: The physical character of the Barwon–Darling system. *Proc. Conf. Researching the Barwon–Darling. Bourke*

166. McLellan J. 1998: Wet or Dry. Social and cultural aspects of the reforms. Case study: CMC approach to water management. Oral presentation to Wet or Dry Conf. Sydney

167. Ollier CD & Pain CF. 1994*:* Landscape evolution and tectonics in south-eastern Australia. *AGSO J. Geol. Geophys.* 15, 335–345

168. Abell R. 1992: Drainage network evolution in SE Australia. Paper given to ANZGG, Port Macquarie

169. Freudenberger D. 1998: *Scoping the management and research needs of the Coolibah woodlands in the Murray–Darling Basin.* CSIRO Canberra

170. Marshall JP. 1993: *Floodplain management for erosion control and high productivity on the Darling Downs.* DPI Qld

171. Date D. 1997: 'Melon-hole' country. *Urimbirra* 31 (5)

172. Wearing J. 1996: Boggomosses. *Urimbirra* April 196

173. Timms BV. 1999: Local runoff, Paroo floods and water extraction impacts on the wetlands of the Currawinya National Park. In Kingsford RT *et al* (eds): *A Free-flowing River.* NPWS. 51–66

174. Kingsford RT (ed.) 1999: *A Free-flowing River: The ecology of the Paroo River.* NSW NPWS

175. Stauffacher M *et al.* 1997*: Salt and water movement in the Liverpool Plains—What's going on?* LWRRDC report

176. Timms W. 1997: Groundwater quality and hydraulic linkages: Interim results. *Proc. Conf. Groundwater & the Liverpool Plains, Gunnedah*

177. Mawhinney W. 1998: *Liverpool Plains water quality project: Nutrients and groundwater quality.* DLWC report

178. Timms W. 1997*: Liverpool Plains water quality project.* DLWC report

179. Mawhinney W. 1998*: Liverpool Plains water quality project. Pesticides Monitoring.* DLWC report

180. Mawhinney W. 1998*: Liverpool Plains water quality project. Land use, pesticide use and their impact on water quality.* DLWC report

181. Smil, V. 1997: Global population and the nitrogen cycle. *Scientific American* July, 58–63

182. Page KJ, Nanson GC & Price D. 1996: Chronology of Murrumbidgee River palaeochannels on the Riverine Plain, southeastern Australia. *J. Quat. Sci.* 11(4), 311–326

183. Page KJ, Nanson GC & Price DM. 1991: Thermoluminescent chronology of Late Quaternary deposition on the Riverine Plain of south-eastern Australia. *Aust. Geographer* 22(1), 14–23

184. Kellett, J.R. 1989*:* The Ivanhoe Block—its structure, hydrogeology and effect on the groundwaters of the Riverine Plain of New South Wales. *BMR J. Aust. Geol. Geophys.* 11(243), 333–353

185. Butler BE. 1950: A theory of prior streams as a causal factor of soil occurrence in the Riverine Plain of south-eastern Australia. *Aust. J. Agric. Res.* 1, 231–252

186. Butler BE. 1958: Depositional systems of the Riverine Plain of south-eastern Australia in relation to soils. *CSIRO Soil publ.* 10

187. Wallbrink PJ, Murray AS & Olley JM. 1998: Determining sources and transit times of suspended sediment in the Murrumbidgee River, New South Wales, Australia, using fallout 137Cs and 210Pb. *Water Resources Res.* 34(4), 870–887

188. Olley JM, Murray AS, Mackenzie DH & Edwards K. 1993: Identifying sediment sources in a gullied catchment using natural and anthropogenic radioactivity. *Water Resources Res.* 29(4), 1037–1043

189. Olley J. (ed.) 1995: Sources of suspended sediment and phosphorus to the Murrumbidgee River. *CSIRO Water Resources* 95–32

191. Starr B. 1998: *The Numeralla: River of change.* Upper Murrumbidgee Soil & Water Management Plan

192. Brooks K. 1998: Rehabilitation at Lake George Mine, Captains Flat. *Groundwork* 1 (2)

193. Norris RH. 1986: Mine waste pollution of the Molonglo River, NSW and the Australian Capital Territory: Effectiveness of remedial works at Captains Flat mining area. *Aust. J. Freshw. Rev.* 37, 147–157

194. Hogg D. 1991: *Evaluation of the remedial works at Captains Flat mine.* David Hogg Pty Ltd, Environmental Consultants

195. Gilmore, Mary. 1934, 1986: *Old Days: Old Ways. A Book of Recollections* Angus & Robertson, Sydney

196. Leitch M. 1985: *Where the Red Gums are Growing.* Oxford Print, Wagga

197. Lawson S & Webb E. 1998*: Review of groundwater use and groundwater level behaviour in the Lower Murrumbidgee Valley.* Tech. Rep. 98/05 DLWC Murrumbidgee Region

198. Murrumbidgee Groundwater MC. 1999: *Management of the alluvial groundwater resources of the Lower Murrumbidgee Valley, NSW.* Discussion paper

199. Lawrie KC, Chan RA, Gibson DL & de Souza Kovacs N. 1999: Alluvial gold potential in buried palaeochannels in the Wyalong district, Lachlan Fold Belt, New South Wales. *AGSO Research Newsletter* 30

200. North Mining Ltd. 1995: *Lake Cowal Gold Project EIS.* NR

Environmental Consultants

201. National Landcare Program booklet: *The Upper Lachlan Catchment*

202. Macumber PG. 1991: *Interaction between groundwater and surface systems in Northern Victoria.* DCE Victoria

203. Robinson D & Mann S. 1996: *Natural values of the public lands along the Broken, Boosey and Nine Mile Creek, north-eastern Victoria.* Goulburn Valley Environment Group

204. Cole LF. 1994: Utilisation of archival maps in mapping and explaining the burial of soils in the valley of the Bendigo Creek, Victoria. *Quaternary Australasia* 12, 35–43

205. Morcom LA & Westbrooke ME. 1998: The Pre-Settlement vegetation of the western and central Wimmera Plains of Victoria, Australia. *Aust. Geogr. Studies* 36(3), 273–288

206. Lee B. 1998: Still life with Platypus. *Groundwork* 2(2), 24–26

207. Bartley R & Rutherfurd I. 1999: Recovery of geomorphic complexity in disturbed streams: using migrating sand slugs as a model. *Proc. 2nd Aust. Stream Management Conf. Adelaide* 39–44

208. Ollier,CD. 1985: Lava flows of Mount Rouse, Western Victoria. *Proc. Roy. Soc. Vic.* 97 (4), 167–174

209. Bowler JM & Magee JW. 1978: Geomorphology of the Mallee region in semi-arid northern Victoria and western New South Wales. *Proc. Roy. Soc. Vic.* 90, 5–26

210. Barnett SR. 1989: The effect of land clearance in the Mallee region on River Murray salinity and land salinisation. *BMR J. Geol. Geophys.* 11, 205–208

211. Leaman D. 1998: A River's Tale: The River Derwent. Unpubl.

212. Locher H. 1999: Changing approaches to river management in the Tasmanian Hydro Electricity System. *Proc. 2nd Aust. Stream Management Conf. Adelaide* 395–400

213. Anderson S. 1999: The science of consultation: A Tasmanian experience. *Proc. 2nd Aust. Stream Management Conf. Adelaide* 9–13

214. Tyler PA, Sherwood JE, Magilton CJ & Hodgson DA. 1996: Limnological and geomorphological considerations underlying Pedder 2000—the campaign to restore Lake Pedder. *Arch. Hydrobiol.* 136(3), 343–361

215. Augustinus P & Colhoun EA. 1986: Glacial history of the upper Pieman and Boco valleys, western Tasmania. *Aust. J. Earth Sci.* 33, 181–191

216. Dept. of Environment. 1976: *Heavy metals in the marine environment of the north west coast of Tasmania.* DE Hobart

217. de Blas A. 1994: *Environmental effects of Mt Lyell operations on Macquarie Harbour and Strahan.* Aust. Centre for Independent Journalism. UTS

218. Needham S & McBride P. 1998: The big one. *Groundwork* 2(1), 20–21

219. Brooks K. 1998: Rehabilitation at Lake George Mine, Captains Flat. *Groundwork* 2(1), 18–19

220. Leaman D. 1998: Hobart's River. Notes for conference excursion, Aust. Soc. Explor. Geophys. Hobart

220A. Leaman D. 1995: A geological appraisal of Mt Wellington. *Proc. AIMM Conf. Tasmania*

221. Murray-Wallace CV & Goede A. 1991: Aminostratigraphy and electron spin resonance studies of the late Quaternary sea level change and coastal neotectonics in Tasmania. *Zeit. Geomorph.* 35(2), 129–149

222. Michael-Leiba MO. 1989: Macroseismic effects, locations and magnitudes of some early Tasmanian earthquakes. *Bur. Miner. Res. J. Geol. Geophys.* 11(1), 89–99

223. Smith MA. 1987: Pleistocene occupation in arid Central Australia. *Nature* 328, 710–711

224. English P. 1998: Palaeodrainage at Uluru-Kata Tjuta National Park and implications for water resources. *Rangel. J.* 20(2), 255–274

225. Stewart AJ, Blake DH & Ollier CD. 1986: Cambrian river terraces and ridgetops in Central Australia: Oldest persisting landforms? *Science* 233, 758–761

226. Gee H. 1998: Draft Lake Pedder restoration plan. Pedder 2000. Unpubl.

227. Gee H. 1995: A cultural appraisal of the restoration proposal for Lake Pedder. *Sci. Symp. Natural History and Restoration of Lake Pedder*

228. Lines-Kelly R. 1995: *Soil Sense. Soil management for NSW North Coast farmers.* North Coast Soil Management WP

229A. NSW Public Works. 1991: *Lower Tweed Estuary, River Management Plan.* Tech. Rep. 3

229B. NSW Public Works. 1991: *Lower Tweed Estuary, River Management Plan*

229C. Cameron McNamara Pty Ltd. 1980: Tweed Valley. *NSW Coastal Rivers Flood Management Studies*

229D. Butler M. 1998: *Tweed River Studies Analysis.* Tweed Shire Council

230. Wasson RJ, Mazari RK, Starr B & Clifton G. 1998: The recent history of erosion and sedimentation on the Southern Tablelands of southeastern Australia: Sediment flux dominated by channel incision. *Geomorphology* 24, 291–308

231. Sippel A. 1999: Oldman Saltbush. The Australian Grazing Plant. Unpubl. booklet.

232. Moller G. 1996: *Herbert River and major tributaries.* State of the Rivers Report DNR Qld

233. Arthington AH. 1995: *State of the rivers in cotton growing areas.* LWRRDC Occ. Pap. 02/95

234. Nanson GC & Price DM. 1998: Quaternary change in the Lake Eyre basin of Australia: An introduction. *Palaeogeog. Palaeoclim. Palaeoecol.* 144, 235–237

235. Burke S, (ed.) 1998: *Bush Lives, Bush Futures.* Historic Houses Trust of NSW

236. Blackburn H. 1998: Wetlands of the Gwydir Watercourse. Unpubl.

237. Tapley I. 1990: Night thermal imagery reveals ancient drainage system of the Canning Basin. *Resource Review* March 1990, 6–8

238. Weaver TR & Lawrence CR, 1998: *Groundwater: Sustainable Solutions.*

Index